Number Theory – Diophantine Problems, Uniform Distribution and Applications

Christian Elsholtz • Peter Grabner

Editors

Number Theory – Diophantine Problems, Uniform Distribution and Applications

Festschrift in Honour of Robert F. Tichy's 60th Birthday

 Springer

Editors
Christian Elsholtz
Institute of Analysis and Number Theory
Graz University of Technology
Graz, Austria

Peter Grabner
Institute of Analysis and Number Theory
Graz University of Technology
Graz, Austria

ISBN 978-3-319-55356-6   ISBN 978-3-319-55357-3   (eBook)
DOI 10.1007/978-3-319-55357-3

Library of Congress Control Number: 2017942736

Printed on acid-free paper

This Springer imprint is published by Springer Nature
The registered company is Springer International Publishing AG
The registered company address is: Gewerbestrasse 11, 6330 Cham, Switzerland

*Dedicated to Robert F. Tichy,
on the occasion of his 60th birthday.
With gratitude and* ad multos annos.

Preface

Without doubt, Robert F. Tichy (see Fig. 1) is one of the most eminent and influential mathematicians in Austria. The breadth and significance of his mathematical œuvre, as well as that of his more than 30 Ph.D. students (and many more co-supervised students), his service to the ÖMG (the Austrian Mathematical Society), his service as editor of scientific journals, his regular advice to many hiring committees, and his long-term service as the representative responsible for mathematics in the board of the FWF (the Austrian science funding body) clearly show his influence.

When addressing technical or administrative matters of any kind, he is an efficient, well-connected, and diplomatic problem solver.

There is no doubt that Austrian mathematics as a whole is deeply indebted to his influential work at all levels of research, education, and administration.

As any description of Robert's mathematical explorations is necessarily incomplete, we would rather like to mention that his non-mathematical explorations as an experienced alpine rock climber (see Fig. 2) are also well known in the community.

When we asked colleagues to contribute to this Festschrift, we received a great number of papers, and we are very grateful to the authors. Following Robert's scientific interests, these papers cover areas ranging from Diophantine problems, asymptotic counting, uniform distribution and discrepancy of sequences (in theory and applications), dynamical systems, and prime numbers to actuarial science. As some papers touch several such topics, it did not seem natural to arrange the papers into sections. We therefore present the papers in alphabetical order.

We are also very grateful to the numerous referees who greatly helped by contributing timely reports and detailed comments. Without their expertise, we could not have edited a volume touching that many topics. Also, we would like thank Springer Verlag (in particular A. Comment, S. Gasser, Dr. J. Holland), and J. Abishag of SPi Technologies for the competent and efficient advice, preparation of the final version, and printing this volume.

Fig. 1 Photo taken by P. Grabner, 9th November 2015

Fig. 2 Photo courtesy of P. Sodamin, taken at the south wall of Hoher Dachstein, 25th August 2016

Acknowledgements The editors Christian Elsholtz and Peter Grabner are supported by the Austrian Science Fund (FWF) Doctoral Program "Discrete Mathematics" (W1230). The editor Peter Grabner is supported by the FWF project F5503 (part of the Special Research Program (SFB) "Quasi-Monte Carlo Methods: Theory and Applications").

Graz, Austria
December 2016

Christian Elsholtz
Peter Grabner

Contents

List of Contributors

Shigeki Akiyama Institute of Mathematics, University of Tsukuba, Tsukuba, Ibaraki, Japan

Hansjörg Albrecher University of Lausanne and Swiss Finance Institute, Quartier UNIL-Dorigny, Lausanne, Switzerland

József Beck Department of Mathematics, Rutgers University, Piscataway, NJ, USA

Michael A. Bennett Department of Mathematics, University of British Columbia, Vancouver, BC, Canada

Vitaly Bergelson Department of Mathematics, Ohio State University, Columbus, OH, USA

István Berkes Institute of Statistics, Graz University of Technology, Graz, Austria

Yuri Bilu Institut de Mathématiques de Bordeaux, Université de Bordeaux and CNRS, Talence, France

Jörg Brüdern Mathematisches Institut, Göttingen, Germany

Arian Cani University of Lausanne, Quartier UNIL-Dorigny, Lausanne, Switzerland

Jean-Marc Deshouillers Institut Mathématiques de Bordeaux, UMR 5251, Bordeaux INP, Université de Bordeaux, CNRS, Talence, France

Michael Drmota Institute of Discrete Mathematics and Geometry, TU Wien, Wien, Austria

Andrej Dujella Department of Mathematics, University of Zagreb, Zagreb, Croatia

Christian Elsholtz Institut für Analysis und Zahlentheorie, Technische Universität Graz, Graz, Austria

Jan-Hendrik Evertse Mathematical Institute, Leiden University, Leiden, The Netherlands

Kálmán Győry Institute of Mathematics, University of Debrecen, Debrecen, Hungary

Philipp Habegger Department of Mathematics and Computer Science, University of Basel, Basel, Switzerland

Laurent Habsieger Institut Camille Jordan, Université de Lyon, CNRS UMR 5208, Université Claude Bernard Lyon 1, Villeurbanne Cedex, France

Clemens Heuberger Institut für Mathematik, Alpen-Adria-Universität Klagenfurt, Klagenfurt, Austria

Richard Hofer Johann Radon Institute for Computational and Applied Mathematics, Austrian Academy of Sciences, Linz, Austria

Roswitha Hofer Institute of Financial Mathematics and Applied Number Theory, Johannes Kepler University Linz, Linz, Austria

Matija Kazalicki Department of Mathematics, University of Zagreb, Zagreb, Croatia

Shanta Laishram Indian Statistical Institute, New Delhi, India

Bernard Landreau LAREMA (Laboratoire Angevin de REcherche en MAthématiques), Université d'Angers, CNRS, Angers, France

Gerhard Larcher Institute of Financial Mathematics and Applied Number Theory, Johannes Kepler University Linz, Linz, Austria

Florian Luca School of Mathematics, University of the Witwatersrand, Johannesburg, South Africa

Manfred G. Madritsch Université de Lorraine, CNRS, Institut Elie Cartan de Lorraine, Vandoeuvre-lès-Nancy, France

László Mérai Johann Radon Institute for Computational and Applied Mathematics, Austrian Academy of Sciences, Linz, Austria

Werner Georg Nowak Institut für Mathematik, Universität für Bodenkultur (BOKU), Wien, Austria

Alina Ostafe School of Mathematics and Statistics, University of New South Wales, Sydney, NSW, Australia

Attila Pethő Department of Computer Science, University of Debrecen, Debrecen, Hungary

János Pintz Rényi Mathematical Institute of the Hungarian Academy of Sciences, Budapest, Hungary

Florian Karl Richter Department of Mathematics, Ohio State University, Columbus, OH, USA

Adrian-Maria Scheerer Institut für Analysis und Zahlentheorie, Technische Universität Graz, Graz, Austria

Andrzej Schinzel Institute of Mathematics, Polish Academy of Sciences, Warsaw, Poland

Igor E. Shparlinski School of Mathematics and Statistics, University of New South Wales, Sydney, NSW, Australia

Paul Surer Institut für Mathematik, Universität für Bodenkultur (BOKU), Wien, Austria

Terence Tao Department of Mathematics, UCLA, Los Angeles, CA, USA

Jörg M. Thuswaldner Chair of Mathematics and Statistics, University of Leoben, Leoben, Austria

Stephan Wagner Department of Mathematical Sciences, Stellenbosch University, Stellenbosch, South Africa

Arne Winterhof Johann Radon Institute for Computational and Applied Mathematics, Austrian Academy of Sciences, Linz, Austria

Volker Ziegler Institute of Mathematics, University of Salzburg, Salzburg, Austria

On Nearly Linear Recurrence Sequences

Shigeki Akiyama, Jan-Hendrik Evertse, and Attila Pethő

To Professor Robert Tichy on the occasion of his 60th birthday

Abstract A nearly linear recurrence sequence (nlrs) is a complex sequence (a_n) with the property that there exist complex numbers $A_0,\ldots, A_{d-1}$ such that the sequence $\left(a_{n+d} + A_{d-1}a_{n+d-1} + \cdots + A_0 a_n\right)_{n=0}^{\infty}$ is bounded. We give an asymptotic Binet-type formula for such sequences. We compare (a_n) with a natural linear recurrence sequence (lrs) $(\tilde{a}_n)$ associated with it and prove under certain assumptions that the difference sequence $(a_n - \tilde{a}_n)$ tends to infinity. We show that several finiteness results for lrs, in particular the Skolem-Mahler-Lech theorem and results on common terms of two lrs, are not valid anymore for nlrs with integer terms. Our main tool in these investigations is an observation that lrs with transcendental terms may have large fluctuations, quite different from lrs with algebraic terms. On the other hand, we show under certain hypotheses that though there may be infinitely many of them, the common terms of two nlrs are very sparse. The proof of this result combines our Binet-type formula with a Baker type estimate for logarithmic forms.

2010 Mathematics Subject Classification: 11B65

S. Akiyama
Institute of Mathematics, University of Tsukuba, 1-1-1 Tennodai, Tsukuba, Ibaraki 350-0006, Japan
e-mail: akiyama@math.tsukuba.ac.jp

J.-H. Evertse
Mathematical Institute, Leiden University, P.O. Box 9512, 2300 RA Leiden, The Netherlands
e-mail: evertse@math.leidenuniv.nl

A. Pethő (✉)
Department of Computer Science, University of Debrecen, P.O. Box 12, H-4010 Debrecen, Hungary
e-mail: Petho.Attila@inf.unideb.hu

© Springer International Publishing AG 2017
C. Elsholtz, P. Grabner (eds.), *Number Theory – Diophantine Problems, Uniform Distribution and Applications,* DOI 10.1007/978-3-319-55357-3_1

1 Introduction

This paper was motivated by the investigations on shift radix systems, defined in [2]. For real numbers $S_0, \ldots, S_{d-1}$ and initial values $s_0, \ldots, s_{d-1} \in \mathbb{Z}$, the inequality

$$0 \le s_{n+d} + S_{d-1} s_{n+d-1} + \cdots + S_0 s_n < 1, \ n \ge 0, \tag{1}$$

uniquely defines a sequence of integers (s_n). If $S_0, \ldots, S_{d-1} \in \mathbb{Z}$, then (s_n) is a linear recurrence sequence. However, if some of the coefficients are non-integers, then we get sequences of a different nature. In earlier papers [1, 3, 4, 10] the case $d = 2, S_0 = 1$, and $|S_1| < 2$ was investigated, as a model of discretized rotation in the plane. In that case it is conjectured that the sequence (s_n) is always periodic.

In this paper, we largely generalize the sequences given by shift radix systems. Let $A_0, \ldots, A_{d-1} \in \mathbb{C}$. Let (a_n) be a sequence of complex numbers and define the error sequence (e_n) by the initial terms $e_0 = \cdots = e_{d-1} = 0$ and by the equations

$$e_{n+d} = a_{n+d} + A_{d-1} a_{n+d-1} + \cdots + A_0 a_n \tag{2}$$

for $n \ge 0$. We call (a_n) a *nearly linear recurrence sequence*, in shortcut nlrs, if for some choice of d and $A_0, \ldots, A_{d-1}$, the sequence $(|e_n|)$ is bounded. The sequence (s_n) from (1) is obviously an nlrs because in that case the terms of the error sequence lie in the interval $[0, 1)$. An interesting number theoretical example is when a_n lies in the integer ring R of an imaginary quadratic field and e_n is chosen to be in a fundamental region of the lattice associated with R, see [15].

It is easily shown that for a given nlrs (a_n), the set of polynomials $B_t x^t + B_{t-1} x^{t-1} + \cdots + B_0$ with complex coefficients such that the sequence $(\sum_{i=0}^{t} B_i a_{n+i})$ is bounded is an ideal of the polynomial ring $\mathbb{C}[x]$, called the *ideal of* (a_n). There is a unique, monic polynomial generating the ideal of (a_n), called the *characteristic polynomial of* (a_n). This corresponds to the necessarily unique relation (2) of minimal length for which (e_n) is bounded. For instance, the characteristic polynomial of a bounded sequence is the polynomial 1.

We mention here that the characteristic polynomial of a linear recurrence sequence (lrs) (a_n) may be different from the characteristic polynomial of (a_n) when viewed as an nlrs. For instance, the Fibonacci sequence (a_n) given by $a_0 = 0, a_1 = 1$ and $a_{n+2} = a_{n+1} + a_n$ for $n \ge 0$ has characteristic polynomial $x^2 - x - 1$ when viewed as an lrs, but characteristic polynomial $x - \theta$ with $\theta = \frac{1}{2}(1 + \sqrt{5})$ when viewed as an nlrs, since the sequence $(a_{n+1} - \theta a_n)$ is bounded. Indeed we will see in Lemma 2.1 (i) in Sect. 2, that the characteristic polynomial of an nlrs does not have roots of modulus < 1, and the characteristic polynomial of an lrs and that of the sequence viewed as an nlrs differ only by factors of the form $x - \alpha$ with $|\alpha| < 1$.

Let (a_n) be an nlrs and

$$P(x) = x^d + A_{d-1} x^{d-1} + \cdots + A_0$$

its characteristic polynomial. Further, let (e_n) be the error sequence from (2). Define the generating function

$$c(z) = \sum_{j=1}^{\infty} e_{d+j-1} z^{-j}.$$

Since (e_n) is bounded, $c(z)$ is convergent for all complex z with $|z| > 1$. If, moreover, (e_n) is a sequence of real numbers, then we have $c(\bar{z}) = \overline{c(z)}$ for all $z \in \mathbb{C}, |z| > 1$, where $\bar{z}$ denotes the complex conjugate of z.

To (a_n) we associate two lrs $(\hat{a}_n)$ and $(\tilde{a}_n)$, as follows. Let $(\hat{a}_n)$ denote the lrs having the initial terms $\hat{a}_0 = \cdots = \hat{a}_{d-2} = 0, \hat{a}_{d-1} = 1$ and satisfying the recursion

$$\hat{a}_{n+d} + A_{d-1}\hat{a}_{n+d-1} + \cdots + A_0\hat{a}_n = 0. \tag{3}$$

The lrs $(\tilde{a}_n)$ is defined by the same recursion (3) with different initial terms $\tilde{a}_j = a_j$ $(j = 0, \ldots, d-1)$.

For the distinct roots $\alpha_1, \ldots, \alpha_h$ of $P(x)$ denote by $m_1, \ldots, m_h$ their respective multiplicities. Although in this paper we are mainly interested in the separable case, where all multiplicities are equal to 1, we recall the so-called *Binet formula*

$$\hat{a}_n = \hat{g}_1(n)\alpha_1^n + \cdots + \hat{g}_h(n)\alpha_h^n \tag{4}$$

in general form. Here the polynomials $\hat{g}_j(x)$ are of degree at most $m_j - 1$ and with coefficients from the field $\mathbb{Q}(\alpha_1, \ldots, \alpha_h)$ for $j = 1, \ldots, h$. For $\tilde{a}_n$ we have a similar expression, with polynomials $\tilde{g}_j(x)$ instead of $\hat{g}_j(x)$. In the case that $P(x)$ is separable, i.e., that all its roots are simple, the polynomials $\hat{g}_j(x)$, $\tilde{g}_j(x)$ are just constants and we write $\hat{g}_j$, $\tilde{g}_j$ for them. With these notions we will prove the following theorem, the essential part of which is a Binet-type expression for nlrs.

Theorem 1.1 *Assume that the characteristic polynomial of the nlrs (a_n) is separable and its zeros are ordered as*

$$|\alpha_1| \geq \cdots \geq |\alpha_{r_1}| > 1 = |\alpha_{r_1+1}| = \cdots = |\alpha_{r_1+r_2}|,$$

where $r_1 + r_2 = d$. Denote by $\tilde{g}_j, \hat{g}_j$ the (constant) coefficients of $\alpha_j^n, j = 1, \ldots, d$ in the expression (4) of $\tilde{a}_n$ and $\hat{a}_n$, respectively. Then

(i) *if $r_1 > 0$ and $r_2 = 0$, then*

$$a_n = (\tilde{g}_1 + \hat{g}_1 c(\alpha_1))\alpha_1^n + \cdots + (\tilde{g}_{r_1} + \hat{g}_{r_1} c(\alpha_{r_1}))\alpha_{r_1}^n + O(1)$$

and $\tilde{g}_i + \hat{g}_i c(\alpha_i) \neq 0$ for $i = 1, \ldots, r_1$;

(ii) *if $r_1 > 0$ and $r_2 > 0$, then*

$$a_n = (\tilde{g}_1 + \hat{g}_1 c(\alpha_1))\alpha_1^n + \cdots + (\tilde{g}_{r_1} + \hat{g}_{r_1} c(\alpha_{r_1}))\alpha_{r_1}^n + O(n)$$

and $\tilde{g}_i + \hat{g}_i c(\alpha_i) \neq 0$ for $i = 1, \ldots, r_1$;

(iii) *and if $r_1 = 0$ and $r_2 > 0$, then*

$$a_n = O(n).$$

We prove this theorem in Sect. 2. It is easy to show that the converse of Theorem 1.1 (i) is also true, that is, if $a_n = \beta_1 \alpha_1^n + \cdots + \beta_{r_1} \alpha_{r_1}^n + O(1)$ for certain constants $\beta_1, \ldots, \beta_{r_1}$ then (a_n) is an nlrs. We will present the simple proof in Sect. 2. Moreover, at the end of Sect. 2 we give examples showing that the $O(n)$-term in Theorem 1.1 (ii), (iii) cannot be improved.

In Sect. 3 we prove first that the fluctuation of an lrs can be extremely large, then we analyze the distance between an nlrs and a naturally chosen lrs. We also deduce some other consequences for nlrs. First, we show that if (a_n) is an nlrs with separable characteristic polynomial and $\alpha_1, \ldots, \alpha_{r_1}$ are as in Theorem 1.1, then the constants $c_1, \ldots, c_{r_1}$ such that $a_n = c_1 \alpha_1^n + \cdots + c_{r_1} \alpha_{r_1}^n + O(n)$ are unique. Second, we prove that the analogue of the Skolem-Mahler-Lech theorem, see, e.g., [9, 18], does not hold generally for nlrs with at least two dominating roots with equal absolute values.

In the last Sect. 4 we investigate the common terms of nlrs, i.e., the solutions (k, m) in non-negative integers of the equation

$$a_k = b_m \tag{5}$$

for two nlrs (a_n), (b_n). We consider the case that the characteristic polynomials of (a_n), (b_n) have multiplicatively independent, real algebraic dominating roots of modulus larger than 1. For lrs (a_n), (b_n) we know that in that case (5) has only finitely many solutions. We give an example, showing that for nlrs this is no longer true. On the other hand, we show that the solutions of (5) are very sparse. More precisely, we show that if (k_1, m_1), (k_2, m_2) are any two distinct solutions of (5) with $\max(k_2, m_2) \geq \max(k_1, m_1)$, then in fact $\max(k_2, m_2)$ exceeds an exponential function of $\max(k_1, m_1)$.

2 Proof of Theorem 1.1

We start with a lemma which imposes some restrictions on the characteristic polynomial of an nlrs.

Lemma 2.1 *Let (a_n) be an nlrs with characteristic polynomial $P(x)$.*

(i) *The roots of $P(x)$ all have modulus ≥ 1.*
(ii) *Assume that $a_n = O(n)$ holds for all n. Then the roots of $P(x)$ all have modulus equal to 1.*

Proof If (a_n) is bounded then this is a void statement because the characteristic polynomial has no roots. Hence in the sequel we assume that (a_n) is unbounded. Let $Q(x) = \sum_{i=0}^{t} Q_i x^i$ be in the ideal of (a_n). Let α be a zero of $Q(x)$ and write $Q(x) = (x - \alpha)R(x)$, $R(x) = \sum_{i=0}^{t-1} R_i x^i$. Define the sequences (q_n), (r_n) by

$$q_{n+t} = \sum_{i=0}^{t} Q_i a_{n+i}, \quad r_{n+t-1} = \sum_{i=0}^{t-1} R_i a_{n+i} \ \text{ for } n \geq 0. \tag{6}$$

By assumption, the sequence (q_n) is bounded. Putting $R_{-1} = R_t = 0$, we have $Q_i = R_{i-1} - \alpha R_i$ for $i = 0, \ldots, t$, hence

$$q_{n+t} = \sum_{i=0}^{t} (R_{i-1} - \alpha R_i) a_{n+i} = \sum_{i=1}^{t} R_{i-1} a_{n+i} - \alpha \sum_{i=0}^{t-1} R_i a_{n+i} \tag{7}$$

$$= r_{n+t} - \alpha r_{n+t-1}.$$

(i) We prove that if $|\alpha| < 1$, then the sequence (r_n) is also bounded, i.e., $R(x) = Q(x)/(x - \alpha)$ is in the ideal of (a_n). By repeatedly applying this, we see that the ideal of (a_n) contains a polynomial all whose zeros have modulus ≥ 1. In particular, the characteristic polynomial of (a_n), being a divisor of this polynomial, cannot have zeros of modulus < 1.

Let $C := \max(|r_{t-1}|, |q_t|, |q_{t+1}|, \ldots)$. By (7) we have

$$|r_{n+t}| \leq C + |\alpha| \cdot |r_{n+t-1}| \ \text{ for all } n \geq 0,$$

implying

$$|r_{n+t}| \leq C \cdot \left(1 + |\alpha| + |\alpha|^2 + \cdots + |\alpha|^{n+1}\right) \ \text{ for all } n \geq 0.$$

This shows that $|r_{n+t}| \leq C/(1 - |\alpha|)$ for all $n \geq 0$, i.e., the sequence (r_n) is bounded.

(ii) We now prove that (r_n) is bounded if $|\alpha| > 1$ when $a_n = O(n)$. Then similarly as above we can deduce that the characteristic polynomial of (a_n) has no roots of modulus > 1. Assume that the sequence (r_n) is not bounded. Let $C := \max(|q_t|, |q_{t+1}|, \ldots)$. There is n_0 such that $|r_{n_0+t}| > 1 + C/(|\alpha| - 1)$. By (7) we have $|r_{n+1+t}| \geq |\alpha| \cdot |r_{n+t}| - C$ for $n = n_0, n_0 + 1, \ldots$ and this implies, by induction on t,

$$|r_{n+t}| \geq |r_{n_0+t}| \cdot |\alpha|^{n-n_0} - C(1 + |\alpha| + \cdots + |\alpha|^{n-n_0-1})$$

$$= |r_{n_0+t}| \cdot |\alpha|^{n-n_0} - C \frac{|\alpha|^{n-n_0} - 1}{|\alpha| - 1}.$$

So $|r_{n+t}| \geq |\alpha|^{n-n_0}$ for $n \geq n_0$. This shows that for $n \geq n_0 + t$, the sequence (r_n) grows exponentially. On the other hand, from our assumption $a_n = O(n)$ and (6), it follows that $r_n = O(n)$. Thus, our assumption that the sequence (r_n) is unbounded leads to a contradiction. $\qquad\square$

We now turn to the proof of Theorem 1.1. We keep the notation from the statement of that theorem. We need a technical lemma, originally given in the context of shift radix systems, Lemma 2 of [14].

Lemma 2.2 *We have*

$$a_n = \tilde{a}_n + \sum_{j=1}^{n-d+1} \hat{a}_{n-j} e_{d-1+j}.$$

Proof By the definition of $(\hat{a}_n)$ and $(\tilde{a}_n)$, it is clearly true for $n \leq d - 1$. Assume that it is true for $n \leq m + d - 1$ with $m \geq 0$. Then

$$a_{m+d} - \tilde{a}_{m+d} = e_{m+d} - \sum_{j=1}^{d} A_{d-j}(a_{m+d-j} - \tilde{a}_{m+d-j})$$

$$= e_{m+d} - \sum_{j=1}^{d} A_{d-j} \sum_{k=1}^{m-j+1} \hat{a}_{m+d-j-k} e_{d-1+k}$$

$$= e_{m+d} - \sum_{k=1}^{m} e_{d-1+k} \sum_{j=1}^{\min(d,m+1-k)} A_{d-j}\hat{a}_{m+d-j-k}.$$

Using the definition of $(\hat{a}_n)$ we have

$$a_{m+d} - \tilde{a}_{m+d} = e_{m+d} - \sum_{k=1}^{m} e_{d-1+k}(-\hat{a}_{m+d-k})$$

$$= \sum_{k=1}^{m+1} \hat{a}_{m+d-k} e_{d-1+k}$$

which finishes the induction. $\qquad\square$

Proof of Theorem 1.1 By the assumptions the characteristic polynomial of (a_n) has a root of modulus at least one, thus (a_n) is an unbounded sequence. Both sequences $(\tilde{a}_n), (\hat{a}_n)$ can be written in the form (4) with $\hat{g}_j(x) = \hat{g}_j$, $\tilde{g}_j(x) = \tilde{g}_j$ constants and $h = d$. Lemma 2.2 implies

$$
\begin{aligned}
a_n &= \sum_{i=1}^{d} \tilde{g}_i \alpha_i^n + \sum_{j=1}^{n-d+1} \sum_{i=1}^{d} \hat{g}_i e_{d-1+j} \alpha_i^{n-j} \\
&= \sum_{i=1}^{d} \left(\tilde{g}_i \alpha_i^n + \hat{g}_i \sum_{j=1}^{n-d+1} e_{d-1+j} \alpha_i^{n-j} \right) \\
&= \sum_{i=1}^{d} \alpha_i^n \left(\tilde{g}_i + \hat{g}_i \sum_{j=1}^{n-d+1} e_{d-1+j} \alpha_i^{-j} \right).
\end{aligned}
$$

If $r_1 = 0$, then the bases of all exponential terms lie in the closed unit disk. Thus all summands are bounded. Further the number of summands is bounded by nd. Thus we proved the theorem for $r_1 = 0$.

The function $c(z)$ is well defined outside the closed unit disk, among others for all $\alpha_1, \ldots, \alpha_{r_1}$. Thus if $r_1 > 0$, then put

$$
\begin{aligned}
b_n &= \sum_{i=1}^{r_1} \left(\tilde{g}_i + \hat{g}_i c(\alpha_i) \right) \alpha_i^n + \sum_{i=r_1+1}^{d} \left(\tilde{g}_i \alpha_i^n + \hat{g}_i \sum_{j=1}^{n-d+1} e_{d-1+j} \alpha_i^{n-j} \right) \\
&= \sum_{i=1}^{r_1} \left(\tilde{g}_i + \hat{g}_i c(\alpha_i) \right) \alpha_i^n + O(r_2 n + 1).
\end{aligned}
$$

Using this notation we obtain

$$
\begin{aligned}
b_n - a_n &= \sum_{i=1}^{r_1} \hat{g}_i \alpha_i^n \left(c(\alpha_i) - \sum_{j=1}^{n-d+1} e_{d-1+j} \alpha_i^{-j} \right) \\
&= \sum_{i=1}^{r_1} \hat{g}_i \alpha_i^n \left(\sum_{j=n-d+2}^{\infty} e_{d-1+j} \alpha_i^{-j} \right) \\
&= O(|\alpha_1|^d) = O(1).
\end{aligned}
$$

From the above observations we immediately deduce (i)–(iii), except that in (i), (ii) we still have to verify that $\tilde{g}_i + \hat{g}_i c(\alpha_i) \neq 0$ for $i = 1, \ldots, r_1$. Let $I \subseteq \{1, \ldots, r_1\}$ be the set of indices i with $\beta_i := \tilde{g}_i + \hat{g}_i c(\alpha_i) \neq 0$, and put

$$
c_n := a_n - \sum_{i \in I} \beta_i \alpha_i^n.
$$

Then (c_n) is an nlrs with $c_n = O(n)$ for all n. By Lemma 2.1 (ii), the characteristic polynomial $g(x)$ of (c_n) has only roots of modulus 1. In general, if (u_n), (v_n) are two nlrs with characteristic polynomials $P_1(x)$, $P_2(x)$, then $u_n + v_n$ is an nlrs, and $P_1(x)P_2(x)$ is in the ideal of $(u_n + v_n)$. In particular, $g(x) \prod_{i \in I}(x - \alpha_i)$ is in the ideal of (a_n). But since the characteristic polynomial of (a_n) has zeros $\alpha_1, \ldots, \alpha_{r_1}$, we must have $I = \{1, \ldots, r_1\}$. $\qquad\square$

Remark 2.1 The assertion (iii) of Theorem 1.1 remains true with simple modifications for nlrs with inseparable characteristic polynomial, but with remainder term $O(n^\kappa)$, where κ is the maximum of the multiplicities of the roots of the characteristic polynomial of (a_n). As in our Diophantine application we cannot deal with this case, we postpone the study of the inseparable case.

Remark 2.2 The error term $O(n)$ in Theorem 1.1 (ii), (iii) is best possible. For instance, let $\alpha_1, \ldots, \alpha_{r_1}, \beta_1, \ldots, \beta_{r_1}$ be as above, let γ be a non-zero complex number, and let (a_n) be a sequence of complex numbers such that

$$a_n = \beta_1 \alpha_1^n + \cdots + \beta_{r_1} \alpha_{r_1}^n + \gamma n + O(1)$$

holds for all $n \geq 1$. Then (a_n) is an nlrs with characteristic polynomial $(x - \alpha_1) \cdots (x - \alpha_{r_1})(x - 1)$. Similarly, if $a_n = \gamma n + O(1)$ holds for all $n \geq 1$, then (a_n) is an nlrs with characteristic polynomial $x - 1$.

Remark 2.3 It is easy to see that the converse of Theorem 1.1 (i) is also true. Indeed, let (a_n) be a sequence of complex numbers such that

$$a_n = \beta_1 \alpha_1^n + \cdots + \beta_{r_1} \alpha_{r_1}^n + O(1)$$

holds for all $n \geq 1$ with non-zero complex numbers $\alpha_1, \ldots, \alpha_{r_1}, \beta_1, \ldots, \beta_{r_1}$ satisfying $|\alpha_j| > 1, j = 1, \ldots, r_1$. Then (a_n) is an nlrs with characteristic polynomial $(x - \alpha_1) \cdots (x - \alpha_{r_1})$. In general, we can show that if there exist non-zero polynomials $g_1, \ldots, g_h$ and complex numbers $\alpha_1, \ldots, \alpha_h$ with $|\alpha_i| \geq 1$ such that

$$a_n = g_1(n)\alpha_1^n + \cdots + g_h(n)\alpha_h^n + O(1), \tag{8}$$

then (a_n) is an nlrs with characteristic polynomial

$$\prod_{i=1}^{h}(x - \alpha_i)^{1 + \deg g_i}.$$

This is not true anymore if in (8) we replace the error term $O(1)$ by $O(n^\kappa)$ with a positive integer κ. We give a counterexample in the simplest case when $a_n = O(n^\kappa)$. Take a sequence (b_n) which is not eventually periodic, taking two values $\{-1, 1\}$. Then the sequence $(n^\kappa b_n)$ cannot be an nlrs. Indeed, if there are $A_0, \ldots A_{\nu-1}$ such that $a_n = n^\kappa b_n$ satisfies

$$a_{n+v} + A_{v-1}a_{n+v-1} + \cdots + A_0 a_n = O(1),$$

then by non-periodicity, we can find two increasing sequences of integers (N_j) and (M_j) for $j = 1, 2, \ldots$ such that

$$(N_j + v)^\kappa b_{N_j+v} + A_{v-1}(N_j + v - 1)^\kappa b_{N_j+v-1} + \cdots + A_0 N_j^\kappa b_{N_j} = O(1),$$

$$(M_j + v)^\kappa b_{M_j+v} + A_{v-1}(M_j + v - 1)^\kappa b_{M_j+v-1} + \cdots + A_0 M_j^\kappa b_{M_j} = O(1)$$

with $b_{N_j+k} = b_{M_j+k}$ for $k = 0, \ldots v-1$ and $b_{N_j+v} + b_{M_j+v} = 0$. Dividing by $(N_j+v)^\kappa$ and $(M_j + v)^\kappa$, respectively, and taking the difference gives an impossibility:

$$2b_{N_j+v} = o(1) \ \text{ as } j \to \infty.$$

3　On the Growth of nlrs

Combining Theorem 1.1 with some Diophantine approximation arguments we are able to prove lower and upper estimates for the growth of nlrs. Specializing our results for lrs we get surprising facts in this case too. Moreover we can estimate the growth of the difference sequence $(a_n - \tilde{a}_n)$. We start with the analysis of a special case.

The main result of this section is the following theorem.

Theorem 3.1 *Assume that $r \geq 2$.*

(i) *Let $\eta_1, \ldots, \eta_r$ be any pairwise distinct complex numbers lying on the unit circle and $\gamma_1, \ldots, \gamma_r$ any non-zero complex numbers. Then there exists a constant $d_1 > 0$ such that*

$$|\gamma_1 \eta_1^n + \cdots + \gamma_r \eta_r^n| > d_1 \tag{9}$$

holds for infinitely many positive integers n.

(ii) *Let $\eta_1, \ldots, \eta_r$ be any pairwise distinct complex numbers lying on the unit circle such that at least one of the quotients η_j/η_r, $1 \leq j < r$ is not a root of unity and $\gamma_1, \ldots, \gamma_{r-1}$ any non-zero complex numbers. Then for all $d_2 > 1$ there exists γ_r such that the inequality*

$$|\gamma_1 \eta_1^n + \cdots + \gamma_r \eta_r^n| < d_2^{-n} \tag{10}$$

holds for infinitely many positive integers n.

Remark 3.1 In relation to (ii), we should remark here that as a consequence of the p-adic Subspace Theorem of Schmidt and Schlickewei, if $\gamma_1, \ldots, \gamma_r, \eta_1, \ldots, \eta_r$ are all algebraic and $|\eta_1| = \cdots = |\eta_r| = 1$, then for every $d_2 > 1$ there are only finitely many positive integers n with (10), see [16] or [5].

In fact, one can show that if $\gamma_1, \ldots, \gamma_{r-1}$ are any non-zero complex numbers and $\eta_1, \ldots, \eta_r$ any complex numbers on the unit circle, then for almost all complex γ_r in the sense of Lebesgue measure, we have that for every $d_2 > 1$, inequality (10) holds for only finitely many positive integers n. To see this, let S be the set of $\gamma_r \in \mathbb{C}$ for which there exists $d_2 > 1$ such that (10) holds for infinitely many n. Then $S = \bigcup_{k=1}^{\infty} S_k$, where S_k is the set of $\gamma_r \in \mathbb{C}$ such that (10) with $d_2 = 1 + k^{-1}$ holds for infinitely n. For fixed n, k, let $B_{n,k}$ be the set of $\gamma_r \in \mathbb{C}$ satisfying (10) with $d_2 = 1 + k^{-1}$. Then $B_{n,k}$ has Lebesgue measure $\lambda(B_{n,k}) = \pi(1 + k^{-1})^{-2n}$, the measure of a ball in $\mathbb{C}$ of radius d_2^{-n}. Thus, S_k is the set of $\gamma_r \in \mathbb{C}$ that are contained in $B_{n,k}$ for infinitely many n. We have $\sum_{n=1}^{\infty} \lambda(B_{n,k}) < \infty$ so by the Borel–Cantelli Lemma, S_k has Lebesgue measure 0. But then, S must have Lebesgue measure 0.

The proof of the second assertion of Theorem 3.1 is based on the following Diophantine approximation result.

Lemma 3.1 *Let $\eta_1, \ldots, \eta_r$ be any pairwise distinct complex numbers lying on the unit circle, at least one of which is not a root of unity. Then for every $d > 0$ there are infinitely many n such that $|\eta_j^n - 1| < d$ holds for $j = 1, \ldots, r$.*

Proof We use the inequality

$$|e^z - 1| = |z| \cdot \left| \sum_{n=1}^{\infty} z^{n-1}/n! \right| < |z| \cdot e^{|z|},$$

which holds for all complex z.

Let $0 < d < 1$. Write $\eta_j = e^{2\pi i u_j}$ with real numbers u_j for $j = 1, \ldots, r$. Since by assumption not all η_j are roots of unity, at least one of the u_j is irrational. By Dirichlet's approximation theorem (see, e.g., [6, Chap. XI, Theorem 200]), there are infinitely many integers n for which there exist integers $m_j = m_j(n)$ such that $|n u_j - m_j| < d/c$ for $j = 1, \ldots, r$, where $c = 2\pi \cdot e^{2\pi}$. For these n,

$$|\eta_j^n - 1| = |e^{2\pi i(n u_j - m_j)} - 1| < e^{2\pi |n u_j - m_j|} 2\pi |n u_j - m_j| < d.$$

$\square$

The second lemma holds under more general assumptions. Its proof was inspired by an idea we found in the Hungarian lecture notes of P. Turán [7, pp. 361–362].

Lemma 3.2 *Let $\eta_1, \ldots, \eta_r$ pairwise different and lying on the unit circle and $\gamma_1(x), \ldots, \gamma_r(x) \in \mathbb{C}[x]$ non-zero. Let $g(n) = \gamma_1(n)\eta_1^n + \cdots + \gamma_r(n)\eta_r^n$ for $n \in \mathbb{Z}$. Assume that $|g(n)| \leq G$ for all $n \geq n_0$. Then for $j = 1, \ldots, r$, $\gamma_j(n)$ is a constant, say γ_j, satisfying $|\gamma_j| \leq G$.*

Proof For every real $\alpha \geq 0$, complex number $\xi \neq 1$ with $|\xi| = 1$ and integer $n_1 \geq n_0$ we have

$$\lim_{T \to \infty} \frac{1}{T} \sum_{n=n_1}^{n_1+T-1} \frac{\xi^n}{n^\alpha} = 0, \tag{11}$$

which follows from Abel summation.

Let n^α be the highest power of n occurring in $\gamma_1(n), \ldots, \gamma_r(n)$. It may occur in various $\gamma_i(n)$. Suppose, for instance, that it occurs in $\gamma_r(n)$ and that the corresponding coefficient is b. Then for any $n_1 \geq n_0$,

$$\lim_{T \to \infty} \frac{1}{T} \sum_{n=n_1}^{n_1+T-1} \frac{g(n)}{n^\alpha \eta_r^n} = \sum_{j=1}^{r-1} \lim_{T \to \infty} \frac{1}{T} \sum_{n=n_1}^{n_1+T-1} \frac{\gamma_j(n)}{n^\alpha} \left(\frac{\eta_j}{\eta_r} \right)^n$$

$$+ \sum_{j=1}^{r-1} \lim_{T \to \infty} \frac{1}{T} \sum_{n=n_1}^{n_1+T-1} \frac{\gamma_r(n)}{n^\alpha} = b$$

by (11). So $|b| \leq G/n_1^\alpha$ for all $n_1 \geq n_0$, implying $\alpha = 0$. Hence $\gamma_1(n), \ldots, \gamma_r(n)$ are all constants, say $\gamma_1, \ldots, \gamma_r$ respectively.

Let $1 \leq j \leq r$. Then applying (11) with $\alpha = 0$ we obtain

$$\lim_{T \to \infty} \frac{1}{T} \sum_{n=n_0}^{n_0+T-1} \frac{g(n)}{\eta_j^n} = \lim_{T \to \infty} \frac{1}{T} \sum_{n=n_0}^{n_0+T-1} \gamma_j = \gamma_j.$$

On the other hand, the modulus of the left-hand side is clearly not greater than G.
$\square$

Now we are in the position to prove Theorem 3.1.

Proof of Theorem 3.1

(i) Let $\Gamma := \max\{|\gamma_j| : j = 1, \ldots, r\}$ and let $0 < d_1 < \Gamma$. If (9) holds for only finitely many integers n, then there exists an n_0 such that

$$|\gamma_1 \eta_1^n + \cdots + \gamma_r \eta_r^n| \leq d_1$$

holds for all $n > n_0$. Then by Lemma 3.2 $|\gamma_j| \leq d_1 < \Gamma$ for all $j = 1, \ldots, r$, which is a contradiction.

(ii) Dividing $\gamma_1 \eta_1^n + \cdots + \gamma_r \eta_r^n$ by η_r^n, we see that without loss of generality we may assume that $\eta_r = 1$ and at least one of $\eta_1, \ldots, \eta_{r-1}$ is not a root of unity. We take any non-zero $\gamma_1, \ldots, \gamma_{r-1}$ and construct γ_r.

Let $u_n := \gamma_1 \eta_1^n + \cdots + \gamma_{r-1} \eta_{r-1}^n$. We construct a sequence (n_k). Let $n_1 := 1$. For $k \geq 1$, given n_k, choose $n_{k+1} > n_k$ such that

$$|\eta_j^{n_{k+1}-n_k} - 1| < (2d_2)^{-n_k}/|rB|, \quad j = 1, \ldots, r-1,$$

where $B > \max(|\gamma_1|, \ldots, |\gamma_{r-1}|)$. This is possible by Lemma 3.1. Then

$$|u_{n_{k+1}} - u_{n_k}| \leq \sum_{j=1}^{r-1} |\gamma_j \eta_j^{n_k}| \cdot |\eta_j^{n_{k+1}-n_k} - 1|$$

$$< (2d_2)^{-n_k}$$

for $k = 1, 2, \ldots$

Now let

$$\gamma_r := -u_{n_1} - \sum_{k \geq 1} (u_{n_{k+1}} - u_{n_k}).$$

This is a convergent series, and for $l \geq 1$,

$$|\gamma_1 \eta_1^{n_l} + \cdots + \gamma_{r-1} \eta_{r-1}^{n_l} + \gamma_r| = |u_{n_l} + \gamma_r|$$

$$= \left| \sum_{k \geq l} (u_{n_{k+1}} - u_{n_k}) \right|$$

$$< (2d_2)^{-n_l} + (2d_2)^{-n_{l+1}} + \cdots$$

$$\leq \frac{2d_2}{2d_2 - 1} (2d_2)^{-n_l} < d_2^{-n_l},$$

completing the proof. $\square$

Theorem 3.1 implies that general linear recurrence sequences may have surprisingly big fluctuation.

Corollary 3.1 *Let $r \geq 2$ be an integer and $h > 1$ be a real number. There exists a lrs u_n of degree r such that $u_n \neq 0$ for all n, $|u_n| \gg h^n$ for infinitely many n and $|u_n| \ll h^{-n}$ for infinitely many n.*

Note that the non-zero assumption expels trivial "degenerate" sequences like $u_n = h^{2n}(1 + \gamma^n + \gamma^{2n} + \cdots + \gamma^{(r-1)n})$ for a primitive rth root of unity γ.

Proof Take distinct algebraic numbers $\eta_1, \ldots, \eta_{r-1}$ that lie on the unit circle and are not roots of unity and set $\eta_r = 1$. Let $D \geq h$ be an integer and put $\alpha_j = D\eta_j$ for $j = 1, \ldots, r$. Finally let $\gamma_j, j = 1, \ldots, r-1$ be non-zero integers. Taking $d_2 = D^2$ there exists by Theorem 3.1 (ii) a complex number γ_r such that

$$|\gamma_1 \eta_1^n + \cdots + \gamma_r \eta_r^n| \ll D^{-2n}$$

holds for infinitely many n. Let $u_n = \gamma_1\alpha_1^n + \cdots + \gamma_r\alpha_r^n$. Then u_n satisfies a linear recursive recursion, for which we have

$$|u_n| = D^n \cdot |\gamma_1\eta_1^n + \cdots + \gamma_r\eta_r^n| \ll D^{-n} \ll h^{-n}$$

for infinitely many n.

We claim that γ_r is transcendental. Indeed, assume that γ_r is algebraic. Then by [16] for every $\varepsilon > 0$ we have $|u_n| \gg D^{n(1-\varepsilon)}$ for sufficiently large n, which is a contradiction. Thus γ_r is transcendental and as $\eta_1,\ldots,\eta_r$ and $\gamma_1,\ldots,\gamma_{r-1}$ are algebraic, we have $u_n \neq 0$ for all n. By using Theorem 3.1 (i), $|u_n| \gg D^n$ for infinitely many n. $\qquad\square$

We deduce some consequences for nlrs.

Corollary 3.2 *Let (a_n) be an nlrs with separable characteristic polynomial and assume that $\alpha_1,\ldots,\alpha_{r_1}$ are its zeros of modulus > 1 with $r_1 \geq 1$. Then there are unique complex numbers $\beta_1,\ldots,\beta_{r_1}$ such that*

$$a_n = \beta_1\alpha_1^n + \cdots + \beta_{r_1}\alpha_{r_1}^n + O(n)$$

holds for all $n \geq 1$.

Proof Such $\beta_1,\ldots,\beta_{r_1}$ exist by Theorem 1.1. Suppose there is also a tuple of complex numbers $(\gamma_1,\ldots,\gamma_{r_1}) \neq (\beta_1,\ldots,\beta_{r_1})$ such that $a_n = \sum_{i=1}^{r_1}\gamma_i\alpha_i^n + O(n)$ for all n. Let k be an index i for which $\gamma_i \neq \beta_i$ and $|\alpha_i|$ is maximal. Then

$$\sum_{i=1}^{r_1}(\gamma_i - \beta_i)(\alpha_i/\alpha_k)^n = O(n \cdot |\alpha_k|^{-n}) \text{ as } n \to \infty.$$

But this clearly contradicts Theorem 3.1 (i). $\qquad\square$

Recall that the Skolem-Mahler-Lech theorem, see, e.g., [9] or [18], asserts that if (a_n) is an lrs, then the set of n with $a_n = 0$ is either finite or contains an infinite arithmetic progression. We show that there is no analogue for nlrs.

Corollary 3.3 *There exists an nlrs with integer terms (a_n) such that* $\limsup_{n\to\infty} |a_n| = \infty$, *but* $a_n = 0$ *for infinitely many n and the set of n with* $a_n = 0$ *does not contain an infinite arithmetic progression.*

Proof Let $\alpha_1,\ldots,\alpha_r$ $(r \geq 2)$ be complex numbers such that

$$|\alpha_1| = \cdots = |\alpha_r| > 1,$$

none of the quotients α_i/α_j $(1 \leq i < j \leq r)$ is a root of unity, and $\overline{\alpha_i} \in \{\alpha_1,\ldots,\alpha_r\}$ for $i = 1,\ldots,r$. Choose non-zero $\gamma_1,\ldots,\gamma_{r-1} \in \mathbb{C}$. Let $C > 1$. By Theorem 3.1

(ii) there exists $\gamma_r \in \mathbb{C}$ such that

$$|\gamma_1\alpha_1^n + \cdots + \gamma_r\alpha_r^n| < C^{-n}$$

for infinitely many n. Let t_n denote the real part of $\sum_{i=1}^r \gamma_i\alpha_i^n$ for all $n \geq 0$ or, in case this is identically 0, $t_n = \frac{1}{2\sqrt{-1}} \cdot \sum_{i=1}^r \gamma_i\alpha_i^n$ for all n. Then t_n is real for all n and $|t_n| < C^{-n}$ for infinitely many n, and by our assumption on the α_i-s, there are $\delta_1, \ldots, \delta_r$, not all 0 such that $t_n = \sum_{i=1}^r \delta_i\alpha_i^n$ for all n. Now we take $a_n := \lfloor t_n \rceil$, where $\lfloor x \rceil := [x + 1/2]$ for $x \in \mathbb{R}$. Then clearly, (a_n) is an nlrs in $\mathbb{Z}$ and $a_n = 0$ for infinitely many n.

It remains to prove that the set of n with $a_n = 0$ does not contain an arithmetic progression. Consider the arithmetic progression $u, u + v, u + 2v, \ldots$. By Theorem 3.1 (i), there are a constant $c > 0$ and infinitely many integers m such that

$$|t_{u+mv}| = |(\delta_1\alpha_1^u)(\alpha_1^v)^m + \cdots + (\delta_r\alpha_r^u)(\alpha_r^v)^m| > c|\alpha_1^v|^m.$$

This implies that $|a_{u+mv}| > c'|\alpha_1^v|^m$ for infinitely many m, where $0 < c' < c$, so in particular, $a_{u+mv} \neq 0$ for infinitely many m. This shows at the same time that $\limsup_{n\to\infty} |a_n| = \infty$. □

In the next corollaries, we compare the nlrs (a_n) and its corresponding lrs analogue $(\tilde{a}_n)$. Although $a_n = \tilde{a}_n$ for $0 \leq n < d$, we can show under a mild condition that the difference $a_n - \tilde{a}_n$ cannot be bounded.

Corollary 3.4 *Under the same assumptions as in Theorem 1.1 set*

$$R = \{\alpha_i \mid i = 1, \ldots, r_1 \text{ and } c(\alpha_i) \neq 0\}.$$

Assume that $R \neq \emptyset$. If among the elements of R there is exactly one of maximum modulus, then $\lim_{n\to\infty} |a_n - \tilde{a}_n| = \infty$, otherwise

$$\limsup_{n\to\infty} |a_n - \tilde{a}_n| = \infty.$$

Proof Observe that the coefficients $\hat{g}_j$ in (4) are all non-zero. Indeed, otherwise $\hat{a}_i$ would be an lrs of order less than d and hence identically 0, which it isn't.

(*i*) Let α_i be the element of R of maximum modulus. Then by Theorem 1.1, we have

$$a_n - \tilde{a}_n = \hat{g}_i c(\alpha_i)\alpha_i^n + o(|\alpha_i|^n).$$

(*ii*) Let $\alpha_{i_1}, \ldots, \alpha_{i_s}$ be the elements of R of maximum modulus. As in case (*i*) we have

$$a_n - \tilde{a}_n = \hat{g}_{i_1} c(\alpha_{i_1}) \alpha_{i_1}^n + \cdots + \hat{g}_{i_s} c(\alpha_{i_s}) \alpha_{i_s}^n + o(|\alpha_{i_s}|^n)$$
$$= d(n) |\alpha_{i_s}|^n + o(|\alpha_{i_s}|^n),$$

where

$$d(n) = \hat{g}_{i_1} c(\alpha_{i_1}) \left(\frac{\alpha_{i_1}}{|\alpha_{i_s}|} \right)^n + \cdots + \hat{g}_{i_s} c(\alpha_{i_s}) \left(\frac{\alpha_{i_s}}{|\alpha_{i_s}|} \right)^n.$$

As the assumptions of Theorem 3.1 (i) hold, we can ensure that $|d(n)| > d_0 > 0$ for infinitely many n, and the proof is complete. $\square$

In the next corollary we need stronger assumptions on the nlrs.

Corollary 3.5 *Assume that $A_0, \ldots, A_{d-2}$ are real, the terms of the nlrs (a_n) are integers and $e_n \geq 0$ for all n, where (e_n) denotes the corresponding error sequence. Further assume that the characteristic polynomial has a single root of maximum modulus, which is real, greater than one and not an algebraic integer. Then $\lim_{n \to \infty} |a_n - \tilde{a}_n| = \infty$.*

Proof Let α_1 be the root of maximum modulus of the characteristic polynomial of (a_n). We prove that $c(\alpha_1) \neq 0$.

Under our assumptions (e_n) is a sequence of real numbers. By definition of $c(z)$, $c(\alpha_1) = 0$ if and only if $e_n = 0$ for all n. This is equivalent to $a_n = \tilde{a}_n$ for all n. If $\tilde{a}_n = a_n$ for all n, then $(\tilde{a}_n)$ is an integer valued lrs. By a result of Fatou (see, e.g., [17]), we know that the formal power series

$$\sum_{n=0}^{\infty} \tilde{a}_n x^n$$

with integer coefficients represents a rational function $P(x)/Q(x)$ with $P, Q \in \mathbb{Z}[x]$ and $Q(0) = 1$. Putting $Q(x) = \sum_{i=0}^{m} q_i x^i$ with $q_0 = 1$, the sequence $(\tilde{a}_n)$ satisfies a linear recurrence:

$$\tilde{a}_{n+m} + q_1 \tilde{a}_{n+m-1} + \cdots + q_m \tilde{a}_n = 0$$

for a sufficiently large n, as well as (3), i.e.,

$$\tilde{a}_{n+d} + A_{d-1} \tilde{a}_{n+d-1} + \cdots + A_0 \tilde{a}_n = 0.$$

Considering the characteristic polynomials of these two recursions, we have $Q(1/\alpha_1) = 0$ and hence α_1 is an algebraic integer. This is a contradiction and we know that $c(\alpha_1) \neq 0$. From Corollary 3.4, we get the result. $\qquad\square$

Corollary 3.5 has the following immediate consequence.

Corollary 3.6 *If the characteristic polynomial of the nlrs (s_n) from (1) has a single root of maximum modulus, and this is real, greater than one and not an algebraic integer, then $\lim_{n\to\infty} |s_n - \tilde{s}_n| = \infty$.*

4   Common Values

Common values of lrs with algebraic terms are quite well investigated. Thanks to the theory of S-unit equations, developed by Evertse [5] and by van der Poorten and Schlickewei [16], Laurent [8] characterized those pairs of lrs's (a_n), (b_n) for which there are infinitely many pairs of indices (k, m) with $a_k = b_m$. His result is not effective. A particular case of Laurent's result is that if (a_n), (b_n) have separable characteristic polynomials then the set of (k, m) with $a_k = b_m$ is either finite or the union of a finite set and of finitely many rational lines. A rational line is a set of the type $\{(k, m) \in \mathbb{Z}^2 : k, m \geq K_0, Ak + Bm + C = 0\}$, where K_0 is a constant ≥ 0 and A, B, C are rational numbers.

We recall that two non-zero complex numbers α, β are multiplicatively dependent if there are integers m, n, not both zero, with $\alpha^m \beta^n = 1$, and multiplicatively independent otherwise. We say that a root α of a polynomial $P(x)$ with complex coefficients is *dominating* if $|\alpha| > |\beta|$ for every other root β of P.

In the case that the characteristic polynomials of the lrs (a_n), (b_n) both have a dominating root and if these two roots are multiplicatively independent, Mignotte [12] proved that there are only finitely many k, m with $a_k = b_m$ and gave an effective upper bound for them. His result was generalized to sequences with at most three, not necessarily dominating roots by Mignotte et al. [13]. One finds a good overview on effective results concerning common values of lnr's in the book of Shorey and Tijdeman [18]. In the above-mentioned results the Binet formula (4) plays a central role.

Theorem 1.1 gives a Binet-type formula for nlrs's, which suggests to study common values of such sequences. The next result implies that the situation for nlrs is quite different from that of lrs.

Theorem 4.1 *Let α, β be two multiplicatively independent real numbers > 1. Then there exist nlrs (a_n), (b_n) with integer terms, having characteristic polynomials with dominating roots α, β, respectively, such that there are infinitely many pairs of nonnegative integers (k, m) with $a_k = b_m$. This set of pairs (k, m) has finite intersection with every rational line.*

In the proof we need some lemmas.

Lemma 4.1 *Let a, b be positive real numbers with $a/b \notin \mathbb{Q}$ and let $C > 1$. Then there exists $c \in \mathbb{R}$ such that the inequality*

$$|ak - bm - c| < C^{-(k+m)}$$

has infinitely many solutions in non-negative integers k, m.

Proof We construct an infinite sequence of triples $(k_n, m_n, \varepsilon_n)$ $(n = 1, 2, \ldots)$ such that $0 < \varepsilon_n < 1$, k_n, m_n are positive integers with $|ak_n - bm_n| < \varepsilon_n$ for all n, and

$$\varepsilon_{n+1} < \min\left(\frac{1}{2}\varepsilon_n, (2C)^{-(k_1 + \cdots + k_n + m_1 + \cdots + m_n)}\right).$$

The existence of such an infinite sequence follows easily from Dirichlet's approximation theorem or the continued fraction expansion of a/b. Now put $s_n := ak_n - bm_n$ and

$$c := \sum_{n=1}^{\infty} s_n.$$

This series is easily seen to be convergent. Further we have, on putting $k_n' := k_1 + \cdots + k_n$, $m_n' := m_1 + \cdots + m_n$,

$$|ak_n' - bm_n' - c| \leq \sum_{l=n+1}^{\infty} |s_l| < 2(2C)^{-(k_n' + m_n')} < C^{-(k_n' + m_n')}.$$

This clearly proves our lemma. $\qquad\square$

Lemma 4.2 *Let α, β be multiplicatively independent reals > 1 and $C > 1$. Then there exists $\gamma > 1$ such that the inequality*

$$|\alpha^k - \gamma\beta^m| < C^{-(k+m)}$$

has infinitely many solutions in positive integers k, m.

Proof By the previous lemma, there exist $\gamma > 1$ and infinitely many pairs of positive integers (k, m), such that

$$|k \log \alpha - m \log \beta - \log \gamma| < (2\beta C)^{-(k+m)}.$$

Using the inequality $|e^x - 1| \leq 2|x|$ for real x sufficiently close to 0, we infer that there are infinitely many pairs (k, m) of positive integers such that

$$
\begin{aligned}
|\alpha^k - \gamma\beta^m| &= \gamma\beta^m \cdot |\alpha^k \beta^{-m} \gamma^{-1} - 1| \\
&\leq 2\gamma\beta^m |k\log\alpha - m\log\beta - \log\gamma| \\
&\leq 2\gamma\beta^m (2\beta C)^{-(k+m)} < C^{-(k+m)}.
\end{aligned}
$$

$\square$

Proof of Theorem 4.1 The previous lemma implies that there exists $\gamma > 0$ such that $[\alpha^k] - [\gamma\beta^m] \in \{-1, 0, 1\}$ for infinitely many pairs of non-negative integers k, m. This implies that there are $u \in \{-1, 0, 1\}$ and infinitely many pairs of non-negative integers k, m such that $[\alpha^k] - [\gamma\beta^m] = u$. Now define (a_n), (b_n) by $a_n := [\alpha^n]$, $b_n := [\gamma\beta^n + u]$. These are easily seen to be nlrs with dominating roots α, β, respectively, and clearly, $a_k = b_m$ for infinitely many pairs k, m. There is $C > 0$ such that $|k\log\alpha - m\log\beta| \leq C$ for all pairs of non-negative integers k, m with $a_k = b_m$. Since by assumption, $\log\alpha / \log\beta \notin \mathbb{Q}$, only finitely many of these pairs (k, m) can lie on a given rational line. This completes our proof. $\square$

Below, we consider the set of pairs (k, m) satisfying $a_k = b_m$ for two given nlrs (a_n), (b_n) in more detail. One of our results is that if (a_n), (b_n) satisfy the conditions of Theorem 4.1 and if moreover α, β are algebraic, then the set of these pairs (k, m) is very sparse.

The main ingredient of our proof is an effective lower bound for linear forms in logarithms of algebraic numbers. We use here a theorem of Matveev [11]. For our qualitative result below it would be enough to use a less explicit form, but we could save almost nothing with it. Before formulating the theorem we have to define the *absolute logarithmic height*—$h(\beta)$—of an algebraic number β. Let β be an algebraic number of degree n and denote by b_0 the leading coefficient of its defining polynomial. Further, denote by $\beta = \beta^{(1)}, \ldots, \beta^{(n)}$ the (algebraic) conjugates of β. Then

$$
h(\beta) = \frac{1}{n}\left(\log|b_0| + \sum_{j=1}^{n} \log\max\{|\beta^{(j)}|, 1\} \right).
$$

Theorem 4.2 *Let $\gamma_1, \ldots, \gamma_t$ be positive real algebraic numbers in a real algebraic number field K of degree D and $b_1, \ldots, b_t$ rational integers such that*

$$
\Lambda := \gamma_1^{b_1} \cdots \gamma_t^{b_t} - 1 \neq 0.
$$

Then

$$
|\Lambda| > \exp\left(-1.4 \times 30^{t+3} \times t^{4.5} \times D^2 (1 + \log D)(1 + \log B)A_1 \cdots A_t\right),
$$

where

$$B \geq \max\{|b_1|,\ldots,|b_t|\},$$

and

$$A_i \geq \max\{Dh(\gamma_i), |\log \gamma_i|, 0.16\} \text{ for all } i = 1,\ldots,t.$$

Now we are in the position to state and prove our main result of this section.

Theorem 4.3 *Let (a_n) and (b_n) be two nlrs. Assume that the characteristic polynomials of (a_n), (b_n) have dominating roots α, β respectively, and that α, β are real algebraic, have absolute value > 1, and are multiplicatively independent.*

Then there exist effectively computable constants K_0, K_1, K_2 depending only on the characteristic polynomials, the initial values, and the sizes of the error terms of (a_n) and (b_n) such that if $(k_1, m_1), (k_2, m_2) \in \mathbb{Z}^2$ are solutions of the diophantine equation

$$a_k = b_m \tag{12}$$

with $K_0 \leq k_1 < k_2$ then $k_2 > k_1 + K_1 \exp(K_2 k_1)$.

Proof As $|\alpha|, |\beta| > 1$ there exist by Theorem 1.1 non-zero constants γ, δ depending only on the starting terms of the sequences (a_n), (b_n) and the coefficients of their characteristic polynomials, and a constant ε depending only on the second largest zeros such that

$$a_k = \gamma \alpha^k + O(\alpha^{k(1-\varepsilon)}), \quad b_m = \delta \beta^m + O(\beta^{m(1-\varepsilon)}) \tag{13}$$

hold for all large enough k, m. Thus if Eq. (12) holds then we have

$$\gamma \alpha^k - \delta \beta^m = O(|\alpha|^{k(1-\varepsilon)}) + O(|\beta|^{m(1-\varepsilon)}).$$

For fixed m this equation has finitely many solutions in k. Let K_0 be a large enough constant, which we will specify later, and assume that (12) has at least one solution (m, k) with $k > K_0$.

We may assume $\alpha, \beta > 0$ without loss of generality. Indeed, otherwise we consider the cases k, m odd and even separately. Moreover we may also assume $\alpha^k > \beta^m$ (equality cannot occur as α and β are multiplicatively independent). Then the last inequality implies

$$\left| \frac{\delta}{\gamma} \frac{\beta^m}{\alpha^k} - 1 \right| < C_1 \alpha^{-k\varepsilon} + C_2 \beta^{-m\varepsilon}.$$

Here and in the sequel the constants $C_1, C_2, \ldots$ are effectively computable and depend on the parameters of the sequences, i.e. on their initial terms and the heights of the coefficients and their characteristic polynomials, on ε and on the upper bound of the terms of the error sequences only.

We now assume that K_0 is large enough so that the right-hand side of the last inequality is less than $1/2$. Then

$$\beta^m > \left| \frac{\gamma}{2\delta} \right| \cdot \alpha^k,$$

thus

$$\left| \frac{\delta}{\gamma} \frac{\beta^m}{\alpha^k} - 1 \right| < C_3 \alpha^{-k\varepsilon}. \tag{14}$$

This inequality seems to have already the form for which Matveev's Theorem 4.2 could be applied. Unfortunately we are not yet so far because we know nothing about the arithmetic nature of γ and δ. They can be (and are probably usually) transcendental numbers. Thus Theorem 4.2 is not applicable and we cannot deduce an upper bound for $\max\{k, m\}$. On the other hand, inequality (14) is strong enough to allow us to prove that the sequence of solutions of (12) is growing very fast.

Indeed, let $(k_1, m_1), (k_2, m_2)$ be solutions of (12) such that $K_0 < k_1 < k_2$. Then (14) holds for both solutions and we get

$$\left| \frac{\delta}{\gamma} \frac{\beta^{m_1}}{\alpha^{k_1}} - \frac{\delta}{\gamma} \frac{\beta^{m_2}}{\alpha^{k_2}} \right| < 2C_3 \alpha^{-k_1 \varepsilon}.$$

Dividing this inequality by the first term, which lies by (14) in the interval $\left(\frac{1}{2}, \frac{3}{2} \right)$ we get

$$|\Lambda| < C_4 \alpha^{-k_1 \varepsilon}, \tag{15}$$

where

$$\Lambda = \beta^{m_2 - m_1} \alpha^{k_1 - k_2} - 1.$$

As α and β are positive real numbers and multiplicatively independent we have $\Lambda \neq 0$, thus we may apply Theorem 4.2 to it, with $t = 2$. In our situation, D is the degree of the number field $\mathbb{Q}(\alpha, \beta)$, A_1, A_2 are constants depending only on the coefficients of the characteristic polynomials of the sequences. Further $b_1 = m_2 - m_1$ and $b_2 = k_1 - k_2$. We proved above that if k, m are integers with $a_k = b_m$ and $k > K_0$ then either

$$\alpha^k > \beta^m > \left| \frac{\gamma}{2\delta} \right| \alpha^k$$

or

$$\beta^m > \alpha^k > \left|\frac{\delta}{2\gamma}\right|\beta^m$$

holds. We have in both cases

$$\left|m - \frac{\log \alpha}{\log \beta}k\right| < C_5.$$

This implies

$$\left|\left|m_1 - m_2\right| - \frac{\log \alpha}{\log \beta}\left|k_1 - k_2\right|\right| < C_6. \tag{16}$$

Thus $|b_1| < C_7|b_2| + C_6$ and Theorem 4.2 implies

$$|\Lambda| > \exp(-C_8 D^2(1 + \log D)A_1 A_2(2 + \log C_7 + \log(k_2 - k_1))).$$

Comparing this inequality with (15) we obtain

$$C_9 \log(k_2 - k_1) + C_{10} > C_{11}\varepsilon k_1 - \log C_4,$$

which implies

$$k_2 > k_1 + K_2 \exp(K_1 k_1), \tag{17}$$

with $K_1 = C_{11}\varepsilon/C_9$ and $K_2 = \exp(-C_{10}/C_9 - (\log C_4)/C_9)$. $\qquad\square$

We now consider the case that the (a_n), (b_n) have characteristic polynomials with multiplicatively dependent dominant roots. We show that in this case, if the number of pairs (k, m) with $a_k = b_m$ is infinite, then apart from at most finitely many exceptions they lie on a rational line.

Theorem 4.4 *Let (a_n) and (b_n) be nlrs's. Assume that the characteristic polynomials of both sequences are separable, and have dominating roots α, β with $|\alpha| > 1$, $|\beta| > 1$ which are multiplicatively dependent. If the equation*

$$a_k = b_m \tag{18}$$

has infinitely many solutions in non-negative integers k, m then there exist integers u, v, w such that for all but finitely many solutions we have $k = um/v + w/v$.

Proof Like in the proof of Theorem 4.3 we write

$$a_k = \gamma\alpha^k + O(|\alpha|^{k(1-\varepsilon)}) \quad \text{and} \quad b_m = \delta\beta^m + O(|\beta|^{m(1-\varepsilon)}),$$

with $\gamma, \delta \neq 0$. As α and β are multiplicatively dependent, there exist positive integers u, v such that

$$\alpha^u = \beta^v,$$

i.e., there exists a vth root of unity ζ with

$$\beta = \zeta \alpha^{u/v}.$$

If $a_k = b_m$, then

$$\gamma \alpha^k - \zeta^m \delta \alpha^{um/v} = O(|\alpha|^{k(1-\varepsilon)}) + O(|\alpha|^{um(1-\varepsilon)/v}). \tag{19}$$

Assume that $k - um/v > \ell_1$, where the integer ℓ_1 is so large that

$$|\delta \alpha^{-\ell_1}| < \left|\frac{\gamma}{3}\right|.$$

If, moreover, k is large enough, then dividing (19) by α^k would make the absolute value of the right-hand side smaller than $\left|\frac{\gamma}{3}\right|$ too, which is impossible. Thus if (18) has infinitely many solutions k, m, then $k - um/v \leq \ell_1$. Similarly, if $um/v - k > \ell_2$, where the integer ℓ_2 is so large that

$$|\gamma \alpha^{-\ell_2}| < \left|\frac{\delta}{3}\right|,$$

then repeating the former argument we get again a contradiction. Thus setting $\ell = \max\{\ell_1, \ell_2\}$ we must have $|k - um/v| \leq \ell$ for all but finitely many solutions of (18).

Thus we have shown that for all but finitely many solutions (k, m) of (18) there is $w \in [-v\ell, v\ell] \cap \mathbb{Z}$ such that $k - um/v = w/v$. We have to show that w is independent of the choice of (k, m). Clearly, there is w such that $k - um/v = w/v$ holds for infinitely many solutions (k, m) of (18). Dividing (19) by $|\alpha|^k$ we see that

$$|\gamma - \zeta^m \delta \alpha^{w/v}| = O(|\alpha|^{-k\varepsilon})$$

for these solutions (k, m). Since the left-hand side of this inequality assumes only finitely many values and the right-hand side can become arbitrarily small, there must be an integer r such that

$$\gamma = \zeta^r \delta \alpha^{w/v}.$$

We show that this uniquely determines w. Indeed, suppose we have $\gamma = \zeta^{r_1} \delta \alpha^{w_1/v} = \zeta^{r_2} \delta \alpha^{w_2/v}$ for two tuples of integers $(r_1, w_1), (r_2, w_2)$. Then $\alpha^{(w_1 - w_2)/v}$ is a root of unity, which implies $w_1 = w_2$ since by assumption, α is not a root of unity. This shows that $k = um/v + w/v$ holds for all but finitely many solutions (k, m) of (18). $\qquad\square$

A nearly immediate consequence of Theorem 4.4 is the following assertion.

Corollary 4.1 *Let (a_n) be an nlrs. Assume that its characteristic polynomial is separable, and has a dominant root α with $|\alpha| > 1$. Then the equation*

$$a_k = a_m \tag{20}$$

has only finitely many solutions with $k \neq m$.

Proof We apply Theorem 4.4 in the situation that the sequences under consideration are equal. We have plainly $u = v = 1$, i.e. $k = m + w$ holds with a fixed integer w for all but finitely many solutions of (20). Then

$$\gamma \alpha^k (\alpha^{m-k} - 1) = O(|\alpha|^{k(1-\varepsilon)}),$$

which is absurd, if $m - k = w \neq 0$. $\qquad\square$

Acknowledgements Research supported in part by the OTKA grants NK104208, NK115479.

References

1. S. Akiyama, A. Pethő, Discretized rotation has infinitely many periodic orbits. Nonlinearity **26**, 871–880 (2013)
2. S. Akiyama, T. Borbély, H. Brunotte, A. Pethő, J. Thuswaldner, Generalized radix representations and dynamical systems I. Acta Math. Hungar. **108**(3), 207–238 (2005)
3. S. Akiyama, H. Brunotte, A. Pethő, W. Steiner, Remarks on a conjecture on certain integer sequences. Period. Math. Hung. **52**, 1–17 (2006)
4. S. Akiyama, H. Brunotte, A. Pethő, W. Steiner, Periodicity of certain piecewise affine planar maps. Tsukuba J. Math. **32**(1), 1–55 (2008)
5. J.-H. Evertse, On sums of S-units and linear recurrences. Compos. Math. **53**, 225–244 (1984)
6. G.H. Hardy, E.M. Wright, *An Introduction to the Theory of Numbers*, 4th edn. (with corrections) (Clarendon Press, Oxford, 1975)
7. I. Lánczi, P. Turán, *Számelmélet* (Number Theory) (Tankönyvkiadó, Budapest, 1969) [in Hungarian]
8. M. Laurent, Equations exponentielles polynômes et suites récurrentes linéaires II. J. Number Theory **31**, 24–53 (1989)
9. C. Lech, A note on recurring series. Ark. Mat. **2**, 417–421 (1953)
10. J.H. Lowenstein, S. Hatjispyros, F. Vivaldi, Quasi-periodicity, global stability and scaling in a model of Hamiltonian round-off. Chaos **7**, 49–56 (1997)
11. E.M. Matveev, An explicit lower bound for a homogeneous rational linear form in the logarithms of algebraic numbers, II. Izv. Ross. Akad. Nauk Ser. Mat. **64**(6), 125–180 (2000). Translation in Izv. Math. **64**(6), 1217–1269 (2000)
12. M. Mignotte, Intersection des images de certaines suites récurrentes linéaires. Theor. Comput. Sci. **7**, 117–122 (1978)
13. M. Mignotte, T.N. Shorey, R. Tijdeman, The distance between terms of an algebraic recurrence sequence. J. Reine Angew. Math. **349**, 63–76 (1984)

14. A. Pethő, Notes on *CNS* polynomials and integral interpolation, in *More Sets, Graphs and Numbers*, ed. by E. Győry, G.O.H. Katona, L. Lovász. Bolyai Society Mathematical Studies, vol. 15 (Springer, Berlin, 2006), pp. 301–315
15. A. Pethő, P. Varga, Canonical number systems over imaginary quadratic euclidean domains. Colloq. Math. **146**, 165–186 (2017)
16. A.J. van der Poorten, H.P. Schlickewei, The Growth Conditions for Recurrence Sequences. Macquarie University, NSW, Australia, Report 82.0041 (1982)
17. R. Salem, *Algebraic Numbers and Fourier Analysis* (D. C. Heath and Co., Boston, MA, 1963)
18. T.N. Shorey, R. Tijdeman, *Exponential Diophantine Equations*. Cambridge Tracts in Mathematics, vol. 87 (Cambridge University Press, Cambridge, 1986)

Risk Theory with Affine Dividend Payment Strategies

Hansjörg Albrecher and Arian Cani

Dedicated to Robert F. Tichy at the occasion of his 60th birthday

Abstract We consider a classical compound Poisson risk model with affine dividend payments. We illustrate how both by analytical and probabilistic techniques closed-form expressions for the expected discounted dividends until ruin and the Laplace transform of the time to ruin can be derived for exponentially distributed claim amounts. Moreover, numerical examples are given which compare the performance of the proposed strategy to classical barrier strategies and illustrate that such affine strategies can be a noteworthy compromise between profitability and safety in collective risk theory.

1 Introduction

The question of how to pay dividends from a surplus process of an insurance portfolio has a long tradition in collective risk theory. The classical criterion to measure the performance of such a dividend strategy is the expected sum of discounted dividend payments over the lifetime of the process, where typically the discount rate is assumed to be positive and constant over time. In this case the optimal strategy is a balance between paying out dividends early (in view of the discounting) and paying dividends later (so that due to the typically positive drift of the process the lifetime (and hence the time span of dividend payments) is prolongated). This criterion was first proposed by De Finetti [11], who proved for a simple random walk model that the optimal strategy is a barrier strategy, that is, dividends are paid out whenever the surplus process

H. Albrecher (✉)
University of Lausanne and Swiss Finance Institute, Quartier UNIL-Dorigny, Bâtiment Extranef, 1015 Lausanne, Switzerland
e-mail: hansjoerg.albrecher@unil.ch

A. Cani
University of Lausanne, Quartier UNIL-Dorigny, Bâtiment Extranef, 1015 Lausanne, Switzerland
e-mail: arian.cani@unil.ch

© Springer International Publishing AG 2017
C. Elsholtz, P. Grabner (eds.), *Number Theory – Diophantine Problems, Uniform Distribution and Applications*, DOI 10.1007/978-3-319-55357-3_2

exceeds a threshold value (the horizontal dividend barrier), and no dividends are paid out below that level (i.e., the process is reflected at this barrier). Later, Gerber [13] proved that for a Cramér–Lundberg risk process, a so-called band strategy is optimal, which simplifies to a barrier strategy in some particular cases (including the one with exponential claim size distribution). More recently, this stochastic control problem was embedded in modern control theory, which led to surprisingly challenging mathematical problems (see, e.g., Schmidli [23] and Azcue and Muler [7]). The optimal dividend problem was also studied intensively in many different variants, including model variations, transaction costs, as well as other objective functions and constraints, see [2] and [5] for an overview.

One disadvantage of the classical criterion of maximizing the expected sum of discounted dividend payments until ruin is that it focuses on profitability only, and does not consider the lifetime of the controlled process (in particular, under the optimal band strategy, the process will be ruined with probability 1, and if the barrier is at level 0, then it is even optimal to pay out all the surplus immediately and get ruined at the occurrence of the first claim payment; we refer to [4] for an overview of the ruin concept and its many mathematical implications). In [27], a variant of the dividend problem was studied, where the objective function is a weighted sum of expected discounted dividend payments until ruin and expected ruin time. It turns out that in such a setting, again a band strategy (respectively, barrier strategy) is optimal, albeit with modified parameters. This approach was then extended to more general models in [18].

The criterion of maximizing the expected sum of discounted dividend payments until ruin may be considered as a somewhat natural target, which also has economic motivation in terms of valuating a company on the basis of this quantity (starting with [15] and later variants within the corporate finance literature). However, if a barrier strategy is optimal, in addition to the solvency aspect mentioned above, this strategy does not pay any dividends whenever the surplus is below the barrier and it pays the maximal feasible amount above the barrier, so that the dividend stream may be very uneven over time. At the same time, empirical research suggests that companies typically strive for a smooth dividend distribution over time with the incentive to gradually move towards a long-term payout ratio (see, e.g., Lintner [17] for a pioneering study on this topic). This goes in line with the observation that dividend payments in practice often adjust to changes in earnings only slowly (indicating that the management exhibits some reluctance to either increase or decrease established dividend levels unless there is sufficient confidence that the new levels are justified for the future, not the least to avoid psychological effects entailed by dividend reductions), see also Brav et al. [8].

In view of these aspects, in this paper we propose a dividend strategy that secures a continuous dividend payment stream, the rate of which is adjusted according to the present surplus value in an affine way. We will study such a strategy for a compound Poisson surplus model. Our approach is in part inspired by Avanzi and Wong [6] who studied a related strategy for a diffusion process and also gave an extensive numerical study of its performance. Mathematically, our model in the Cramér–Lundberg framework will lead to an Ornstein–Uhlenbeck process driven by the compound Poisson subordinator. For such a setup we will derive equations for the expected discounted dividend payments until ruin as well as for the Laplace transform of the time of ruin. These equations turn out to be challenging in their own right, and various different approaches to solve them will lead to interesting relations between special functions of hypergeometric type.[1]

An interesting consequence of the numerical results at the end of the paper is that utilizing such an affine dividend strategy leads to almost the same performance as the barrier strategy in terms of expected sum of discounted dividend payments, but has—in many different parameter settings—a considerably longer lifetime. Consequently, in view of a compromise between profitability and safety, such an affine strategy is certainly an interesting alternative. In fact, such a strategy is known to be optimal in a somewhat different context of linear quadratic optimal control problems, where quadratic deviations of a target "dividend" rate are punished in the objective function, see Steffensen [26] for an application in the control of pension funds and Parlar [21] for a model in forest management systems.

The rest of the paper is organized as follows. In Sect. 2, we introduce the model and discuss some basic properties. Section 3 then derives the integro-differential equation for the expected discounted dividend payments and studies its solution for the case of exponentially distributed claim amounts. In Sect. 4 we pursue an alternative approach for the solution of the latter equation via Laplace transforms, leading to a rather intricate study of certain special functions and suggesting an identity that seems to be new and non-obvious. In Sect. 5 we adapt the calculations of Sect. 3 to study the Laplace transform of the time to ruin. In order to retrieve a concrete formula for the expected ruin time from the Laplace transform, we then employ an approach based on digamma functions and another one based on Kampé de Fériet functions. Section 6 gives a simple and intuitive probabilistic view to connect the quantities of Sects. 3 and 5. Finally, Sect. 7 provides detailed numerical illustrations to test the proposed strategy and determines optimal parameters. The results are then compared to the optimal barrier strategies showing that affine strategies can be a competitive alternative to barrier strategies when paying dividends.

---

[1]In this way, a practically motivated question of insurance risk theory leads to non-trivial mathematical problems and relations, a connection which is also in the tradition of Robert Tichy's work, to whom this paper is dedicated. For the application of Quasi-Monte Carlo results to risk theory by Robert Tichy, see, e.g., [3, 28].

2 The Model

In the classical Cramér–Lundberg risk model, the surplus process of an insurance company $(R_t)_{t\geq 0}$ is described by

$$R_t = x + ct - \sum_{i=1}^{N_t} Y_i, \qquad t \geq 0, \tag{1}$$

where $x = R_0$ is the initial capital, $c > 0$ is the constant premium rate and the claims $\{Y_i\}_{i\in\mathbb{N}}$ are a sequence of independent and identically distributed positive random variables with distribution function F_Y, bounded density f_Y, and finite mean μ. The number of claims up to time $t \geq 0$ is assumed to be a homogeneous Poisson process N_t with intensity $\lambda > 0$, independent of $\{Y_i\}_{i\in\mathbb{N}}$.

Let D_t denote the accumulated dividends paid up to time t, so $X_t := R_t - D_t$ is the surplus process after dividend payments. Assume now that dividends are paid according to an affine strategy, i.e.

$$dD_t = (qX_t + \beta)dt, \tag{2}$$

where $q > 0$ is a fixed proportionality constant and $0 \leq \beta \leq c$ is a constant rate. Then

$$dX_t = (c - (qX_t + \beta))\,dt - dS_t, \tag{3}$$

which identifies X_t as a Lévy-driven Ornstein–Uhlenbeck process (in the present case, the driving Lévy process is the compound Poisson process $S_t = \sum_{i=1}^{N_t} Y_i$). The unique solution to (3) is given in terms of the stochastic integral

$$X_t = \frac{c - \beta}{q} + \left(x - \frac{c - \beta}{q}\right) e^{-qt} - \int_0^t e^{-q(t-u)} dS_u, \tag{4}$$

i.e.,

$$X_t = \frac{c - \beta}{q} + \left(x - \frac{c - \beta}{q}\right) e^{-qt} - \sum_{i=1}^{N_t} e^{-q(t-T_i)} Y_i,$$

which embeds X_t into the class of shot-noise processes. One sees that the process X_t behaves like an exponentially decaying function between the claim occurrences, and the influence of past claims on the value of X_t also decays exponentially in time (see Fig. 1 for a sample path of X_t). Let

$$\tau_x := \inf\{t \geq 0 : X_t < 0 \mid X_0 = x\}$$

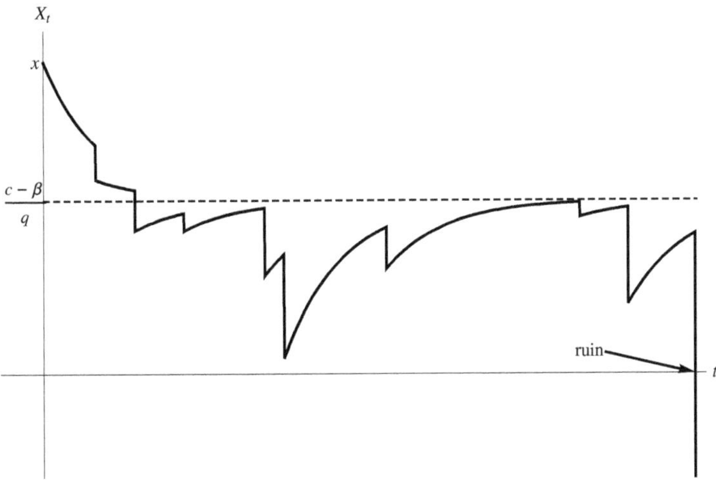

Fig. 1 Sample path of X_t

denote the time of ruin of X_t and note that $P(\tau_x < \infty) = 1$ for all $x \geq 0$ (i.e., ruin is certain). The latter holds true, since the process X_t is upper-bounded by $\max\{x, (c-\beta)/q\}$ (above $(c-\beta)/q$ there is a negative drift down to this level and below it is bounded by this level).

If X_t is not stopped at ruin, then as $t \to \infty$

$$X_t \xrightarrow{a.s.} X_\infty := \frac{c-\beta}{q} - \int_0^\infty e^{-qu} dS_u, \tag{5}$$

see, e.g., [22]. If the claim sizes Y_i are $\mathrm{Exp}(\alpha)$-distributed, then the self-decomposable limit random variable X_∞ simplifies further to a shifted Gamma random variable $X_\infty = (c-\beta)/q - Z$ with $Z \sim \Gamma(\lambda/q, \alpha)$, see also [9, 24].

3 Expected Discounted Dividend Payments

We are now interested in the expected value of the sum of the discounted dividend payments up to the time of ruin

$$V(x) := \mathbb{E}_x\left[\int_0^{\tau_x} e^{-\delta t}(qX_t + \beta)\,dt\right], \tag{6}$$

where $\delta \geq 0$ is a force of interest for valuation. Let us first consider some elementary, but general properties of the function $V(x)$ regarding bounds and growth rate.

Proposition 3.1 *For $x \geq 0$, the function $V(x)$ satisfies the following bounds:*

$$\underline{l} + \frac{qx}{q+\delta} \leq V(x) \leq \bar{l} + \frac{qx}{q+\delta}, \tag{7}$$

where $\underline{l} = \frac{\lambda(q(c-\lambda\mu)+\delta\beta)}{(q+\delta)(\delta+\lambda)}$ and $\bar{l} = \frac{cq+\delta\beta}{\delta(q+\delta)}$.

Proof For any $t \geq 0$, the process X_t in (3) satisfies $X_t \leq \left(x - \frac{c-\beta}{q}\right)e^{-qt} + \frac{c-\beta}{q} := \tilde{X}_t$. Then, clearly,

$$V(x) \leq \mathbb{E}_x\left[\int_0^\infty e^{-\delta t}\left(q\tilde{X}_t + \beta\right)dt\right] = \frac{qx - c + \beta}{q+\delta} + \frac{c}{\delta},$$

which yields the upper bound.

For the lower bound, define $h(x) := \frac{qx}{q+\delta}\mathbb{1}_{\{x\geq 0\}}$ and let M be an operator acting on h defined as

$$(Mh)(x) := \mathcal{L}h(x) - \delta h(x) + qx + \beta, \tag{8}$$

for $x \geq 0$, where $\mathcal{L}h(x) := (c - (qx + \beta))h'(x) + \lambda\left(\int_0^x h(x-y)dF_Y(y) - h(x)\right)$ is the infinitesimal generator of the process (3). More concretely, (8) can be rewritten as

$$(Mh)(x) = (c - (qx + \beta))\frac{q}{q+\delta} + \lambda\left(\int_0^x \frac{q(x-y)}{q+\delta}dF_Y(y) - \frac{qx}{q+\delta}\right)$$
$$- \frac{\delta qx}{q+\delta} + qx + \beta,$$
$$= \frac{cq + \delta\beta}{q+\delta} + \frac{\lambda qx}{q+\delta}(F_Y(x) - 1) - \frac{\lambda q}{q+\delta}\int_0^x y\,dF_Y(y).$$

Observe that $(Mh)'(x) = \frac{\lambda q}{q+\delta}(F_Y(x) - 1) \leq 0$ with boundary values $(Mh)(0) = \frac{cq+\delta\beta}{q+\delta} > 0$ and $\lim_{x\to\infty}(Mh)(x) = \frac{q(c-\lambda\mu)+\delta\beta}{q+\delta} > 0$. Thus, $(Mh)(x)$ is strictly positive and monotone decreasing, bounded from below by $\frac{q(c-\lambda\mu)+\delta\beta}{q+\delta}$.

In view of the Dynkin formula applied to the function $e^{-\delta t}h(X_t)$, the process

$$e^{-\delta t}h(X_t) - h(x) - \int_0^t e^{-\delta s}\left[\mathcal{L}h(X_s) - \delta h(X_s)\right]ds$$

is a zero-expectation martingale. Bearing in mind that the stopped process $X_{t\wedge\tau}$ is also a martingale, we obtain

$$\mathbb{E}_x\left(e^{-\delta t\wedge\tau}h(X_{t\wedge\tau})\right) = h(x) + \mathbb{E}_x\left(\int_0^{t\wedge\tau} e^{-\delta s}\left[\mathcal{L}h(X_s) - \delta h(X_s)\right]ds\right).$$

From the properties of M, we get that the integrand on the right-hand side is bounded from below by $-(qX_s + \beta) + \frac{q(c-\lambda\mu)+\delta\beta}{q+\delta}$. Furthermore, since $h(X_{t\wedge\tau})$ is linearly bounded in t, an application of the monotone convergence theorem implies that as $t \to \infty$, the right-hand side converges to 0. Combining the above and rearranging terms yields

$$\mathbb{E}_x\left[\int_0^\tau e^{-\delta s}(qX_s + \beta)\,ds\right] \geq h(x) + \mathbb{E}_x\left[\int_0^\tau e^{-\delta s}\frac{q(c-\lambda\mu)+\delta\beta}{q+\delta}\,ds\right],$$

$$\geq h(x) + \mathbb{E}_x\left[\int_0^{T_1} e^{-\delta s}\frac{q(c-\lambda\mu)+\delta\beta}{q+\delta}\,ds\right],$$

$$= h(x) + \frac{\lambda\left(q(c-\lambda\mu)+\delta\beta\right)}{(q+\delta)(\delta+\lambda)},$$

which gives the result. $\qquad\square$

Proposition 3.2 *For* $0 \leq y < x$ *and* $\overline{f_Y} := \max_x f_Y(x) < \infty$, *the following inequality holds*

$$\frac{\lambda q(x-y)}{q+\lambda+\delta} \leq V(x) - V(y) \leq \frac{q(x-y)}{q+\delta}\left(1 + \left(x - y + \frac{c}{\delta} + \frac{\beta}{q}\right)\overline{f_Y}\right).$$

Proof Let $0 \leq y < x$ and let X_t^y and X_t^x be the processes in (4) started in y and x with respective times of ruin τ_y and τ_x. Additionally, define $\mathcal{M} = \{\omega \in \Omega \mid \tau_x(\omega) = \tau_y(\omega)\}$ and denote by $\mathcal{M}^c$ its complementary set. A pathwise comparison of both processes on $\mathcal{M}$ gives $X_t^x(\omega) - X_t^y(\omega) = (x-y)e^{-qt}$. We have

$$V(x) - V(y) = \mathbb{E}\left[\int_0^{\tau_y} e^{-\delta t}qX_t^x dt\right] - \mathbb{E}\left[\int_0^{\tau_y} e^{-\delta t}qX_t^y dt\right]$$

$$+ \mathbb{E}\left[\mathbb{1}_{\mathcal{M}^c}\int_{\tau_y}^{\tau_x} e^{-\delta t}\left(qX_t^x + \beta\right)dt\right],$$

$$= \mathbb{E}\left[\int_0^{\tau_y} e^{-(q+\delta)t}q(x-y)dt\right] + \mathbb{E}\left[\mathbb{1}_{\mathcal{M}^c}\int_{\tau_y}^{\tau_x} e^{-\delta t}\left(qX_t^x + \beta\right)dt\right],$$

$$\geq \mathbb{E}\left[\int_0^{T_1} e^{-(q+\delta)t}q(x-y)dt\right],$$

$$= \frac{\lambda q(x-y)}{q+\lambda+\delta}.$$

For the reverse direction, we can write

$$V(x) - V(y) = \mathbb{E}\left[\int_0^{\tau_y} e^{-(q+\delta)t} q(x-y)dt\right] + \mathbb{E}\left[\mathbb{1}_{\mathcal{M}^c}\int_{\tau_y}^{\tau_x} e^{-\delta t} X_t^x dt\right],$$

$$\leq \int_0^{\infty} e^{-(q+\delta)t} q(x-y)dt + V(x-y)\mathbb{E}\left[\mathbb{1}_{\mathcal{M}^c}\right]. \tag{9}$$

The last inequality follows from the a.s. finiteness of τ_y in the first integral combined with the strong Markov property of the process X^x and observing that on $\mathcal{M}^c$, $X_{\tau_y}^x(\omega) \leq (x-y)e^{-q\tau_y(\omega)} \leq x - y$ in the second integral. By definition, $\mathcal{M}^c$ comprises all paths ω such that $\tau_x(\omega) > \tau_y(\omega)$, therefore $\mathbb{E}[\mathbb{1}_{\mathcal{M}^c}] = P(\tau_x > \tau_y)$. Writing $X_{\tau_y-}^y$ for the surplus immediately prior to ruin of the surplus started in y and conditioning on the latter leads to

$$P(\tau_x > \tau_y) = \int_0^{\max\left(y, \frac{c-\beta}{q}\right)} P(\tau_x > \tau_y \mid X_{\tau_y-}^y = z)P(X_{\tau_y-}^y \in dz)$$

$$= \int_0^{\max\left(y, \frac{c-\beta}{q}\right)} P(z < Y \leq z + x - y)\, P(X_{\tau_y-}^y \in dz),$$

$$= \int_0^{\max\left(y, \frac{c-\beta}{q}\right)} \int_z^{z+x-y} f_Y(w)\, dw\, P(X_{\tau_y-}^y \in dz) \leq (x-y)\overline{f_Y}.$$

Substituting the last result in (9) and explicitly evaluating the first integral in the aforementioned expression gives

$$V(x) - V(y) \leq \frac{q(x-y)}{q+\delta} + V(x-y)(x-y)\overline{f_Y}.$$

Combining this with the upper bound obtained in Proposition 3.1 establishes the result.   □

Hence $V(x)$ is locally Lipschitz and differentiable almost everywhere. If the derivative exists, then using the typical infinitesimal generator arguments for X_t and in view of (7), one gets that $V(x)$ is characterized as a solution to the integro-differential equation (IDE)

$$(c - (qx + \beta))\, V'(x) - (\lambda + \delta)V(x) + \lambda\int_0^x V(x-y)dF_Y(y) = -(qx+\beta), \qquad x \geq 0. \tag{10}$$

3.1 Constructing an Exact Solution for Exponential Claims

We now assume that the claims are exponentially distributed with rate $\alpha > 0$. Then, applying the operator $\left(\frac{d}{dx} + \alpha\right)$ to both sides of (10) leads to the second-order differential equation

$$(c - (qx + \beta))\, V''(x) + [\alpha\,(c - (qx + \beta)) - (q + \lambda + \delta)]\, V'(x) - \alpha\delta V(x)$$

$$= -q(1 + \alpha x) - \alpha\beta. \tag{11}$$

Let V_h be the solution to the related homogeneous differential equation of (11). Choosing $f(z) := V_h(x)$ associated to the change of variable $z := z(x) = \frac{\alpha(c-(qx+\beta))}{q}$ produces *Kummer's* confluent hypergeometric equation

$$zf''(z) + (b - z)f'(z) - af(z) = 0, \qquad z \le \frac{\alpha(c - \beta)}{q}, \tag{12}$$

with parameters

$$a = \frac{\delta}{q}, \qquad b = 1 + \frac{\lambda + \delta}{q},$$

which has a regular singular point at $z = 0$ and an irregular singular point at $z = -\infty$ (which in the original coordinates correspond to $x = (c-\beta)/q \ge 0$ and $x = \infty$, respectively). This gives

$$V_h(x) = f(z)$$

$$= \begin{cases} A_1\, M\left(\frac{\delta}{q}, 1 + \frac{\lambda+\delta}{q}, z(x)\right) + A_2\, U\left(\frac{\delta}{q}, 1 + \frac{\lambda+\delta}{q}, z(x)\right), & 0 \le x \le \frac{c-\beta}{q}, \\[2mm] A_3\, M\left(\frac{\delta}{q}, 1 + \frac{\lambda+\delta}{q}, z(x)\right) + A_4\, e^{z(x)} U\left(1 + \frac{\lambda}{q}, 1 + \frac{\lambda+\delta}{q}, -z(x)\right), & x > \frac{c-\beta}{q}, \end{cases}$$

$$\tag{13}$$

for arbitrary constants $A_i, i = 1, \ldots 4$. Here

$$M(a, b, z) = {}_1F_1(a, b, z) = \sum_{n=0}^{\infty} \frac{(a)_n}{(b)_n} \frac{z^n}{n!} \tag{14}$$

denotes the Kummer confluent hypergeometric function with the Pochhammer symbol $(a)_n = \Gamma(a + n)/\Gamma(n)$, and

$$U(a, b, z) = \begin{cases} \frac{\Gamma(1-b)}{\Gamma(1+a-b)} M(a, b, z) + \frac{\Gamma(b-1)}{\Gamma(a)} z^{1-b} M(1 + a - b, 2 - b, z) & b \notin \mathbb{Z}, \\[2mm] \lim_{\theta \to b} U(a, \theta, z) & b \in \mathbb{Z}, \end{cases}$$

$$\tag{15}$$

is the Tricomi's confluent hypergeometric function. The piecewise construction of V_h originates from the fact that Tricomi's function $U(a, b, z)$ is in general complex-valued when its argument z is negative, that is, when $x > (c - \beta)/q$. Since we are looking for a real-valued solution V over the entire domain $x \geq 0$, another independent pair of solutions to (12), here, $M(a, b, z)$ and $e^z U(b - a, b, -z)$ needs to be chosen for $z < 0$, namely, $x > (c - \beta)/q$.

The general solution to (11) can then be written as

$$V(x) = V_h(x) + V_p(x),$$

where $V_p(x)$ is a particular solution to (11). Looking for a form $V_p(x) = Ax + B$, one finds

$$V_p(x) = \frac{1}{q + \delta}\left(qx + \beta + \frac{q}{\delta}\left(c - \frac{\lambda}{\alpha}\right)\right), \qquad x \geq 0.$$

To determine the constant coefficients $A_i, i = 1, \ldots 4$, we first investigate the components of V_h involving the Tricomi function U. For $a = \delta/q$ and $b = 1 + \frac{\lambda + \delta}{q} > 1$, $U(a, b, z)$ is singular at $z = 0$. Linear boundedness of V established in Proposition 3.1 then leads to the requirement $A_2 = 0$. Next, we focus on A_4: one has (cf. Olver [20])

$$\lim_{z \to 0} U(a, b, z) = \begin{cases} \frac{\Gamma(b-1)}{\Gamma(a)} z^{1-b} + \frac{\Gamma(1-b)}{\Gamma(a-b+1)} + \mathcal{O}\left(z^{2-\Re(b)}\right) & \text{if } 1 \leq \Re(b) < 2, \\ \frac{1}{\Gamma(a)} z^{-1} + \mathcal{O}(\log z) & \text{if } b = 2, \\ \frac{\Gamma(b-1)}{\Gamma(a)} z^{1-b} + \frac{\Gamma(1-b)}{\Gamma(a-b+1)} + \mathcal{O}\left(z^{2-\Re(b)}\right) & \text{if } \Re(b) \geq 2, b \neq 2. \end{cases}$$

In the original coordinates, this translates to

$$\lim_{x \to \frac{c-\beta}{q}+} e^{z(x)} U\left(1 + \frac{\lambda}{q}, 1 + \frac{\lambda + \delta}{q}, -z(x)\right)$$

$$= \begin{cases} \frac{\Gamma\left(\frac{\lambda+\delta}{q}\right)}{\Gamma\left(1+\frac{\lambda}{q}\right)}(-z(x))^{-\frac{\lambda+\delta}{q}} + \frac{\Gamma\left(-\frac{\lambda+\delta}{q}\right)}{\Gamma\left(1-\frac{\delta}{q}\right)} + +\mathcal{O}\left(x - \frac{c-\beta}{q}\right)^{1-\frac{\lambda+\delta}{q}} & \text{if } \frac{\lambda+\delta}{q} < 1, \\ \frac{1}{\Gamma\left(1+\frac{\lambda}{q}\right)}(-z(x))^{-1} + \mathcal{O}\left(\log\left(x - \frac{c-\beta}{q}\right)\right) & \text{if } \frac{\lambda+\delta}{q} = 1, \\ \frac{\Gamma\left(\frac{\lambda+\delta}{q}\right)}{\Gamma\left(1+\frac{\lambda}{q}\right)}(-z(x))^{-\frac{\lambda+\delta}{q}} + \frac{\Gamma\left(-\frac{\lambda+\delta}{q}\right)}{\Gamma\left(1-\frac{\delta}{q}\right)} + \mathcal{O}\left(x - \frac{c-\beta}{q}\right)^{1-\frac{\lambda+\delta}{q}} & \text{if } \frac{\lambda+\delta}{q} > 1. \end{cases}$$

The latter expression is unbounded for all choices of $(\lambda + \delta)/q$, so that by the linear boundedness of V we can also conclude $A_4 = 0$. On the other hand, the Kummer function M is analytic over the entire domain $x \geq 0$.

Next, the constant A_1 is determined by setting $x = 0$ in (10) which yields $(c - \beta)V'(0) = (\lambda + \delta)V(0) = -\beta$. Using the differentiation property

$$\frac{d}{dz}M(a, b, z) = \frac{a}{b}M(a + 1, b + 1, z) \tag{16}$$

(see [1]), this translates into

$$(c - \beta)\left[\frac{-\alpha\delta}{q + \lambda + \delta}A_1 M\left(1 + \frac{\delta}{q}, 2 + \frac{\lambda + \delta}{q}, z(0)\right) + \frac{q}{q + \delta}\right]$$

$$+ (\lambda + \delta)A_1 M\left(\frac{\delta}{q}, 1 + \frac{\lambda + \delta}{q}, z(0)\right)$$

$$+ (\lambda + \delta)\left[\frac{q}{\alpha\delta(q + \delta)}(\alpha(c - \beta) - (q + \lambda + \delta)) + \frac{q + \alpha\beta}{\alpha\delta}\right] = -\beta.$$

Solving for A_1 gives

$$A_1 = \frac{\beta + \frac{q(c-\beta)}{q+\delta} - (\lambda + \delta)\left[\frac{q}{\alpha\delta(q+\delta)}(\alpha(c - \beta) - (q + \lambda + \delta)) + \frac{q+\alpha\beta}{\alpha\delta}\right]}{\frac{\alpha\delta(c-\beta)}{q+\lambda+\delta}M\left(1 + \frac{\delta}{q}, 2 + \frac{\lambda+\delta}{q}, z(0)\right) + (\lambda + \delta)M\left(\frac{\delta}{q}, 1 + \frac{\lambda+\delta}{q}, z(0)\right)}.$$

Finally, by the continuity of V at $x = (c - \beta)/q$ (which follows from Proposition 3.2), we get $A_3 = A_1$, so that we arrive at the following result.

Proposition 3.3 *For any $x \geq 0$, the sum of the expected discounted dividend payments up to the time of ruin in a Cramér–Lundberg model with affine dividend strategy (2) and $Exp(\alpha)$-distributed claims is given by*

$$V(x) = \frac{\beta + \frac{q(c-\beta)}{q+\delta} - \frac{\lambda+\delta}{q+\delta}\left(\beta + \frac{q}{\delta}\left(c - \frac{\lambda}{\alpha}\right)\right)}{\frac{\alpha\delta(c-\beta)}{q+\lambda+\delta}M\left(1 + \frac{\delta}{q}, 2 + \frac{\lambda+\delta}{q}, z(0)\right) + (\lambda + \delta)M\left(\frac{\delta}{q}, 1 + \frac{\lambda+\delta}{q}, z(0)\right)}$$

$$M\left(\frac{\delta}{q}, 1 + \frac{\lambda + \delta}{q}, z(x)\right) + \frac{1}{q+\delta}\left(qx + \beta + \frac{q}{\delta}\left(c - \frac{\lambda}{\alpha}\right)\right), \tag{17}$$

where $z(x) = \frac{\alpha(c-(qx+\beta))}{q}$.

Remark 3.1 For $q \to \infty$ (i.e., infinite dividend rate), $V(x)$ in (17) tends to $x + \frac{c}{\lambda+\delta}$. Note that an infinite rate q instantaneously drives the process X_t to 0 implying an immediate lump sum dividend payment of size x. From then on, all incoming premium at rate c is immediately paid out as dividends (of which a magnitude $c - \beta$ is due to the proportional factor and β is due to the constant part), and the process X_t is continuously pushed back towards 0. The first claim will then lead to ruin and stops the dividend payments. That is, $q \to \infty$ corresponds to a horizontal dividend barrier strategy with barrier $b = 0$.

4 A Laplace Transform Approach

The structure of Eq. (10) suggests that a Laplace transform approach could in general also be a feasible tool to determine V. Indeed, denote by

$$\widetilde{V}(s) := \int_0^\infty e^{-sx} V(x)\, dx, \quad \widetilde{f_Y}(s) := \int_0^\infty e^{-sx} f_Y(x)\, dx$$

the corresponding Laplace transforms. Then (10) turns into a first-order differential equation for $\widetilde{V}(s)$:

$$\widetilde{V}'(s) = \widetilde{V}(s) \frac{(\beta - c)s - q + \lambda + \delta - \lambda \widetilde{f_Y}(s)}{qs} + \frac{(c - \beta)V(0) - \frac{q}{s^2} - \frac{\beta}{s}}{qs}.$$

It has the solution

$$\widetilde{V}(s) = e^{-\frac{(c-\beta)}{q}s} s^{\frac{\lambda+\delta}{q}-1} e^{-\frac{\lambda}{q}\int \frac{\widetilde{f_Y}(s)}{s} ds} \times$$

$$\left(\int \frac{(c-\beta)V(0) - \frac{q}{s^2} - \frac{\beta}{s}}{qs} e^{\frac{(c-\beta)}{q}s} s^{1-\frac{\lambda+\delta}{q}} e^{\frac{\lambda}{q}\int \frac{\widetilde{f_Y}(s)}{s} ds}\, ds + C \right)$$

for some constant C. In addition to the algebraic manipulations required in the Laplace transform domain, the inversion of $\widetilde{V}(s)$ is another intricate problem, see Sect. 4.1.

4.1 Exponential Claims

It is instructive to see how for exponential claims with rate α the above expression simplifies to the explicit solution derived in the previous section. While it will become clear that for this case the approach of Sect. 3.1 leads to the result with considerably less effort, a comparison of the two approaches gives rise to identities between special functions which are interesting in their own right.

From $\widetilde{f_Y}(s) = \alpha/(s + \alpha)$ one gets after standard algebraic manipulations

$$\widetilde{V}(s) = e^{-\frac{(c-\beta)}{q}s} s^{\frac{\delta}{q}-1} (s + \alpha)^{\frac{\lambda}{q}}$$

$$\times \left(C + \int \left(\frac{(c-\beta)V(0) - \frac{q}{s^2} - \frac{\beta}{s}}{q} \right) e^{\frac{(c-\beta)}{q}s} s^{-\frac{\delta}{q}} (s + \alpha)^{-\frac{\lambda}{q}}\, ds \right).$$

Expanding the exponential term inside the integral gives

$$\widetilde{V}(s) = e^{-\frac{(c-\beta)}{q}s} s^{\frac{\delta}{q}-1}(s+\alpha)^{\frac{\lambda}{q}}\left(C + \frac{(c-\beta)}{q}V(0)\sum_{n=0}^{\infty}\frac{\left(\frac{c-\beta}{q}\right)^n}{n!}\int s^{n-\frac{\delta}{q}}(s+\alpha)^{-\frac{\lambda}{q}}\,ds \right.$$

$$\left. -\sum_{n=0}^{\infty}\frac{\left(\frac{c-\beta}{q}\right)^n}{n!}\int s^{n-\frac{\delta}{q}-2}(s+\alpha)^{-\frac{\lambda}{q}}\,ds - \frac{\beta}{q}\sum_{n=0}^{\infty}\frac{\left(\frac{c-\beta}{q}\right)^n}{n!}\int s^{n-\frac{\delta}{q}-1}(s+\alpha)^{-\frac{\lambda}{q}}\,ds \right).$$

$$(18)$$

The following lemma establishes a connection between the integrals in (18) and the Gauss hypergeometric function $_2F_1$. Recall (also for later use) that the generalized hypergeometric function $_pF_q$ is defined through

$$_pF_q(a_1,\ldots,a_p;b_1,\ldots,b_q;z) = \sum_{n=0}^{\infty}\frac{(a_1)_n\cdots(a_p)_n}{(b_1)_n\cdots(b_q)_n}\frac{z^n}{n!}.$$

Lemma 4.1 *For $(n,k) \in \mathbb{N}_0 \times \mathbb{N}_0$ and $n - \frac{\delta}{q} - k \notin \mathbb{Z}^-$, one has*

$$\int s^{n-\frac{\delta}{q}-k}(s+\alpha)^{-\frac{\lambda}{q}}\,ds$$

$$= \frac{\alpha^{-\frac{\lambda}{q}}}{\left(n-\frac{\delta}{q}-k+1\right)}s^{n-\frac{\delta}{q}-k+1}\,_2F_1\left(\frac{\lambda}{q};n-\frac{\delta}{q}-k+1,n-\frac{\delta}{q}-k+2;-\frac{s}{\alpha}\right),$$

for $s \in \mathfrak{S} = \{s : |\frac{s}{\alpha}| < 1, s \neq 0\}$.

Proof Let $(n,k) \in \mathbb{N}_0 \times \mathbb{N}_0$ and define the new variable $\xi = s/\alpha$, i.e.

$$\int s^{n-\frac{\delta}{q}-k}(s+\alpha)^{-\frac{\lambda}{q}}\,ds = \alpha^{n-\frac{\delta+\lambda}{q}-k+1}\int \xi^{n-\frac{\delta}{q}-k}(1+\xi)^{-\frac{\lambda}{q}}\,d\xi$$

$$= \alpha^{-\frac{\lambda}{q}}(\alpha\xi)^{n-\frac{\delta}{q}-k+1}\sum_{j=0}^{\infty}\frac{\left(\frac{\lambda}{q}\right)_j}{\left(n-\frac{\delta}{q}-k+1+j\right)}\frac{(-\xi)^j}{j!}.$$

In terms of the original variable s, this can be recast into the form

$$\frac{\alpha^{-\frac{\lambda}{q}}}{\left(n-\frac{\delta}{q}-k+1\right)}s^{n-\frac{\delta}{q}-k+1}\sum_{j=0}^{\infty}\frac{\left(\frac{\lambda}{q}\right)_j\left(n-\frac{\delta}{q}-k+1\right)_j}{\left(n-\frac{\delta}{q}-k+2\right)_j}\frac{\left(-\frac{s}{\alpha}\right)^j}{j!}.$$

$$\qquad\qquad\qquad\qquad\qquad\qquad\qquad\qquad\qquad\qquad\qquad\qquad\qquad\square$$

Using Lemma 4.1, we can rewrite (18) as

$$
\widetilde{V}(s) = e^{-\frac{(c-\beta)}{q}s} \left(C\, s^{\frac{\delta}{q}-1}(s+\alpha)^{\frac{\lambda}{q}} + \frac{(c-\beta)}{q} V(0) \left(1+\frac{s}{\alpha}\right)^{\frac{\lambda}{q}} \right.
$$

$$
\times \sum_{n=0}^{\infty} \frac{\left(\frac{c-\beta}{q}\right)^n}{n!} \frac{s^n}{(n-\frac{\delta}{q}+1)} \, {}_2F_1\left(\frac{\lambda}{q}, n-\frac{\delta}{q}+1; n-\frac{\delta}{q}+2; -\frac{s}{\alpha}\right)
$$

$$
- \left(1+\frac{s}{\alpha}\right)^{\frac{\lambda}{q}} \frac{1}{s^2} \sum_{n=0}^{\infty} \frac{\left(\frac{c-\beta}{q}\right)^n}{n!} \frac{s^n}{(n-\frac{\delta}{q}-1)} \, {}_2F_1\left(\frac{\lambda}{q}, n-\frac{\delta}{q}-1; n-\frac{\delta}{q}; -\frac{s}{\alpha}\right)
$$

$$
\left. - \left(1+\frac{s}{\alpha}\right)^{\frac{\lambda}{q}} \frac{\beta}{qs} \sum_{n=0}^{\infty} \frac{\left(\frac{c-\beta}{q}\right)^n}{n!} \frac{s^n}{(n-\frac{\delta}{q})} \, {}_2F_1\left(\frac{\lambda}{q}, n-\frac{\delta}{q}; n-\frac{\delta}{q}+1; -\frac{s}{\alpha}\right) \right).
$$

$$\tag{19}$$

Denote

$$
{}_2\mathbf{F_1}(a, b; \nu; z) := \frac{{}_2F_1(\nu-a, \nu-b; \nu; z)}{\Gamma(\nu)}.
$$

Successively using the transformation formulas

$$
{}_2\mathbf{F_1}(a, b; \nu; z) = (1-z)^{\nu-a-b} \, {}_2\mathbf{F_1}(\nu-a, \nu-b; \nu; z),
$$

and

$$
{}_2\mathbf{F_1}(a, b; \nu; z) = \frac{\pi}{\sin(\pi(b-a))} \left[\frac{(-z)^{-a}}{\Gamma(b)\Gamma(\nu-a)} \, {}_2\mathbf{F_1}\left(a, a-\nu+1; a-b+1; \frac{1}{z}\right) \right.
$$

$$
\left. - \frac{(-z)^{-a}}{\Gamma(b)\Gamma(\nu-a)} \, {}_2\mathbf{F_1}\left(a, a-\nu+1; a-b+1; \frac{1}{z}\right) \right],
$$

(cf. [20]), we can then rewrite (19) as

$$
\widetilde{V}(s) = e^{\frac{(c-\beta)}{q}s} \left[\left(C + \frac{(c-\beta)}{q} V(0)\, \alpha^{-\frac{\lambda+\delta}{q}+1} \right. \right.
$$

$$
\times \sum_{n=0}^{\infty} \frac{(z(0))^n}{n!} \frac{\Gamma\left(-(n+1)+\frac{\lambda+\delta}{q}\right) \Gamma\left(n-\frac{\delta}{q}+1\right)}{\Gamma\left(\frac{\lambda}{q}\right)}
$$

$$
-\alpha^{-\frac{\lambda+\delta}{q}-1}\sum_{n=0}^{\infty}\frac{(z(0))^n}{n!}\frac{\Gamma\left(-n+1+\frac{\lambda+\delta}{q}\right)\Gamma\left(n-\frac{\delta}{q}-1\right)}{\Gamma\left(\frac{\lambda}{q}\right)}
$$

$$
-\frac{\beta}{q}\alpha^{-\frac{\lambda+\delta}{q}}\sum_{n=0}^{\infty}\frac{(z(0))^n}{n!}\frac{\Gamma\left(-n+\frac{\lambda+\delta}{q}\right)\Gamma\left(n-\frac{\delta}{q}\right)}{\Gamma\left(\frac{\lambda}{q}\right)}\Bigg)s^{\frac{\delta}{q}-1}(s+\alpha)^{\frac{\lambda}{q}}
$$

$$
-\frac{(c-\beta)}{q}V(0)\left(1+\frac{\alpha}{s}\right)\sum_{n=0}^{\infty}\frac{\left(\frac{(c-\beta)}{q}\right)^n}{n!}\frac{s^n}{\left(-(n+1)+\frac{\lambda+\delta}{q}\right)}
$$

$$
\times{}_2F_1\left(1,\frac{\delta}{q}-n;\frac{\lambda+\delta}{q}-n;-\frac{\alpha}{s}\right)+\frac{\left(1+\frac{\alpha}{s}\right)}{s^2}\sum_{n=0}^{\infty}\frac{\left(\frac{(c-\beta)}{q}\right)^n}{n!}
$$

$$
\times\frac{s^n}{\left(-n+1+\frac{\lambda+\delta}{q}\right)}{}_2F_1\left(1,\frac{\delta}{q}-n+2;\frac{\lambda+\delta}{q}-n+2;-\frac{\alpha}{s}\right)+\frac{\beta}{q}\frac{\left(1+\frac{\alpha}{s}\right)}{s}
$$

$$
\times\sum_{n=0}^{\infty}\frac{\left(\frac{(c-\beta)}{q}\right)^n}{n!}\frac{s^n}{\left(-n+\frac{\lambda+\delta}{q}\right)}{}_2F_1\left(1,\frac{\delta}{q}-n+1;\frac{\lambda+\delta}{q}-n+1;-\frac{\alpha}{s}\right)\Bigg].
$$

$$
\tag{20}
$$

Since

$$
\sum_{n=0}^{\infty}\frac{z^n}{n!}\frac{\Gamma\left(-n+\kappa+\frac{\lambda+\delta}{q}\right)\Gamma\left(n-\frac{\delta}{q}-\kappa\right)}{\Gamma\left(\frac{\lambda}{q}\right)}
$$

$$
=\frac{\Gamma\left(-\frac{\delta}{q}-\kappa\right)\Gamma\left(\kappa+\frac{\lambda+\delta}{q}\right)}{\Gamma\left(\frac{\lambda}{q}\right)}M\left(-\frac{\delta}{q}-\kappa,1-\kappa-\frac{\lambda+\delta}{q},-z\right),
$$

Eq. (20) simplifies to

$$
\widetilde{V}(s)\;=e^{-\frac{(c-\beta)}{q}s}\left[s^{\frac{\delta}{q}-1}(s+\alpha)^{\frac{\lambda}{q}}(C+D)-\frac{(c-\beta)}{q}V(0)\left(1+\frac{\alpha}{s}\right)\right.
$$

$$
\times\sum_{n=0}^{\infty}\frac{\left(\frac{(c-\beta)}{q}\right)^n}{n!}\frac{s^n}{\left(-(n+1)+\frac{\lambda+\delta}{q}\right)}{}_2F_1\left(1,\frac{\delta}{q}-n;\frac{\lambda+\delta}{q}-n;-\frac{\alpha}{s}\right)
$$

$$
+ \frac{\left(1+\frac{\alpha}{s}\right)}{s^2} \sum_{n=0}^{\infty} \frac{\left(\frac{(c-\beta)}{q}\right)^n}{n!} \frac{s^n}{\left(-n+1+\frac{\lambda+\delta}{q}\right)} \, {}_2F_1\left(1, \frac{\delta}{q}-n+2; \frac{\lambda+\delta}{q}-n+2; -\frac{\alpha}{s}\right)
$$

$$
+ \frac{\beta}{q} \frac{\left(1+\frac{\alpha}{s}\right)}{s} \sum_{n=0}^{\infty} \frac{\left(\frac{(c-\beta)}{q}\right)^n}{n!} \frac{s^n}{\left(-n+\frac{\lambda+\delta}{q}\right)} \, {}_2F_1\left(1, \frac{\delta}{q}-n+1; \frac{\lambda+\delta}{q}-n+1; -\frac{\alpha}{s}\right) \Bigg]
$$

$$
\tag{21}
$$

where

$$
D = \frac{(c-\beta)}{q} V(0)\, \alpha^{-\frac{\lambda+\delta}{q}+1} \frac{\Gamma\left(1-\frac{\delta}{q}\right)\Gamma\left(\frac{\lambda+\delta}{q}-1\right)}{\Gamma\left(\frac{\lambda}{q}\right)} M\left(1-\frac{\delta}{q}, 2-\frac{\lambda+\delta}{q}, -\frac{\alpha(c-\beta)}{q}\right)
$$

$$
- \alpha^{-\frac{\lambda+\delta}{q}-1} \frac{\Gamma\left(-1-\frac{\delta}{q}\right)\Gamma\left(\frac{\lambda+\delta}{q}+1\right)}{\Gamma\left(\frac{\lambda}{q}\right)} M\left(-1-\frac{\delta}{q}, -\frac{\lambda+\delta}{q}, -\frac{\alpha(c-\beta)}{q}\right)
$$

$$
- \frac{\beta}{q} \alpha^{-\frac{\lambda+\delta}{q}} \frac{\Gamma\left(-\frac{\delta}{q}\right)\Gamma\left(\frac{\lambda+\delta}{q}\right)}{\Gamma\left(\frac{\lambda}{q}\right)} M\left(-\frac{\delta}{q}, 1-\frac{\lambda+\delta}{q}, -\frac{\alpha(c-\beta)}{q}\right).
$$

From

$$
\frac{1}{\Gamma(b)} \int_0^{\infty} e^{-sx} x^{b-1} M(a,b,x)\, dx = s^{a-b}(s-1)^{-a}, \qquad \Re(b) > 0, \Re(s) > 1,
$$

one deduces that for $\lambda + \delta < q$ and $\Re(s) > 0$, the inverse Laplace transform of $s^{\frac{\delta}{q}-1}(s+\alpha)^{\frac{\lambda}{q}}$ is given by

$$
\Re\left\{ \left(\frac{\Gamma\left(\frac{\delta}{q}\right)}{\Gamma\left(\frac{\lambda+\delta}{q}\right)} U\left(\frac{\delta}{q}, 1+\frac{\lambda+\delta}{q}, -\alpha x\right) \right.\right.
$$

$$
\left.\left. - \frac{\Gamma\left(\frac{\delta}{q}\right)}{\Gamma\left(\frac{\lambda+\delta}{q}\right)} \frac{\Gamma\left(-\frac{\lambda+\delta}{q}\right)}{\Gamma\left(-\frac{\lambda}{q}\right)} M\left(\frac{\delta}{q}, 1+\frac{\lambda+\delta}{q}, -\alpha x\right) \right) \frac{(-\alpha)^{\frac{\lambda+\delta}{q}}}{\Gamma\left(1-\frac{\lambda+\delta}{q}\right)} \right\}.
$$

For the sake of brevity, define $g(x) := x^{-\frac{\lambda+\delta}{q}} M\left(-\frac{\lambda}{q}, 1-\frac{\lambda+\delta}{q}, -\alpha x\right) / \Gamma(1-\frac{\lambda+\delta}{q})$. The first term of (21) can be interpreted as

$$
e^{-\frac{(c-\beta)}{q}s} \tilde{g}(s) = \mathcal{L}\left\{ u\left(x - \frac{c-\beta}{q}\right) g\left(x - \frac{c-\beta}{q}\right) \right\}.
$$

where u is the Heaviside function. Since $u\left(x - \frac{c-\beta}{q}\right) g\left(x - \frac{c-\beta}{q}\right)$ becomes unbounded as x approaches $(c - \beta)/q$ from the right, the linear boundedness of $V(x)$ imposes $C = -D$.

The three hypergeometric functions ${}_2F_1$ in (21) have parameters differing by an integer. A connection between those terms is given by the identity

$$
{}_2F_1\left(1, \frac{\delta}{q} - n; \frac{\lambda + \delta}{q} - n; -\frac{\alpha}{s}\right) = \sum_{k=0}^{m-1} \frac{\left(n + 1 - k - \frac{\delta}{q}\right)_k}{\left(n + 1 - k - \frac{\lambda+\delta}{q}\right)_k} \left(-\frac{\alpha}{s}\right)^k
$$

$$
+ \left(-\frac{\alpha}{s}\right)^m \frac{\left(1 - \frac{\delta}{q} + n - m\right)_m}{\left(1 - \frac{\lambda+\delta}{q} + n - m\right)_m} {}_2F_1\left(1, \frac{\delta}{q} - n + m; \frac{\lambda + \delta}{q} - n + m; -\frac{\alpha}{s}\right)
$$

for $(n, m) \in \mathbb{N}_0 \times \mathbb{N}$. Respective substitution in (21) gives

$$
\widetilde{V}(s) = e^{-\frac{(c-\beta)}{q}s}\left[-\frac{(c - \beta)}{q} V(0) \left(1 + \frac{\alpha}{s}\right) \sum_{n=0}^{\infty} \frac{\left(\frac{(c-\beta)}{q}\right)^n}{n!} \frac{s^n}{\left(-(n + 1) + \frac{\lambda+\delta}{q}\right)} \times \right.
$$

$$
\times {}_2F_1\left(1, \frac{\delta}{q} - n; \frac{\lambda + \delta}{q} - n; -\frac{\alpha}{s}\right)
$$

$$
+ \frac{\left(1 + \frac{\alpha}{s}\right)}{\alpha^2} \sum_{n=0}^{\infty} \frac{\left(\frac{(c-\beta)}{q}\right)^n}{n!} \frac{s^n \left(1 - {}_2F_1\left(1, \frac{\delta}{q} - n; \frac{\lambda+\delta}{q} - n; -\frac{\alpha}{s}\right)\right)\left(n - \frac{\lambda+\delta}{q}\right)}{\left(n - \frac{\delta}{q} - 1\right)\left(n - \frac{\delta}{q}\right)}
$$

$$
- \frac{\left(1 + \frac{\alpha}{s}\right)}{\alpha s} \sum_{n=0}^{\infty} \frac{\left(\frac{(c-\beta)}{q}\right)^n}{n!} \frac{s^n}{\left(n - \frac{\delta}{q} - 1\right)}
$$

$$
\left. - \frac{\beta}{q} \frac{\left(1 + \frac{\alpha}{s}\right)}{\alpha} \sum_{n=0}^{\infty} \frac{\left(\frac{(c-\beta)}{q}\right)^n}{n!} \frac{s^n \left(1 - {}_2F_1\left(1, \frac{\delta}{q} - n; \frac{\lambda+\delta}{q} - n; -\frac{\alpha}{s}\right)\right)}{\left(n - \frac{\delta}{q}\right)} \right].
$$

With considerable effort, the latter expression can be represented as

$$
\widetilde{V}(s) = e^{-\frac{(c-\beta)}{q}s}\left(1 + \frac{\alpha}{s}\right)\left[B \cdot {}_2F_1\left(1, \frac{\delta}{q}; \frac{\lambda + \delta}{q}; -\frac{\alpha}{s}\right)\right.
$$

$$
\left. + \frac{(c - \beta)}{q} V(0) \sum_{n=1}^{\infty} \frac{\left(\frac{(c-\beta)}{q}\right)^n s^n}{n!\left(n + 1 - \frac{\lambda+\delta}{q}\right)}\right.
$$

$$\times {}_2F_2\left(1, n+1-\frac{\delta}{q}; n+1, n+2-\frac{\lambda+\delta}{q}; -\frac{\alpha(c-\beta)}{q}\right)$$

$$-\left(\frac{c-\beta}{q}\right)^2 \sum_{n=1}^{\infty} \frac{\left(\frac{(c-\beta)}{q}\right)^n s^n}{(n+2)!\left(n+1-\frac{\lambda+\delta}{q}\right)}$$

$$\times {}_2F_2\left(1, n+1-\frac{\delta}{q}; n+3, n+2-\frac{\lambda+\delta}{q}; -\frac{\alpha(c-\beta)}{q}\right)$$

$$-\frac{\beta}{q}\frac{(c-\beta)}{q} \sum_{n=1}^{\infty} \frac{\left(\frac{(c-\beta)}{q}\right)^n s^n}{(n+1)!\left(n+1-\frac{\lambda+\delta}{q}\right)}$$

$$\times {}_2F_2\left(1, n+1-\frac{\delta}{q}; n+2, n+2-\frac{\lambda+\delta}{q}; -\frac{\alpha(c-\beta)}{q}\right)$$

$$\left. +\frac{1}{\alpha s\left(1+\frac{\delta}{q}\right)} - \frac{q(\lambda+\delta)}{\alpha^2\delta(q+\delta)} + \frac{c}{\alpha\delta}\right], \tag{22}$$

where

$$B = \frac{(c-\beta)}{q}V(0)\frac{M\left(1-\frac{\delta}{q}, 2-\frac{\lambda+\delta}{q}, -\frac{\alpha(c-\beta)}{q}\right)}{\left(1-\frac{\lambda+\delta}{q}\right)}$$

$$+\frac{q(\lambda+\delta)}{\alpha^2\delta(q+\delta)}M\left(-1-\frac{\delta}{q}, -\frac{\lambda+\delta}{q}, -\frac{\alpha(c-\beta)}{q}\right)$$

$$-\frac{\beta}{\alpha\delta}M\left(-\frac{\delta}{q}, 1-\frac{\lambda+\delta}{q}, -\frac{\alpha(c-\beta)}{q}\right).$$

Using the contiguous relation

$$v(1-z)\,{}_2F_1(a, b; v; z) = v\,{}_2F_1(a-1, b; v; z) - (b-v)z\,{}_2F_1(a, b; v+1; z)$$

for our context, we obtain

$${}_2F_1\left(1, \frac{\delta}{q}; \frac{\lambda+\delta}{q}; -\frac{\alpha}{s}\right) = \frac{1}{1+\frac{\alpha}{s}}\left(1 + \frac{\alpha\lambda}{\lambda+\delta}\frac{{}_2F_1\left(1, \frac{\delta}{q}; 1+\frac{\lambda+\delta}{q}; -\frac{\alpha}{s}\right)}{s}\right),$$

which transforms (22) into

$$\widetilde{V}(s) = e^{-\frac{(c-\beta)}{q}s}\frac{B\alpha\lambda}{(\lambda+\delta)}\frac{{}_2F_1\left(1,\frac{\delta}{q};1+\frac{\lambda+\delta}{q};-\frac{\alpha}{s}\right)}{s} + e^{-\frac{(c-\beta)}{q}s}\left\{B + \left(1+\frac{\alpha}{s}\right)\right.$$

$$\left[\frac{(c-\beta)}{q}V(0)\sum_{n=1}^{\infty}\frac{\left(\frac{(c-\beta)}{q}\right)^n s^n}{n!\left(n+1-\frac{\lambda+\delta}{q}\right)}\right.$$

$$\times {}_2F_2\left(1,n+1-\frac{\delta}{q};n+1,n+2-\frac{\lambda+\delta}{q};-\frac{\alpha(c-\beta)}{q}\right)$$

$$-\left(\frac{c-\beta}{q}\right)^2\sum_{n=1}^{\infty}\frac{\left(\frac{(c-\beta)}{q}\right)^n s^n}{(n+2)!\left(n+1-\frac{\lambda+\delta}{q}\right)}$$

$$\times {}_2F_2\left(1,n+1-\frac{\delta}{q};n+3,n+2-\frac{\lambda+\delta}{q};-\frac{\alpha(c-\beta)}{q}\right).$$

$$-\frac{\beta}{q}\frac{(c-\beta)}{q}\sum_{n=1}^{\infty}\frac{\left(\frac{(c-\beta)}{q}\right)^n s^n}{(n+1)!\left(n+1-\frac{\lambda+\delta}{q}\right)}$$

$$\times {}_2F_2\left(1,n+1-\frac{\delta}{q};n+2,n+2-\frac{\lambda+\delta}{q};-\frac{\alpha(c-\beta)}{q}\right)$$

$$\left.\left.+\frac{1}{\alpha s\left(1+\frac{\delta}{q}\right)} - \frac{q(\lambda+\delta)}{\alpha^2\delta(q+\delta)} + \frac{c}{\alpha\delta}\right]\right\}. \tag{23}$$

Utilizing the relationship (cf. Olver [20])

$$\int_0^\infty e^{-sx}x^{b-1}\mathbf{M}(a,v,kx)\,dx$$

$$= \frac{\Gamma(b)}{s^b}\,{}_2\mathbf{F}_1\left(a,b;v;\frac{k}{s}\right), \qquad \Re(b) > 0, \Re(s) > \max\left(\Re(k),0\right),$$

gives the inverse Laplace transform of the first term of (23):

$$e^{-\frac{(c-\beta)}{q}s}\frac{B\alpha\lambda}{(\lambda+\delta)}\frac{{}_2F_1\left(1,\frac{\delta}{q};1+\frac{\lambda+\delta}{q};-\frac{\alpha}{s}\right)}{s}$$

$$= \frac{B\alpha\lambda}{(\lambda+\delta)}\mathcal{L}\left\{u\left(x-\frac{c-\beta}{q}\right)M\left(\frac{\delta}{q},1+\frac{\lambda+\delta}{q},-\alpha\left(x-\frac{c-\beta}{q}\right)\right)\right\}.$$

Since we know by different means from Sect. 3 the expression for $V(x)$, we can take the (direct) Laplace transform of (17) in order to compare it with the above expression. After some efforts, one obtains from (17)

$$\tilde{V}(s) = e^{-\frac{(c-\beta)}{q}s}\tilde{C}\left[H(s) + \frac{{}_2F_1\left(\frac{\delta}{q}, 1; 1 + \frac{\lambda+\delta}{q}; -\frac{\alpha}{s}\right)}{s}\right]$$

$$+ \frac{1}{q+\delta}\left(\frac{q}{s^2} + \frac{\beta}{s} + \frac{q\left(c - \frac{\lambda}{\alpha}\right)}{\delta s}\right), \tag{24}$$

where

$$H(s) := \int_{-\frac{(c-\beta)}{q}}^{0} e^{-sy}M\left(\frac{\delta}{q}, 1 + \frac{\lambda+\delta}{q}, -\alpha y\right)dy.$$

One can show that $\frac{B\alpha\lambda}{(\lambda+\delta)} = \tilde{C}$ and

$$H(s) = \sum_{n=0}^{\infty} \frac{\left(\frac{c-\beta}{q}\right)^{n+1}s^n}{(n+1)!} {}_2F_2\left(n+1, \frac{\delta}{q}; 2+n, 1+\frac{\lambda+\delta}{q}; \frac{\alpha(c-\beta)}{q}\right);$$

so the first term coincides with the one in (23). Since Expressions (24) and (23) have to coincide altogether, this leads to the identity

$$e^{-\frac{(c-\beta)}{q}s}\frac{B\alpha\lambda}{(\lambda+\delta)}H(s) + \frac{1}{q+\delta}\left(\frac{q}{s^2} + \frac{\beta}{s} + \frac{q\left(c - \frac{\lambda}{\alpha}\right)}{\delta s}\right) = e^{-\frac{(c-\beta)}{q}s}\left\{B + \left(1 + \frac{\alpha}{s}\right)\right.$$

$$\left[\frac{(c-\beta)}{q}V(0)\sum_{n=1}^{\infty}\frac{\left(\frac{(c-\beta)}{q}\right)^n s^n}{n!\left(n+1-\frac{\lambda+\delta}{q}\right)}\right.$$

$$\times {}_2F_2\left(1, n+1-\frac{\delta}{q}; n+1, n+2-\frac{\lambda+\delta}{q}; -\frac{\alpha(c-\beta)}{q}\right)$$

$$-\left(\frac{c-\beta}{q}\right)^2\sum_{n=1}^{\infty}\frac{\left(\frac{(c-\beta)}{q}\right)^n s^n}{(n+2)!\left(n+1-\frac{\lambda+\delta}{q}\right)}$$

$$\times {}_2F_2\left(1, n+1-\frac{\delta}{q}; n+3, n+2-\frac{\lambda+\delta}{q}; -\frac{\alpha(c-\beta)}{q}\right)$$

$$-\frac{\beta}{q}\frac{(c-\beta)}{q}\sum_{n=1}^{\infty}\frac{\left(\frac{(c-\beta)}{q}\right)^n s^n}{(n+1)!\left(n+1-\frac{\lambda+\delta}{q}\right)}$$

$$\times {}_2F_2\left(1, n+1-\frac{\delta}{q}; n+2, n+2-\frac{\lambda+\delta}{q}; -\frac{\alpha(c-\beta)}{q}\right)$$

$$+\frac{1}{\alpha s\left(1+\frac{\delta}{q}\right)} - \frac{q(\lambda+\delta)}{\alpha^2\delta(q+\delta)} + \frac{c}{\alpha\delta}\bigg]\bigg\}. \tag{25}$$

While it is far from obvious to show analytically that (25) holds true, it is indeed the case, as numerical verifications show. In fact, the two alternative approaches of Sects. 3 and 4—through identity (25)—suggest new relations between ${}_2F_2$-hypergeometric functions of argument $\pm z$. A detailed study of such relations is, however, beyond the scope of this paper.

5 The Time of Ruin

Let us now study the effect of the proposed dividend strategy on the distribution of the ruin time τ_x. For this purpose, consider the expected discounted penalty at ruin

$$m_\delta(x) := \mathbb{E}\left[e^{-\delta\tau_x}w\left(|X_{\tau_x}|\right)\right],$$

where w is a non-negative penalty function of the deficit at ruin. Given differentiability of $m_\delta(x)$, the standard arguments based on the infinitesimal generator then lead to the integro-differential equation

$$(c-(qx+\beta))\, m_\delta'(x) - (\lambda+\delta)m_\delta(x) + \lambda \int_0^x m_\delta(x-y)dF_Y(y) = -\lambda A(x), \quad x \geq 0, \tag{26}$$

where

$$A(x) := \int_x^\infty w(y-x)\, dF_Y(y).$$

We will again restrict our considerations to exponentially distributed claims with parameter $\alpha > 0$. In this case, $|X_{\tau_x}| \sim \mathrm{Exp}(\alpha)$ due to lack of memory, so that we can focus on the (Laplace transform of the) time of ruin, i.e. $w(x) = 1$.

Similarly to Sect. 3.1, this leads to the second-order homogeneous differential equation

$$(c-(qx+\beta))\, m_\delta''(x) + [\alpha\,(c-(qx+\beta)) - (q+\lambda+\delta)]\, m_\delta'(x) - \alpha\delta m_\delta(x) = 0, \quad x \geq 0. \tag{27}$$

Because of the linear boundedness of $m_\delta(x)$ in x, the solution to (27) matches the homogeneous solution $V_h(x) := V(x) - V_p(x)$ to (11) up to a constant factor. That is, for $x \geq 0$, we can write $m_\delta(x) = B V_h(x)$ for some constant B. Letting $x = 0$ in (26) yields

$$(c - \beta)m'_\delta(0) - (\lambda + \delta)m_\delta(0) = -\lambda,$$

that is

$$(c - \beta)B\left(V'(0) - V'_p(0)\right) - (\lambda + \delta)B\left(V(0) - V_p(0)\right) = -\lambda$$

and hence

$$B = \frac{\lambda}{\beta + \frac{q(c-\beta)}{q+\delta} - \frac{\lambda+\delta}{q+\delta}\left(\beta + \frac{q}{\delta}\left(c - \frac{\lambda}{\alpha}\right)\right)}.$$

Proposition 5.1 *For any $x \geq 0$, the Laplace transform of the ruin time in a Cramér–Lundberg model with affine dividend strategy (2) and $Exp(\alpha)$-distributed claims is given by*

$$m_\delta(x) = \frac{\lambda M\left(\frac{\delta}{q}, 1 + \frac{\lambda+\delta}{q}, z(x)\right)}{\frac{\alpha\delta(c-\beta)}{q+\lambda+\delta}M\left(1 + \frac{\delta}{q}, 2 + \frac{\lambda+\delta}{q}, z(0)\right) + (\lambda + \delta)M\left(\frac{\delta}{q}, 1 + \frac{\lambda+\delta}{q}, z(0)\right)}, \quad x \geq 0,$$

$$(28)$$

where $z(x) = \frac{\alpha(c-(qx+\beta))}{q}$.

One particular quantity of interest is the expected ruin time. While it can be simply obtained by taking the derivative

$$\mathbb{E}[\tau_x] = -\frac{d}{d\delta}m_\delta(x)\bigg|_{\delta=0},$$

the concrete calculation is considerably involved. It requires differentiating M with respect to its first two parameters:

$$M^{(a)} = \frac{d}{da}M(a,b,z), \qquad M^{(b)} = \frac{d}{db}M(a,b,z).$$

The so-called Kummer transformation $M(a,b,z) = e^z M(b - a, b, -z)$ will facilitate the mathematical tractability, under which (28) reads

$$m_\delta(x) = \frac{\lambda e^{-\alpha x}M\left(a, a + \frac{\delta}{q}, -z(x)\right)}{\frac{\alpha\delta(c-\beta)}{q+\lambda+\delta}M\left(a, a + 1 + \frac{\delta}{q}, -z(0)\right) + (\lambda + \delta)M\left(a, a + \frac{\delta}{q}, -z(0)\right)}, \quad x \geq 0,$$

$$(29)$$

where $a = 1 + \frac{\lambda}{q}$.

We now discuss two possible representations for the derivative of M:

5.1 Digamma Functions

A first and customary approach to calculate $M^{(a)}$ makes use of the derivative of the Pochhammer symbol $a_{(n)} := \Gamma(a + n)/\Gamma(n)$, which is given by

$$\frac{d(a)_n}{da} = (a)_n \left[\psi(a + n) - \psi(a) \right],$$

where $\psi(a) = \frac{d}{da} \log \Gamma(a) = \frac{\Gamma'(a)}{\Gamma(a)}$ is the logarithmic derivative of the gamma function, known as the digamma function. Next, using the power series definition of M given in (14) leads to the representations

$$M^{(a)} = \sum_{n=0}^{\infty} \left[\psi(a + n) - \psi(a) \right] \frac{(a)_n}{(b)_n} \frac{z^n}{n!},$$

and

$$M^{(b)} = \sum_{n=0}^{\infty} \left[\psi(b) - \psi(b + n) \right] \frac{(a)_n}{(b)_n} \frac{z^n}{n!}. \tag{30}$$

With that in mind, the derivative of (29) with respect to δ (together with the property $M(a, a, z) = e^z$) yields

$$\frac{d}{d\delta} m_\delta(x) \bigg|_{\delta=0} = \frac{e^{z(x)}}{q} \sum_{n=0}^{\infty} \left[\psi(a) - \psi(a + n) \right] \frac{(-z(x))^n}{n!} - \frac{\alpha(c - \beta)}{\lambda(q + \lambda)} e^{z(0)} M(a, a + 1, -z(0))$$

$$- \frac{1}{\lambda} - \frac{e^{z(0)}}{q} \sum_{n=0}^{\infty} \left[\psi(a) - \psi(a + n) \right] \frac{(-z(0))^n}{n!}.$$

But for series of the above type there are expressions in terms of the generalized hypergeometric function $_2F_2$ available (cf. [10]):

$$\sum_{n=0}^{\infty} \left[\psi(a + n) - \psi(a) \right] \frac{(z(x))^n}{n!} = \frac{z(x)}{a} e^{z(x)} \, _2F_2 \left(1, 1; 2, a + 1; -z(x) \right).$$

Using this last result together with the relation $M(a, a + 1, -z) = az^{-a}\gamma(a, z)$ (where $\gamma(a, z)$ denotes the lower incomplete gamma function) leads to the following formula:

Proposition 5.2 *For any $x \geq 0$, the expected time of ruin under the proposed dividend strategy in model (1) with exponentially distributed claims with parameter α is given by*

$$\mathbb{E}[\tau_x] = \frac{1 + e^{z(0)}\gamma\left(1 + \frac{\lambda}{q}, z(0)\right) z(0)^{-\frac{\lambda}{q}}}{\lambda}$$

$$+ \frac{{}_2F_2\left(1, 1; 2, 2 + \frac{\lambda}{q}; z(0)\right) z(0) - {}_2F_2\left(1, 1; 2, 2 + \frac{\lambda}{q}; z(x)\right) z(x)}{q + \lambda},$$

$$\tag{31}$$

where $z(x) = \frac{\alpha(c - (qx + \beta))}{q}$.

5.2 Kampé de Fériet Functions

As an alternative, one can express the derivatives of the Kummer function also in terms of the bivariate Kampé de Fériet function

$$F_{R,S,U}^{A,B,D}\left(\begin{array}{c} a_1, \ldots, a_A \; ; b_1, \ldots, b_B \; ; d_1, \ldots, d_D \; ; \\ r_1, \ldots, r_R \; ; s_1, \ldots, s_S \; ; u_1, \ldots, u_U \; ; \end{array} x, y\right)$$

$$= \sum_{m=0}^{\infty} \sum_{n=0}^{\infty} \frac{\prod_{j=1}^{A}(a_j)_{m+n} \prod_{j=1}^{B}(b_j)_m \prod_{j=1}^{D}(d_j)_n}{\prod_{j=1}^{R}(r_j)_{m+n} \prod_{j=1}^{S}(s_j)_m \prod_{j=1}^{U}(u_j)_n} \frac{x^m y^n}{m! \, n!},$$

see, e.g., [12, 25]. The concrete connection is

$$M^{(a)} = \frac{z}{b} F_{2,1,0}^{1,2,1}\left(\begin{array}{c} a+1 \; ; \; 1, a \; ; 1 \; ; \\ 2, b+1 \; ; a+1 \; ; - \; ; \end{array} z, z\right),$$

$$M^{(b)} = -\frac{a}{b}\frac{z}{b} F_{2,1,0}^{1,2,1}\left(\begin{array}{c} a+1 \; ; \; 1, b \; ; 1 \; ; \\ 2, b+1 \; ; b+1 \; ; - \; ; \end{array} z, z\right),$$

where the empty product indicated by the the solid horizontal line is interpreted to be unity. Employing this formula and proceeding similarly as in Sect. 5.1, one then obtains

$$\mathbb{E}[\tau_x] = -\frac{d}{d\delta}m_\delta(x)\Big|_{\delta=0} = \frac{1 + e^{z(0)}\gamma\left(1 + \frac{\lambda}{q}, z(0)\right)z(0)^{-\frac{\lambda}{q}}}{\lambda}$$

$$+ \frac{\left(e^{z(0)}z(0)F_{1,1,0}^{0,2,1}\left(\begin{matrix} -\,; & 1,1+\frac{\lambda}{q}\,; & 1\,; \\ 2,\,; & 2+\frac{\lambda}{q} & ;-\,; \end{matrix} - z(0), -z(0)\right)\right.}{q+\lambda}$$
$$\left.\frac{-e^{z(x)}z(x)F_{1,1,0}^{0,2,1}\left(\begin{matrix} -\,; & 1,1+\frac{\lambda}{q}\,; & 1\,; \\ 2,\,; & 2+\frac{\lambda}{q} & ;-\,; \end{matrix} - z(x), -z(x)\right)\right)}{q+\lambda}.$$

The equivalency of this expression with (31) follows from the reduction formula (cf. [19])

$$F_{1,1,0}^{0,2,1}\left(\begin{matrix} -\,; & a,b\,; & d-a\,; \\ d\,; & f\,; & -\,; \end{matrix} z, z\right) = e^z \,{}_2F_2\left(a, f-b; d, f; -z\right),$$

with $a = 1, b = 1 + \frac{\lambda}{q}, d = 2$ and $f = 2 + \frac{\lambda}{q}$.

6 A Probabilistic Argument

Inspired by Avanzi and Wong [6], we also present here a probabilistic argument to connect the function $V(x)$ of Sect. 3 with the Laplace transform of the time to ruin in the previous section. To that end, consider first a surplus process of the form (4), but let it continue after ruin (i.e., whenever $X_t < -\frac{\beta}{q}$, negative dividends are paid, which could be interpreted as capital injections).[2] Denote with

$$\overline{V}(x) := \mathbb{E}_x\left[\int_0^\infty e^{-\delta t}(qX_t + \beta)\,dt\right],$$

the respective expected discounted dividend payments (or, more precisely, the difference between expected discounted dividend payments and expected discounted amount of such capital injections).

Proposition 6.1 *For $x \geq 0$,*

$$\overline{V}(x) = \frac{c - \lambda\mu}{\delta} + \frac{x - \frac{c-\beta-\lambda\mu}{q}}{1 + \delta/q}.$$

---

[2]In fact, there is some methodological link to a calculation in Tichy [29], where for a horizontal dividend barrier an explicit calculation for V without ruin was given.

Proof

$$\overline{V}(x) = \mathbb{E}_x\left[\int_0^\infty e^{-\delta t}\left[q\left(\left(x - \frac{c-\beta}{q}\right)e^{-qt} + \frac{c-\beta}{q} - \int_0^t e^{-q(t-u)}dS_u\right) + \beta\right]dt\right],$$

$$= \frac{qx}{q+\delta} + \frac{(c-\beta)q}{\delta(q+\delta)} + \frac{\beta}{\delta} - \mathbb{E}\left[\int_0^\infty qe^{qu}\int_u^\infty e^{-(q+\delta)t}dt\,dS_u\right],$$

$$= \frac{qx}{q+\delta} + \frac{(c-\beta)q}{\delta(q+\delta)} + \frac{\beta}{\delta} - \frac{q}{q+\delta}\mathbb{E}\left[\sum_{i=1}^\infty e^{-\delta T_i}Y_i\right],$$

where the last equality follows from $\lim_{t\to\infty} N_t = \infty$ a.s. Since the *i*th claim arrival time T_i in the Poisson model is independent of Y_i and $\Gamma(i, \lambda)$-distributed, we then have

$$\overline{V}(x) = \frac{qx}{q+\delta} + \frac{(c-\beta-\lambda\mu)q}{\delta(q+\delta)} + \frac{\beta}{\delta}, \qquad x \geq 0.$$

$\square$

Due to the strong Markov property of X_t, we can now deduce

$$V(x) = \overline{V}(x) - \mathbb{E}\left[e^{-\delta\tau_x}\overline{V}(X_{\tau_x})\right], \qquad x \geq 0.$$

For exponentially distributed claims with mean $\mu = 1/\alpha$, the lack-of-memory property implies that the deficit at $t = \tau_x$ is again exponentially distributed and independent of the time of ruin. This leads to

$$V(x) = \overline{V}(x) - \mathbb{E}\left[e^{-\delta\tau_x}\right]\overline{V}\left(-\frac{1}{\alpha}\right).$$

Combining Proposition 6.1 with the expression for the Laplace transform of the time of ruin derived in (28) then again leads to the formula given in Proposition 3.3. In particular, this approach gives a complementary probabilistic interpretation of the particular solution $V_p(X) = \overline{V}(x)$ in Sect. 3 as well as the relation between the time of ruin and amount of dividend payments, in a certain sense akin to the dividends-penalty identity of Gerber et al. [14] in the model with horizontal dividend barrier.

7 Numerical Illustrations

7.1 General Considerations

In this section we analyze the effects of the affine dividend strategy numerically. In particular, we are interested in the influence of the parameters q and β on the

Table 1 Expected present value of dividends for different rates q with $\alpha = 1/3$, $\lambda = 1$, $c = 3.5$, $\beta = 1.5$, and $\delta = 0.05$

$V(x; q, 1.5)$

x	$q = 0.1$	$q = 0.2$	$q = 0.3$	$q = 0.5$	$q = 1$	$q = 10$
0	3.385	3.403	3.406	3.403	3.389	3.344
0.5	3.896	3.919	3.923	3.920	3.903	3.846
1	4.401	4.430	4.436	4.433	4.414	4.349
2	5.396	5.440	5.452	5.451	5.430	5.352
3	6.371	6.435	6.454	6.459	6.440	6.354
4	7.327	7.415	7.445	7.458	7.443	7.356
5	8.268	8.384	8.426	8.450	8.442	8.356
10	12.763	13.079	13.213	13.321	13.381	13.352
20	21.052	22.007	22.433	22.818	23.117	23.324

expected discounted dividend payments. Assume that $\alpha = 1/3$, $\lambda = 1$, $c = 3.5$ and $\delta = 0.05$. With each pair $(q, \beta) \in \mathbb{R}_+ \times [0, c]$, we associate $V(x; q, \beta) := V(x)$. Table 1 shows the influence of q on $V(x)$ for $\beta = 1.5$. We observe that $V(x; q, 1.5)$ increases in q up to a certain value and decreases thereafter. This demonstrates the compromise between paying larger amounts early (which is preferable due to discounting) and maintaining a longer survival in order to receive more payments later, i.e. too large proportions q (in addition to the constant rate β) reduce the lifetime of the process too much. One observes that this turning point appears for larger values of q the larger the initial surplus value x is (and for $x = 20$, this turning point is not yet visible for the depicted range of q). Incidentally, for $x = 0$ one sees that the choices $q = 0.2$ and $q = 0.5$ lead to roughly the same total expected dividend payouts (where for the larger q, more dividends are collected earlier and over a shorter portfolio's lifetime compared to the case with the smaller q, i.e. different time patterns of dividend payments here lead to the same aggregate payout in expectation).

Next, let us consider the effect of β on $V(x)$ for a given level of q. Table 2 illustrates a qualitatively similar pattern: the larger x is, the higher constant rate β can be afforded, and for $x \geq 10$, it is preferable to pay out the entire premium rate c as dividends (in addition to the proportional payments).

In order to better understand the contribution of the proportional rate q and the constant rate β to the overall value of $V(x)$, we now decompose (6) as $V(x) := V_q(x) + V_\beta(x)$, where

$$V_q(x) := \mathbb{E}_x\left[\int_0^{\tau_x} e^{-\delta t} q X_t \, dt\right], \quad \text{and} \quad V_\beta(x) := \mathbb{E}_x\left[\int_0^{\tau_x} e^{-\delta t} \beta \, dt\right].$$

Table 2 Expected present value of dividends for different rates β with $\alpha = 1/3$, $\lambda = 1$, $c = 3.5$, $q = 0.3$, and $\delta = 0.05$

$V(x; 0.3, \beta)$						
x	$\beta = 0$	$\beta = 0.5$	$\beta = 1$	$\beta = 2$	$\beta = 3$	$\beta = 3.5$
0	3.354	3.394	3.409	3.394	3.355	3.333
0.5	3.855	3.903	3.922	3.913	3.876	3.854
1	4.352	4.407	4.432	4.428	4.393	4.372
2	5.336	5.405	5.440	5.449	5.419	5.399
3	6.307	6.390	6.435	6.457	6.434	6.415
4	7.267	7.363	7.418	7.453	7.438	7.422
5	8.217	8.326	8.391	8.440	8.433	8.420
10	12.863	13.028	13.139	13.258	13.298	13.302
20	21.860	22.108	22.294	22.537	22.675	22.721

Along the same line of arguments as in Sect. 3 one can then derive

$$V_q(x) = A_q M\left(\frac{\delta}{q}, 1 + \frac{\lambda + \delta}{q}, z(x)\right) + V_{qp}(x), \qquad x \geq 0, \tag{32}$$

and

$$V_\beta(x) = A_\beta M\left(\frac{\delta}{q}, 1 + \frac{\lambda + \delta}{q}, z(x)\right) + V_{qp}(x), \qquad x \geq 0, \tag{33}$$

for respective constants A_q and A_β. We now proceed with an example to discuss the influence of the expected claim size $1/\alpha$ on the relative contribution of (32) and (33) to the total expected dividend payouts (6). Consider the following constellation of parameters: $\lambda = 1, c = 5, q = 0.5, \beta = 1$, and $\delta = 0.05$ so that the dividend rate at time t is given by $0.5X_t + 1$. Hence, the linear term qX_t constitutes the dominant term in the dividend rate if $X_t > 2$. Figure 2 displays the ratio $V_q(x)/V(x)$ for $\alpha = 1/3$ (solid line) and $\alpha = 1/4$ (dashed line). First, we observe that the ratio $V_q(x)/V(x)$ is increasing in x. This is in line with intuition since for larger initial capital x, the proportion of dividends from the linear term is larger; the concrete value of that ratio, however, reflects the occupation time of the various levels of the process. This is also illustrated by the different response of the ratio $V_q(x)/V(x)$ on changing the average claim size, depending on the range of x.

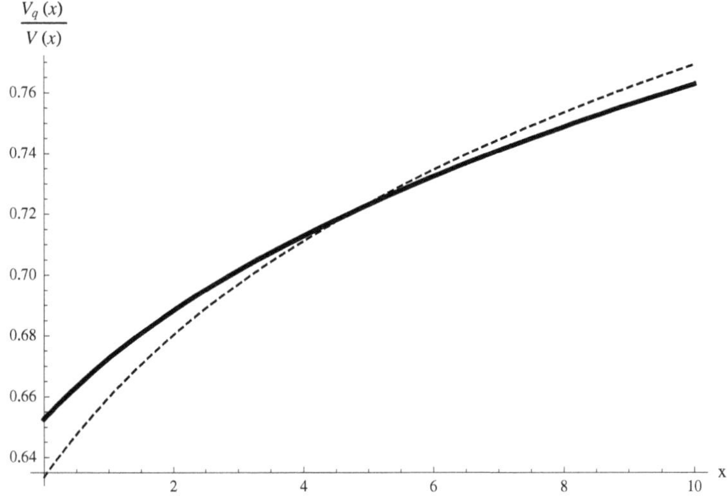

Fig. 2 Ratio $V_q(x)/V(x)$ for $\alpha = 1/3$ (*solid line*) and $\alpha = 1/4$ (*dashed line*)

7.2 Optimal Parameters

In a next step, it is natural to ask which combination of parameters q and β maximizes the expected present value of dividends until ruin. Let $\Theta = \{(q,\beta) : q > 0, \beta \in [0, c]\}$. With each pair $(q, \beta) \in \Theta$, we associate $V(x; q, \beta) := V(x)$. The optimization problem then consists in finding a pair (q^*, β^*) such that

$$V(x; q^*, \beta^*) = \max_{(q,\beta)\in\Theta} V(x; q, \beta), \tag{34}$$

for a given initial capital $x \geq 0$. In view of (17), such an optimization problem has to be approached numerically (here we used respectives routines in Mathematica). Table 3 displays the optimal parameters and resulting optimal dividend values for the case $\alpha = 1/3, \lambda = 1, c = 3.5$, and $\delta = 0.05$ and different initial capital values x. For illustration purposes we also depict the corresponding values for $\delta = 0.07$ in parentheses. One can observe that $q^*(x)$ increases in x (note that $q^*(x)$ is chosen as a function of initial capital x, but by construction then kept fixed throughout the lifetime of the process, i.e. not a function of current surplus value). Furthermore, we always have $\beta^*(x) = 0$, i.e. a constant dividend rate does not contribute favorably to the compromise between profitability and length (lifetime) of the payments (this is also the case for other parameter values in numerical experiments). An interpretation of the latter is that since such a constant rate would be applied at all capital levels, the survival when close to ruin is more important than the payment of immediate dividends, which is somewhat in line with the philosophy behind dividend barrier strategies, cf. Sect. 7.3. For the higher discount rate $\delta = 0.07$, $q^*(x)$ changes drastically and even becomes infinity for large x, mimicking lump

Table 3 Maximal expected present value of dividends and optimal pairs (q^*, β^*) for $\alpha = 1/3, \lambda = 1, c = 3.5$ and $\delta = 0.05$ $(\delta = 0.07)$

x	$V(x; q^*, \beta^*)$	$q^*(x)$	$\beta^*(x)$
0	3.426 (3.279)	0.751 (3.789)	0.000 (0.000)
0.5	3.939 (3.780)	0.756 (3.871)	0.000 (0.000)
1	4.449 (4.280)	0.768 (4.088)	0.000 (0.000)
2	5.461 (5.279)	0.806 (4.896)	0.000 (0.000)
3	6.466 (6.276)	0.860 (6.413)	0.000 (0.000)
4	7.465 (7.274)	0.927 (9.502)	0.000 (0.000)
5	8.460 (8.272)	1.008 (18.227)	0.000 (0.000)
10	13.406 (13.271)	1.719 (∞)	0.000 (0.000)
20	23.334 (23.271)	31.623 (∞)	0.000 (0.000)

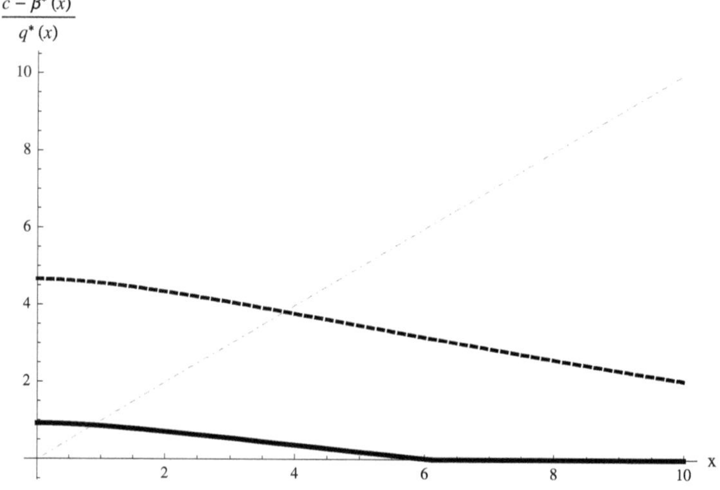

Fig. 3 $\frac{c - \beta^*(x)}{q^*(x)}$ for $\delta = 0.05$ (*dashed line*) and $\delta = 0.07$ (*solid line*)

sum payments of a barrier strategy with barrier at level zero. Note from (4) that for $(c - \beta^*(x))/q^*(x) \geq x$, the drift of the process X_t will never be positive, cf. Fig. 3.

7.3 Comparison with the Optimal Barrier Strategy

From classical results one knows that the optimal dividend strategy in a compound Poisson model with exponential claims is a barrier strategy (cf. Gerber [13]), if the sole criterion is the profitability. It is hence instructive to compare our optimal expected dividend payouts $V(x; q^*, \beta^*)$ according to the affine dividend strategy with $V_{b^*}(x)$, the one under the optimal barrier strategy b^*. Recall that

$$V_{b^*}(x) = \begin{cases} \frac{h(x)}{h'(b^*)}, & 0 \le x \le b^*, \\ x - b + \frac{h(b^*)}{h'(b^*)}, & x > b^*, \end{cases}$$

where $h(u) = (r + \alpha)e^{rx} - (s + \alpha)e^{sx}$, $r > 0$ and $s < 0$ are the roots of the characteristic equation

$$c\xi^2 + (\alpha c - (\lambda + \delta))\,\xi - \alpha\delta = 0,$$

and

$$b^* = \frac{1}{r - s}\ln\frac{s^2(s + \alpha)}{r^2(r + \alpha)}.$$

Table 4 compares the resulting $V_{b^*}(x)$ to the dividend payout of the optimal affine strategy $V(x; q^*, \beta^*)$ for the case $\alpha = 1/3, \lambda = 1, c = 3.5$, and $\delta = 0.05$ (in which case $b^* = 3.26$). Knowing that the barrier strategy is optimal among all admissible strategies, it is quite remarkable to observe how close one gets to this optimal value $V_{b^*}(x)$ by the best affine strategy.

A next question in this context is then how sensitive the performance of the affine dividend strategy is when varying $(q, \beta) \in \Theta$. To get an impression on that, Fig. 4 depicts the contour lines $\{(q, \beta) \in \Theta : V(x; q, \beta) = aV_{b^*}(x)\}$, i.e. those parameter values for which we achieve a certain percentage a of the optimal dividend barrier strategy. Here $x = 10$ and $a \in [0.98, 0.9942]$. The red area, for instance, consists of all pairs (q, β) leading to at least 99.4% of $V_{b^*}(10)$. The size of that area is quite remarkable, showing that one can achieve quite convincing performance for a variety of (q, β)-values. The pair $(q, \beta) = (1.719, 0)$ maximizes $V(10; q, \beta)$ and yields 99.5% of $V_{b^*}(10)$. One also sees that the gradient becomes larger as one moves towards smaller q-values indicating that $V(10; q, \beta)$ is sensitive to changes in q for smaller q.

Table 4 Comparison of the expected present value of dividends under affine and barrier dividend strategies for initial capitals x with $\alpha = 1/3, \lambda = 1, c = 3.5$, $q = 0.3$ and $\delta = 0.05$

x	$V_{b^*}(x)$	$V(x; q^*, \beta^*)$	$q^*(x)$	$\beta^*(x)$
0	3.437	3.426	0.751	0
$0.5\,b^*$	5.232	5.223	0.795	0
b^*	7.000	6.994	0.893	0
$1.5\,b^*$	8.764	8.749	1.034	0
$2\,b^*$	10.527	10.496	1.226	0
$3\,b^*$	14.055	13.981	1.854	0
$5\,b^*$	21.110	20.977	7.668	0

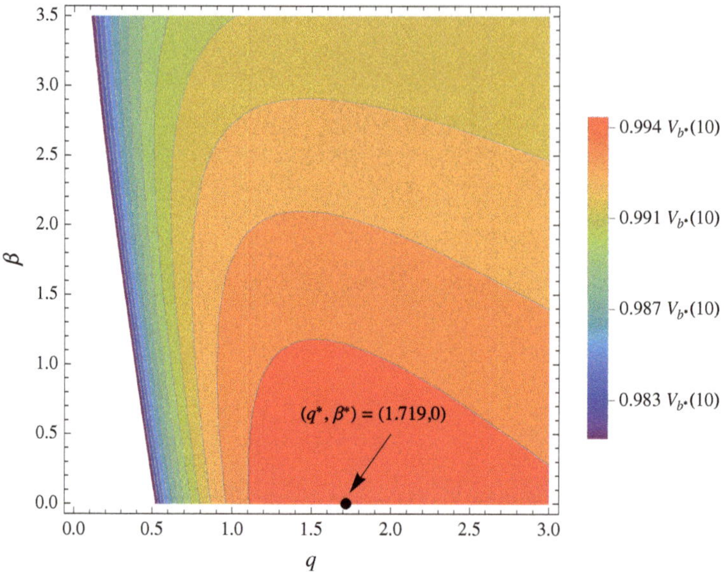

Fig. 4 Contour lines for $x = 10, \alpha = 1/3, \lambda = 1, c = 3.5, q = 0.3$, and $\delta = 0.05$

7.4 Dividend Payments Versus Expected Ruin Time

Since one motivation to introduce an affine dividend strategy was the increased lifetime of the process, and since we have seen in the previous section that the performance of this strategy gets quite close to the one of optimal barrier strategies, it is now interesting to see to what extent the expected ruin time is improved using such an affine dividend strategy.

Let T_x^b the time to ruin with a barrier at level b. The Laplace transform of T_x^b for exponentially distributed claim amounts in the compound Poisson model is then given by

$$
\mathbb{E}\left[e^{-\delta T_x^b}\right] = \begin{cases} \frac{(s+\alpha)(r+\alpha)}{\alpha}\left[\frac{re^{rb}e^{sx}-se^{sb}e^{rx}}{r(r+\alpha)e^{rb}-s(s+\alpha)e^{sb}}\right] & 0 \leq x \leq b, \\ \frac{(s+\alpha)(r+\alpha)}{\alpha}\left[\frac{e^{(r+s)b}(r-s)}{r(r+\alpha)e^{rb}-s(s+\alpha)e^{sb}}\right] & x > b, \end{cases}
$$

see, e.g., [16, Eq. (6.3)]. We then have

$$
\mathbb{E}\left[T_x^b\right] = -\frac{d}{d\delta}\mathbb{E}\left[e^{-\delta T_x^b}\right]\bigg|_{\delta=0}.
$$

Let $\mathbb{E}\left[\tau_{x;q,\beta}\right]$ be the expected ruin time under the affine dividend strategy $(q,\beta) \in \Theta$ (cf. Proposition 5.2) and consider the following constrained optimization problem:

$$\max_{(q,\beta)\in\Theta} \quad \mathbb{E}\left[\tau_{x;q,\beta}\right]$$

$$\text{subject to} \quad V(x;q,\beta) = aV_{b*}(x),$$

where $a \in (0,1)$. Note that the resulting optimal values q^* are not the same as the ones in the previous section, whereas β^* turns out to be again always zero. Figure 5 depicts the ratio $\mathbb{E}\left[\tau_{x;q^*,\beta^*}\right]/\mathbb{E}\left[T_x^{b^*}\right]$ for different initial capital values x as function of the performance factor a, for the same parameters $\alpha = 1/3, \lambda = 1, c = 3.5$, and $\delta = 0.05$. To match a higher required performance level a, one has to select larger values in the set Θ, which causes a reduction of the expected time to ruin $\mathbb{E}\left[\tau_{x;q^*,\beta^*}\right]$. Hence, the ratio $\mathbb{E}\left[\tau_{x;q^*,\beta^*}\right]/V_{b*}(x)$ is monotone decreasing in a. For $x = 2$, we have that for $a = 0.95$, selecting the best pair (q^*, β^*) roughly doubles the expected lifetime of the portfolio, whilst for $a = 0.99$, the improvement factor is still 1.33. For $x = 4$, it is worth noticing that under the horizontal dividend strategy (with $b^* = 3.257$), an immediate dividend payment occurs, leading to a reduced expected ruin time $\mathbb{E}\left[T_{b*}^{b^*}\right]$. Here the affine dividend strategy then compares even more favorably in terms of lifetime of the process. However, this trend is not preserved for even higher values of x. For illustrative purposes, we consider $x = 50$, where clearly there is a considerable initial dividend payment under the

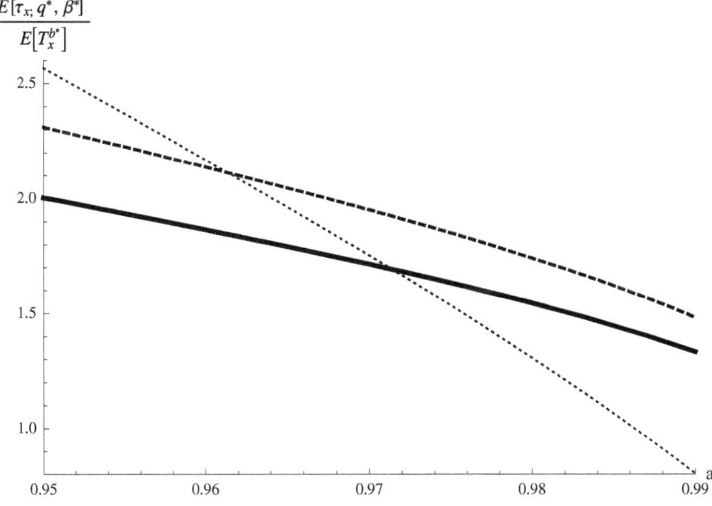

Fig. 5 Ratio $\mathbb{E}\left[\tau_{x;q^*,\beta^*}\right]/\mathbb{E}\left[T_x^{b^*}\right]$ for $x = 2$ (*solid line*), $x = 4$ (*dashed line*), and $x = 50$ (*dotted line*)

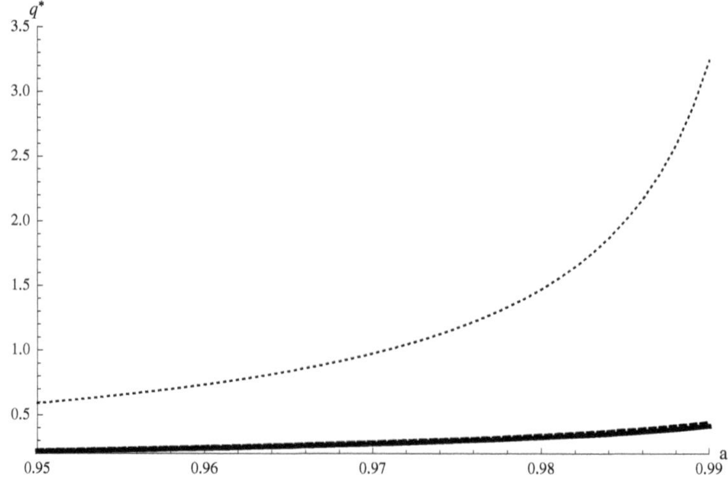

Fig. 6 Maximizer q^* for $x = 2$ (*solid line*), $x = 4$ (*dashed line*), and $x = 50$ (*dotted line*)

optimal barrier strategy, contributing to a major extent to the overall value $V_{b*}(50)$. Matching this performance under an affine dividend strategy for some large factor a, say 0.99, requires to increase (q^*, β^*) to an extent that the ratio $\mathbb{E}\left[\tau_{x;q*,\beta*}\right]/\mathbb{E}\left[T_x^b\right]$ is then even below 1 (see dotted line at $a = 0.99$).

Figure 6 depicts the resulting maximizer q^* for $x = 2$ (solid line), $x = 4$ (dashed line) and $x = 50$ (dotted line) as a function of a. For $x = 2$ and $x = 4$, the choice of q^* is almost identical. However, for $x = 50$, a significant non-linear increase of q^* is needed for higher values of a to make up for the large initial lump sum payment under the optimal barrier strategy.

Altogether, one sees that a suitably chosen affine dividend strategy can lead to almost as large values for the expected discounted dividend payments, while leading to considerably improved safety, measured in terms of expected ruin time of the portfolio. Note that the chosen numerical values of the discount rate δ are quite high, and smaller values of δ can lead to an even better performance of the affine strategy relative to the optimal barrier strategy.

Acknowledgements Financial support by the Swiss National Science Foundation Project 200020 143889 is gratefully acknowledged.

References

1. M. Abramowitz, I.A. Stegun, *Handbook of Mathematical Functions with Formulas, Graphs, and Mathematical Tables*. National Bureau of Standards Applied Mathematics Series, vol. 55 (Dover, New York, 1964)

2. H. Albrecher, S. Thonhauser, Optimality results for dividend problems in insurance. Rev. R. Acad. Cienc. Exactas Fís. Nat. Ser. A Math. RACSAM **103**(2), 295–320 (2009)
3. H. Albrecher, R. Kainhofer, R.F. Tichy, Simulation methods in ruin models with non-linear dividend barriers. Math. Comput. Simul. **62**(3–6), 277–287 (2003). 3rd IMACS Seminar on Monte Carlo Methods—MCM 2001 (Salzburg)
4. S. Asmussen, H. Albrecher, *Ruin Probabilities*. Advanced Series on Statistical Science and Applied Probability, vol. 14, 2nd edn. (World Scientific, Hackensack, NJ, 2010)
5. B. Avanzi, Strategies for dividend distribution: a review. N. Am. Actuar. J. **13**(2), 217–251 (2009)
6. B. Avanzi, B. Wong, On a mean reverting dividend strategy with Brownian motion. Insur. Math. Econ. **51**(2), 229–238 (2012)
7. P. Azcue, N. Muler, *Stochastic Optimization in Insurance: A Dynamic Programming Approach*. Springer Briefs in Quantitative Finance (Springer, New York, 2014)
8. A. Brav, J.R. Graham, C.R. Harvey, R. Michaely, Payout policy in the 21st century. J. Financ. Econ. **77**(3), 483–527 (2005)
9. R. Cont, P. Tankov, *Financial Modelling with Jump Processes*. Chapman & Hall/CRC Financial Mathematics Series (Chapman & Hall/CRC, Boca Raton, FL, 2004)
10. D. Cvijović, Closed-form summations of certain hypergeometric-type series containing the digamma function. J. Phys. A Math. Theor. **41**(45), 455205 (2008)
11. B. De Finetti, Su un'impostazione alternativa della teoria collettiva del rischio, in *Transactions of the XVth International Congress of Actuaries*, vol. 2 (1957), pp. 433–443
12. H. Exton, *Multiple Hypergeometric Functions and Applications* (Wiley, Chichester, 1976)
13. H.-U. Gerber, Entscheidungskriterien für den zusammengesetzten Poisson-Prozess. PhD thesis, 1969
14. H.U. Gerber, S. Lin, H. Yang, A note on the dividends-penalty identity and the optimal dividend barrier. ASTIN Bull. **36**(2), 489–503 (2006)
15. M.J. Gordon, Dividends, earnings and stock prices. Rev. Econ. Stat. **41**, 99–105 (1959)
16. X.S. Lin, G.E. Willmot, S. Drekic, The classical risk model with a constant dividend barrier: analysis of the Gerber-Shiu discounted penalty function. Insur. Math. Econ. **33**(3), 551–566 (2003)
17. J. Lintner, Distribution of incomes of corporations among dividends, retained earnings, and taxes. Am. Econ. Rev. **46**(2), 97–113 (1956)
18. R.L. Loeffen, J.-F. Renaud, De Finetti's optimal dividends problem with an affine penalty function at ruin. Insur. Math. Econ. **46**(1), 98–108 (2010)
19. A.R. Miller, Summations for certain series containing the digamma function. J. Phys. A **39**(12), 3011–3020 (2006)
20. F.W.J. Olver, D.W. Lozier, R.F. Boisvert, C.W. Clark (eds.), *NIST Handbook of Mathematical Functions* (U.S. Department of Commerce, National Institute of Standards and Technology, Washington, DC; Cambridge University Press, Cambridge, 2010)
21. M. Parlar, Use of stochastic control theory to model a forest management system. Appl. Math. Model. **9**(2), 125–130 (1985)
22. K. Sato, *Lévy Processes and Infinitely Divisible Distributions* (Cambridge University Press, Cambridge, 2013)
23. H. Schmidli, *Stochastic Control in Insurance*. Probability and Its Applications (New York) (Springer, London, 2008)
24. W. Schoutens, *Lévy Processes in Finance* (Wiley, New York, 2003)
25. H.M. Srivastava, P.W. Karlsson, *Multiple Gaussian Hypergeometric Series* (Wiley, Chichester, 1985)
26. M. Steffensen, Quadratic optimization of life and pension insurance payments. ASTIN Bull. **36**(01), 245–267 (2006)
27. S. Thonhauser, H. Albrecher, Dividend maximization under consideration of the time value of ruin. Insur. Math. Econ. **41**(1), 163–184 (2007)

28. R.F. Tichy, Über eine zahlentheoretische Methode zur numerischen Integration und zur Behandlung von Integralgleichungen. Österreich. Akad. Wiss. Math. Natur. Kl. Sitzungsber. II **193**(4–7), 329–358 (1984)
29. R.F. Tichy, Bemerkung zu einem versicherungsmathematischen Modell. Mitt. Verein. Schweiz. Versicherungsmath. **2**, 237–241 (1987)

A Discrepancy Problem: Balancing Infinite Dimensional Vectors

József Beck

Abstract As a corollary of a general balancing result, we prove that there exists a balanced "2-coloring" g of the set of natural numbers $\mathbb{N}$ such that simultaneously for all integers $d \geq 1$, every (finite) arithmetic progression of difference d has discrepancy $D_g(d) \leq d^{8+\varepsilon}$, independently of the starting point and the length of the arithmetic progression. Formally, for every $\varepsilon > 0$ there exists a function $g : \mathbb{N} \to \{-1, 1\}$ such that

$$D_g(d) = \max_{a \geq 1, m \geq 1} \left| \sum_{i=0}^{m-1} g(a + id) \right| \leq d^{8+\varepsilon}$$

for all sufficiently large $d \geq d_0(\varepsilon)$. This reduces an old superexponential upper bound $\leq d!$ of Cantor, Erdős, Schreiber, and Straus to a polynomial upper bound. Note that the polynomial range is the correct range, since a well known result of Roth implies the lower bound $D_g(d) \geq \sqrt{d}/20$ for every $g : \mathbb{N} \to \{-1, 1\}$.

We derive this concrete number theoretic upper bound result about arithmetic progressions from a very general vector balancing result. It is about balancing infinite dimensional vectors in the maximum norm, and it is interesting in its own right (possibly, more interesting than the special case above).

1 Introduction

Our starting point is the following trivial observation: given an arbitrary infinite sequence of real numbers $x_1, x_2, x_3, \ldots$ with $|x_i| \leq 1$, we can always find an infinite sequence of "balancing" signs $\delta_1, \delta_2, \delta_3, \ldots$ with $\delta_i \in \{-1, 1\}$ such that

$$\max_{n \geq 1} \left| \sum_{i=1}^{n} \delta_i x_i \right| \leq 1. \tag{1}$$

J. Beck (✉)

Department of Mathematics, Rutgers University, 110 Frelinghuysen Road, Piscataway, NJ 08854-8019, USA

e-mail: jbeck@math.rutgers.edu

© Springer International Publishing AG 2017

C. Elsholtz, P. Grabner (eds.), *Number Theory – Diophantine Problems, Uniform Distribution and Applications*, DOI 10.1007/978-3-319-55357-3_3

Indeed, proceed by induction, and choose δ_{n+1} in such a way that $\delta_{n+1}x_{n+1}$ and $\sum_{i=1}^{n}\delta_i x_i$ have opposite signs.

This trivial observation motivates the following non-trivial question: does there exist a *simultaneous* generalization of (1)? More precisely, does there exist a function $f(k)$ such that, given an arbitrary infinite sequence of infinite dimensional vectors $\mathbf{v}_1, \mathbf{v}_2, \mathbf{v}_3, \ldots$ with $\|\mathbf{v}_i\|_\infty \leq 1$ (where $\|.\|_\infty$ denotes the maximum norm, i.e., every coordinate $v_{i,k}$ of $\mathbf{v}_i$ has absolute value $|v_{i,k}| \leq 1$, where $i \geq 1, k \geq 1$), we can always find an infinite sequence of "balancing" signs $\delta_1, \delta_2, \delta_3, \ldots$ with $\delta_i \in \{-1, 1\}$ such that

$$\max_{n \geq 1} \left| \sum_{i=1}^{n} \delta_i v_{i,k} \right| \leq f(k) \text{ for all } k \geq 1? \tag{2}$$

Note that the sum

$$\sum_{i=1}^{n} \delta_i v_{i,k}$$

represents the kth coordinate of the vector sum

$$\sum_{i=1}^{n} \delta_i \mathbf{v}_i.$$

In 1981 we gave an affirmative answer to question (2) by proving the upper bound (see Beck [1])

$$f(k) = O\left(k^{(2+\varepsilon)\log k}\right).$$

The goal of this paper is to prove the following upgraded version, where we replace the superpolynomial upper bound above with the polynomial upper bound

$$f(k) = O\left(k^{4+\varepsilon}\right).$$

Theorem 1 *Let $\varepsilon > 0$ be arbitrarily small but fixed. Let $\mathbf{v}_1, \mathbf{v}_2, \mathbf{v}_3, \ldots$ be an infinite sequence of infinite dimensional vectors $\mathbf{v}_i \in \mathbb{R}^\infty$ with $\|\mathbf{v}_i\|_\infty \leq 1, 1 \leq i < \infty$. Then there exists an infinite sequence of balancing signs $\delta_1, \delta_2, \delta_3, \ldots, \delta_i \in \{-1, 1\}$ such that the kth coordinate of the sum*

$$\sum_{i=1}^{n} \delta_i \mathbf{v}_i \text{ has absolute value less than } k^{4+\varepsilon} \text{ for all } k \geq c'(\varepsilon) \tag{3}$$

and for all $1 \leq n < \infty$, and for $1 \leq k < c'(\varepsilon)$ the kth coordinate of the sum

$$\sum_{i=1}^{n} \delta_i \mathbf{v}_i \text{ has absolute value less than } c_0'(\varepsilon) \tag{3'}$$

for all $1 \leq n < \infty$, where $c'(\varepsilon)$ and $c_0'(\varepsilon)$ are finite constants depending only on ε.

We do not know how close (3) is to be optimal. We do know, however, that

$$\limsup_{K \to \infty} \frac{\max_{1 \leq k \leq K} f(k)}{\sqrt{K}} > 0. \tag{4}$$

The message of (3)–(4) is that in (2) the polynomial range for $f(k)$ is the correct range, but we do not know where the right exponent is between $1/2$ and $4 + \varepsilon$.

One way to prove (4) is to use Hadamard matrices. Indeed, a Hadamard matrix of order k is a k by k matrix with entries ± 1 such that any two columns are orthogonal (i.e., the inner product is zero). It is customary to assume that the first row and the first column consist entirely of $+1$'s (which can be always achieved if we permute rows or columns or if we multiply some rows or columns by -1). Then the remaining rows (or columns) have as many $+1$'s as -1's (so $k > 1$ has to be even). Let $\mathbf{v}_i$, $1 \leq i \leq k$ be the k column vectors; they all have maximum norm one. By using orthogonality, for any sequence of signs $\delta_1, \delta_2, \ldots, \delta_k, \delta_i \in \{-1, 1\}$ we can compute the l_2-norm $\|.\|_2$ of the sum

$$\left(\left\| \sum_{i=1}^{k} \delta_i \mathbf{v}_i \right\|_2 \right)^2 = \sum_{i=1}^{k} \|\mathbf{v}_i\|_2^2 + 2 \sum_{1 \leq i < j \leq k} \mathbf{v}_i \cdot \mathbf{v}_j = \sum_{i=1}^{k} \|\mathbf{v}_i\|_2^2 = k^2.$$

It follows that the l_2-norm of the sum is k, and so

$$\text{maximum norm of } \sum_{i=1}^{k} \delta_i \mathbf{v}_i \text{ is } \geq \sqrt{k}. \tag{5}$$

Equation (5) implies (4), since starting with

$$H_1 = \begin{pmatrix} 1 & 1 \\ 1 & -1 \end{pmatrix} \text{ and using the pattern } H_{i+1} = \begin{pmatrix} H_i & H_i \\ H_i & -H_i \end{pmatrix},$$

we obtain a growing sequence of Hadamard matrices. (It follows, as a by product, that for every power of two $k = 2^s$ there exists a Hadamard matrix of order k.)

To prove Theorem 1 we use the proof technique developed in Beck [1], but we put the same ingredients together in a more efficient way. This is how we reduce the superpolynomial upper bound in [1] to the polynomial upper bound in (3).

We derive Theorem 1—a result about balancing vectors with arbitrary real coordinates—from a similar result about *integer* coordinates. Working with integers makes it much simpler to execute the basic idea of the proof: to force as many perfect cancellations as possible (based on the simple fact that zeros do not accumulate, but many small real numbers together may give a huge error term).

We slightly change the setup: Theorem 2 is about a matrix, where the columns play the role of the vectors in Theorem 1.

Theorem 2 *Let $\varepsilon > 0$ be arbitrarily small but fixed. Given an infinite matrix $\mathcal{U} = \{u_{k,j}\}$, $1 \leq k, j < \infty$ of entries $u_{k,j} \in \{-1, 0, 1\}$, there exists an infinite sequence $\delta_1, \delta_2, \delta_3, \ldots, \delta_i \in \{-1, 1\}$ such that all partial sums of the kth row have the upper bound*

$$\max_{j \geq 1} \left| \sum_{i=1}^{j} \delta_i u_{k,i} \right| \leq k^{4+\varepsilon} \text{ for } k \geq \hat{c}(\varepsilon), \tag{6}$$

and for $1 \leq k < \hat{c}(\varepsilon)$ all partial sums of the kth row have the upper bound

$$\max_{j \geq 1} \left| \sum_{i=1}^{j} \delta_i u_{k,i} \right| \leq \hat{c}_0(\varepsilon), \tag{6'}$$

where $\hat{c}(\varepsilon)$ and $\hat{c}_0(\varepsilon)$ are finite constants depending only on ε.

The main difficulty is to prove Theorem 2; see Sects. 2 and 3. The deduction of Theorem 1 from Theorem 2 is quite simple; we postpone it to the end of the paper.

Instead of flipping the sign $\pm$ of vectors, we may achieve that all partial sums be small just by *rearranging* them (assuming of course that the total sum is small). A remarkably simple and elegant result, called Chobayan's transference lemma, shows that the two kinds of balancing problems are strongly related.

Chobayan's transference lemma *Let $\mathbf{v}_j \in \mathbb{R}^d$, $1 \leq j \leq n$ be n vectors such that $\sum_{j=1}^{n} \mathbf{v}_j = \mathbf{0}$, and let $\|.\|$ be an arbitrary norm in $\mathbb{R}^d$. Suppose that for every permutation $\pi_n = (i_1, i_2, i_3, \ldots, i_n)$ of $1, 2, 3, \ldots, n$ there exists a sign sequence $\delta_j \in \{-1, 1\}$, $1 \leq j \leq n$ (depending on π_n) such that*

$$\max_{1 \leq t \leq n} \left\| \sum_{j=1}^{t} \delta_j \mathbf{v}_{i_j} \right\| \leq A.$$

Then there is a permutation $\pi^* = (\ell_1, \ell_2, \ell_3, \ldots, \ell_n)$ of $1, 2, 3, \ldots, n$ such that

$$\max_{1 \leq t \leq n} \left\| \sum_{j=1}^{t} \mathbf{v}_{\ell_j} \right\| \leq A.$$

For a proof of Chobayan's transference lemma; see, e.g., Theorem 4.8 in [3].

Combining Theorem 1 with Chobayan's transference lemma, we obtain the following result.

Corollary *Let* $\mathbf{v}_j \in \mathbb{R}^\infty$, $1 \le j \le n$ *be* n *infinite dimensional vectors satisfying* $\|\mathbf{v}_j\|_\infty \le 1$ *and* $\sum_{j=1}^n \mathbf{v}_j = \mathbf{0}$. *Then for every* $\varepsilon > 0$ *there is a permutation* $\pi^* = (\ell_1, \ell_2, \ell_3, \ldots, \ell_n)$ *of* $1, 2, 3, \ldots, n$ *such that the kth coordinate of the sum*

$$\sum_{j=1}^t \mathbf{v}_{\ell_j} \text{ has absolute value } \le c_0^*(\varepsilon) k^{4+\varepsilon}$$

for every $k \ge 1$ *and* $1 \le t \le n$, *where* $c_0^*(\varepsilon)$ *is a constant that depends only on* $\varepsilon > 0$.

Note that earlier I. Bárány proved a weaker version of the Corollary with the double-exponential upper bound 2^{3^k} (oral communication).

We conclude Sect. 1 with a number-theoretic application. Cantor, Erdős, Schreiber, and Straus ([5]; see also [6]) studied the 2-coloring discrepancy of the family of all arithmetic progressions of the integers. They asked the question: Does there exist a balanced "2-coloring" of the set of natural numbers $\mathbb{N}$ (where a "2-coloring" simply means a function $g : \mathbb{N} \to \{-1, 1\}$) such that, simultaneously for all $d \ge 1$, every (finite) arithmetic progression of difference d has discrepancy $\le h(d)$ for a certain function h, independently of the starting point and the length of the arithmetic progression? Formally, for every $g : \mathbb{N} \to \{-1, 1\}$ write

$$D_g(d) = \max_{a \ge 1, m \ge 1} \left| \sum_{i=0}^{m-1} g(a + id) \right|.$$

Does there exist a "2-coloring" $g : \mathbb{N} \to \{-1, 1\}$ such that

$$D_g(d) < \infty \text{ for every } d? \tag{7}$$

Cantor, Erdős, Schreiber, and Straus gave an affirmative answer to question (7) by constructing a 2-coloring $g : \mathbb{N} \to \{-1, 1\}$ such that $D_g(d) \le d!$.

First we replace the superexponential upper bound $d!$ with an exponential upper bound; see (15) below.

We define the sign function $g(n) = \pm 1$, $n \ge 1$ by induction. Let $g(1) = 1$, $g(2) = -1$, and write $M(1) = 2$. Now let $r \ge 2$, and assume that $g(n) \in \{-1, 1\}$ is already defined for $1 \le n \le M(r-1)$. We distinguish two cases.

Case 1: $r = p^\nu$, $\nu \ge 1$ is a prime power
Then we extend the definition of $g(n)$ for the first $2pM(r-1) = M(r)$ natural numbers by using the rule "double by the negative, and then repeat p times":

$$g(iM(r-1)+j) = (-1)^i g(j) \text{ for all } 1 \le i \le 2p-1 \text{ and } 1 \le j \le M(r-1). \tag{8}$$

Case 2: r is not a prime power

Then $M(r) = M(r-1)$, i.e., we do nothing.

For illustration note that $g(3) = -1, g(4) = 1, g(5) = 1, g(6) = -1, g(7) = -1, g(8) = 1, g(9) = -1, g(10) = 1$, and so on; and $M(2) = 2 \cdot 2M(1) = 8$, $M(3) = 2 \cdot 3M(2) = 48$, $M(4) = 2 \cdot 2M(3) = 192$, $M(5) = 2 \cdot 5M(4) = 1920$, $M(6) = M(5) = 1920$, and so on.

For later application note that $M(r)$ is divisible by r. It follows from $M(d) = 2pM(d-1)$ if d is a p-power; see Case 1 and Case 2 above.

Construction rule (8) implies that the ±1-sequence $g(n), n \in \mathbb{N}$ has the following key property:

$$g(2\ell M(d) + j) = -g((2\ell + 1)M(d) + j) \text{ for every } d \geq 1, \ 1 \leq j \leq M(d), \ \ell \geq 1. \tag{9}$$

By using (9) we give an upper bound to

$$\left| \sum_{i=0}^{m-1} g(a + id) \right|$$

as follows. Let

$$a \leq 2\ell_1 M(d) \text{ and } 2\ell_2 M(d) < a + (m-1)d, \tag{10}$$

where ℓ_1 is the smallest integer satisfying the first inequality in (10), and ℓ_2 is the largest integer satisfying the second inequality. We recall that $M(d)$ is divisible by d, and combining this fact with (9) we have for every integer $\ell_1 \leq \ell < \ell_2$

$$\sum_{i:\ 2\ell M(d)<a+id\leq(2\ell+1)M(d)} g(a + id) = - \sum_{i:\ (2\ell+1)M(d)<a+id\leq(2\ell+2)M(d)} g(a + id). \tag{11}$$

By (10) and (11)

$$\left| \sum_{i=0}^{m-1} g(a + id) \right| \leq \left| \sum_{0\leq i\leq m-1:\ a+id\leq 2\ell_1 M(d)} g(a + id) \right|$$

$$+ \left| \sum_{0\leq i\leq m-1:\ 2\ell_2 M(d)<a+id} g(a + id) \right|$$

$$\leq 2M(d) + 2M(d) = 4M(d). \tag{12}$$

Next we estimate $M(r)$ from above in terms of r. Let $L(r)$ denote the least common multiple of the odd natural numbers $\leq r$. Clearly $L(1) = L(2) = 1$ and for $r \geq 3$

$$L(r) = \prod_{3 \leq p \leq r} p^{\alpha(r;p)} \text{ where } p^{\alpha(r;p)} \leq r \text{ but } p^{1+\alpha(r;p)} > r,$$

and $p \geq 3$ denotes odd primes. Since there are relatively few prime-squares (and even fewer prime-cubes, etc.), it easily follows from the prime number theorem that

$$L(r) \leq e^{(1+o(1))r}. \tag{13}$$

Let $\pi(r)$ denote the number of primes $\leq r$, and let $\pi^*(r)$ denote the number of prime-powers $\leq r$. We have

$$\pi^*(r) = \pi(r) + \pi\left(\sqrt{r}\right) + \pi\left(\sqrt[3]{r}\right) + \cdots = (1 + o(1))r/\log r,$$

where in the last step we used the prime number theorem. We also need the trivial fact that the number of two-powers $\leq r$ is $\leq 1 + \log_2 r$ (binary logarithm). Combining these facts with the definitions of $M(r)$ and $L(r)$, we have

$$M(r) \leq 2 \cdot 2^{1+\log_2 r} \cdot 2^{\pi^*(r)} \cdot L(r),$$

where the first factor 2 comes from $r = 1$ and the second factor comes from the contribution of the two-powers. Using (13) we obtain

$$M(r) \leq 2 \cdot 2^{1+\log_2 r} \cdot 2^{\pi^*(r)} \cdot e^{(1+o(1))r}$$

$$= 4r \cdot 2^{(1+o(1))r/\log r} \cdot e^{(1+o(1))r} = e^{(1+o(1))r}.$$

Using this in (12), we conclude

$$\left| \sum_{i=0}^{m-1} g(a + id) \right| \leq 4M(d) \leq e^{(1+o(1))d},$$

which implies the desired exponential upper bound

$$D_g(d) < (e + \varepsilon)^d \tag{15}$$

for any $\varepsilon > 0$ if d is sufficiently large depending on $\varepsilon > 0$.

Next we replace the exponential upper bound in (15) with a polynomial upper bound; see (17) below. It is based on the observation that question (7) is a special case of question (2). To apply Theorem 1 we make some preparation: we arrange the residue classes $b \pmod{r}$ with $0 \le b < r$ in linear order as follows:

$$(0,1), (0,2), (1,2), (0,3), (1,3), (2,3), \ldots, (0,r), \ldots, (b,r), \ldots, (r-1,r), \ldots$$

Note that there are $1 + 2 + 3 + \cdots + d = d(d+1)/2$ residue classes $b \pmod{r}$ with $0 \le b < r \le d$.

Let $a \ge 0$, and write $a = b + hr$ with $0 \le b < r$; then we have the simple inequality

$$\left| \sum_{i=0}^{m-1} g(a+ir) \right| \le \left| \sum_{j=0}^{h+m-1} g(b+jr) \right| + \left| \sum_{j=0}^{h-1} g(b+jr) \right|.$$

Combining these facts with (2) and (7), we obtain that there exists a 2-coloring $g : \mathbb{N} \to \{-1, 1\}$ such that

$$D_g(d) \le 2f\left(\frac{d(d+1)}{2} \right), \tag{16}$$

and because we have the upper bound

$$f(k) \le k^{4+\varepsilon}$$

for all sufficiently large $k \ge k_0(\varepsilon)$ (see Theorem 1; in fact, it suffices to apply Theorem 2), (16) implies

$$D_g(d) \le \left(d^2 \right)^{4+\varepsilon} = d^{8+2\varepsilon} \tag{17}$$

for all sufficiently large $d \ge d_0(\varepsilon)$.

Note that there is an interesting difference between the upper bounds (15) and (17): (15) is *constructive* and (17) is *non-constructive*. Indeed, for (15) we defined the balancing sign function $g : \mathbb{N} \to \{-1, 1\}$ by the explicit rule (8) (see Case 1 and Case 2 above), which makes it constructive. In sharp contrast, the proof of Theorem 2 contains two non-constructive ingredients: the pigeonhole principle (see Lemma 1 in Sect. 2) and the compactness argument (see Sect. 3). This is why we call (17) a "pure existence" result.

We also point out that (15) heavily uses the fact that the residue classes are "homogeneous," but the proof of (17)—as a direct application of the general Theorem 2—does not make any use of this special property of the residue classes.

Equation (17) is the best upper bound that we know (and (15) is the best *constructive* upper bound that we know). To get a lower bound, we can use a well-known result of Roth [8] on the discrepancy of integer sequences relative to arithmetic progressions. Let $g(n) \in \{-1, 1\}$, $1 \le n \le N$ be an arbitrary "2-coloring" of the interval $[1, N] = \{1, 2, 3, \ldots, N\}$. It follows from Roth's general l_2-norm lower bound that there exists an arithmetic progression $a + id$, $0 \le i < m$ contained in the interval $[1, N]$ such that $d \le N^{1/2}$ and

$$\left| \sum_{i=0}^{m-1} g(a + id) \right| \ge \frac{1}{20} N^{1/4}. \tag{18}$$

Since $d \le N^{1/2}$, (18) implies the lower bound $D_g(d) \ge c\sqrt{d}$ with the absolute constant $c = 1/20$. (Note that Roth's lower bound is sharp apart from the value of c; see Beck [2] and Matoušek–Spencer [7].)

We can say, therefore, that in (7) the polynomial range for $D_g(d)$ is the correct range, but we do not know where the right exponent is between $1/2$ and $8 + \varepsilon$. It is well possible that the right exponent is $1/2$, i.e., the lower bound is sharp. Eliminating the $\log n$-power in the following result would imply that $1/2$ is the right exponent.

Theorem A (Beck and Spencer [4]) *Let n be a positive integer. Then there exists a sign function $g(k) \in \{-1, 1\}$, $k \in \mathbb{N}$ such that for any arithmetic progression*

$$P = P(d) = \{a, a + d, a + 2d, a + 3d, \ldots\}$$

of difference $d \le n$ and of arbitrary finite length,

$$\left| \sum_{k \in P(d)} g(k) \right| < c \cdot \sqrt{d}(\log n)^{7/2}, \quad 1 \le d \le n,$$

where c is an absolute constant.

Finally, we mention a somewhat related recent breakthrough of Terence Tao [9]: the solution of the well-known 500-dollar Erdős discrepancy problem. Roth's lower bound result is about the discrepancy relative to arithmetic progressions; the Erdős discrepancy problem is about a special class of arithmetic progressions: $\{d, 2d, 3d, \ldots, nd\}$, called *homogeneous* (or "starting from zero"). In the 1930s Erdős conjectured that every function $g(n) \in \{-1, 1\}$, $n \in \mathbb{N}$ has infinite discrepancy relative to homogeneous arithmetic progressions, i.e.,

$$\sup_{n, d \in \mathbb{N}} \left| \sum_{j=1}^{n} g(jd) \right| = \infty. \tag{19}$$

This long-standing open problem was very recently solved by Terence Tao [9] in a remarkable paper using deep analytic number theory [in fact, Tao proved a vector-valued generalization of (19)].

Here we just briefly point out two basic differences between Roth's lower bound result and Tao's lower bound result. First, the Erdős conjecture is false if we weaken the hypothesis to functions "$g(n) \in \{-1, 0, 1\}, n \in \mathbb{N}$, where the non-zero values have positive density." Indeed, let χ_3 be the non-principal Dirichlet character modulo 3 [i.e., $\chi_3(n) \in \{-1, 1\}$ if $n \equiv \pm 1 \pmod 3$ and $\chi_3(n) = 0$ if $n \equiv 0 \pmod 3$]. Since χ_3 is completely multiplicative, we have

$$\left| \sum_{j=1}^{n} \chi_3(jd) \right| = |\chi_3(d)| \cdot \left| \sum_{j=1}^{n} \chi_3(j) \right| \leq \left| \sum_{j=1}^{n} \chi_3(j) \right| \leq 1,$$

and so the homogeneous discrepancy is bounded.

In sharp contrast, Roth's discrepancy theorem does work for functions "$g(n) \in \{-1, 0, 1\}, n \in \mathbb{N}$, where the non-zero values have positive density" (and even far beyond that). We may say that Roth's theorem works for a substantially larger class of functions.

The second basic difference is the quantitative aspect. In Roth's theorem the discrepancy function is *polynomial* (and in general we know the exact order of magnitude). The homogeneous discrepancy function, on the other hand, is much smaller: we know that it is *at most logarithmic*. Indeed, we can easily eliminate the zero values in the character χ_3 mentioned above: let $\widetilde{\chi_3}(p) = \chi_3(p)$ for primes $p \neq 3$, let $\widetilde{\chi_3}(3) = 1$, and extend $\widetilde{\chi_3}(n)$ as a completely multiplicative function. It is easy to see that

$$\left| \sum_{j=1}^{n} \widetilde{\chi_3}(j) \right|$$

is equal to the number of digits 1 in the base 3 expansion of n, i.e., the homogeneous discrepancy function of $\widetilde{\chi_3}$ is only logarithmic in the parameter n.

Tao's paper [9] does not give any estimation on how fast the homogeneous discrepancy function tends to infinity. So, we do not know whether the logarithmic upper bound $O(\log n)$ is close to the truth, or perhaps the truth is much smaller like $O(\log \log n)$, or $O(\log \log \log n)$, and so on. The homogeneous discrepancy problem turned out to be much harder, and it is not surprising that we know much less about its quantitative aspect.

2 Proof of Theorem 2 (I): Main Idea

Let C_j denote the jth column vector of matrix $\mathcal{U}$. Theorem 2 can be restated as follows: there exist $\delta_j = \pm 1, j \geq 1$ such that the kth coordinate of the sum $\sum_{j=1}^{m} \delta_j C_j$ has absolute value $\leq k^{4+\varepsilon}$ for any $m \geq 1$ and $k \geq \hat{c}(\varepsilon)$.

We will start by showing that if a ± 1-0-matrix is finite, with a sufficient number of columns, we can always find $\delta_j = \pm 1$'s that cause some of the columns to cancel. The first lemma is an application of the pigeonhole principle.

Lemma 1 *Let $\{V_j\}_{j=1}^{t}$ be a finite set of d-dimensional integer vectors with maximum norm $\|V_j\|_\infty \le M$, and also $d \ge 17$ and $t \ge 2d\log(dM)$. Then there exist a nonempty subset $H \subset \mathbb{N}$ of the natural numbers and a sequence $\{\delta_j\}_{j=1}^{t}$ where $\delta_j \in \{-1,1\}$ such that*

$$\sum_{j\in H} \delta_j V_j = \mathbf{0} \text{ (=zero vector)}.$$

Proof Consider the 2^t vector sums of the form

$$\sum_{j\in I} V_j \text{ where } I \subset \{1,2,\ldots,t\}. \tag{20}$$

Each coordinate of a vector sum $\sum_{j\in I} V_j$ lies in the interval $[-tM, tM]$, and so there are at most $(2tM+1)^d$ distinct vector sums among (20). But for $M \ge 1, d \ge 17, t \ge 2d\log(dM)$ we have the inequality $2^t > (2tM+1)^d$, so the pigeonhole principle applies, and yields that there exist two different subsets $I_1, I_2 \subset \{1,2,\ldots,t\}$ such that

$$\sum_{j\in I_1} V_j = \sum_{j\in I_2} V_j, \text{ implying } \sum_{j\in I_1} V_j - \sum_{j\in I_2} V_j = \mathbf{0}. \tag{21}$$

We may assume that I_1 and I_2 are disjoint, since the identical terms cancel out in the last equality in (21). So if we define $H = I_1 \cup I_2$ (where $I_1 \cap I_2 = \emptyset$), and $\delta_j = 1$ if $j \in I_1$ and $\delta_j = -1$ if $j \in I_2$, we get

$$\sum_{j\in H} \delta_j V_j = \mathbf{0},$$

completing the proof of the lemma. $\qquad\qquad\square$

We may say, intuitively speaking, that in the argument below Lemma 1 replaces the perfect cancellation in (11) that was enforced by the rule "double by the negative." The reason why (17) is much better than (15) is that Lemma 1 is much more efficient than the rule "double by the negative."

By repeated application of Lemma 1 we can handle the case of infinitely many columns. The next lemma shows how we can keep the partial sums of the column vectors bounded by choosing the $\delta_j = \pm 1$'s to cancel some sets of column vectors. It is the key lemma: the proof of Theorem 2 is basically a repeated application of Lemma 2; see the "Nutshell Summary of the Proof of Theorem 2" at the end of Sect. 2. Section 3 is just a routine execution of this plan.

Lemma 2 *Let $\{V_j\}_{j=1}^{\infty}$ be an infinite set of d-dimensional integer vectors with maximum norm $\|V_j\|_\infty \le M$ and $d \ge 17$. Then there exist disjoint nonempty subsets $H_\ell \subset \mathbb{N}$ of the natural numbers and a sequence $\{\delta_j\}_{j=1}^{\infty}$ where $\delta_j \in \{-1,1\}$ with the* *following properties:*

$$|H_\ell| \leq 2d \log(dM), \tag{22}$$

for every natural number $\ell \in \mathbb{N}$,

$$\sum_{j \in H_\ell} \delta_j V_j = \mathbf{0}, \tag{23}$$

moreover, for every $m \geq 1$, the interval $\{1, 2, \ldots, m\}$ can be expressed in the form

$$\{1, 2, \ldots, m\} = \left(\bigcup_{\ell=1}^{n} H_\ell \right) \cup R_m, \text{ where } |R_m| \leq 2d \log(dM) - 1, \tag{24}$$

and

$$\left\| \sum_{j=1}^{m} \delta_j V_j \right\|_\infty \leq 2dM \log(dM), \tag{25}$$

and finally, the modified sequence, $\delta_j^* = -\delta_j$ if $j \in H_\ell$ for some ℓ and $\delta_j^* = \delta_j$ if $j \notin H_\ell$ for every ℓ, also satisfies properties (23) and (25).

Proof We will prove (22)–(24) by a simple induction on m. Assume

$$\{1, 2, \ldots, m - 1\} = \left(\bigcup_{\ell=1}^{n} H_\ell \right) \cup R_{m-1},$$

where the H_ℓ, δ_j and R_{m-1} satisfy the first three properties (for appropriate values of ℓ and j).

If $|R_{m-1}| \leq 2d \log(dM) - 2$, then we can set $R_m = R_{m-1} \cup \{m\}$. Otherwise $|R_{m-1} \cup \{m\}| = 2d \log(dM)$, and Lemma 1 can be applied to this set of vectors to obtain $H_{n+1} \subset R_{m-1} \cup \{m\}$ and $\delta_j = \pm 1$ for $j \in H_{n+1}$ such that

$$\sum_{j \in H_{n+1}} \delta_j V_j = \mathbf{0},$$

which satisfies (23).

Clearly

$$|H_{n+1}| \leq |R_{m-1} \cup \{m\}| = 2d \log(dM),$$

and if we set

$$R_m = (R_{m-1} \cup \{m\}) \setminus H_{n+1},$$

we get

$$|R_m| \leq |R_{m-1}| + 1 - |H_{n+1}| \leq |R_{m-1}| \leq 2d \log(dM) - 1,$$

proving (22) and (24).

Since the above induction does not define δ_j for

$$j \notin \bigcup_{\ell=1}^{\infty} H_\ell,$$

we can arbitrarily assign a value of ± 1 to each of these.

To prove (25), we just apply the triangle inequality

$$\left\| \sum_{j=1}^{m} \delta_j V_j \right\|_\infty \leq \left\| \sum_{j \in R_m} \delta_j V_j \right\|_\infty + \sum_{\ell=1}^{n} \left\| \sum_{j \in H_\ell} \delta_j V_j \right\|_\infty$$

$$\leq \sum_{j \in R_m} M + \sum_{\ell=1}^{n} 0 = M|R_m| \leq 2dM \log(dM).$$

Finally, it should be noted that the only important property of the ± 1-sequence $\{\delta_j\}_{j=1}^{\infty}$ is the fact

$$\sum_{j \in H_\ell} \delta_j V_j = \mathbf{0}.$$

If the sign of each δ_j is changed for every $j \in H_\ell$, the above property still holds, proving the last statement of Lemma 2 about the modified sequence δ_j^*. This completes the induction proof of Lemma 2. $\qquad\square$

The proof of Theorem 2 is basically a repeated application of Lemma 2 as follows.

Nutshell Summary of the Proof of Theorem 2 Fix an infinite parameter sequence $\{d_i\}_{i=1}^{\infty}, d_i \in \mathbb{N}, d_i \geq 17$, and let

$$k_i = \sum_{\tau=1}^{i} d_\tau. \tag{26}$$

(We will optimize the choice of the sequence $\{d_i\}_{i=1}^{\infty}$ at the end of the proof of Theorem 2.) If we apply Lemma 2 to the first d_1 rows of matrix $\mathcal{U}$, we obtain a sequence of H_ℓ and δ_j with the property that

$$\sum_{j=1}^{m} \delta_j C_j$$

is bounded in the first d_1 rows for every m (where C_j denotes the jth column vector of matrix $\mathcal{U}$). Our aim is to modify the signs δ_j's in order to bound the sum of the next d_2 rows of matrix $\mathcal{U}$, and so on. The last statement of Lemma 2 suggests that if the sign of δ_j is changed, where $j \in H_\ell$, then all the δ_v where $v \in H_\ell$ should be changed [this is similar to the proof of (15)]. To facilitate this, we collect all the column vectors C_j, where $j \in H_\ell$, into a single unit. We will do this by creating a new matrix $\mathcal{U}^{(1)}$, where the ℓth column $C_\ell^{(1)}$ is defined as

$$C_\ell^{(1)} = \sum_{j \in H_\ell} \delta_j C_j.$$

Note that changing the sign of one column $C_\ell^{(1)}$ has the same effect on the sum as changing the signs of all δ_j for $j \in H_\ell$. While we have little control of the sums over the H_ℓ in the lower rows, we can try to keep the sums of the next d_2 rows small by *reapplying* Lemma 2, and then to continue on inductively. So far this is the same as the old proof in Beck [1].

The novelty is the last step where we optimize the choice of the parameter sequence $\{d_i\}_{i=1}^{\infty}$. This is not completely trivial: this is where we make a better choice here than what we did in [1]. The better choice of the parameter sequence $\{d_i\}_{i=1}^{\infty}$ leads to the polynomial upper bound; see (6).

3   Proof of Theorem 2 (II): Technical Details

Here we turn the informal Nutshell Summary into a formal proof. It consists of several small steps, each proved by an easy induction. We conclude Sect. 3 by a routine calculation.

First we fix an infinite parameter sequence $\{d_i\}_{i=1}^{\infty}, d_i \in \mathbb{N}, d_i \geq 17$ (to be specified later), and let

$$k_i = \sum_{\tau=1}^{i} d_\tau \tag{27}$$

(let $k_0 = 0$). We also need a universal bound which we will define by induction as follows: let

$$M_0 = 1 \text{ and } M_i = 2d_i M_{i-1} \log(d_i M_{i-1}) \text{ for } i \geq 1. \tag{28}$$

Let $\mathcal{U}^{(0)} = \mathcal{U} = \{u_{k,j}\}\, 1 \le k,j < \infty$ (i.e., the given infinite matrix in Theorem 2 with entries $u_{k,j} \in \{-1,0,1\}$), and define $\mathcal{U}^{(i)}$ by induction as follows. Apply Lemma 2 to the submatrix of $\mathcal{U}^{(i-1)}$ formed by its rows indexed between $k_{i-1} + 1$ and k_i (see (27); d_i rows altogether) with parameters $d = d_i$, $M = M_{i-1}$ [see (28)]. This creates $\{H_\ell^{(i)}\}_{\ell=1}^{\infty}$ and $\{\delta_j^{(i)}\}_{j=1}^{\infty}$, from which we define column $C_\ell^{(i)}$ of $\mathcal{U}^{(i)}$ by induction as $C_\ell^{(0)} = C_\ell$ and

$$C_\ell^{(i)} = \sum_{j \in H_\ell^{(i)}} \delta_j^{(i)} C_j^{(i-1)}. \tag{29}$$

Note that the underlying structure of this construction is an infinite (locally finite) directed tree—we hope this information helps the reader to visualize the formal definitions.

As we group the columns and their signs, we will need to know what has happened to the original column vectors of $\mathcal{U}$. In order to do this, we define by induction

$$S_\ell^{(0)} = \{C_\ell\}, \text{ and let } S_\ell^{(i)} = \bigcup_{j \in H_\ell^{(i)}} S_j^{(i-1)} \text{ for } i \ge 1, \tag{30}$$

and

$$\Delta_j^{(0)} = 1, \text{ and for } i \ge 1 \text{ let } \Delta_j^{(i)} = \delta_\ell^{(i)} \Delta_j^{(i-1)} \tag{31}$$

where $C_j \in S_\ell^{(i-1)}$.

We claim

$$C_\ell^{(i)} = \sum_{C_j \in S_\ell^{(i)}} \Delta_j^{(i)} C_j. \tag{32}$$

Indeed, the case $i = 0$ is trivial:

$$\sum_{C_j \in S_\ell^{(0)}} \Delta_j^{(0)} C_j = 1 C_\ell = C_\ell^{(0)},$$

and for $i \ge 1$ we proceed by induction: by (29)–(31) we have

$$\sum_{C_j \in S_\ell^{(i)}} \Delta_j^{(i)} C_j = \sum_{C_j \in \bigcup_{v \in H_\ell^{(i)}} S_v^{(i-1)}} \delta_v^{(i)} \Delta_j^{(i-1)} C_j$$

$$= \sum_{v \in H_\ell^{(i)}} \delta_v^{(i)} \sum_{C_j \in S_v^{(i-1)}} \Delta_j^{(i-1)} C_j = \sum_{v \in H_\ell^{(i)}} \delta_v^{(i)} C_v^{(i-1)} = C_\ell^{(i)},$$

where at the end we used the induction hypothesis.

We claim [see (28) and (30)]

$$|S_\ell^{(i)}| \leq M_i. \tag{33}$$

We prove (33) by induction. When $i = 0$ the statement is obvious since $|S_\ell^{(0)}| = 1 = M_0$. For $i \geq 1$ we apply induction, use (22) in Lemma 2 (with $d = d_i$ and $M = M_{i-1}$), and also (28), (30):

$$|S_\ell^{(i)}| \leq \sum_{j \in H_\ell^{(i)}} |S_j^{(i-1)}| \leq |H_\ell^{(i)}| M_{i-1} \leq 2d_i \log(d_i M_{i-1}) \cdot M_{i-1} = M_i,$$

as we claimed.

Lemma 3 *We have*

$$\left\| C_\ell^{(i)} \right\|_\infty \leq M_i,$$

moreover, the kth row of column $C_\ell^{(i)}$ is $\mathbf{0}$ for all i where $k_i \geq k$.

Proof By (32)

$$\left\| C_\ell^{(i)} \right\|_\infty \leq \sum_{C_j \in S_\ell^{(i)}} \left\| \Delta_j^{(i)} C_j \right\|_\infty \leq |S_\ell^{(i)}| \cdot 1 \leq M_i,$$

where in the last step we used (33).

For the second part let i be the smallest integer such that $k_i \geq k$. Then the kth row is one of the d_i rows of $\mathcal{U}^{(i-1)}$ to which Lemma 2 was applied to in order to construct matrix $\mathcal{U}^{(i)}$. The $\{H_\ell^{(i)}\}_{\ell=1}^\infty$ and $\{\delta_j^{(i)}\}_{j=1}^\infty$ were selected (via Lemma 2) so that in these d_i rows

$$\sum_{C_j \in H_\ell^{(i)}} \delta_j^{(i)} C_j^{(i-1)} = \mathbf{0},$$

but this is precisely $C_\ell^{(i)}$. Applying induction for larger i we get

$$C_\ell^{(i)} = \sum_{C_j \in H_\ell^{(i)}} \delta_j^{(i)} C_j^{(i-1)} = \sum_{C_j \in H_\ell^{(i)}} \delta_j^{(i)} \mathbf{0} = \mathbf{0},$$

completing the proof of Lemma 3. $\qquad\square$

Lemma 4 *For every m and i we can express the first m column vectors of matrix $\mathcal{U}$ in the form*

$$\{C_j\}_{j=1}^m = \left(\bigcup_{\ell=1}^n S_\ell^{(i)}\right) \cup \tilde{R}_m^{(i)}$$

for some appropriate integer n and $\tilde{R}_m^{(i)}$ such that

$$|\tilde{R}_m^{(i)}| \le M_i - 1.$$

Proof We prove it by induction on i. When $i = 0$, we can choose $\tilde{R}_m^{(i)} = \emptyset$, since

$$\{C_j\}_{j=1}^m = \bigcup_{\ell=1}^m \{C_\ell\} = \left(\bigcup_{\ell=1}^m S_\ell^{(0)}\right) \cup \emptyset.$$

For $i \ge 1$ we use the induction hypothesis

$$\{C_j\}_{j=1}^m = \left(\bigcup_{\ell=1}^n S_\ell^{(i-1)}\right) \cup \tilde{R}_m^{(i-1)}. \tag{34}$$

By Lemma 2 we know that

$$\{1, 2, \ldots, n\} = \left(\bigcup_{\ell'=1}^{n'} H_{\ell'}^{(i)}\right) \cup R_n^{(i)}. \tag{35}$$

Combining (34) and (35),

$$\{C_j\}_{j=1}^m = \left(\left(\bigcup_{\ell'=1}^{n'} \bigcup_{j \in H_{\ell'}^{(i)}} S_j^{(i-1)}\right) \cup \left(\bigcup_{\ell \in R_n^{(i)}} S_\ell^{(i-1)}\right)\right) \cup \tilde{R}_m^{(i-1)}$$

$$= \left(\bigcup_{\ell'=1}^{n'} S_{\ell'}^{(i)}\right) \cup \left(\left(\bigcup_{\ell \in R_n^{(i)}} S_\ell^{(i-1)}\right) \cup \tilde{R}_m^{(i-1)}\right). \tag{36}$$

If we choose

$$\tilde{R}_m^{(i)} = \left(\bigcup_{\ell \in R_n^{(i)}} S_\ell^{(i-1)}\right) \cup \tilde{R}_m^{(i-1)}, \tag{37}$$

then

$$|\tilde{R}_m^{(i)}| \leq \left(\sum_{\ell \in R_n^{(i)}} |S_\ell^{(i-1)}|\right) + |\tilde{R}_m^{(i-1)}|$$

$$\leq |R_n^{(i)}| \max_{\ell \in R_n^{(i)}} |S_\ell^{(i-1)}| + M_{i-1} - 1$$

$$\leq (2d_i \log(d_i M_{i-1}) - 1)M_{i-1} + M_{i-1} - 1 = M_i - 1, \tag{38}$$

where at the end we used (24) in Lemma 2, (28) and (33). Combining (36)–(38), the induction step is complete, and Lemma 4 follows. $\qquad\square$

Lemma 5 *Let $k \leq k_i$ [see (27)] and $v \geq i$, then*

$$\left|\sum_{j=1}^{m} \Delta_j^{(v)} u_{k,j}\right| \leq M_i.$$

Proof By the definition of $\Delta_j^{(i)}$ [see (31)] it is easy to see that $\Delta_j^{(i)} = \pm\Delta_j^{(v)}$, and that this sign remains fixed for all j where $C_j \in S_\ell^{(i)}$. First consider the whole column sums: by Lemma 4,

$$\sum_{j=1}^{m} \Delta_j^{(v)} C_j = \sum_{C_j \in \left(\bigcup_{\ell=1}^{n} S_\ell^{(i)}\right) \cup \tilde{R}_m^{(i)}} \pm\Delta_j^{(i)} C_j$$

$$= \sum_{\ell=1}^{n} (\pm 1) \sum_{C_j \in S_\ell^{(i)}} \Delta_j^{(i)} C_j + \sum_{C_j \in \tilde{R}_m^{(i)}} \pm\Delta_j^{(i)} C_j$$

$$= \sum_{\ell=1}^{n} \pm C_\ell^{(i)} + \sum_{C_j \in \tilde{R}_m^{(i)}} \pm\Delta_j^{(i)} C_j.$$

By Lemma 4 for $k \leq k_i$ the kth row of this becomes bounded by

$$\sum_{\ell=1}^{n} 0 + |\tilde{R}_m^{(i)}||u_{k,j}| \leq M_i. \tag{39}$$

Equation (39) completes the proof of Lemma 5. $\qquad\square$

Let Δ be the set of sequences

$$\Delta = \left\{\left(\Delta_1^{(i)}, \Delta_2^{(i)}, \Delta_3^{(i)}, \ldots\right)\right\}_{i=1}^{\infty}. \tag{40}$$

By Lemma 5 for each i we have the upper bound

$$\left| \sum_{j=1}^{m} \Delta_j^{(i)} u_{k,j} \right| \leq M_i$$

for rows $1 \leq k \leq k_i$. By using compactness we can easily create a sequence $\{\Delta_j\}_{j=1}^{\infty}$ which bounds all row sums. Indeed, each $\Delta_1^{(i)}$ is ± 1, so one must occur an infinite number of times. Let Δ_1 be this sign, and remove from Δ in (40) all sequences where $\Delta_1^{(i)} \neq \Delta_1$. For the remaining infinitely many indexes i we have $\Delta_2^{(i)} = \pm 1$, so one must occur an infinite number of times. Let Δ_2 be this sign, and again remove all sequences where $\Delta_2^{(i)} \neq \Delta_2$. Continue inductively, creating each Δ_j in the same fashion. Clearly $\{\Delta_j\}_{j=1}^{\infty}$ satisfies

$$\left| \sum_{j=1}^{m} \Delta_j u_{k,j} \right| \leq M_i \text{ for all rows } k_{i-1} + 1 \leq k \leq k_i, \tag{41}$$

since we can find a v where $v \geq i$ and $\Delta_j = \Delta_j^{(v)}$ for all $1 \leq j \leq m$.

The last step in the proof of Theorem 2 is optimization. We need to choose the parameter sequence $\{d_i\}_{i=1}^{\infty}$ to minimize the upper bounds M_i for k where $k_{i-1} + 1 \leq k \leq k_i$ [see (41)].

Fix $\varepsilon > 0$ and define $\{d_i\}_{i=1}^{\infty}$ as follows. Choose d_1 such that

$$d_1 \geq 17, \quad n^{\varepsilon/2} \geq (4 + \varepsilon) \log n \text{ for all } n \geq d_1, \quad 2d_1 \log d_1 \leq d_1^2. \tag{42}$$

All three conditions in (42) are satisfied for sufficiently large d_1, so such an integer clearly exists.

Next let

$$k_1 = d_1, \quad d_2 = k_1^2, \quad k_2 = d_1 + d_2, \quad d_3 = k_2^2, \tag{43}$$

and in general, for $i \geq 2$, let

$$d_{i+1} = k_i^2 \text{ where } k_i = d_1 + \cdots + d_i. \tag{44}$$

By (42)–(44),

$$k_i \geq d_i = k_{i-1}^2 \text{ and } k_i^{\varepsilon/2} \geq (4 + \varepsilon) \log k_i. \tag{45}$$

We claim

$$M_i \leq k_i^{2+\varepsilon} \text{ for all } i \geq 1. \tag{46}$$

We prove (46) by induction. When $i = 1$, we get

$$M_1 = 2d_1 \log d_1 \le d_1^2 = k_1^2 \le k_1^{2+\varepsilon}.$$

For $i \ge 2$ we apply (42)–(45) and the induction hypothesis:

$$M_i = 2d_i M_{i-1} \log(d_i M_{i-1}) \le 2k_{i-1}^2 k_{i-1}^{2+\varepsilon} \log(k_{i-1}^2 k_{i-1}^{2+\varepsilon})$$

$$\le \left(k_{i-1}^2\right)^{2+(\varepsilon/2)} (4 + \varepsilon) \log(k_{i-1}^2) \le k_i^{2+(\varepsilon/2)} (4 + \varepsilon) \log(k_i)$$

$$\le k_i^{2+(\varepsilon/2)} k_i^{\varepsilon/2} = k_i^{2+\varepsilon},$$

completing the induction proof of (46).

The last technical detail is that in view of (41) we need to bound M_i in terms of k with $k_{i-1} + 1 \le k \le k_i$, and not in terms of k_i as we did in (46). We use the simple fact

$$(x + x^2)^{1/2} \le x + 1 \quad \text{for } x \ge 1,$$

which implies

$$k_i^{1/2} = \left(k_{i-1} + k_{i-1}^2\right)^{1/2} \le k_{i-1} + 1. \tag{47}$$

Combining (46) and (47) we conclude

$$k^{4+2\varepsilon} \ge (k_{i-1} + 1)^{4+2\varepsilon} \ge \left(k_i^{1/2}\right)^{4+2\varepsilon} = k_i^{2+\varepsilon} \ge M_i. \tag{48}$$

Now (41) and (48) imply Theorem 2 for $k \ge k_1 = d_1$. For $k < k_1 = d_1$ it suffices to combine (41) and (46), and thus the proof of Theorem 2 is complete. $\square$

4 Proof of Theorem 1

We deduce Theorem 1 from Theorem 2. Let

$$\mathbf{v}_j = (v_{j,1}, v_{j,2}, v_{j,3}, \ldots), \ j \ge 1.$$

Since $-1 \le v_{j,k} \le 1$, we can write it in the form (base 3 representation)

$$v_{j,k} = \sum_{i=0}^{\infty} v(j, k; i) 3^{-i}, \quad \text{where } v(j, k; i) \in \{-1, 0, 1\}. \tag{49}$$

The following infinite sequence has the property that every positive integer shows up infinitely many times (the five times repetition in the pattern is explained by the fact that $5 > 4 + \varepsilon$, where $4 + \varepsilon$ is the exponent in Theorem 2)

$$1, 2, 3, 1, 2, 3, 1, 2, 3, 1, 2, 3, 1, 2, 3,$$

$$1, 2, \ldots, 9, 1, 2, \ldots, 9, 1, 2, \ldots, 9, 1, 2, \ldots, 9, 1, 2, \ldots, 9,$$

$$1, 2, \ldots, 27, 1, 2, \ldots, 27, 1, 2, \ldots, 27, 1, 2, \ldots, 27, 1, 2, \ldots, 27, \ldots$$

$$1, 2, \ldots, 3^\ell, 1, 2, \ldots, 3^\ell, 1, 2, \ldots, 3^\ell, 1, 2, \ldots, 3^\ell, 1, 2, \ldots, 3^\ell, \ldots \tag{50}$$

where the last line of (50) clearly describes the simple rule of how the sequence is generated by increasing blocks of integers between 1 and 3^ℓ, $\ell \geq 1$. Using sequence (50) we define a bijection

$$\beta : \mathbb{N} \times (\mathbb{N} \cup \{0\}) \to \mathbb{N} \tag{51}$$

as follows: let $\beta(k, i)$ denote the position of natural number k where it shows up the $(1 + i)$th time in (50) (for example, $\beta(4, 0) = 19$ and $\beta(4, 1) = 28$). It follows from the construction of (50) that

$$\beta(k, i) < 10k3^{i/5}. \tag{52}$$

Let $u_{t,j} = v(j, k; i)$ where $(k, i) = \beta^{-1}(t)$ (and of course β^{-1} denotes the inverse of bijection β). Consider the matrix $\mathcal{U} = \{u_{t,j}\}$, $1 \leq t, j < \infty$ with entries $-1, 0, 1$. By Theorem 2 there exists an infinite sequence $\delta_j \in \{-1, 1\}, j \geq 1$ such that

$$\max_{j \geq 1} \left| \sum_{\ell=1}^{j} \delta_\ell u_{t,\ell} \right| = O\left(t^{4+\varepsilon}\right). \tag{53}$$

(53) is equivalent to

$$\max_{j \geq 1} \left| \sum_{\ell=1}^{j} \delta_\ell v(\ell, k; i) \right| = O\left(\beta(k, i)^{4+\varepsilon}\right). \tag{54}$$

Multiplying (54) by 3^{-i}, adding them up for $i = 0, 1, 2, \ldots$, and using (49), we obtain

$$\max_{j \geq 1} \left| \sum_{\ell=1}^{j} \delta_\ell v_{\ell, k} \right| = \sum_{i=0}^{\infty} O\left(\beta(k, i)^{4+\varepsilon}\right) 3^{-i}. \tag{55}$$

Applying (52) in (55), we have

$$\max_{j \geq 1} \left| \sum_{\ell=1}^{j} \delta_\ell v_{\ell,k} \right| = \sum_{i=0}^{\infty} O\left(\left(10k3^{i/5}\right)^{4+\varepsilon}\right) 3^{-i}$$

$$= O\left(k^{4+\varepsilon}\right) \sum_{i=0}^{\infty} 3^{-(1-\varepsilon)i/5} = O\left(k^{4+\varepsilon}\right), \tag{56}$$

if $0 < \varepsilon < 1$. Equation (56) completes the proof of Theorem 1. $\qquad\square$

Acknowledgements I am very grateful to I. Bárány for his remarks and suggestions, and to D. Reimer for his help in formulating the proof of Theorem 2.

References

1. J. Beck, Balancing families of integer sequences. Combinatorica **1**, 209–216 (1981)
2. J. Beck, Roth's estimate of discrepancy of integer sequences is nearly sharp. Combinatorica **1**, 319–325 (1981)
3. J. Beck, V.T. Sós, Discrepancy theory, Chap. 26, in *Handbook of Combinatorics*, ed. by R. Graham, M. Grőtschel, L. Lovász (Elsevier, Amsterdam, 1995), pp. 1405–1446
4. J. Beck, J. Spencer, Well-distributed 2-colorings of integers relative to long arithmetic progressions. Acta Arith. **43**, 287–294 (1984)
5. P. Erdős, Extremal problems in number theory II (in Hungarian). Mat. Lapok **17**, 135–155 (1966)
6. P. Erdős, Problems and results on combinatorial number theory, in *A Survey of Combinatorial Theory*, ed. by J.N. Srivastava, et al. (North-Holland, Amsterdam, 1973), pp. 117–138
7. J. Matoušek, J. Spencer, Discrepancy in arithmetic progressions. J. Am. Math. Soc. **9**(1), 195–204 (1996)
8. K.F. Roth, Remark concerning integer sequences. Acta Arith. **9**, 257–260 (1964)
9. T. Tao, The Erdős discrepancy problem. arXiv: 1509.05363v5, see also the new journal Discrete Analysis

Squares with Three Nonzero Digits

Michael A. Bennett and Adrian-Maria Scheerer

Abstract We determine all integers n such that n^2 has at most three base-q digits for $q \in \{2, 3, 4, 5, 8, 16\}$. More generally, we show that all solutions to equations of the shape

$$Y^2 = t^2 + M \cdot q^m + N \cdot q^n,$$

where q is an odd prime, $n > m > 0$ and $t^2, |M|, N < q$, either arise from "obvious" polynomial families or satisfy $m \leq 3$. Our arguments rely upon Padé approximants to the binomial function, considered q-adically.

1991 *Mathematics Subject Classification.* Primary 11D61, Secondary 11A63, 11J25

1 Introduction

Let us suppose that $q > 1$ is an integer. A common way to measure the lacunarity of the base-q expansion of a positive integer n is through the study of functions we will denote by $N_q(n)$ and $S_q(n)$, the number of and sum of the nonzero digits in the base-q expansion of n, respectively. Our rough expectation is that, if we restrict n to lie in a subset $S \subset \mathbb{N}$, these quantities should behave in essentially the same way as for unrestricted integers, at least provided the subset is not too "thin." Actually quantifying such a statement can be remarkably difficult; particularly striking successes along these lines, for S the sets of primes and squares can be found in the work of Mauduit and Rivat [16] and [17].

M.A. Bennett (✉)
Department of Mathematics, University of British Columbia, Vancouver, BC, Canada
e-mail: bennett@math.ubc.edu

A.-M. Scheerer
Institute of Analysis and Number Theory, Graz University of Technology, Graz, Austria
e-mail: scheerer@math.tugraz.at

In this paper, we will restrict our attention to the case where S is the set of integer squares. Since (see [12])

$$\sum_{n<N} S_q(n) \sim \frac{1}{2} \sum_{n<N} S_q(n^2) \sim \frac{q-1}{2\log q} N \log N,$$

it follows that the ratios

$$\frac{S_q(n^2)}{S_q(n)} \quad \text{and} \quad \frac{N_q(n^2)}{N_q(n)}$$

are infrequently "small." On the other hand, in the case $q = 2$ (where $S_q(n)$ and $N_q(n)$ coincide), Stolarsky [19] proved that, for infinitely many n,

$$\frac{N_2(n^2)}{N_2(n)} \leq \frac{4 \left(\log \log n\right)^2}{\log n},$$

a result that was subsequently substantially sharpened and generalized by Hare et al. [13]. Further developments are well described in [14] where, in particular, one finds that

$$\# \left\{ n < N \ : \ N_2(n) = N_2(n^2) \right\} \gg N^{1/19}$$

and that the set

$$\left\{ n \in \mathbb{N}, \, n \text{ odd} \ : \ N_2(n) = N_2(n^2) = k \right\}$$

is finite for $k \leq 8$ and infinite for $k \in \{12, 13\}$ or $k \geq 16$.

In what follows, we will focus our attention on integers n with the property that $N_q(n^2) = k$, for small fixed positive integer k. Classifying those integers n in the set

$$B_k(q) = \left\{ n \in \mathbb{N} \ : \ n \not\equiv 0 \mod q \ \text{and} \ N_q(n) \geq N_q(n^2) = k \right\}$$

is, apparently, a rather hard problem, even for the case $k = 3$ (on some level, this is the smallest "nontrivial" situation as those n with $N_q(n^2) < 3$ are readily understood). There are infinitely many squares, coprime to q with precisely three nonzero digits base-q, as evidenced by the identity

$$\left(1 + q^b\right)^2 = 1 + 2 \cdot q^b + q^{2b}. \tag{1}$$

There are, however, other squares with three nonzero digits, arising more subtly. For example, if $n = 10{,}837$, then, base $q = 8$, we have

$$10837 = 2 \cdot 8^4 + 5 \cdot 8^3 + 1 \cdot 8^2 + 2 \cdot 8 + 5$$

while

$$10837^2 = 7 \cdot 8^8 + 7 \cdot 8 + 1.$$

On the other hand, a result of Corvaja and Zannier [10] implies that all but finitely many squares with three base-q digits arise from polynomial identities like (1), and, further, that $B_3(q)$ is actually finite. The proof of this in [10], however, depends upon Schmidt's Subspace Theorem and is thus ineffective (in that it does not allow one to precisely determine $B_3(q)$—it does, however, lead to an algorithmic determination of all relevant polynomial identities, if any). Analogous questions for $B_k(q)$ with $k \geq 4$ are, as far as we are aware, unsettled, except for the case of $B_4(2)$ (see [11]).

In this paper, we will explicitly determine $B_3(q)$ for certain fixed values of q. We prove the following theorem.

Theorem 1.1 *The only positive integers n for which n^2 has at most three nonzero digits base q for $q \in \{2, 3, 4, 5, 8, 16\}$ and $n \not\equiv 0 \mod q$ are as follows:*

$$q = 2 \; : \; n \in \{1, 5, 7, 23\} \; or \; n = 2^b + 1,$$

$$q = 3 \; : \; n \in \{1, 5, 8, 13\} \; or \; n = 3^b + 1,$$

$$q = 4 \; : \; n = t \; or \; 2t \; for \; t \in \{1, 7, 15, 23, 31, 111\}, \; or \; t = 4^b + 1 \; or \; 2 \cdot 4^b + 1,$$

$$q = 5 \; : \; n \in \{1, 4, 8, 9, 12, 16, 23, 24, 56, 177\} \; or \; n = 5^b + 1, 2 \cdot 5^b + 1 \; or \; 5^b + 2,$$

$$q = 8 \; : \; n \leq 63, \; n \in \{92, 111, 124, 126, 158, 188, 316, 444, 479, 508, 10837\}$$
$$or \; n = r \cdot 8^b + s \; for \; r, s \in \{1, 2, 4\}$$

and

$$q = 16 \; : \; n = t, 2t \; or \; 4t \; for \; t \leq 100, \; t \in \{111, 125, 126, 127\}$$
$$or \; t = r \cdot 16^b + s \; where \; either \; r, s \in \{1, 2, 4, 8\} \; or \; the \; set$$
$$\{r, s\} \; is \; one \; of \; \{1, 3\}, \{2, 3\}, \{3, 8\}, \{2, 12\}, \{4, 12\} \; or \; \{8, 12\}.$$

Here, b is a nonnegative integer.

This immediately implies

Corollary 1.2 *We have*

$$B_3(2) = \{7, 23\}, \; B_3(3) = \{13\}, \; B_3(4) = \{23, 30, 31, 46, 62, 111, 222\},$$

$$B_3(5) = \{56, 177\}, \; B_3(8) = \{92, 111, 124, 126, 158, 188, 316, 444, 479, 508, 10837\}$$

and

$$B_3(16) = \{364, 444, 446, 500, 504, 508, 574, 628, 680, 760, 812, 888, 924, 958,$$
$$1012, 1016, 1020, 1022, 1784, 2296, 3832, 3966, 4088, 10837, 15864, 43348\}.$$

We note that the case $q = 2$ of Theorem 1.1 was originally proved by Szalay [20] in 2002, through appeal to a result of Beukers [7]. This latter work was based upon Padé approximation to the binomial function (as are the results of the paper at hand, though our argument is quite distinct). In 2012, the first author [2] treated the case $q = 3$ in Theorem 1.1. We should point out that there are computational errors in the last two displayed equations on page 4 of [2] that require repair; we will do this in the current paper.

Our main result which leads to Theorem 1.1 is actually rather more general—we state it for a prime base, though our arguments extend to more general q with the property that q has a prime-power divisor p^α with $p^\alpha > q^{3/4}$. We prove

Theorem 1.3 *If q is an odd prime, if we have a solution to the equation*

$$Y^2 = t^2 + Mq^m + Nq^n, \tag{2}$$

in integers Y, t, M, N, m and n satisfying

$$t, Y, N \geq 1, \quad |M|, N, t^2 \leq q - 1 \quad and \quad 1 \leq m < n, \tag{3}$$

then either $n = 2m$ and $Y = q^m \cdot Y_0 \pm t$, for integers t and Y_0 with $\max\{Y_0^2, 2tY_0\} < q$, or we have $m \leq 3$.

In the special case $t = 1, M = \pm 1, N = 1$, a sharper version of this result already appears as the main theorem of Luca [15]; the proof of this result relies upon primitive divisors in binary recurrence sequences and does not apparently generalize. It seems likely that the last upper bound in Theorem 1.3 can be replaced by $m \leq 2$; indeed our argument can be sharpened to prove this for "many" pairs (m, n), though not all. We know of a number of families of solutions to (2), with, for instance, $(m, n) = (2, 6)$, $q = r^2 + 1$ prime, $r \in \mathbb{Z}$:

$$\left(\frac{1}{2}r(r^6 + 5r^4 + 7r^2 + 5)\right)^2 = r^2 + (r^2 - 1)q^2 + \left(\frac{r^2 + 4}{4}\right)q^6 \tag{4}$$

and $(m, n) = (1, 5)$, for $q = 64r^2 + 1$, corresponding to the identity

$$\left(r(32768r^4 + 1280r^2 + 15)\right)^2 = 9r^2 - (40r^2 + 1)q + q^5.$$

Further families with $(m, n) = (1, 3)$, $(2, 3)$, and $(1, 4)$ are readily observed [as are many more examples with $(m, n) = (1, 5)$]. Beyond these, we also know a few (possibly) sporadic examples, with $(m, n) = (1, 6), (1, 7)$, and $(2, 7)$:

$$430683365^2 = 9^2 - 51 \cdot 311 + 205 \cdot 311^6,$$

$$6342918641^2 = 25^2 - 97 \cdot 673 + 433 \cdot 673^6,$$

$$49393781643^2 = 34^2 - 875 \cdot 1229 + 708 \cdot 1229^6,$$

$$559^2 = 1^2 - 4 \cdot 5 + 4 \cdot 5^7,$$

$$574588^2 = 3^2 + 13 \cdot 31 + 12 \cdot 31^7,$$

$$1815^2 = 2^2 + 7^2 + 4 \cdot 7^7$$

and

$$20958^2 = 2^2 - 11 \cdot 13^2 + 7 \cdot 13^7.$$

For a fixed odd prime q, Theorem 1.3 provides an effective way to completely solve Eq. (2) under the conditions of (3). Indeed, given an upper bound upon m, say m_0, solving (2) with (3) amounts to treating at most $O(q^{5/2} m_0)$ "Ramanujan-Nagell" equations of the shape

$$Y^2 + D = Nq^n \quad \text{where} \quad D = -(t^2 + Mq^m). \tag{5}$$

These can be handled efficiently via algorithms from Diophantine approximation; see Pethő and de Weger [18] or de Weger [21] for details. Alternatively, if $n \equiv n_0$ mod 3, where $n_0 \in \{0, 1, 2\}$, we may rewrite (2) as

$$U^2 = V^3 + k, \tag{6}$$

where

$$U = Nq^{n_0} Y, \quad V = q^{\frac{n+2n_0}{3}} N \quad \text{and} \quad k = N^2 q^{2n_0} \left(t^2 + Mq^m \right). \tag{7}$$

We can therefore solve Eq. (2) if we are able to find the "integer points" on at most $O(q^{5/2} m_0)$ "Mordell curves" of the shape (6), where we may subsequently check to see if any solutions encountered satisfy (7). The integer points on these curves are known for $|k| \leq 10^7$ (see [4]) and are listed at http://www.math.ubc.ca/~bennett/BeGa-data.html. For larger values of $|k|$, one can, in many cases, employ Magma or a similar computational package to solve equations of the shape (6). For our purposes, however, we are led to consider a number of values of k for which approaches to solving (6) reliant upon computation of a full Mordell-Weil basis (as Magma does) for the corresponding curve are extremely time-consuming. We instead choose to solve a number of equations of the form (5), via lower bounds for linear forms in p-adic logarithms and reduction techniques from Diophantine approximation, as in [18]. An alternative approach, at least for the equations we encounter, would be to appeal to strictly elementary properties of the corresponding binary recurrences, as in a paper of Bright [9] on the Ramanujan-Nagell equation.

It is probably worth mentioning that similar problems to those discussed in this paper, only for higher powers with few digits, are treated in a series of papers by the first author, together with Yann Bugeaud [3] and with Bugeaud and Maurice Mignotte [5, 6]. The results therein require rather different techniques than those employed here, focussing on lower bounds for linear forms in logarithm, p-adic and complex.

2 Three Digits, Without Loss of Generality

Suppose that $q > 1$ is an integer and that we have a square y^2 with (at most) three nonzero base-q digits. If q is either squarefree or a square, it follows that y is necessarily a multiple by some power of q (or $\sqrt{q}$ if q is a square) of an integer Y satisfying a Diophantine equation of the shape

$$Y^2 = C + M \cdot q^m + N \cdot q^n, \tag{8}$$

where C, M, N, m, and n are nonnegative integers with

$$C, M, N \leq q - 1 \quad \text{and} \quad 1 \leq m < n. \tag{9}$$

If q is neither a square nor squarefree, we may similarly reduce to consideration of Eq. (8), only with weaker bounds for M and N.

The machinery we will employ to prove Theorems 1.1 and 1.3 requires that, additionally, the integer C in Eq. (8) is square. Whilst this is certainly without loss of generality if every quadratic residue modulo q in the range $1 \leq C < q$ is itself a square, it is easy to show that such a condition is satisfied only for $q \in \{2, 3, 4, 5, 8, 16\}$. If we have the somewhat weaker constraint upon q that every least positive quadratic residue C modulo q is either a square or has the property that it fails to be a quadratic residue modulo q^k for some exponent $k > 1$, then we may reduce to consideration of (8) with either C square or m bounded. This weaker condition is satisfied for the following q:

$$q = 2, 3, 4, 5, 6, 8, 10, 12, 14, 15, 16, 18, 20, 21, 22, 24, 28, 30, 36, 40, 42, 44, 48, 54, 56,$$
$$60, 66, 70, 72, 78, 84, 88, 90, 102, 120, 126, 140, 150, 156, 168, 174, 180, 210, 240,$$
$$330, 390, 420, 462, 630, 660, 840, 2310.$$

Of these, the only ones with a prime-power divisor p^α with $p^\alpha > q^{3/4}$ (another requirement for our techniques to enable the complete determination of squares with three base-q digits) are

$$q = 2, 3, 4, 5, 8, 16, 18, 22 \text{ and } 54.$$

The principal reason we restrict our attention to Eq. (8) with C square is to guarantee that the exponent n is relatively large compared to m, enabling us to employ machinery from Diophantine approximation (this is essentially the content of Sect. 3). This might not occur if C is nonsquare, as examples like

$$45454^2 = 13 + 22 \cdot 23^5 + 13 \cdot 23^6$$

and

$$9730060^2 = 46 + 96 \cdot 131^5 + 18 \cdot 131^6$$

illustrate.

3 Three Digits: Gaps Between Exponents

For the next few sections, we will restrict attention to the case where the base q is an odd prime. Let us now suppose that we have a solution to (2) with (3). In this section, we will show that necessarily the ratio n/m is not too small, except when $Y = q^m \cdot Y_0 \pm t$ for small Y_0. Specifically, we will prove the following result.

Lemma 3.1 *If there exists a solution to Eq. (2) with (3) and $m \geq 4$, then either $n = 2m$ and $Y = q^m \cdot Y_0 \pm t$, for integers t and Y_0 with $\max\{Y_0^2, 2tY_0\} < q$, or we have $n \geq 10m - 10$.*

Let us begin by considering the case where $M = 0$ (where we will relax the condition that $n \geq 2$). Since q is an odd prime, we may write

$$Y = q^n \cdot Y_0 + (-1)^\delta t,$$

for some positive integer Y_0 and $\delta \in \{0, 1\}$, whence

$$N = q^n \cdot Y_0^2 + (-1)^\delta 2t \cdot Y_0.$$

Since $1 \leq N, t^2 \leq q - 1$, if $n \geq 2$, it follows that

$$q - 1 \geq q^2 - 2\sqrt{q - 1},$$

a contradiction since $q \geq 3$. We thus have $n = 1$, so that

$$N = q \cdot Y_0^2 + (-1)^\delta 2t \cdot Y_0,$$

whence $N < q$ implies that $Y_0 = \delta = 1$, corresponding to the identities

$$(q - t)^2 = t^2 + (q - 2t)q.$$

It is worth observing that whilst there are no solutions to (8) with (9), q an odd prime and $M = 0$, provided C is square, this is not true without this last restriction, as the identity

$$3233069 1^2 = 182 + 157 \cdot 367^5$$

illustrates.

We may thus, without loss of generality, suppose that $M \neq 0$ in what follows and write

$$Y = q^m \cdot Y_0 + (-1)^\delta t,$$

for some positive integer Y_0 and $\delta \in \{0, 1\}$, so that

$$q^m Y_0^2 + 2(-1)^\delta t \cdot Y_0 = M + N q^{n-m}. \tag{10}$$

We thus have

$$q^m - 2q^{1/2} < q^{n-m+1} - q^{n-m} + q.$$

If $n \leq 2m - 2$ (so that $m \geq 3$), it follows that $q^m - 2q^{1/2} < q^{m-1} - q^{m-2} + q$, an immediate contradiction. If $n = 2m - 1$, then

$$q^{m-1} < q + 2q^{1/2},$$

and so $m = 2$, $n = 3$, whereby (10) becomes

$$q^2 Y_0^2 + 2(-1)^\delta t \cdot Y_0 = M + Nq \leq (q-1)q + q - 1 = q^2 - 1.$$

We thus have $Y_0 = 1$ and $\delta = 1$. Since $q \mid M - 2(-1)^\delta t = M + 2t$, it follows that either $M = -2t$ or $M = q - 2t$. In the first case, we have that $q \mid N$, a contradiction. The second corresponds to the identity

$$(q^2 - t)^2 = t^2 + (q - 2t)q^2 + (q-1)q^3. \tag{11}$$

Otherwise, we may suppose that $n \geq 2m$. From the series expansion

$$(t^2 + x)^{1/2} = t + \frac{x}{2t} - \frac{x^2}{8t^3} + \frac{x^3}{16t^5} - \frac{5x^4}{128t^7} + \frac{7x^5}{256t^9} - \frac{21x^6}{1024t^{11}} + \frac{33x^7}{2048t^{13}} - \frac{429x^8}{32768t^{15}} + \cdots,$$

and (2), it follows that

$$Y \equiv (-1)^\delta \left(t + \frac{Mq^m}{2t} \right) \mod q^{2m},$$

so that

$$2tY \equiv (-1)^\delta \left(2t^2 + Mq^m\right) \mod q^{2m}.$$

If $2tY = (-1)^\delta \left(2t^2 + Mq^m\right)$, then

$$n = 2m, \quad \frac{M^2}{4t^2} = N \quad \text{and} \quad |Y_0| = \left|\frac{M}{2t}\right|,$$

corresponding to the identity

$$\left(q^m \cdot Y_0 + (-1)^\delta t\right)^2 = t^2 + ((-1)^\delta t 2 Y_0) \cdot q^m + Y_0^2 \cdot q^{2m}, \tag{12}$$

where $\max\{t^2, Y_0^2, 2tY_0\} < q$.

If we are not in situation (12), we may write

$$2tY = \kappa q^{2m} + (-1)^\delta (Mq^m + 2t^2), \tag{13}$$

for some positive integer κ, so that

$$4t^2 \cdot N \cdot q^{n-2m} = \kappa^2 q^{2m} + 2\kappa(-1)^\delta \left(Mq^m + 2t^2\right) + M^2. \tag{14}$$

We rewrite this as

$$4t^2 \cdot N \cdot q^{n-2m} = \left(\kappa q^m + (-1)^\delta M\right)^2 + \kappa(-1)^\delta 4t^2. \tag{15}$$

If $n = 2m$, this becomes

$$4t^2 \cdot N = \left(\kappa q^m + (-1)^\delta M\right)^2 + \kappa(-1)^\delta 4t^2,$$

the left-hand side of which is at most $4(q-1)^2$. Since the right-hand side is at least

$$(\kappa q^m - q + 1)^2 - 4(q-1),$$

it follows that $m = 1$ and $\kappa \in \{1, 2\}$. If $\kappa = 1$, we have

$$q + (-1)^\delta M \equiv 0 \mod 2t,$$

say $q = 2tq_0 - (-1)^\delta M$, for q_0 a positive integer with $N = q_0^2 + (-1)^\delta$, with corresponding identity

$$\left(q_0 q + (-1)^\delta t\right)^2 = t^2 + (-1)^\delta (2tq_0 - q)q + (q_0^2 + (-1)^\delta)q^2, \tag{16}$$

where $t, q_0 < \sqrt{q}$. If $\kappa = 2$, then M is necessarily even, say $M = 2M_0$, and

$$q + (-1)^\delta M_0 \equiv 0 \mod t,$$

say $q = tq_0 - (-1)^\delta M_0$. This corresponds to

$$\left(q_0 q + (-1)^\delta t\right)^2 = t^2 + (-1)^\delta 2(tq_0 - q)q + (q_0^2 + 2(-1)^\delta)q^2, \tag{17}$$

where we require that $q/2 < tq_0 < 3q/2, t < \sqrt{q}$ and $q_0 < \sqrt{q - 2(-1)^\delta}$.

With these families excluded, we may thus assume that $n \geq 2m + 1$ and that (15) is satisfied. For the remainder of this section, we will suppose that $m \geq 4$. Then, since the right-hand side of (14) is

$$\kappa^2 q^{2m}\left(1 + (-1)^\delta\left(\frac{2M}{\kappa}q^{-m} + \frac{4t^2}{\kappa}q^{-2m}\right) + \frac{M^2}{\kappa^2}q^{-2m}\right), \tag{18}$$

and we assume that $|M| < q$ and $t < \sqrt{q}$, we have

$$N \cdot q^{n-2m} > \frac{1 - 2q^{1-m} - 4q^{1-2m}}{4}q^{2m-1}. \tag{19}$$

Since $N < q, m \geq 4$ and $q \geq 3$ this implies that

$$q^{n-2m+1} > \frac{2021}{8748}q^{2m-1}$$

and hence $n \geq 4m - 3 \geq 3m + 1$. We thus have

$$Y \equiv (-1)^\delta\left(t + \frac{Mq^m}{2t} - \frac{M^2 q^{2m}}{8t^3}\right) \mod q^{3m},$$

whence

$$8t^3 Y \equiv (-1)^\delta\left(8t^4 + 4t^2 Mq^m - M^2 q^{2m}\right) \mod q^{3m}.$$

If

$$8t^3 Y = (-1)^\delta\left(8t^4 + 4t^2 Mq^m - M^2 q^{2m}\right),$$

then

$$64t^6 \cdot N \cdot q^{n-3m} = M^4 q^m - 8t^2 \cdot M^3,$$

an immediate contradiction, since $n \geq 3m + 1$ and q is coprime to tM.

We may thus assume that

$$8t^3 Y = \kappa_1 q^{3m} + (-1)^\delta \left(-M^2 q^{2m} + 4t^2 M q^m + 8t^4 \right),$$

for a positive integer κ_1, whereby

$$\begin{aligned}
64t^6 N q^{n-3m} = {} & \kappa_1^2 q^{3m} + M^4 q^m - 8t^2 M^3 \\
& + (-1)^\delta \left(-2\kappa_1 M^2 q^{2m} + 8t^2 \kappa_1 M q^m + 16t^4 \kappa_1 \right)
\end{aligned} \tag{20}$$

and so

$$64t^6 N q^{n-3m} > q^{3m} - 2M^2 q^{2m} - 8t^2 |M| q^m. \tag{21}$$

This implies that

$$64 q^{n-3m+4} > q^{3m} \left(1 - 2q^{2-m} - 8q^{2-2m} \right). \tag{22}$$

For $q \geq 7$, we therefore have

$$q^{n-3m+4} > \frac{1}{67} q^{3m},$$

so that $n \geq 6m - 4$ if $q \geq 67$. If $q = 3$, we obtain the inequality $n \geq 6m - 4$ directly from (21). For each $5 \leq q \leq 61$, (22) implies that $n \geq 6m - 6$. In every case, we may thus assume that $n \geq 6m - 6 > 4m$, so that

$$Y \equiv (-1)^\delta \left(t + \frac{M q^m}{2t} - \frac{M^2 q^{2m}}{8t^3} + \frac{M^3 q^{3m}}{16t^5} \right) \mod q^{4m}$$

and hence

$$16t^5 Y = \kappa_2 q^{4m} + (-1)^\delta \left(16t^6 + 8t^4 M q^m - 2t^2 M^2 q^{2m} + M^3 q^{3m} \right)$$

for a nonnegative integer κ_2, whence

$$\begin{aligned}
256t^{10} N q^{n-4m} = {} & \kappa_2^2 q^{4m} + (-1)^\delta \left(32\kappa_2 t^6 + 16\kappa_2 t^4 M q^m \right. \\
& \left. - 4\kappa_2 t^2 M^2 q^{2m} + 2\kappa_2 M^3 q^{3m} \right) + 20t^4 M^4 - 4t^2 M^5 q^m + M^6 q^{2m}.
\end{aligned} \tag{23}$$

If $\kappa_2 = 0$,

$$256t^{10} N q^{n-4m} = 20t^4 M^4 - 4t^2 M^5 q^m + M^6 q^{2m},$$

contradicting the fact that $q \nmid tM$. We therefore have that

$$256t^{10}Nq^{n-4m} > q^{4m} - 2|M|^3q^{3m} - 4t^2M^2q^{2m} \tag{24}$$

and so

$$q^{n-4m+6} > \frac{1}{263}q^{4m}, \tag{25}$$

whence $n \geq 8m - 8$ unless, possibly, $q \in \{3, 5\}$. If $q = 3$, since $t = 1$ and $|M|, N \leq 2$, inequality (24) implies a stronger inequality. If $q = 5$, $t \leq 2$, $|M|, N \leq 4$ and inequality (24) again yield $n \geq 8m - 8$ and hence we may conclude, in all cases that, provided $m \geq 4$, we have $n \geq 8m - 8 \geq 6m$.

From (23), we have

$$(-1)^\delta 8\kappa_2 t^2 + 5M^4 \equiv 0 \mod q^m. \tag{26}$$

If this is equality, we must have $\delta = 1$ and so (23) becomes

$$256t^{10}Nq^{n-5m} = \kappa_2^2 q^{3m} - 16\kappa_2 t^4 M + 4\kappa_2 t^2 M^2 q^m - 2\kappa_2 M^3 q^{2m} - 4t^2 M^5 + M^6 q^m.$$

It follows that

$$4\kappa_2 t^2 + M^4 \equiv 0 \mod q^m. \tag{27}$$

Combining (26) and (27), we thus have

$$7M^4 \equiv 0 \mod q^m,$$

contradicting the fact that $m \geq 4$, while $0 < |M| < q$.

We thus have

$$(-1)^\delta 8\kappa_2 t^2 + 5M^4 = \upsilon q^m \tag{28}$$

for some nonzero integer υ. If υ is negative, necessarily $\kappa_2 > \frac{q^m}{8t^2}$. If $\upsilon \geq 6$, we have that, again, $\kappa_2 > \frac{q^m}{8t^2}$. Let us therefore assume that $1 \leq \upsilon \leq 5$. Now (23) is

$$256t^{10}Nq^{n-5m} = \kappa_2^2 q^{3m} + 4t^4\upsilon + (-1)^\delta \left(16\kappa_2 t^4 M \right. \\ \left. -4\kappa_2 t^2 M^2 q^m + 2\kappa_2 M^3 q^{2m}\right) - 4t^2 M^5 + M^6 q^m$$

and so, since $n \geq 6m$,

$$t^2\upsilon + (-1)^\delta 4\kappa_2 t^2 M - M^5 \equiv 0 \mod q^m.$$

From (26), we therefore have

$$5\upsilon + (-1)^{\delta}28\kappa_2 M \equiv 0 \mod q^m. \tag{29}$$

Since $1 \le \upsilon \le 5$, the left-hand side here is nonzero and so

$$28\kappa_2|M| \ge q^m - 25.$$

For $q^m \ge 375$, it follows immediately that

$$\kappa_2 > \frac{q^{m-1}}{30}, \tag{30}$$

whilst the inequality is trivial if $q = 3$ and $m = 4$. If $q = 3$ and $m = 5$, we check that for $|M| \in \{1, 2\}$ and $1 \le \upsilon \le 5$, the smallest positive solution to the congruence (29) has $\kappa_2 \ge 17$, whereby (30) is again satisfied.

Combining this with (23), we have that

$$256 t^{10} N q^{n-4m} > \frac{1}{900} q^{6m-2} - \frac{1}{15}|M|^3 q^{4m-1} - \frac{2}{15} t^2 M^2 q^{3m-1} - \frac{8}{15} t^4 |M| q^{2m-1}, \tag{31}$$

whence

$$q^{n-4m+6} > \frac{1}{480^2} q^{6m-2} \left(1 - 60 q^{4-2m} - 120 q^{4-3m} - 480 q^{4-4m}\right).$$

It follows that

$$n \ge 10m - 10 \tag{32}$$

if $q \ge 23$.

We note that, combining (28) and (29), we have

$$2\upsilon t^2 \equiv 7M^5 \mod q^{m-\delta_5}, \tag{33}$$

where $\delta_5 = 1$ if $q = 5$ and 0 otherwise. For $q = 3$, we have $t = 1$, $M = \pm 1, \pm 2$, and find that $\upsilon \equiv \pm 37 \mod 81$ if $|M| = 1$ and $\upsilon \equiv \pm 31 \mod 81$ if $|M| = 2$. In all cases, from (28), we have

$$\kappa_2 \ge \frac{1}{8}\left(31 \cdot 3^m - 80\right) > \frac{15}{4} 3^m.$$

Together with (23), we find, after a little work, that, again, $n \geq 10m - 10$. If $q = 5$, congruence (33) implies that $|v| \geq 13$, so that (28) yields, crudely,

$$\kappa_2 \geq \frac{1}{32}(13 \cdot 5^m - 1280) > \frac{1}{3}5^m,$$

which again, with (23), implies (32). Arguing similarly for the remaining values of q with $7 \leq q \leq 19$, enables us to conclude that inequality (32) holds for all $q \geq 3$ and $m \geq 4$. This concludes the proof of Lemma 3.1.

4 Padé Approximants to the Binomial Function

We now consider Padé approximants to $(1+x)^{1/2}$, defined, for n_1 and n_2 nonnegative integers, via

$$P_{n_1,n_2}(x) = \sum_{k=0}^{n_1}\binom{n_2 + 1/2}{k}\binom{n_1 + n_2 - k}{n_2}x^k \tag{34}$$

and

$$Q_{n_1,n_2}(x) = \sum_{k=0}^{n_2}\binom{n_1 - 1/2}{k}\binom{n_1 + n_2 - k}{n_1}x^k. \tag{35}$$

As in [1], we find that

$$P_{n_1,n_2}(x) - (1 + x)^{1/2}\, Q_{n_1,n_2}(x) = x^{n_1+n_2+1}\, E_{n_1,n_2}(x), \tag{36}$$

where (see, e.g., Beukers [7])

$$E_{n_1,n_2}(x) = \frac{(-1)^{n_2}\,\Gamma(n_2 + 3/2)}{\Gamma(-n_1 + 1/2)\Gamma(n_1 + n_2 + 1)}F(n_1 + 1/2, n_1 + 1, n_1 + n_2 + 2, -x), \tag{37}$$

for F the hypergeometric function given by

$$F(a, b, c, -x) = 1 - \frac{a \cdot b}{1 \cdot c}x + \frac{a \cdot (a + 1) \cdot b \cdot (b + 1)}{1 \cdot 2 \cdot c \cdot (c + 1)}x^2 - \cdots.$$

Appealing twice to (36) and (37) and eliminating $(1 + x)^{1/2}$, the quantity

$$P_{n_1+1,n_2}(x)Q_{n_1,n_2+1}(x) - P_{n_1,n_2+1}(x)Q_{n_1+1,n_2}(x)$$

is a polynomial of degree $n_1 + n_2 + 2$ with a zero at $x = 0$ of order $n_1 + n_2 + 2$ (and hence is a monomial). It follows that we may write

$$P_{n_1+1,n_2}(x)Q_{n_1,n_2+1}(x) - P_{n_1,n_2+1}(x)Q_{n_1+1,n_2}(x) = cx^{n_1+n_2+2}. \tag{38}$$

Here, we have

$$c = (-1)^{n_2+1} \frac{(2n_1 - 2n_2 - 1)\Gamma(n_2 + 3/2)}{2(n_1 + 1)!\,(n_2 + 1)!\,\Gamma(-n_1 + 1/2)} \neq 0.$$

We further observe that

$$\binom{n + \frac{1}{2}}{k} 4^k \in \mathbb{Z},$$

so that, in particular, if $n_1 \geq n_2$, $4^{n_1} P_{n_1,n_2}(x)$ and $4^{n_1} Q_{n_1,n_2}(x)$ are polynomials with integer coefficients.

4.1 Choosing n_1 and n_2

For our purposes, optimal choices for n_1 and n_2 are as follows (we denote by $[x]$ the greatest integer not exceeding a real number x and set $x = [x] + \{x\}$).

Definition 1 Define

$$(n_1, n_2) = \left(\left[\frac{3n}{4m}\right] + \delta - \Delta_1, \left[\frac{n}{4m}\right] - \delta + \Delta_2 \right)$$

where $\delta \in \{0, 1\}$,

$$\Delta_1 = \begin{cases} 1 & \text{if } \left\{\frac{n}{4m}\right\} \in [0, 1/4] \cup [1/3, 1/2] \cup [2/3, 3/4] \\ 0 & \text{if } \left\{\frac{n}{4m}\right\} \in (1/4, 1/3) \cup (1/2, 2/3) \cup (3/4, 1), \end{cases}$$

and

$$\Delta_2 = \begin{cases} 1 & \text{if } \left\{\frac{n}{4m}\right\} > 0 \\ 0 & \text{if } \left\{\frac{n}{4m}\right\} = 0. \end{cases}$$

Note that for these choices of n_1 and n_2, we may check that

$$(n_1 + n_2 + 1)m = n + \left(\Delta_2 - \Delta_1 + 1 - \left\{\frac{n}{4m}\right\} - \left\{\frac{3n}{4m}\right\} \right) m \geq n.$$

Further, we have

$$n_1(m+1) = \frac{3n}{4} + \frac{3n}{4m} + \kappa_1(m, n, \delta)$$

and

$$n_2(m+1) + n_1 - n_2 + \frac{n}{2} = \frac{3n}{4} + \frac{3n}{4m} + \kappa_2(m, n, \delta),$$

where

$$\kappa_1(m, n, \delta) = (m+1)\left(\left[3\left\{\frac{n}{4m}\right\}\right] + \delta - \Delta_1 - 3\left\{\frac{n}{4m}\right\}\right)$$

and

$$\kappa_2(m, n, \delta) = -(m+3)\left\{\frac{n}{4m}\right\} + \left[3\left\{\frac{n}{4m}\right\}\right] + (\Delta_2 - \delta)m + \delta - \Delta_1.$$

A short calculation ensures that, in every situation, we have

$$\max\{n_1(m+1), n_2(m+1) + n_1 - n_2 + n/2\} \leq \frac{3n}{4} + \frac{3n}{4m} + m - \frac{5}{4}, \qquad (39)$$

where the right-hand side is within $O(1/m)$ of the "truth" for $\delta = 0$, $\Delta_1 = \Delta_2 = 1$. Note that the fact that $n \geq 10m - 10$ implies that we have $n_2 \geq 2$, unless

$$(m, n) \in \{(4, 30), (4, 31), (4, 32), (5, 40)\},$$

where we might possibly have $n_2 = 1$. In all cases, we also have

$$|n_1 - 3n_2| \leq 3. \qquad (40)$$

4.2   Bounds for $|P_{n_1,n_2}(x)|$ and $|Q_{n_1,n_2}(x)|$

We will have need of the following result.

Lemma 4.1 *If n_1 and n_2 are as given in Definition 1, where $m \geq 4$ and $n \geq 10m - 10$ are integers, then we have*

$$|P_{n_1,n_2}(x)| \leq 2|x|^{n_1} \quad and \quad |Q_{n_1,n_2}(x)| \leq 2^{n_1+n_2-1}\left(1 + \frac{|x|}{2}\right)^{n_2},$$

for all real numbers x with $|x| \geq 16$.

Proof Arguing as in the proof of Lemma 1 of Beukers [8], we have that

$$|Q_{n_1,n_2}(x)| \leq \sum_{k=0}^{n_2} \binom{n_1}{k}\binom{n_1 + n_2 - k}{n_1} |x|^k = \sum_{k=0}^{n_2} \binom{n_2}{k}\binom{n_1 + n_2 - k}{n_2} |x|^k.$$

Since $n_1 > n_2$ and $\binom{n_1+n_2-k}{n_2} \leq 2^{n_1+n_2-k-1}$, it follows that

$$|Q_{n_1,n_2}(x)| \leq 2^{n_1+n_2-1}\left(1 + \frac{|x|}{2}\right)^{n_2}.$$

Next, note that, since $n_1 > n_2$, $|P_{n_1,n_2}(x)|$ is bounded above by

$$\sum_{k=0}^{n_2+1} \binom{n_2 + 1}{k}\binom{n_1 + n_2 - k}{n_2} |x|^k + \sum_{k=n_2+2}^{n_1} \frac{(n_2+1)!(k - n_2 - 1)!}{k!}\binom{n_1+n_2 - k}{n_2} |x|^k.$$

The first sum here is, arguing as previously, at most

$$2^{n_1+n_2-1}\left(1 + \frac{|x|}{2}\right)^{n_2+1}.$$

For the second, we split the summation into the ranges $n_2 + 2 \leq k \leq \left[\frac{n_1+n_2}{2}\right]$ and $\left[\frac{n_1+n_2}{2}\right] + 1 \leq k \leq n_1$. In the second of these, we have $n_1 + n_2 - k < k$ and so

$$\binom{n_1 + n_2 - k}{n_2} < \binom{k}{n_2},$$

whence

$$\sum_{k=\left[\frac{n_1+n_2}{2}\right]+1}^{n_1} \frac{(n_2 + 1)!(k - n_2 - 1)!}{k!}\binom{n_1 + n_2 - k}{n_2} |x|^k < \sum_{k=\left[\frac{n_1+n_2}{2}\right]+1}^{n_1} \frac{n_2 + 1}{k - n_2} |x|^k.$$

Appealing to Definition 1, we may show that $2n_2 \leq \left[\frac{n_1+n_2}{2}\right]+2$ and hence $\frac{n_2+1}{k-n_2} \leq 1$, so that

$$\sum_{k=\left[\frac{n_1+n_2}{2}\right]+1}^{n_1} \frac{n_2 + 1}{k - n_2} |x|^k \leq \sum_{k=\left[\frac{n_1+n_2}{2}\right]+1}^{n_1} |x|^k < \frac{|x|}{|x| - 1} |x|^{n_1},$$

provided $|x| > 1$. Since

$$\sum_{k=n_2+2}^{\left[\frac{n_1+n_2}{2}\right]} \frac{(n_2+1)!(k-n_2-1)!}{k!} \binom{n_1+n_2-k}{n_2} |x|^k < \sum_{k=n_2+2}^{\left[\frac{n_1+n_2}{2}\right]} \binom{n_1+n_2-k}{n_2} |x|^k$$

and

$$\sum_{k=n_2+2}^{\left[\frac{n_1+n_2}{2}\right]} \binom{n_1+n_2-k}{n_2} |x|^k \leq \sum_{k=n_2+2}^{\left[\frac{n_1+n_2}{2}\right]} 2^{n_1+n_2-k-1} |x|^k < \sum_{k=n_2+2}^{\left[\frac{n_1+n_2}{2}\right]} |2x|^k,$$

we may conclude that $|P_{n_1,n_2}(x)|$ is bounded above by

$$2^{n_1+n_2-1} \left(1 + \frac{|x|}{2}\right)^{n_2+1} + \frac{|x|}{|x|-1} |x|^{n_1} + \frac{|2x|}{|2x|-1} |2x|^{\frac{n_1+n_2}{2}}.$$

Since $|x| \geq 16$ and, via (40), $n_1 \geq 3n_2 - 3$, checking values with $n_2 \leq 10$ separately, we may conclude that

$$|P_{n_1,n_2}(x)| < 2 |x|^{n_1}.$$

This concludes our proof. $\square$

5   Proof of Theorem 1.3

To prove Theorem 1.3, we will, through the explicit Padé approximants of the preceding section, construct an integer that is nonzero and, in archimedean absolute value "not too big," while, under the assumptions of the theorem, being divisible by a very large power of our prime q. With care, this will lead to the desired contradiction.

Setting $\eta = \sqrt{t^2 + Mq^m}$, since $(1+x)^{1/2}$, $P_{n_1,n_2}(x)$ and $Q_{n_1,n_2}(x)$ have q-adic integral coefficients, the same is also true of $E_{n_1,n_2}(x)$ and so, via Eq. (36),

$$\left| t P_{n_1,n_2}\left(\frac{Mq^m}{t^2}\right) - \eta Q_{n_1,n_2}\left(\frac{Mq^m}{t^2}\right) \right|_q \leq q^{-n}.$$

On the other hand, from the fact that $\eta^2 \equiv Y^2 \mod q^n$, we have

$$\eta \equiv (-1)^{\delta_1} Y \mod q^n,$$

for some $\delta_1 \in \{0, 1\}$, and hence

$$\left| t P_{n_1, n_2}\left(\frac{Mq^m}{t^2}\right) - (-1)^{\delta_1} Y Q_{n_1, n_2}\left(\frac{Mq^m}{t^2}\right) \right|_q \leq q^{-n}.$$

Equation (38) implies that for at least one of our two pairs (n_1, n_2), we must have

$$t P_{n_1, n_2}\left(\frac{Mq^m}{t^2}\right) \neq (-1)^{\delta_1} Y Q_{n_1, n_2}\left(\frac{Mq^m}{t^2}\right)$$

and hence, for the corresponding pair (n_1, n_2), we have that

$$(2t)^{2n_1} P_{n_1, n_2}\left(\frac{Mq^m}{t^2}\right) - (-1)^{\delta_1} Y \, 2^{2n_1} \, t^{2n_1 - 1} \, Q_{n_1, n_2}\left(\frac{Mq^m}{t^2}\right)$$

is a nonzero integer, divisible by q^n, and so, in particular,

$$\left| (2t)^{2n_1} P_{n_1, n_2}\left(\frac{Mq^m}{t^2}\right) - (-1)^{\delta_1} Y \, 2^{2n_1} \, t^{2n_1 - 1} \, Q_{n_1, n_2}\left(\frac{Mq^m}{t^2}\right) \right| \geq q^n. \tag{41}$$

From Lemma 4.1 and the fact that $Y < q^{(n+1)/2}$, we thus have

$$q^n \leq 2^{2n_1 + 1} |M|^{n_1} q^{mn_1} + 2^{3n_1 + n_2 - 1} q^{(n+1)/2} t^{2n_1 - 1} \left(1 + \frac{|M| q^m}{2t^2}\right)^{n_2}. \tag{42}$$

From the inequalities

$$|M|, t^2 < q \quad \text{and} \quad \frac{|M| q^m}{2t^2} \geq \frac{81}{2},$$

it follows from (42) that

$$q^n \leq 2^{2n_1 + 1} \cdot q^{(m+1)n_1} + 2^{3n_1 - 1} q^{n/2 + (m+1)n_2 + n_1 - n_2} (83/81)^{n_2}, \tag{43}$$

and hence, since $n \geq 10m - 10$ and $m \geq 4$, we may argue rather crudely to conclude that

$$q^n < 9^{n_1} \cdot q^{\max\{n_1(m+1), n_2(m+1) + n_1 - n_2 + n/2\}}. \tag{44}$$

Inequality (39) thus implies

$$q^n < 9^{\frac{3n}{4m} + 1} \cdot q^{\frac{3n}{4} + \frac{3n}{4m} + m - \frac{5}{4}},$$

whence

$$q^{1-\frac{3}{m}-\frac{4m}{n}+\frac{5}{n}} < 9^{\frac{3}{m}+\frac{4}{n}}. \tag{45}$$

Since $m \geq 4$, if n is suitably large, this provides an upper bound upon q. In particular, if

$$n > \frac{4m^2 - 5m}{m - 3}, \tag{46}$$

then

$$q < 3^{\frac{6n+8m}{mn-3n-4m^2+5m}}. \tag{47}$$

Since $m \geq 4$ and $n \geq 10m - 10$, (46) is satisfied unless we have $m = 4$ and $30 \leq n \leq 44$. Excluding these values for the moment, we thus have

$$q < 3^{\frac{68m-60}{6m^2-35m+30}}.$$

Since $q \geq 3$, it follows, therefore, that, in all cases, $m \leq 16$. If $q \geq 5$, we have the sharper inequality $m \leq 12$.

5.1 Small Values of m

To treat the remaining values of m, we argue somewhat more carefully. For fixed q and m, Eq. (2) under the conditions in (3) can, in many cases, be shown to have no solutions via simple local arguments. In certain cases, however, when the tuple (t, M, N, m) matches up with an actual solution, we will not be able to find such local obstructions. For example, the identities

$$(q^m \cdot Y_0 \pm t)^2 = t^2 \pm 2tY_0q^m + Y_0^2q^{2m}$$

imply that we cannot hope, through simple congruential arguments, to eliminate the cases (here $n \equiv n_0 \mod 3$)

$$(t, M, N, n_0) = (t, \pm 2tY_0, Y_0^2, 2m \mod 3), \tag{48}$$

where $\max\{t^2, Y_0^2, 2tY_0\} < q$. For even values of m, we are also unable to summarily dismiss tuples like

$$(t, M, N, n_0) = (t, Y_0^2, 2tY_0, m/2 \mod 3). \tag{49}$$

Additionally, the "trivial" identity

$$t^2 = t^2 - M \cdot q^m + M \cdot q^m$$

leaves us with the necessity of treating tuples

$$(t, M, N, n_0) = (t, -N, N, m \mod 3) \tag{50}$$

via other arguments. By way of example, if $q = m = 5$, sieving by primes p with the property that the smallest positive t with $5^t \equiv 1 \mod p$ divides 300, we find that all tuples (t, M, N, n_0) are eliminated except for

$$(1, -2, 1, 1), (1, -1, 1, 2), (1, 2, 1, 1), (1, -2, 2, 2), (1, 1, 2, 1), (1, -3, 3, 2),$$
$$(1, -4, 4, 1), (1, -4, 4, 2), (1, 4, 4, 1), (2, -4, 1, 1), (2, -1, 1, 2), (2, 4, 1, 1),$$
$$(2, -2, 2, 2), (2, -3, 3, 2) \text{ and } (2, -4, 4, 2).$$

These all correspond to (48) or (50), except for $(t, M, N, n_0) = (1, 1, 2, 1)$ which arises from the identity $56^2 = 1^2 + 2 \cdot 5 + 5^5$.

For the cases where we fail to obtain a local obstruction, we can instead consider Eq. (6), with the conditions (7). Our expectation is that, instead of needing to treat roughly $6(q-1)^{5/2}$ such equations (for a fixed pair (q, m)), after local sieving we will be left with on the order of $O(q)$ Mordell curves to handle.

By way of example, let us begin with the case where $q = 3$. Here, from (42),

$$3^n \leq 2^{3n_1+1} 3^{mn_1} + 2^{3n_1+n_2-1} (82/81)^{n_2} 3^{mn_2+(n+1)/2}.$$

Since $\max\{mn_1, mn_2 + (n+1)/2\} \leq \frac{3n}{4} + m + \frac{1}{4}$, and $n_2 \geq 2$ (provided $n > 40$), we thus have

$$3^n \leq 2^{3n_1+n_2} (82/81)^{n_2} 3^{\frac{3n}{4}+m+\frac{1}{4}},$$

so that

$$3^{n/4-m-1/4} \leq 2^{3n_1+n_2} (82/81)^{n_2}.$$

We check that $n_2 \leq \frac{n}{4m} + 1$ and $3n_1 + n_2 \leq \frac{5n}{2m} + \frac{3}{2}$, whence either $n \leq 40$, or we have

$$3^{\frac{n}{4}-m-\frac{1}{4}} \leq 2^{\frac{5n}{2m}+\frac{3}{2}} (82/81)^{\frac{n}{4m}+1}.$$

In this latter case, if $m \geq 12$, the fact that $n \geq 10m - 10$ leads to a contradiction, whilst, for $8 \leq m \leq 11$, we have that $n \leq 157$. A short calculation ensures that there are no solutions to Eq. (2) with (3), if $q = 3$, $8 \leq m \leq 11$ and $10m - 10 \leq n \leq 157$. For $q = 3$ and $4 \leq m \leq 7$, we are led to equation of the shape (6), where now

$|k| \leq 324\,(1 + 2 \cdot 3^m) \leq 1417500$. As noted previously, the integer points on the corresponding Mordell curves are known (see [4]) and listed at http://www.math. ubc.ca/~bennett/BeGa-data.html. We check that no solutions exist with U and V as in (7).

We may thus suppose that $q \geq 5$ and hence it remains to treat the values of m with $4 \leq m \leq 12$. If $m = 12$, appealing to (47), we have, from the fact that $n \geq 110$, necessarily $110 \leq n \leq 118$ and $q = 5$. A short calculation ensures that there are no corresponding solutions to Eq. (2) with (3). Similarly, if $m = 11$, we have that either $q = 5$ and $100 \leq n \leq 125$, or $q = 7$, $100 \leq n \leq 103$. If $m = 10$, $q = 5$ and $90 \leq n \leq 139$, or $q = 7$ and $90 \leq n \leq 109$, or $q = 11$ and $n = 90$. For $m = 9$ we have, in all cases, $n \leq 172$ and $q \leq 19$. For $m = 8$, $n \leq 287$ and $q \leq 47$. A modest computation confirms that we have no new solutions to the equation of interest and hence we may suppose that $4 \leq m \leq 7$ (and that $q \geq 5$).

For small values of q, each choice of m leads to at most $2q^{5/2}$ Ramanujan–Nagell equations (5) which we can solve as in [18]. In practice, the great majority of these are eliminated by local sieving. By way of example, if $q = 5$, after local sieving, we are left to treat precisely 32 pairs (D, N) in Eq. (5), corresponding to

$$D \in \{-312498, -15624, -15623, -12498, -2498, -1249,$$
$$-624, 1251, 2502, 6251, 12502, 31251, 312502\}, \ \text{if } N = 1,$$

$$D \in \{-156248, -31248, 3126, 15626\}, \ \text{if } N = 2,$$

$$D \in \{-234374, -234373, -46874, -46873, -1873, 31252\}, \ \text{if } N = 3$$

and

$$D \in \{-312499, -62498, -2499, -2498, 627, 2501, 12501, 15627, 62501\} \ \text{if } N = 4.$$

For these values of (D, N), we find that Eq. (5) has precisely solutions as follows

D	N	n	D	N	n
-312499	4	7	2501	4	2
-312499	4	14	2501	4	8
-234374	3	7	3126	2	1
-46874	3	6	6251	1	10
-15624	1	6	12501	4	10
-2499	4	4	15626	2	3
-2499	4	8	31251	1	12
-1249	1	8	62501	4	3
-624	1	4	62501	4	12
1251	1	8			

In all cases, these solutions correspond to values of m that have either $m \geq n$ or $n = 2m$. More generally, implementing a "Ramanujan-Nagell" solver as in [18], in conjunction with local sieving, we completely solve Eq. (2) with (3), for $m \in \{4, 5, 6, 7\}$ and $5 \leq q \leq 31$. No new solutions accrue. If we appeal again to inequality (47), using that $q \geq 37$, we find that $60 \leq n \leq 81$ (if $m = 7$), $50 \leq n \leq 109$ (if $m = 6$) and $40 \leq n \leq 499$ (if $m = 5$). After a short computation, we are left to consider the cases with $m = 4$ and $q \geq 37$.

For the value $m = 4$, proceeding in this manner would entail an extremely large computation, without additional ingredients. By way of example, in case $m = 4$ and $n = 45$, inequality (47) implies an upper bound upon q that exceeds 10^{144} (and no upper bound whatsoever for $30 \leq n \leq 44$). To sharpen this and related inequalities, we will argue as follows. Notice that if we have

$$t P_{n_1, n_2}\left(\frac{M q^m}{t^2}\right) = (-1)^{\delta_1} Y Q_{n_1, n_2}\left(\frac{M q^m}{t^2}\right),$$
(51)

then

$$t^2 P_{n_1, n_2}^2\left(\frac{M q^m}{t^2}\right) - (t^2 + M q^m + N q^n) Q_{n_1, n_2}^2\left(\frac{M q^m}{t^2}\right) = 0.$$

From our construction, it follows that

$$\left| t^2 P_{n_1, n_2}^2\left(\frac{M q^m}{t^2}\right) - (t^2 + M q^m) Q_{n_1, n_2}^2\left(\frac{M q^m}{t^2}\right)\right|_q \leq q^{-m(n_1 + n_2 + 1)}.$$

and hence, if $(n_1 + n_2 + 1)m > n$ and (51), then

$$q^{(n_1 + n_2 + 1)m - n} \text{ divides } Q_{n_1, n_2}^2(0) = \binom{n_1 + n_2}{n_2}^2.$$
(52)

In particular, if $m = 4$ and $30 \leq n \leq 32$, then we have $(n_1, n_2) \in \{(5, 2), (6, 1)\}$ and hence, since $q \geq 37$, (52) fails to hold. We thus obtain inequality (41) for both pairs (n_1, n_2), rather than just for one of them, provided $n \in \{30, 31\}$ (if $n = 32$, we have $(n_1 + n_2 + 1)m = n$). Choosing $(n_1, n_2) = (5, 2)$, it follows from (44) that, if $n = 30$, we have $q^2 < 3^{10}$, so that $q \leq 241$, while $n = 31$ implies $q^{5/2} < 3^{10}$, i.e. $q \leq 79$. If $n = 32$, the worse case corresponds to $(n_1, n_2) = (6, 1)$, where we find, again from (44), that $q^2 < 3^{12}$ and so $q \leq 727$. Continuing in this fashion, observing that the greatest prime factor $\binom{n_1 + n_2}{n_2}$ is bounded above by roughly $n/4$, and that $4(n_1 + n_2 + 1) = n$ precisely when $4 \mid n$, we have, via (44), an upper bound upon q of the shape $q < \min_{\delta \in \{0,1\}}\{3^{2n_1/(n-\mu)}\}$, if $4 \nmid n$, and $q < \max_{\delta \in \{0,1\}}\{3^{2n_1/(n-\mu)}\}$, if $4 \mid n$, where

$$\mu = \max\{n_1(m + 1), n_2(m + 1) + n_1 - n_2 + n/2\}.$$

Here, we exclude the cases where $\mu \geq n$, corresponding to $(n_1, n_2) = (5, 3)$ if $n = 33$ or 34 and $(n_1, n_2) = (9, 2)$ if $n = 45$; in each of these, the other choice of (n_1, n_2) leads to a bound upon q. For $n \leq 1000$, we find that $q < 3^{10}$, in case $n = 36$, $q < 3^{28/3}$ (if $q = 41$), $q < 3^8$ (if $n = 52$ or $n = 57$) and otherwise $q < 3155$. A painful but straightforward computation finds that we have no additional solutions to Eq. (2) with (3) for $n \leq 1000$. Applying once again inequality (47), we may thus assume that $q \leq 1021$. After local sieving and solving corresponding equations of the shape (5), we verify that Eq. (2) has no unexpected solutions with (3), for $m = 4$ and $37 \leq q \leq 1021$. This completes the proof of Theorem 1.3.

Full details of our computations are available from the authors upon request.

6 Proof of Theorem 1.1

For $q \in \{3, 5\}$, we may apply Theorem 1.3 to conclude that either $n = 3^b + 1$ (in case $q = 3$) or that $n \in \{5^b + 1, 2 \cdot 5^b + 1, 5^b + 2\}$ (if $q = 5$), for some positive integer b, or that we have either

$$n^2 = 1 + M \cdot 3^m + N \cdot 3^n, \quad n^2 = 1 + M \cdot 5^m + N \cdot 5^n \quad \text{or} \quad n^2 = 4 + M \cdot 5^m + N \cdot 5^n, \tag{53}$$

with $m \in \{1, 2, 3\}$, $n > m$ and $1 \leq M, N \leq q - 1$. Checking the corresponding solutions to (6) (all available at http://www.math.ubc.ca/~bennett/BeGa-data.html), we find that the only solutions to (53) are with

$$n \in \{4, 5, 8, 9, 12, 13, 16, 23, 24, 56, 177\},$$

as claimed. Adding in the "trivial" solutions with $n \in \{1, 2\}$ completes the proof of Theorem 1.1 in case $q \in \{3, 5\}$.

Our argument for $q \in \{2, 4, 8, 16\}$ follows along very similar lines to the proof of Theorem 1.3, only with slight additional complications, arising from the fact that none of $(1 + x)^{1/2}$, $P_{n_1, n_2}(x)$ or $Q_{n_1, n_2}(x)$ have 2-adic integral coefficients. On the other hand, $(1 + 4x)^{1/2}$, $P_{n_1, n_2}(4x)$, and $Q_{n_1, n_2}(4x)$ do have 2-adic integral coefficients and so we can proceed as in Sect. 5, taking $x = Mq^m/t^2$, where now $q = 2^\alpha$ for $\alpha \in \{1, 2, 3, 4\}$. Under mild assumptions upon m ($m \geq 5$ is satisfactory), the arguments of Sects. 3 and 5 go through with essentially no changes. We are left to treat a number of equations of the shape (5), to complete the proof of Theorem 1.1. We suppress the details.

7 Concluding Remarks

In this paper, we have focussed our attention on Eq. (8) in case C is square and q is prime. Even in this very restricted situation, we have been able to use our results to completely determine $B_3(q)$ only for $q \in \{2, 3, 5\}$. We conclude with some speculations upon the structure of the sets $B_3(q)$. Let us write

$$B_k(q) = \bigcup_{j=k}^{\infty} B_{k,j}(q),$$

where

$$B_{k,j}(q) = \left\{ n \in \mathbb{N} : n \not\equiv 0 \mod q, \ N_q(n) = j \ \text{and} \ N_q(n^2) = k \right\}.$$

If $q = r^2 + 1$ is prime for r an integer, since we have

$$\frac{1}{2} r(r^6 + 5r^4 + 7r^2 + 5) = r + r \cdot q^2 + \frac{r}{2} \cdot q^3,$$

identity (4) implies that $B_{3,3}(q)$ is nonempty for such q. Further, for odd prime q, we can find examples to verify that $B_{3,4}(q)$ is nonempty for (at least)

$$q = 7, 11, 17, 23, 31, 47, 101, 131, 151,$$

amongst the primes up to 200. We observe that

$$35864 \in B_{3,5}(11).$$

We know of no other value in $B_{3,j}(q)$ for $j \geq 5$ and q prime. Perhaps there are none.

Acknowledgements The authors are grateful to the referees for pointing out a number of errors, typographical, and otherwise. The authors were supported in part by grants from NSERC. The second author (Adrian-Maria Scheerer) was supported by the Austrian Science Fund (FWF): I 1751-N26; W1230, Doctoral Program "Discrete Mathematics"; and SFB F 5510-N26.

References

1. M. Bauer, M. Bennett, Application of the hypergeometric method to the generalized Ramanujan-Nagell equation. Ramanujan J. **6**, 209–270 (2002)
2. M. Bennett, Perfect powers with few ternary digits. Integers **12**(6), 1159–1166 (2012)
3. M. Bennett, Y. Bugeaud, Perfect powers with three digits. Mathematika **60**, 66–84 (2014)
4. M. Bennett, A. Ghadermarzi, Mordell's equation: a classical approach. LMS J. Comput. Math. **18**, 633–646 (2015)

5. M. Bennett, Y. Bugeaud, M. Mignotte, Perfect powers with few binary digits and related Diophantine problems II. Math. Proc. Camb. Philos. Soc. **153**, 525–540 (2012)
6. M. Bennett, Y. Bugeaud, M. Mignotte, Perfect powers with few binary digits and related Diophantine problems. Ann. Sc. Norm. Super. Pisa Cl. Sci. **XII**, 941–953 (2013)
7. F. Beukers, On the generalized Ramanujan–Nagell equation. I. Acta Arith. **38**, 389–410 (1980/1981)
8. F. Beukers, On the generalized Ramanujan–Nagell equation. II. Acta Arith. **39**, 113–123 (1981)
9. C. Bright, *Solving Ramanujan's Square Equation Computationally* (2007), pp. 1–4, https://cs.uwaterloo.ca/~cbright/nsra/ramanujans-square-equation.pdf
10. P. Corvaja, U. Zannier, On the Diophantine equation $f(a^m, y) = b^n$. Acta Arith. **94**, 25–40 (2000)
11. P. Corvaja, U. Zannier, Finiteness of odd perfect powers with four nonzero binary digits. Ann. Inst. Fourier (Grenoble) **63**(2), 715–731 (2013)
12. H. Delange, Sur la fonction sommatoire de la fonction "somme des chiffres". Enseign. Math. **21**, 31–47 (1975)
13. K. Hare, S. Laishram, T. Stoll, Stolarsky's conjecture and the sum of digits of polynomial values. Proc. Am. Math. Soc. **139**(1), 39–49 (2011)
14. K. Hare, S. Laishram, T. Stoll, The sum of digits of n and n^2. Int. J. Number Theory **7**(7), 1737–1752 (2011)
15. F. Luca, The diophantine equation $x^2 = p^a \pm p^b + 1$. Acta. Arith. **112**, 87–101 (2004)
16. C. Mauduit, J. Rivat, La somme des chiffres des carrés. Acta Math. **203**(1), 107–148 (2009)
17. C. Mauduit, J. Rivat, Sur un problème de Gelfond: la somme des chiffres des nombres premiers. Ann. Math. (2) **171**(3), 1591–1646 (2010)
18. A. Pethő, B.M.M. de Weger, Products of prime powers in binary recurrence sequences. I. The hyperbolic case, with an application to the generalized Ramanujan-Nagell equation. Math. Comput. **47**(176), 713–727 (1986)
19. K. Stolarsky, The binary digits of a power. Proc. Am. Math. Soc. **71**, 1–5 (1978)
20. L. Szalay, The equations $2^n \pm 2^m \pm 2^l = z^2$. Indag. Math. (N.S.) **13**, 131–142 (2002)
21. B.M.M. de Weger, *Algorithms for Diophantine Equations*. CWI-Tract No. 65 (Centre for Mathematics and Computer Science, Amsterdam, 1989)

On the Density of Coprime Tuples of the Form $(n, \lfloor f_1(n) \rfloor, \ldots, \lfloor f_k(n) \rfloor)$, Where $f_1, \ldots, f_k$ Are Functions from a Hardy Field

Vitaly Bergelson and Florian Karl Richter

It is a well-known theorem of Čebyšev[1] that the probability of the relation $\gcd(n, m) = 1$ is $\frac{6}{\pi^2}$. One can expect this still to remain true if $m = g(n)$ is a function of n, provided that $g(n)$ does not preserve arithmetic properties of n.

P. Erdős and G. Lorentz

Abstract Let $k \in \mathbb{N}$ and let $f_1, \ldots, f_k$ belong to a Hardy field. We prove that under some natural conditions on the k-tuple $(f_1, \ldots, f_k)$ the density of the set

$$\{n \in \mathbb{N} : \gcd(n, \lfloor f_1(n) \rfloor, \ldots, \lfloor f_k(n) \rfloor) = 1\}$$

exists and equals $\frac{1}{\zeta(k+1)}$, where ζ is the Riemann zeta function.

1 Introduction

The above epigraph is a quote from the introduction to a paper by Erdős and Lorentz [12], which establishes sufficient conditions for a differentiable function $f : [1, \infty) \to \mathbb{R}$ of sub-linear growth to satisfy

$$d\big(\{n \in \mathbb{N} : \gcd(n, \lfloor f(n) \rfloor) = 1\}\big) = \frac{6}{\pi^2}; \tag{1}$$

here $d(A)$ denotes the natural density of a set $A \subset \mathbb{N}$.

---

[1] The attribution of this result to Čebyšev (Чебышёв) seems not to be justified; see, however, the very interesting recent preprint [1] where Čebyšev's role in the popularization of this theorem is traced and analyzed. The result itself goes back to Dirichlet (see [10, pp. 51–66] where the equivalent statement $\sum_{n=1}^{N} \phi(n) \sim \frac{3}{\pi^2} n^2$ is proven) and was rediscovered multiple times—see, for example, [6, 7, 21, 23, 24]. It is worth noting that it was Cesàro who formulated this result in probabilistic terms [6] and also gave a probabilistic, though not totally rigorous, proof in [7].

V. Bergelson (✉) • F.K. Richter
Department of Mathematics, The Ohio State University, Columbus, OH 43210, USA
e-mail: bergelson.1@osu.edu; richter.109@osu.edu

© Springer International Publishing AG 2017

C. Elsholtz, P. Grabner (eds.), *Number Theory – Diophantine Problems, Uniform Distribution and Applications*, DOI 10.1007/978-3-319-55357-3_5

Perhaps the earliest result of this kind is due to Watson [28], who showed that (1) holds for $f(n) = n\alpha$, where α is an irrational number (see also [13, 22]). Other examples of functions for which (1) holds are $f(n) = n^c$, where $c > 0$, $c \notin \mathbb{N}$, (see [20] for the case $0 < c < 1$ and [9] for the general case) and $f(n) = \log^r(n)$ for all $r > 1$ (see [20] for the case $r > 2$ and [12] for the general case).

The purpose of this paper is to establish (1) for a large class of smooth functions that naturally includes examples such as $f(n) = n^c$ or $f(n) = \log^r(n)$; this is the class of functions belonging to a Hardy field.

Let G denote the set of all germs[2] at ∞ of real valued functions defined on the half-line $[1, \infty)$. Note that G forms a ring under pointwise addition and multiplication, which we denote by $(G, +, \cdot)$. Any subfield of the ring $(G, +, \cdot)$ that is closed under differentiation is called a *Hardy field*. By abuse of language, we say that a function $f : [1, \infty) \to \mathbb{R}$ belongs to some Hardy field $\mathcal{H}$ (and write $f \in \mathcal{H}$) if its germ at ∞ belongs to $\mathcal{H}$. See [3–5] and some references therein for more information on Hardy fields.

A classical example of a Hardy field is the class of logarithmico-exponential functions[3] introduced by Hardy in [17, 18]; we denote this class by $\mathcal{L}$. It is worth noting that for any Hardy field $\mathcal{H}$ there exists a Hardy field $\mathcal{H}'$ such that $\mathcal{H}' \supset \mathcal{L} \cup \mathcal{H}$.

If $\mathcal{H}$ is a Hardy field, then one has the following basic properties:

- If $f \in \mathcal{H}$, then $\lim_{t \to \infty} f(t)$ exists (as an element in $\mathbb{R} \cup \{-\infty, \infty\}$);
- Any non-constant $f \in \mathcal{H}$ is eventually either strictly increasing or strictly decreasing; any non-linear $f \in \mathcal{H}$ is eventually either strictly concave or strictly convex.
- If $f \in \mathcal{H}$, $g \in \mathcal{L}$, and $\lim_{t \to \infty} g(t) = \infty$, then there exists a Hardy field $\mathcal{H}'$ containing $f(g(t))$.
- If $f \in \mathcal{H}$, $g \in \mathcal{L}$, and $\lim_{t \to \infty} f(t) = \infty$, then there exists a Hardy field $\mathcal{H}'$ containing $g(f(t))$.

Some well-known examples of functions coming from Hardy fields are:

$$t^c \ (\forall c \in \mathbb{R}), \ \log(t), \ \exp(t), \ \Gamma(t), \ \zeta(t), \ \mathrm{Li}(t), \ \sin\left(\frac{1}{t}\right), \ \text{etc.}$$

Before formulating our main results, we introduce some convenient notation. We use $\log_n(t)$ to abbreviate the n-th iteration of logarithms, that is, $\log_2(t) = \log\log(t)$, $\log_3(t) = \log\log\log(t)$, and so on. Also, given two functions $f, g : [1, \infty) \to \mathbb{R}$ we will write $f(t) \prec g(t)$ if $\frac{g(t)}{f(t)} \to \infty$ as $t \to \infty$.

---

[2] We define a *germ at ∞* to be any equivalence class of functions under the equivalence relationship $(f \sim g) \Leftrightarrow (\exists t_0 > 0 \text{ such that } f(t) = g(t) \text{ for all } t \in [t_0, \infty))$.

[3] By a *logarithmico-exponential function* we mean any function $f : (0, \infty) \to \mathbb{R}$ that can be obtained from constants, $\log(t)$ and $\exp(t)$ using the standard arithmetical operations $+, -, \cdot, \div,$ and the operation of composition.

Let $\mathcal{H}$ be a Hardy field and let $f \in \mathcal{H}$. Consider the following two conditions:

(A) $\log(t) \log_4(t) \prec f(t)$;
(B) There exists $j \in \mathbb{N}$ such that $t^{j-1} \prec f(t) \prec t^j$.

We have the following theorem.

Theorem 1 *Let $\mathcal{H}$ be a Hardy field and assume that $f \in \mathcal{H}$ satisfies conditions (A) and (B). Then the natural density of the set*

$$\left\{ n \in \mathbb{N} : \gcd(n, \lfloor f(n) \rfloor) = 1 \right\}$$

exists and equals $\frac{6}{\pi^2}$.

Examples of sequences $(f(n))_{n \in \mathbb{N}}$ to which Theorem 1 applies are n^c (with $c \notin \mathbb{N}$), $\log^2(n)$, $n^{\sqrt{3}} \log(n)$, $\frac{n}{\log_2(n)}$, $\log(n!)$, $\mathrm{Li}(n)$, $\log(|B_{2n}|)$ (where B_n denotes the n-th Bernoulli number), and many more.

We remark that condition (A) is sharp. Indeed, it is shown in [12, Sect. 3] that Theorem 1 does not hold for the function $f(t) = \log(t) \log_4(t)$, as well as for many other functions that grow slower than $\log(t) \log_4(t)$.

As for condition (B), it can perhaps be replaced by the following:

(B′) There exists $j \in \mathbb{N}$ such that $f(t) \prec t^j$ and for all polynomials $p(t) \in \mathbb{Q}[t]$ we have $|f(t) - p(t)| \succ \log(t)$.

Condition (B′) is inspired by a theorem of Boshernitzan (cf. [5, Theorem 1.3]). However, proving Theorem 1 under conditions (A) and (B′) would certainly necessitate introduction of new ideas.

We actually prove a multi-dimensional generalization of Theorem 1. Let $\mathcal{H}$ be a Hardy field and assume $f_1, \ldots, f_k \in \mathcal{H}$. In addition to conditions (A) and (B) consider the following:

(C) $\frac{f_{i+1}}{f_i} \succ \log_2^4(t)$ for all $i = 1, \ldots, k - 1$.

Theorem 2 *Let $\mathcal{H}$ be a Hardy field and assume $f_1, \ldots, f_k \in \mathcal{H}$ satisfy conditions (A)–(C). Then the natural density of the set*

$$\left\{ n \in \mathbb{N} : \gcd(n, \lfloor f_1(n) \rfloor, \ldots, \lfloor f_k(n) \rfloor) = 1 \right\}$$

exists and equals $\frac{1}{\zeta(k+1)}$, where ζ is the Riemann zeta function.

We would like to remark that our proof of Theorem 2 works for (a larger class of) functions which have sufficiently many derivatives and possess some other natural regularity properties. We decided in favor of dealing with Hardy fields since they (a) provide an ample supply of interesting examples and (b) allow for, so to say, cleaner proofs.

The structure of the paper is as follows. In Sect. 2 we prove some differential inequalities for functions from a Hardy field; these inequalities will play a crucial role in the later sections. In Sect. 3 we briefly recall van der Corput's method for estimating exponential sums. In Sect. 4 we apply van der Corput's method to derive useful estimates for exponential sums involving functions from a Hardy field and in Sect. 5 we use a higher dimensional version of the Erdős-Turán inequality to convert these estimates into discrepancy estimates. In Sect. 6 we use the estimates derived in the previous sections to give a proof of Theorem 2. Finally, in Sect. 7, we formulate some natural open questions.

2   Differential Inequalities for Functions from a Hardy Field

In this section we derive some differential inequalities for functions belonging to a Hardy field. Similar inequalities can be found in [14, Sect. 2.1] and in [2, Sect. 2.1].

Given two functions $f, g : [1, \infty) \to \mathbb{R}$ we write $f(t) \ll g(t)$ if there exist $C > 0$ and $t_0 \geq 1$ such that $f(t) \leq Cg(t)$ for all $t \geq t_0$. Also, for $\ell \in \mathbb{N}$ we use $f^{(\ell)}(t)$ to denote the ℓ-th derivative of $f(t)$.

The following lemma appears in [14].

Proposition 3 (See [14, Corollary 2.3])   *Let $\mathcal{H}$ be a Hardy field. Suppose $f \in \mathcal{H}$ satisfies condition (B). Then, for all $\ell \in \mathbb{N}$, we have*

$$\frac{f(t)}{t^\ell \log^2(t)} \prec |f^{(\ell)}(t)| \ll \frac{f(t)}{t^\ell}.$$

Next, we derive a series of lemmas (Lemmas 4–7) which are needed for the proof of the main result of this section, Proposition 8.

Lemma 4   *Let $m \in \mathbb{N}$ and let $\mathcal{H}$ be a Hardy field. Suppose $f, g \in \mathcal{H}$ satisfy $|f(t)| \prec |g(t)| \prec |f(t)| \log^m(t)$ and $|\log(|f(t)|)| \succ \log_2(t)$. Then*

$$\frac{f'(t)}{f(t)} \sim \frac{g'(t)}{g(t)}.$$

Proof Our goal is to show that

$$\frac{\frac{g'(t)}{g(t)}}{\frac{f'(t)}{f(t)}} \xrightarrow{t \to \infty} 1.$$

First we note that since $\mathcal{H}$ is a field closed under differentiation, the function $\frac{g'(t)/g(t)}{f'(t)/f(t)}$ is contained in $\mathcal{H}$. From this it follows that $\lim_{t \to \infty} \frac{g'(t)/g(t)}{f'(t)/f(t)}$ exists (as a number in $\mathbb{R} \cup \{-\infty, \infty\}$). From L'Hospital's rule we now obtain

$$\lim_{t \to \infty} \frac{\frac{g'(t)}{g(t)}}{\frac{f'(t)}{f(t)}} = \lim_{t \to \infty} \frac{\log(|g(t)|)}{\log(|f(t)|)}.$$

To finish the proof we distinguish between the cases $|f(t)| \succ 1$ and $|f(t)| \prec 1$. If $|f(t)| \succ 1$ then, using $|f(t)| \prec |g(t)| \prec |f(t)| \log^m(t)$ and $|\log(|f(t)|)| \succ \log_2(t)$, we deduce that

$$1 \le \lim_{t\to\infty} \frac{\log(|g(t)|)}{\log(|f(t)|)} \le \lim_{t\to\infty} \frac{\log(|f(t)|) + m\log_2(t)}{\log(|f(t)|)} = 1.$$

Likewise, if $|f(t)| \prec 1$, then we have

$$1 \ge \lim_{t\to\infty} \frac{\log(|g(t)|)}{\log(|f(t)|)} \ge \lim_{t\to\infty} \frac{\log(|f(t)|) + m\log_2(t)}{\log(|f(t)|)} = 1.$$

This finishes the proof. $\qquad\qquad\square$

Lemma 5 *Let $\mathcal{H}$ be a Hardy field and suppose $f \in \mathcal{H}$ satisfies condition (B). Then $f^{(\ell)}$ satisfies either $f^{(\ell)}(t) \succ 1$ or $f^{(\ell)}(t) \prec 1$ for all $\ell \in \mathbb{N}$.*

Proof By way of contradiction, let us assume that there exist $\ell \in \mathbb{N}$ and $c \in \mathbb{R}$ such that $f^{(\ell)}(t) \sim c$. Observe that $c \ne 0$, because otherwise $f(t)$ is a polynomial, which contradicts condition (B).

Using Proposition 3 we deduce that

$$\frac{f(t)}{t^\ell \log^2(t)} \prec |c| \ll \frac{f(t)}{t^\ell},$$

which is equivalent to

$$t^\ell \ll f(t) \prec t^\ell \log^2(t).$$

It follows from condition (B) that we can replace $t^\ell \ll f(t)$ with $t^\ell \prec f(t)$. Therefore, we have

$$t^\ell \prec f(t) \prec t^\ell \log^2(t).$$

By using induction on i and by repeatedly applying Lemma 4 to the functions $f^{(i)}(t)$ and t^i, we conclude that for all $i \in \{0, 1, \ldots, \ell - 1\}$,

$$\frac{f^{(i+1)}(t)}{f^{(i)}(t)} \sim \frac{(\ell - 1)t^{\ell - i - 1}}{t^{\ell - i}} = \frac{\ell - i}{t}.$$

In particular, this shows that

$$\frac{f^{(\ell)}(t)}{f(t)} \sim \frac{\ell!}{t^\ell}.$$

Finally, combing $t^\ell \prec f(t)$ and $\frac{f^{(\ell)}(t)}{f(t)} \sim \frac{\ell!}{t^\ell}$ yields $f^{(\ell)}(t) \succ 1$, which contradicts $f^{(\ell)}(t) \sim c$. $\qquad\square$

Lemma 6 *Let $\mathcal{H}$ be a Hardy field and suppose that $f \in \mathcal{H}$ satisfies either $f(t) \succ 1$ or $f(t) \prec 1$. Also, assume $|\log(|f(t)|)| \ll \log_2(t)$. Then*

$$\frac{|f(t)|}{t \log(t) \log_2^2(t)} \prec |f'(t)| \ll \frac{|f(t)|}{t \log(t)}.$$

Proof (cf. The Proof of Lemma 2.1 in [14]) By L'Hospital's rule we get

$$\lim_{t\to\infty} \frac{\frac{f'(t)}{f(t)}}{\frac{1}{t\log(t)}} = \lim_{t\to\infty} \frac{\log(|f(t)|)}{\log_2(t)} \ll 1.$$

This proves that $|f'(t)| \ll \frac{|f(t)|}{t\log(t)}$.

On the other hand, we have

$$\lim_{t\to\infty} \frac{\frac{f'(t)}{f(t)}}{\frac{1}{t\log(t)\log_2^2(t)}} = \lim_{t\to\infty} \log(|f(t)|) \log_2(t) = \pm\infty,$$

which shows that $\frac{|f(t)|}{t\log(t)\log_2^2(t)} \prec |f'(t)|$. $\qquad\square$

Lemma 7 *Let $m \in \mathbb{N}$, let $\mathcal{H}$ be a Hardy field and let $f, g \in \mathcal{H}$. Assume that f satisfies either $f(t) \succ 1$ or $f(t) \prec 1$ and g satisfies either $g(t) \succ 1$ or $g(t) \prec 1$. Also, assume $|f(t)| \prec |g(t)| \prec |f(t)| \log^m(t)$ and $|\log(|f(t)|)| \ll \log_2(t)$. Then*

$$\left|\frac{f'(t)}{f(t)}\right| \frac{1}{\log_2^2(t)} \prec \left|\frac{g'(t)}{g(t)}\right| \prec \left|\frac{f'(t)}{f(t)}\right| \log_2^2(t).$$

Proof It follows from $|f(t)| \prec |g(t)| \prec |f(t)| \log^m(t)$ and $|\log(|f(t)|)| \ll \log_2(t)$ that $|\log(|g(t)|)| \ll \log_2(t)$. Hence we can apply Lemma 6 to both f and g and obtain

$$\frac{1}{t\log(t)\log_2^2(t)} \prec \left|\frac{f'(t)}{f(t)}\right| \ll \frac{1}{t\log(t)}$$

as well as

$$\frac{1}{t\log(t)\log_2^2(t)} \prec \left|\frac{g'(t)}{g(t)}\right| \ll \frac{1}{t\log(t)}.$$

We deduce that

$$\left| \frac{g'(t)}{g(t)} \right| \ll \frac{1}{t \log(t)} = \frac{\log_2^2(t)}{t \log(t) \log_2^2(t)} \prec \left| \frac{f'(t)}{f(t)} \right| \log_2^2(t).$$

Similarly,

$$\left| \frac{g'(t)}{g(t)} \right| \succ \frac{1}{t \log(t) \log_2^2(t)} \gg \left| \frac{f'(t)}{f(t)} \right| \frac{1}{\log_2^2(t)}.$$

$\square$

Proposition 8 *Let $m \in \mathbb{N}$, let $\mathcal{H}$ be a Hardy field, let $f, g \in \mathcal{H}$, and let $F : [1, \infty) \to (0, \infty)$ be an increasing function satisfying $1 \prec F(t) \prec \log^m(t)$. If f and g satisfy condition (B) and $\frac{g(t)}{f(t)} \succ F(t)$, then*

$$\left| \frac{g^{(\ell)}(t)}{f^{(\ell)}(t)} \right| \succ \frac{F(t)}{\log_2^2(t)}, \qquad \forall \ell \in \mathbb{N}.$$

Proof Let $\ell \in \mathbb{N}$ be arbitrary. We distinguish between the following two cases. The first case is $g(t) \succ f(t) \log^{m+2}(t)$ and the second case is $g(t) \ll f(t) \log^{m+2}(t)$.

We start with the proof of the first case. Using Proposition 3 we obtain the estimate

$$\left| \frac{g^{(\ell)}(t)}{f^{(\ell)}(t)} \right| \succ \frac{g(t)}{f(t) \log^2(t)},$$

and, since $g(t) \succ f(t) \log^{m+2}(t)$, we get

$$\frac{g(t)}{f(t) \log^2(t)} \succ \log^m(t) \succ F(t).$$

Therefore $\left| \frac{g^{(\ell)}(t)}{f^{(\ell)}(t)} \right| \succ F(t)$, which concludes the proof of case one.

Next, we deal with the second case. Consider the product

$$\frac{f(t)}{g(t)} \left| \frac{g^{(\ell)}(t)}{f^{(\ell)}(t)} \right| = \prod_{i=0}^{\ell-1} \frac{\left| \frac{g^{(i+1)}(t)}{g^{(i)}(t)} \right|}{\left| \frac{f^{(i+1)}(t)}{f^{(i)}(t)} \right|}.$$

In virtue of Lemma 5, $f^{(i)}$ satisfies either $f^{(i)} \succ 1$ or $f^{(i)} \prec 1$. The same is true for $g^{(i)}$. Also, it follows from Proposition 3 that for at most one i between 1 and ℓ the function $f^{(i)}$ satisfies $|\log(|f^{(i)}(t)|)| \ll \log_2(t)$; for all other i between 1 and

ℓ the function $f^{(i)}$ must satisfy $|\log(|f^{(i)}(t)|)| \succ \log_2(t)$. We can therefore apply Lemmas 4 and 7 to deduce that for at most one i between 1 and ℓ we have

$$\frac{\left|\frac{g^{(i+1)}(t)}{g^{(i)}(t)}\right|}{\left|\frac{f^{(i+1)}(t)}{f^{(i)}(t)}\right|} \succ \log_2^2(t)$$

and for all other i we have

$$\frac{\left|\frac{g^{(i+1)}(t)}{g^{(i)}(t)}\right|}{\left|\frac{f^{(i+1)}(t)}{f^{(i)}(t)}\right|} \xrightarrow{t\to\infty} 1.$$

Therefore

$$\frac{f(t)}{g(t)}\left|\frac{g^{(\ell)}(t)}{f^{(\ell)}(t)}\right| = \prod_{i=0}^{\ell-1} \frac{\left|\frac{g^{(i+1)}(t)}{g^{(i)}(t)}\right|}{\left|\frac{f^{(i+1)}(t)}{f^{(i)}(t)}\right|} \succ \log_2^2(t).$$

This, together with $\frac{g(t)}{f(t)} \succ F(t)$, implies

$$\left|\frac{g^{(\ell)}(t)}{f^{(\ell)}(t)}\right| \succ \frac{F(t)}{\log_2^2(t)}.$$

$\square$

3   van der Corput's Method for Estimating Exponential Sums

We recall three classical theorems on estimating exponential sums. For proofs and more detailed discussion we refer the reader to Sect. 2 in the book of Graham and Kolesnik [16].

We start with the Kusmin–Landau inequality for exponential sums (cf. [16, Theorem 2.1]).

Theorem 9 (Estimate Based on 1st Derivative) *Suppose $I \subset \mathbb{R}$ is an interval, $f \in C^1(I)$ and there exists $\lambda > 0$ such that $\lambda < |f'(t)| < (1 - \lambda)$ for all $t \in I$. Then*

$$\left|\sum_{n\in I} e(f(n))\right| \ll \lambda^{-1}.$$

The next two theorems are due to van der Corput [26, 27].

Theorem 10 (Estimate Based on 2nd Derivative) *Suppose $I \subset \mathbb{R}$ is an interval, $f \in C^2(I)$ and there are $\lambda > 0$ and $\eta \geq 1$ such that $\lambda < |f''(t)| \leq \eta\lambda$ for all $t \in I$. Then*

$$\left| \sum_{n \in I} e(f(n)) \right| \ll |I| \eta \lambda^{\frac{1}{2}} + \lambda^{-\frac{1}{2}}.$$

Theorem 11 (Estimate Based on 3rd and Higher Derivatives) *Suppose $I \subset \mathbb{R}$ is an interval, $\ell \geq 3$, $f \in C^\ell(I)$ and there are $\lambda > 0$ and $\eta \geq 1$ such that $\lambda < \left| f^{(\ell)}(t) \right| \leq \eta\lambda$ for all $t \in I$. Let $Q := 2^{\ell-2}$. Then*

$$\left| \sum_{n \in I} e(f(n)) \right| \ll |I|(\eta^2\lambda)^{\frac{1}{4Q-2}} + |I|^{1-\frac{1}{2Q}}\eta^{\frac{1}{2Q}} + |I|^{1-\frac{2}{Q}+\frac{1}{Q^2}}\lambda^{-\frac{1}{2Q}}.$$

4 Deriving Estimates for Exponential Sums Involving Functions from Hardy Fields

Proposition 12 *Let $\mathcal{H}$ be a Hardy field and assume $f_1, \ldots, f_k : [1, \infty) \to \mathbb{R}$ are in $\mathcal{H}$. For $t \in [1, \infty)$ define $f(t) := (f_1(t), \ldots, f_k(t))$ and $E(t) := \min\{|f_1(t)|, \ldots, |f_k(t)|\}$. Suppose we have*

(i) for all $i \in \{1, \ldots, k\}$ the function f_i satisfies condition (B);
(ii) $\log_2(t) \prec \log(f_i(t))$ for all $i = 1, \ldots, k$;
(iii) after reordering $f_1, \ldots, f_k$ if necessary, we have $\frac{f_{i+1}}{f_i} \succ \log_2^4(t)$ for all $i = 1, \ldots, k-1$;

Then there exists a constant $C > 0$ such that for all $M \in \mathbb{N}$, all $r, s \in \left[1, \min\left\{M\log^2(M), \frac{E(M)}{\log^4(M)}\right\}\right]$ with $r \geq s$ and all $\tau \in [-\log_2^2(M), \log_2^2(M)]^k \cap \mathbb{Z}^k$ with $\tau \neq (0, \ldots, 0)$ we have

$$\left| \sum_{n=M}^{2M} e\left(\frac{1}{r} \langle f(sn), \tau \rangle \right) \right| \leq \frac{CM}{\log^2(M)}.$$

Proof Let $\epsilon > 0$, $r, s \in \left[1, \min\left\{M\log^2(M), \frac{E(M)}{\log^4(M)}\right\}\right]$ with $r \geq s$ and $\tau \in [-\log_2^2(M), \log_2^2(M)]^k \cap \mathbb{Z}^k$ with $\tau = (\tau_1, \ldots, \tau_k) \neq (0, \ldots, 0)$ be arbitrary. Let $b(t) := \frac{1}{r}\langle f(st), \tau \rangle$. Our goal is to estimate

$$\left| \sum_{n=M}^{2M} e(b(t)) \right|$$

by using van der Corput's method of estimating exponential sums. We therefore have to find convenient estimates for the derivatives of $b(t)$ on the interval $[M, 2M]$.

Let us pick $i_0 \in \{1, \ldots, k\}$ such that $\tau_{i_0} \neq 0$ and $\tau_i = 0$ for all $i > i_0$. Define $E_0(t) := |\tau_{i_0} f_{i_0}(t)|$. Using condition (iii) we deduce that $\tau_{i_0} f_{i_0}(t)$ is the dominating term in the sum $\tau_1 f_1(t) + \cdots + \tau_{i_0} f_{i_0}(t)$ and therefore

$$E_0(t) \ll |\langle f(t), \tau \rangle| \ll E_0(t), \tag{2}$$

where the implied constants depend neither on t nor on the value of $\tau_1, \ldots, \tau_{i_0}$. A similar argument also applies to the derivatives of $\langle f(t), \tau \rangle$. Indeed, it follows from condition (iii) and Proposition 8 (with $F(t) = \log_2^2(t)$) that $\tau_{i_0} f_{i_0}^{(\ell)}(t)$ is the dominating term in $\tau_1 f_1^{(\ell)}(t) + \cdots + \tau_{i_0} f_{i_0}^{(\ell)}(t)$ and therefore

$$|\tau_{i_0} f_{i_0}^{(\ell)}(t)| \ll |\langle f^{(\ell)}(t), \tau \rangle| \ll |\tau_{i_0} f_{i_0}^{(\ell)}(t)|. \tag{3}$$

Next let $u := \inf\{c \in [0, \infty) : f_{i_0}(t) \prec t^c\}$ and pick $d \in \mathbb{R}$ such that $s = (M \log^2(M))^d$ and $h \in \mathbb{R}$ such that $r = (M \log^2(M))^h$. From the conditions on r and s we deduce that $d \in [0, \min\{1, u\}]$ and $h \in [d, \min\{1, u\}]$. We now define

$$\ell := \left\lceil u + du - h + \frac{1}{2} \right\rceil$$

and set $x := \ell - u - ud + h$. Note that $x \in \left[\frac{1}{2}, \frac{3}{2}\right)$.

In view of Proposition 3 we have

$$\frac{f_{i_0}(t)}{t^\ell \log^2(t)} \ll |f_{i_0}^{(\ell)}(t)| \ll \frac{f_{i_0}(t)}{t^\ell}. \tag{4}$$

By combining Eqs. (2)–(4) we obtain

$$\frac{E_0(t)}{t^\ell \log^2(t)} \ll |\langle f^{(\ell)}(t), \tau \rangle| \ll \frac{E_0(t)}{t^\ell}.$$

Hence the minimum of the function $b^{(\ell)}(t)$ on the interval $[M, 2M]$ is at least

$$\gg \frac{E_0(2sM)}{r(2M)^\ell \log^2(2sM)}$$

whereas the maximum is at most

$$\ll \frac{E_0(sM)}{rM^\ell}.$$

Since $E_0(t)$ is eventually increasing, we have $E_0(2sM) \gg E_0(sM)$. Also, since $E_0(t)$ has polynomial growth and $s \leq E_0(M)$ we can estimate $\log(2sM) \ll \log(M)$.

Therefore

$$\frac{E_0(2sM)}{r(2M)^\ell \log^2(2sM)} \gg \frac{E_0(sM)}{rM^\ell \log^2(M)}.$$

If we choose

$$\lambda := \frac{E_0(sM)}{rM^\ell \log^2(M)} \qquad \text{and} \qquad \eta := \log^2(M)$$

then it follows that

$$\lambda \ll b^{(\ell)}(t) \ll \eta\lambda, \qquad \forall t \in [M, 2M]. \tag{5}$$

We now distinguish between the cases $\ell = 1$, $\ell = 2$, and $\ell \geq 3$.

The Case $\ell = 1$ The case $\ell = 1$ only occurs if

$$f_{i_0}(t) \prec t^{\frac{1}{2}}.$$

Therefore

$$b'(t) \ll \eta\lambda = \frac{E_0(sM)}{rM} \leq \frac{|\tau_{i_0} f_{i_0}(sM)|}{rM} \leq \frac{\log_2^2(M) f_{i_0}(sM)}{sM} \prec 1.$$

This means we can apply Theorem 9 and obtain

$$\left| \sum_{n=M}^{2M} e(b(n)) \right| \ll \lambda^{-1}.$$

For λ^{-1} we have

$$\lambda^{-1} = \frac{rM \log^2(M)}{E_0(sM)} \leq \frac{rM \log^2(M)}{E_0(M)} \leq \frac{E(M)M}{E_0(M) \log^2(M)}.$$

Finally, since $\frac{E(M)}{E_0(M)} \leq 1$ we have

$$\lambda^{-1} \ll \frac{M}{\log^2(M)}.$$

The Case $\ell = 2$ If $\ell = 2$, then invoking Theorem 10 yields the estimate

$$\left| \sum_{n=M}^{2M} e(b(n)) \right| \ll M\eta\lambda^{\frac{1}{2}} + \lambda^{-\frac{1}{2}}. \tag{6}$$

Using $(sM)^{u-\beta} \prec f_{i_0}(sM) \prec (sM)^{u+\beta}$ for all $\beta > 0$ we can bound λ from above and below,

$$\frac{|\tau_{i_0}|s^{u-\beta}}{rM^{2-u+\beta}\log^2(M)} \ll \lambda \ll \frac{|\tau_{i_0}|s^{u+\beta}}{rM^{2-u-\beta}\log^2(M)}, \tag{7}$$

taking into account that the implied constants in the above equation depend on our choice of β.

Furthermore, since $x = \ell - u - ud + h$, $s = (M\log^2(M))^d$ and $r = (M\log^2(M))^h$ we obtain from (7) that

$$\frac{1}{M^{x+2\beta}\log^q(M)} \ll \lambda \ll \frac{\log^q(M)}{M^{x-2\beta}}. \tag{8}$$

for some sufficiently large constant $q > 1$. We can use (8) to further estimate (6) and obtain

$$M\eta\lambda^{\frac{1}{2}} + \lambda^{-\frac{1}{2}} \ll \frac{\log^{2+\frac{q}{2}}(M)M}{M^{\frac{x-2\beta}{2}}} + \frac{M^{\frac{x+2\beta}{2}}}{\log^{\frac{q}{2}}(M)}.$$

Finally, by choosing β sufficiently small and taking into account that $x \in \left[\frac{1}{2}, \frac{3}{2}\right]$, we have

$$\frac{\log^{2+\frac{q}{2}}(M)M}{M^{\frac{x-2\beta}{2}}} + \frac{M^{\frac{x+2\beta}{2}}}{\log^{\frac{q}{2}}(M)} \ll \frac{M}{\log^2(M)}.$$

The Case $\ell \geq 3$ The case $\ell \geq 3$ can be dealt with analogously to the case $\ell = 2$, only one must use Theorem 11 instead of Theorem 10. With $Q = 2^{\ell-2}$, we have

$$\left|\sum_{n=M}^{2M} e(b(n))\right| \ll M(\eta^2\lambda)^{\frac{1}{4Q-2}} + M^{1-\frac{1}{2Q}}\eta^{\frac{1}{2Q}} + M^{1-\frac{2}{Q}+\frac{1}{Q^2}}\lambda^{-\frac{1}{2Q}},$$

$$\ll \frac{M}{\log^2(M)},$$

which finishes the proof. $\qquad\square$

Theorem 13 *Let $\mathcal{H}$ be a Hardy field and assume $f_1,\ldots,f_k : [1,\infty) \to \mathbb{R}$ are in $\mathcal{H}$. For $t \in [1,\infty)$ define $f(t) := (f_1(t),\ldots,f_k(t))$ and $E(t) := \min\{|f_1(t)|,\ldots,|f_k(t)|\}$. Suppose we have*

(i) *for all $i \in \{1,\ldots,k\}$ the function f_i satisfies condition (B);*
(ii) *$\log_2(t) \prec \log(f_i(t))$ for all $i = 1,\ldots,k$;*
(iii) *after reordering $f_1,\ldots,f_k$ if necessary, we have $\frac{f_{i+1}}{f_i} \succ \log_2^4(t)$ for all $i = 1,\ldots,k-1$;*

Then there exists a constant $C > 0$ such that for all $N \in \mathbb{N}$, all $r, s \in \left[1, \min \left\{ N, \frac{E(N)}{\log^5(N)} \right\} \right]$ with $r \geq s$ and all $\tau \in [-\log_2^2(M), \log_2^2(M)]^k \cap \mathbb{Z}^k$ with $\tau \neq (0, \ldots, 0)$ we have

$$\left| \sum_{n=1}^{N} e\left(\frac{1}{r} \langle f(sn), \tau \rangle \right) \right| \leq \frac{CN}{\log(N)}.$$

Proof First we note that

$$\left| \sum_{n=1}^{N} e\left(\frac{1}{r} \langle f(sn), \tau \rangle \right) \right| \leq \frac{N}{\log(N)} + \left| \sum_{\frac{N}{\log(N)} \leq n \leq N} e\left(\frac{1}{r} \langle f(sn), \tau \rangle \right) \right|,$$

so it suffices to estimate the expression

$$\left| \sum_{\frac{N}{\log(N)} \leq n \leq N} e\left(\frac{1}{r} \langle f(sn), \tau \rangle \right) \right|.$$

Dissect the interval $\left[\frac{N}{\log(N)}, N \right]$ into $\log_2(N)$-many intervals of the form $[M, 2M]$. If $M \in \left[\frac{N}{\log(N)}, N \right]$, then $N < M \log^2(M)$ and $\frac{N}{\log^5(N)} < \frac{M}{\log^4(M)}$ and therefore

$$\left[1, \min \left\{ N, \frac{N}{\log^5(N)} \right\} \right] \subset \left[1, \min \left\{ M \log^2(M), \frac{M}{\log^4(M)} \right\} \right].$$

Hence applying Proposition 12 to each of the $\log_2(N)$-many intervals of the form $[M, 2M]$ we get

$$\left| \sum_{\frac{N}{\log(N)} \leq n \leq N} e\left(\frac{1}{r} \langle f(sn), \tau \rangle \right) \right| \ll \log_2(N) \frac{M}{\log^2(M)}$$

$$\ll \frac{N}{\log(N)}.$$

This finishes the proof. $\qquad\square$

5 Discrepancy Estimates

The following higher dimensional version of the classical Erdős-Turán inequality was discovered by Szüsz [25] and independently by Koksma [19].

Theorem 14 (See [8, 15, 25] or [11, Theorem 1.21]) *Let $k \geq 1$, let $\vartheta_n \in [0, 1)^k$, $n \in \mathbb{N}$, let $N \in \mathbb{N}$, and let $a_1, \ldots, a_k, b_1, \ldots, b_k \in [0, 1)$ with $0 \leq a_i < b_i < 1$. Then*

$$\frac{\left|\{n \leq N : \vartheta_n \in [a_1, b_1] \times \cdots \times [a_k, b_k]\}\right|}{N} = \prod_{i=1}^{k} (b_i - a_i) + R_{N,k}$$

and where for all $H \geq 1$,

$$|R_{N,k}| \leq C_k \left(\frac{1}{H+1} + \sum_{\substack{\tau \in [-H,H]^k, \\ \tau \neq (0,\ldots,0)}} \left(\prod_{i=1}^{k} \frac{1}{1 + |\tau_i|} \right) \left| \frac{1}{N} \sum_{n=1}^{N} e(\langle \vartheta_n, \tau \rangle) \right| \right).$$

Here, C_k is a constant which depends only on k.

Theorem 15 *Suppose $f_1, \ldots, f_k : [1, \infty) \to \mathbb{R}$ and $E : [0, \infty) \to (0, \infty)$ are as in the statement of Theorem 13. Then there exists a constant $C > 0$ such that for all $N \in \mathbb{N}$ and all $d \in \left[1, \min \left\{ N, \frac{E(N)}{\log^5(N)} \right\} \right]$,*

$$\left| \left|\{n \leq N : d \mid \gcd(n, \lfloor f_1(n) \rfloor, \ldots, \lfloor f_k(n) \rfloor)\}\right| - \frac{N}{d^{k+1}} \right| \leq \frac{CN}{d \log_2^2(N)}.$$

Proof Define $\vartheta_{d,n} := \left(\left\{ \frac{f_1(dn)}{d} \right\}, \ldots, \left\{ \frac{f_k(dn)}{d} \right\} \right)$, where $\{x\}$ denotes the fractional part of a real number x. We first observe that

$$\left|\{1 \leq n \leq N : d \mid \gcd(n, \lfloor f_1(n) \rfloor, \ldots, \lfloor f_k(n) \rfloor)\}\right|$$

$$= \sum_{n=1}^{N} 1_{d\mathbb{Z}}(n) 1_{d\mathbb{Z}}(\lfloor f_1(n) \rfloor) \cdots 1_{d\mathbb{Z}}(\lfloor f_k(n) \rfloor)$$

$$= \sum_{n \leq N/d} 1_{d\mathbb{Z}}(\lfloor f_1(dn) \rfloor) \cdots 1_{d\mathbb{Z}}(\lfloor f_k(dn) \rfloor)$$

$$= \left| \left\{ 1 \leq n \leq \frac{N}{d} : \vartheta_{d,n} \in \left[0, \frac{1}{d} \right)^k \right\} \right|.$$

From Theorem 13 we get that

$$\left| \sum_{n=1}^{N} e\left(\frac{1}{d} \langle f(dn), \tau \rangle \right) \right| \leq \frac{CN}{\log(N)},$$

for all $d \in \left[1, \frac{E(N)}{\log^5(N)} \right]$ and $\tau \in [-\log_2^2(N), \log_2^2(N)]^k \cap \mathbb{Z}^k$ with $\tau \neq (0, \ldots, 0)$.

We now apply Theorem 14 with $H = \log_2^2(N)$ and obtain

$$\left| \left| \left\{ 1 \leq n \leq \frac{N}{d} : \vartheta_{d,n} \in \left[0, \frac{1}{d} \right)^k \right\} \right| - \frac{N}{d^{k+1}} \right|$$

$$\leq \frac{C_k}{d} \left(\frac{N}{\log_2^2(N)} + \sum_{\substack{\tau \in [-\log_2^2(N), \log_2^2(N)]^k, \\ \tau \neq (0,\ldots,0)}} \left(\prod_{i=1}^{k} \frac{1}{1 + |\tau_i|} \right) \left| \sum_{n=1}^{N} e(\langle \vartheta_n, \tau \rangle) \right| \right)$$

$$\leq \frac{C_k}{d} \left(\frac{N}{\log_2^2(N)} + \frac{CN}{\log(N)} \sum_{\substack{\tau \in [-\log_2^2(N), \log_2^2(N)]^k, \\ \tau \neq (0,\ldots,0)}} \left(\prod_{i=1}^{k} \frac{1}{1 + |\tau_i|} \right) \right)$$

$$\ll \frac{N}{d \log_2^2(N)}.$$

□

6 Proving Theorem 2

Proposition 16 *Let $k \in \mathbb{N}$. Let $\xi_1, \xi_2, \ldots$ be a sequence of positive integers and let $E : [1, \infty) \to (0, \infty)$ be a function that satisfies $E(N) \geq \max\{\xi_n : 1 \leq n \leq N\}$ and $\log_2(t) \prec \log(E(t))$ and assume that $E(N)$ has polynomial growth (i.e., there exists $j \in \mathbb{N}$ such that $E(t) \prec t^j$). If*

$$\left| \left| \{ n \leq N : d \mid \xi_n \} \right| - \frac{N}{d^{k+1}} \right| \ll \frac{N}{d \log_2^2(N)}, \quad \forall d \in \mathbb{N} \cap \left[1, \frac{E(N)}{\log^5(N)} \right] \tag{9}$$

and

$$\left| \{ n \leq N : p \mid \xi_n \} \right| \ll \frac{N}{p}, \text{ for all primes } p \in \left(\frac{E(N)}{\log^5(N)}, E(N) \right], \tag{10}$$

then the natural density of $\{ n \in \mathbb{N} : \xi_n = 1 \}$ exists and equals $\frac{1}{\zeta(k+1)}$.

Our proof of Proposition 16 is similar to the proof of the main result in [9].

Proof Define $G(N) := \log^4(t)$. Let $D(N)$ be a slow growing function in N and let Π denote the *primorial* of $D(N)$, that is,

$$\Pi := \prod_{\substack{p \text{ prime,} \\ p \leq D(N)}} p.$$

Here, by "slow growing function" we mean that $D(N)$ converges to ∞ as $N \to \infty$, but slowly enough so that the inequality $\Pi \leq \min\left\{\frac{E(N)}{\log^5(N)}, \log_2^2(N)\right\}$ is satisfied for all $N \geq 1$.

Let $\mu(n)$ denote the classical Möbius function: For $n \in \mathbb{N}$ define

$$\mu(n) = \begin{cases} 1 & \text{if } n = 1; \\ (-1)^j & \text{if } n = p_1 \cdot p_2 \cdot \ldots \cdot p_j \text{ where } p_1, \ldots, p_j \text{ are distinct primes;} \\ 0 & \text{otherwise.} \end{cases}$$

Using the identity

$$\sum_{d \mid a} \mu(d) = \begin{cases} 1 & \text{if } a = 1 \\ 0 & \text{otherwise} \end{cases}$$

we obtain

$$|\{n \leq N : \xi_n = 1\}| = \sum_d \mu(d) \, |\{n \leq N : d \mid \xi_n\}|.$$

It will be convenient to decompose the right-hand side of the above equation into two sums $\Sigma_1 + \Sigma_2$ where

$$\Sigma_1 := \sum_{d \mid \Pi} \mu(d) \, |\{n \leq N : d \mid \xi_n\}|$$

and

$$\Sigma_2 := \sum_{d \nmid \Pi} \mu(d) \, |\{n \leq N : d \mid \xi_n\}|.$$

Our goal is to show that $\lim_{N \to \infty} \frac{1}{N} \Sigma_1 = \frac{1}{\zeta(k+1)}$ and that $\lim_{N \to \infty} \frac{1}{N} \Sigma_2 = 0$; this will finish the proof.

First, let us show $\lim_{N \to \infty} \frac{1}{N} \Sigma_1 = \frac{1}{\zeta(k+1)}$. Recall that $\Pi \leq \frac{E(N)}{\log^5(N)}$. Invoking condition (9) it thus follows that

$$\left| \frac{1}{N} \Sigma_1 - \sum_{d \mid \Pi} \frac{\mu(d)}{d^{k+1}} \right| \ll \frac{1}{\log_2^2(N)} \sum_{d \mid \Pi} \frac{1}{d}.$$

Since $\Pi \leq \log_2^2(N)$ it follows that $\sum_{d|\Pi} \frac{1}{d} = \mathcal{O}\left(\log_3(N)\right)$ and hence

$$\frac{1}{\log_2^2(N)} \sum_{d|\Pi} \frac{1}{d} \ll \frac{\log_3(N)}{\log_2^2(N)} \xrightarrow{N \to \infty} 0.$$

Therefore

$$\lim_{N \to \infty} \frac{1}{N} \Sigma_1 = \sum_{d=1}^{\infty} \frac{\mu(d)}{d^{k+1}} = \frac{1}{\zeta(k+1)}.$$

For Σ_2 we obtain the estimate

$$|\Sigma_2| = \left| \sum_{\substack{p \text{ prime,} \\ p > D(N)}} \sum_{\substack{1 \leq d \leq E(N), \\ p|d}} \mu(d) \left| \{ n \leq N : d \mid \xi_n \} \right| \right|$$

$$= \left| \sum_{n=1}^{N} \sum_{\substack{p \text{ prime,} \\ p > D(N)}} \sum_{\substack{d|\xi_n, \\ p|d}} \mu(d) \right|$$

$$\leq \sum_{n=1}^{N} \sum_{\substack{p \text{ prime,} \\ p > D(N)}} \left| \sum_{\substack{d|\xi_n, \\ p|d}} \mu(d) \right|.$$

It is well known (and easy to show) that

$$\sum_{\substack{d|a, \\ p|d}} \mu(d) = \begin{cases} 1 & \text{if } a = p^j \text{ for some } j \in \mathbb{N}, \\ 0 & \text{otherwise} \end{cases}$$

and hence

$$\sum_{n=1}^{N} \sum_{\substack{p \text{ prime,} \\ p > D(N)}} \left| \sum_{\substack{d|\xi_n, \\ p|d}} \mu(d) \right| \leq \sum_{n=1}^{N} \sum_{\substack{p \text{ prime,} \\ p > D(N), \\ p|\xi_n}} 1$$

$$\leq \sum_{\substack{p \text{ prime,} \\ p > D(N)}} \left| \{ n \leq N : p \mid \xi_n \} \right|.$$

Putting everything together we obtain

$$|\Sigma_2| \leq \sum_{\substack{p \text{ prime,} \\ p > D(N)}} \left|\{n \leq N : p \mid \xi_n\}\right|.$$

Again, we split the right-hand side of the above equation into two more manageable sums $\Sigma_{2,1} + \Sigma_{2,2}$, where

$$\Sigma_{2,1} := \sum_{\substack{p \text{ prime,} \\ D(N) < p \leq \frac{E(N)}{\log^5(N)}}} \left|\{n \leq N : p \mid \xi_n\}\right|$$

and

$$\Sigma_{2,2} := \sum_{\substack{p \text{ prime,} \\ \frac{E(N)}{\log^5(N)} < p \leq E(N)}} \left|\{n \leq N : p \mid \xi_n\}\right|.$$

Using condition (9) for the sum $\Sigma_{2,1}$ and using condition (10) for the sum $\Sigma_{2,2}$ we obtain the estimates

$$\frac{1}{N}\Sigma_{2,1} \ll \sum_{\substack{p \text{ prime,} \\ D(N) < p \leq \frac{E(N)}{\log^5(N)}}} \frac{1}{p^2} + \frac{1}{\log_2^2(N)} \sum_{p \leq E(N)} \frac{1}{p}$$

and

$$\frac{1}{N}\Sigma_{2,2} \leq \sum_{\substack{p \text{ prime,} \\ \frac{E(N)}{\log^5(N)} < p \leq E(N)}} \frac{1}{p}.$$

Using

$$\lim_{N \to \infty} \left| \sum_{\substack{p \text{ prime,} \\ p \leq N}} \frac{1}{p} - \log_2 N \right| = M,$$

(where M is the Meissel–Mertens constant), we can estimate

$$\frac{1}{\log_2^2(N)} \sum_{p \leq E(N)} \frac{1}{p} \ll \frac{\log_2(E(N))}{\log_2^2(N)} \ll \frac{1}{\log_2(N)},$$

and

$$\lim_{N \to \infty} \frac{1}{N} \Sigma_{2,2} \ll \log(\log(E(N))) - \log\left(\log\left(\frac{E(N)}{\log^5(N)}\right)\right)$$

$$\ll -\log\left(1 - \frac{\log(\log^5(N))}{\log(E(N))}\right)$$

$$\ll \frac{\log_2(N)}{\log(E(N))}.$$

Therefore

$$\lim_{N \to \infty} \frac{1}{N} |\Sigma_2| \le \lim_{N \to \infty} \frac{1}{N} \Sigma_{2,1} + \lim_{N \to \infty} \frac{1}{N} \Sigma_{2,2} = 0.$$

$\square$

Corollary 17 *Let $\mathcal{H}$ be a Hardy field and suppose $f_1, \ldots, f_k \in \mathcal{H}$ satisfy conditions (A) and (B) and $\log_2(t) \prec \log(f_i(t))$ for all $i = 1, \ldots, k$. Also, assume that $\frac{f_{i+1}}{f_i} \succ \log_2^4(t)$ for all $i = 1, \ldots, k-1$. Then the natural density of the set*

$$\{n \in \mathbb{N} : \gcd(n, \lfloor f_1(n) \rfloor, \ldots, \lfloor f_k(n) \rfloor) = 1\}\big|$$

exists and equals $\frac{1}{\zeta(k+1)}$.

Proof We define

$$\xi_n := \gcd(n, \lfloor f_1(n) \rfloor, \ldots, \lfloor f_k(n) \rfloor)$$

and

$$E(t) := \min\{|f_1(t)|, \ldots, |f_k(t)|\}.$$

Trivially, $(\xi_n)_{n \in \mathbb{N}}$ satisfies condition (10). Moreover, it follows from Theorem 15 that $(\xi_n)_{n \in \mathbb{N}}$ satisfies condition (9). Therefore, in view of Proposition 16, we have $d(\{n \in \mathbb{N} : \xi_n = 1\}) = \frac{1}{\zeta(k+1)}$. $\square$

Lemma 18 *Let $\mathcal{H}$ be a Hardy field and suppose $f \in \mathcal{H}$ satisfies*

$$\log(t) \log_4(t) \prec f(t) \prec \frac{t}{\log_2(t)}.$$

Define $S(N, d) := \big|\{1 \le n \le N : d \mid \gcd(n, \lfloor f(n) \rfloor)\}\big|$. Then

$$\lim_{D \to \infty} \limsup_{N \to \infty} \frac{1}{N} \sum_{\substack{p \ prime \\ D < p \le f(N)}} S(N, p) = 0. \tag{11}$$

We remark that Lemma 18 can be derived from the proof of Theorem 2 in [12]. For the convenience of the reader we include a separate proof here, where we follow the arguments used by Erdős and Lorentz in [12].

Proof Let $N \in \mathbb{N}$ be arbitrary. Define $I_m := \{1 \le t \le N : f(t) \in [m, m+1)\}$ and note that I_m is a finite subinterval of $\mathbb{R}$. Let k_m denote the length of I_m. Note that the contribution of $I_m \cap \mathbb{N}$ to the size of $S(N, p)$ is given by $I_m \cap (p\mathbb{N})$. Hence this contribution is zero if $p \nmid m$ and it does not exceed $\frac{k_m}{p} + 1$ otherwise. It follows that

$$
\sum_{\substack{p \text{ prime} \\ D < p \le f(N)}} S(N, p) \le \sum_{\substack{p \text{ prime} \\ D < p \le f(N)}} \sum_{\substack{1 \le m \le f(N) \\ p \mid m}} \frac{k_m}{p} + 1
$$

$$
\le \sum_{\substack{p \text{ prime} \\ D < p \le f(N)}} \sum_{1 \le m \le \frac{f(N)}{p}} \frac{k_{pm}}{p} + 1
$$

$$
\le \sum_{\substack{p \text{ prime} \\ D < p \le f(N)}} \sum_{1 \le m \le \frac{f(N)}{p}} \frac{k_{pm}}{p} + \sum_{\substack{p \text{ prime} \\ D < p \le f(N)}} \frac{f(N)}{p}.
$$

Define $\ell_0 := \lfloor \frac{f(N)}{p} \rfloor$. We can write

$$
\sum_{\substack{p \text{ prime} \\ D < p \le f(N)}} \sum_{1 \le m \le \frac{f(N)}{p}} \frac{k_{pm}}{p} + \sum_{\substack{p \text{ prime} \\ D < p \le f(N)}} \frac{f(N)}{p} = \Sigma_1 + \Sigma_2 + \Sigma_3
$$

where

$$
\Sigma_1 := \sum_{\substack{p \text{ prime} \\ D < p \le f(N)}} \frac{f(N)}{p},
$$

$$
\Sigma_2 := \sum_{\substack{p \text{ prime} \\ D < p \le f(N)}} \sum_{1 \le m \le \ell_0 - 1} \frac{k_{pm}}{p},
$$

$$
\Sigma_3 := \sum_{\substack{p \text{ prime} \\ D < p \le f(N)}} \frac{k_{p\ell_0}}{p}.
$$

We have to estimate each of the sums Σ_1, Σ_2, and Σ_3 individually. We start with Σ_1. Since $f(t) \prec \frac{t}{\log_2(t)}$, we conclude that

$$
\Sigma_1 \le f(N) \sum_{p \le N} \frac{1}{p} \ll f(N) \log_2(N) \prec N
$$

and therefore

$$\lim_{N \to \infty} \frac{1}{N} \Sigma_1 = 0. \tag{12}$$

Next, we derive an estimate for Σ_2. Since $f(t)$ is eventually strictly increasing, the inverse function of f^{-1} is well defined on some half-line $[t_0, \infty)$. We have the identity $k_m = f^{-1}(m+1) - f^{-1}(m)$. This means that using the inverse function theorem and the mean value theorem we see that there exists a number $\xi_m \in I_m$ such that

$$k_m = \frac{1}{f'(\xi_m)}.$$

In view of Proposition 3, the derivative of f is eventually decreasing and therefore

$$k_m \ll k_l$$

for $l \geq m$. From this we derive that

$$\sum_{1 \leq m \leq \ell_0 - 1} \frac{k_{pm}}{p} \ll \frac{1}{p} \left(\frac{k_p}{p} + \frac{k_{p+1}}{p} + \frac{k_{p+2}}{p} + \cdots + \frac{k_{\ell_0 p - 1}}{p} \right) \ll \frac{N}{p^2}.$$

Hence

$$\Sigma_2 \ll \sum_{p > D} \frac{N}{p^2}$$

which proves that

$$\lim_{D \to \infty} \lim_{N \to \infty} \frac{1}{N} \Sigma_2 = 0. \tag{13}$$

Next, let $\epsilon > 0$ be arbitrary and define $P' := \{p \text{ prime} : D < p < f(N) + 1 - \frac{\log_2(f(N))}{\epsilon}\}$ and $P'' := \{p \text{ prime} : f(N) + 1 - \frac{\log_2(f(N))}{\epsilon} \leq p \leq f(N)\}$. We now split the sum Σ_3 into two more sums Σ_3' and Σ_3'', where

$$\Sigma_3' := \sum_{p \in P'} \frac{k_{p\ell_0}}{p}$$

and

$$\Sigma_3'' := \sum_{p \in P''} \frac{k_{p\ell_0}}{p}.$$

Arguing as before we have for all $p \in P'$,

$$k_{p\ell_0} \ll \frac{1}{f(N) - p\ell_0 + 1}(k_{p\ell_0} + k_{p\ell_0+1} + \cdots + k_{f(N)})$$
$$\ll \frac{N}{f(N) - p\ell_0 + 1}$$
$$= \frac{N\epsilon}{\log_2(f(N))}.$$

This yields the following estimate for Σ_3':

$$\Sigma_3' \ll \sum_{p \leq f(N)} \frac{N\epsilon}{\log_2(f(N))p} \ll N\epsilon.$$

Let $\Pi := \prod_{p \in P''} p$. Certainly,

$$\Pi \leq (f(N))^{\frac{\log_2(N)}{\epsilon}}.$$

Using the well-known estimate

$$\sum_{p|n} \frac{1}{p} \ll \log_3(n)$$

we obtain

$$\Sigma_3'' \leq k_{f(N)} \sum_{p|\Pi} \frac{1}{p} \ll \frac{1}{f'(N)} \log_3(\Pi) \ll \frac{1}{f'(N)} \log_3(f(N)).$$

Using L'Hospital's rule, we see that

$$\frac{tf'(t)}{\log_3(f(t))} \gg \frac{f(t)}{\log(t)\log_3(f(t))} \gg \frac{f(t)}{\log(t)\log_3(\log^k(t))} \gg \frac{f(t)}{\log(t)\log_4(t)},$$

and therefore, using $\log(t)\log_4(t) \prec f(t)$, we get

$$\frac{1}{f'(N)} \log_3(f(N)) \prec N.$$

Putting everything together yields

$$\frac{1}{N}\Sigma_3 = \frac{1}{N}\Sigma_3' + \frac{1}{N}\Sigma_3'' \ll \epsilon.$$

Since ϵ was chosen arbitrarily, we deduce that

$$\lim_{N \to \infty} \frac{1}{N} \Sigma_3 = 0. \tag{14}$$

Finally, combining Eqs. (12)–(14) completes the proof. $\qquad\square$

Lemma 19 *Let $\mathcal{H}$ be a Hardy field and suppose $f_1, \ldots, f_k \in \mathcal{H}$ satisfy condition (B). Also assume that $f_1(t) \succ \log(t)$ and $\frac{f_{i+1}}{f_i} \succ 1$ for all $i = 1, \ldots, k - 1$. For $D \in \mathbb{N}$ define*

$$A_{D,N} := \left\{ 1 \le n \le N : \gcd(n, \lfloor f_1(n) \rfloor, \ldots, \lfloor f_k(n) \rfloor, D!) = 1 \right\}.$$

Then

$$\lim_{N \to \infty} \frac{|A_{D,N}|}{N} = \sum_{d \mid D!} \frac{\mu(d)}{d^{k+1}}.$$

For the proof of Lemma 19 we need the following Proposition, which is an immediate corollary of [5, Theorem 1.8].

Proposition 20 *Let $k \in \mathbb{N}$, let $\mathcal{H}$ be a Hardy field, and let $g_1, \ldots, g_k \in \mathcal{H}$ satisfy $g_1(t) \succ \log(t)$ and $\frac{g_{i+1}}{g_i} \succ 1$ for all $i = 1, \ldots, k - 1$. Then the sequence $(\{g_1(n)\}, \ldots, \{g_k(n)\})$, $n \in \mathbb{N}$, is equidistributed in $[0, 1)^k$.*

Proof of Lemma 19 Define $\vartheta_{d,n} := \left(\left\{ \frac{f_1(dn)}{d} \right\}, \ldots, \left\{ \frac{f_k(dn)}{d} \right\} \right)$, where $\{x\}$ denotes the fractional part of a real number x. We first observe that

$$|A_{D,N}| = \sum_{d \mid D!} \mu(d) \left| \left\{ 1 \le n \le N : d \mid \gcd(n, \lfloor f_1(n) \rfloor, \ldots, \lfloor f_k(n) \rfloor) \right\} \right|$$

$$= \sum_{d \mid D!} \sum_{n=1}^{N} 1_{d\mathbb{Z}}(n) 1_{d\mathbb{Z}}(\lfloor f_1(n) \rfloor) \cdots 1_{d\mathbb{Z}}(\lfloor f_k(n) \rfloor)$$

$$= \sum_{d \mid D!} \sum_{n \le N/d} 1_{d\mathbb{Z}}(\lfloor f_1(dn) \rfloor) \cdots 1_{d\mathbb{Z}}(\lfloor f_k(dn) \rfloor)$$

$$= \sum_{d \mid D!} \left| \left\{ 1 \le n \le \frac{N}{d} : \vartheta_{d,n} \in \left[0, \frac{1}{d} \right)^k \right\} \right|.$$

Applying Proposition 20 to the functions $g_1(t) = \frac{f_1(dt)}{d}, \ldots, g_k(t) = \frac{f_k(dt)}{d}$, we deduce that

$$\lim_{N \to \infty} \left| \left\{ 1 \le n \le \frac{N}{d} : \vartheta_{d,n} \in \left[0, \frac{1}{d} \right)^k \right\} \right| = \frac{1}{d^{k+1}},$$

which proves the claim. $\qquad\square$

Theorem 21 *Let $\mathcal{H}$ be a Hardy field and suppose $f_1, \ldots, f_k \in \mathcal{H}$ satisfy conditions (A) and (B). Also assume that $f_1(t) \prec \frac{t}{\log_2(t)}$ and $\frac{f_{i+1}}{f_i} \succ 1$ for all $i = 1, \ldots, k-1$. Then the natural density of the set*

$$\{n \in \mathbb{N} : \gcd(n, \lfloor f_1(n) \rfloor, \ldots, \lfloor f_k(n) \rfloor) = 1\}\big|$$

exists and equals $\frac{1}{\zeta(k+1)}$.

Proof Let $D \in \mathbb{N}$. We define

$$A_N := \{1 \leq n \leq N : \gcd(n, \lfloor f_1(n) \rfloor, \ldots, \lfloor f_k(n) \rfloor) = 1\}$$

and

$$A_{D,N} := \{1 \leq n \leq N : \gcd(n, \lfloor f_1(n) \rfloor, \ldots, \lfloor f_k(n) \rfloor, D!) = 1\}.$$

Certainly, $A_N \subset A_{D,N}$. It follows directly from Lemma 19 that

$$\lim_{D \to \infty} \lim_{N \to \infty} \frac{|A_{D,N}|}{N} = \frac{1}{\zeta(k+1)}.$$

Hence, it suffices to show that

$$\lim_{D \to \infty} \limsup_{N \to \infty} \frac{|A_{D,N} \setminus A_N|}{N} = 0.$$

Let

$$S'(N, d) := \big|\{1 \leq n \leq N : d \mid \gcd(n, \lfloor f_1(n) \rfloor, \ldots, \lfloor f_k(n) \rfloor)\}\big|$$

and let

$$S(N, d) := \big|\{1 \leq n \leq N : d \mid \gcd(n, \lfloor f_1(n) \rfloor)\}\big|.$$

It is clear that

$$|A_{D,N} \setminus A_N| \leq \sum_{\substack{p \text{ prime} \\ D < p \leq \lfloor f_1(n) \rfloor}} S'(N, p).$$

However, we have that

$$\sum_{\substack{p \text{ prime} \\ D < p \leq \lfloor f_1(n) \rfloor}} S'(N, p) \leq \sum_{\substack{p \text{ prime} \\ D < p \leq \lfloor f_1(n) \rfloor}} S(N, p)$$

and it follows from Lemma 18 that

$$\lim_{D \to \infty} \limsup_{N \to \infty} \frac{1}{N} \sum_{\substack{p \text{ prime} \\ D < p \leq \lfloor f_1(n) \rfloor}} S(N, p) = 0.$$

$\square$

Proof of Theorem 2 Let $f_1, \ldots, f_k \in \mathcal{H}$ satisfy conditions (A)–(C).

If $f_1(t) \prec \frac{t}{\log_2(t)}$, then the conclusion of Theorem 2 follows from Theorem 21. On the other hand, if $f_1(t) \gg \frac{t}{\log_2(t)}$, then the conclusion of Theorem 2 follows from Corollary 17. $\square$

7 Some Open Questions

We end this paper with formulating some open questions.

7.1 The first question concerns a natural extension of Watson's [28] original result.

Question 1 *Let $\alpha_1, \ldots, \alpha_k$ be k irrational numbers. Is it true that the natural density of the set*

$$\{n \in \mathbb{N} : \gcd\left(n, \lfloor n\alpha_1 \rfloor, \lfloor n^2\alpha_2 \rfloor, \ldots, \lfloor n^k\alpha_k \rfloor\right) = 1\}$$

exists and equals $\frac{1}{\zeta(k+1)}$?

7.2 Let $\mathcal{H}$ be a Hardy field, let $f_1, \ldots, f_k \in \mathcal{H}$ and consider the condition:

(C′) $\frac{f_{i+1}}{f_i} \succ 1$ for all $i = 1, \ldots, k - 1$.

Question 2 *In the statement of Theorem 2, can one replace condition (C) with condition (C′)?*

7.3 By slightly generalizing the methods used by Estermann in [13], one can prove the following theorem:

Theorem 22 *For any k-tuple $(\alpha_1, \ldots, \alpha_k)$ of rationally independent irrational numbers the natural density of the set*

$$\{n \in \mathbb{N} : \gcd(\lfloor n\alpha_1 \rfloor, \ldots, \lfloor n\alpha_k \rfloor) = 1\}$$

exists and equals $\frac{1}{\zeta(k)}$ here and in Question 3.

This leads to the following question.

Question 3 *Let $\mathcal{H}$ be a Hardy field. Suppose $f_1,\ldots,f_k \in \mathcal{H}$ satisfy conditions (A), (B), and (C'). Is it true that the natural density of the set*

$$\left\{n \in \mathbb{N} : \gcd(\lfloor f_1(n)\rfloor,\ldots,\lfloor f_k(n)\rfloor) = 1\right\}$$

exists and equals $\frac{1}{\zeta(k)}$?

7.4 We conclude this paper with a question which addresses a possible weakening of condition (B) in Theorems 1 and 2.

Question 4 *Can condition (B) be replaced by condition (B') (see page 111) in Theorems 1 and 2?*

Acknowledgements The authors would like to thank the anonymous referees for their helpful comments and Christian Elsholtz for the efficient handling of the submission process.
The first author gratefully acknowledges the support of the NSF under grant DMS-1500575.

References

1. S. Abramovich, Y.Y. Nikitin, On the probability of co-primality of two natural numbers chosen at random (Who was the first to pose and solve this problem?). arXiv e-prints (2016). http://arxiv.org/abs/1608.05435
2. V. Bergelson, G. Kolesnik, Y. Son, Uniform distribution of subpolynomial functions along primes and applications. J. Anal. Math. arXiv e-prints (2015, to appear). http://arxiv.org/abs/1503.04960
3. M. Boshernitzan, An extension of Hardy's class L of "orders of infinity". J. Anal. Math. **39**, 235–255 (1981)
4. M. Boshernitzan, New "orders of infinity". J. Anal. Math. **41**, 130–167 (1982)
5. M.D. Boshernitzan, Uniform distribution and Hardy fields. J. Anal. Math. **62**, 225–240 (1994)
6. E. Cesàro, Questions proposées, # 75. Mathesis **1**, 184 (1981)
7. E. Cesàro, Solutions de questions proposées, question 75. Mathesis **3**, 224–225 (1983)
8. T. Cochrane, Trigonometric approximation and uniform distribution modulo one. Proc. Am. Math. Soc. **103**, 695–702 (1988)
9. F. Delmer, J.-M. Deshouillers, On the probability that n and $[n^c]$ are coprime. Period. Math. Hungar. **45**, 15–20 (2002)
10. G. Dirichlet, Mathematische Werke. Band II, Herausgegeben auf Veranlassung der Königlich Preussischen Akademie der Wissenschaften von L. Kronecker, Druck und Verlag von Georg Reimer (1897)
11. M. Drmota, R.F. Tichy, *Sequences, Discrepancies and Applications*. Lecture Notes in Mathematics, vol. 1651 (Springer, Berlin, 1997)
12. P. Erdős, G.G. Lorentz, On the probability that n and $g(n)$ are relatively prime. Acta Arith. **5**, 35–44 (1959)
13. T. Estermann, On the number of primitive lattice points in a parallelogram. Can. J. Math. **5**, 456–459 (1953)
14. N. Frantzikinakis, Equidistribution of sparse sequences on nilmanifolds. J. Anal. Math. **109**, 353–395 (2009)
15. P.J. Grabner, Erdős-Turán type discrepancy bounds. Monatsh. Math. **111**, 127–135 (1991)
16. S.W. Graham, G. Kolesnik, *Van der Corput's Method of Exponential Sums*. London Mathematical Society Lecture Note Series, vol. 126 (Cambridge University Press, Cambridge, 1991)

17. G.H. Hardy, Properties of logarithmico-exponential functions. Proc. Lond. Math. Soc. **S2–10**, 54–90 (1912)
18. G.H. Hardy, *Orders of Infinity. The Infinitärcalcül of Paul du Bois-Reymond* (Hafner Publishing, New York, 1971). Reprint of the 1910 edition, Cambridge Tracts in Mathematics and Mathematical Physics, No. 12
19. J.F. Koksma, Some theorems on Diophantine inequalities, Scriptum no. 5, Math. Centrum Amsterdam (1950)
20. J. Lambek, L. Moser, On integers n relatively prime to $f(n)$. Can. J. Math. **7**, 155–158 (1955)
21. F. Mertens, Über einige asymptotische Gesetze der Zahlentheorie. J. Reine Angew. Math. **77**, 289–338 (1874)
22. J. Spilker, Die Fastperiodizität der Watson-Funktion. Arch. Math. (Basel) **74**, 26–29 (2000)
23. J.J. Sylvester, *The Collected Mathematical Papers. Volume III (1870–1883)* (Cambridge University Press, Cambridge, 1909), pp. 672–676
24. J.J. Sylvester, *The Collected Mathematical Papers. Volume IV (1882–1897)* (Cambridge University Press, Cambridge, 1912), pp. 84–87
25. P. Szüsz, Über ein Problem der Gleichverteilung, in *Comptes Rendus du Premier Congrès des Mathématiciens Hongrois*, 27 Août–2 Septembre 1950 (Akadémiai Kiadó, Budapest, 1952), pp. 461–472
26. J.G. van der Corput, Neue zahlentheoretische Abschätzungen. Math. Ann. **89**, 215–254 (1923)
27. J.G. van der Corput, Neue zahlentheoretische Abschätzungen (Zweite Mitteilung). Math. Z. **29**, 397–426 (1929)
28. G.L. Watson, On integers n relatively prime to $[\alpha n]$. Can. J. Math. **5**, 451–455 (1953)

On the Uniform Theory of Lacunary Series

István Berkes

Dedicated to Professor Robert Tichy on the occasion of his 60th birthday

Abstract The theory of lacunary series starts with Weierstrass' famous example (1872) of a continuous, nondifferentiable function and now we have a wide and nearly complete theory of lacunary subsequences of classical orthogonal systems, as well as asymptotic results for thin subsequences of general function systems. However, many applications of lacunary series in harmonic analysis, orthogonal function theory, Banach space theory, etc. require uniform limit theorems for such series, i.e., theorems holding simultaneously for a class of lacunary series, and such results are much harder to prove than dealing with individual series. The purpose of this paper is to give a survey of uniformity theory of lacunary series and discuss new results in the field. In particular, we study the permutation-invariance of lacunary series and their connection with Diophantine equations, uniform limit theorems in Banach space theory, resonance phenomena for lacunary series, lacunary sequences with random gaps, and the metric discrepancy theory of lacunary sequences.

1 Introduction

Let (n_k) be a sequence of positive integers satisfying the Hadamard gap condition

$$n_{k+1}/n_k \geq q > 1 \quad (k = 1, 2, \ldots). \tag{1}$$

Salem and Zygmund [59] proved that $(\cos 2\pi n_k x)$ obeys the central limit theorem

$$\lim_{N \to \infty} \lambda \left\{ x \in (0, 1) : (N/2)^{-1/2} \sum_{k=1}^{N} \cos 2\pi n_k x \leq t \right\} = (2\pi)^{-1/2} \int_{-\infty}^{t} e^{-u^2/2} du \tag{2}$$

I. Berkes (✉)

Graz University of Technology, Institute of Statistics, Kopernikusgasse 24, 8010 Graz, Austria

e-mail: berkes@tugraz.at

© Springer International Publishing AG 2017

C. Elsholtz, P. Grabner (eds.), *Number Theory – Diophantine Problems, Uniform Distribution and Applications*, DOI 10.1007/978-3-319-55357-3_6

where λ denotes the Lebesgue measure. Under the same gap condition Erdős and Gál [33] proved that $(\cos 2\pi n_k x)$ obeys the law of the iterated logarithm

$$\limsup_{N\to\infty} (N\log\log N)^{-1/2} \sum_{k=1}^{N} \cos 2\pi n_k x = 1 \qquad \text{a.e.} \tag{3}$$

These results show that under the gap condition (1) the functions $\cos 2\pi n_k x$ behave like independent random variables. Morgenthaler [53] and Weiss [67] showed that the analogue of (2) and (3) holds for any uniformly bounded orthonormal system except that one needs a much stronger gap condition for (n_k), the Gaussian limit distribution in (2) becomes mixed Gaussian, and the limsup in (3) becomes a nonconstant function of x. As shown by Gaposhkin [41], at the cost of introducing an extra centering factor, even the orthogonality can be dropped here. Specifically, he proved (see also Chatterji [30, 31]) that if a sequence (X_n) of r.v.'s satisfies $\sup_n EX_n^2 < \infty$, then one can find a subsequence (X_{n_k}) and r.v.'s X and $Y \geq 0$ such that

$$\frac{1}{\sqrt{N}} \sum_{k \leq N} (X_{n_k} - X) \xrightarrow{d} N(0, Y) \tag{4}$$

and

$$\limsup_{N\to\infty} \frac{1}{\sqrt{2N\log\log N}} \sum_{k \leq N} (X_{n_k} - X) = Y^{1/2} \qquad \text{a.s.,} \tag{5}$$

where $\xrightarrow{d}$ denotes convergence in distribution and $N(0, Y)$ denotes the distribution of the r.v. $Y^{1/2}\zeta$ where ζ is a standard normal r.v. independent of Y. Komlós [50] proved that under $\sup_n E|X_n| < \infty$ there exist a subsequence (X_{n_k}) and an integrable r.v. X such that

$$\lim_{N\to\infty} \frac{1}{N} \sum_{k=1}^{N} X_{n_k} = X \qquad \text{a.s.} \tag{6}$$

and Chatterji [28] showed that under $\sup_n E|X_n|^p < \infty, 0 < p < 2$ the conclusion of the previous theorem can be changed to

$$\lim_{N\to\infty} \frac{1}{N^{1/p}} \sum_{k=1}^{N} (X_{n_k} - X) = 0 \qquad \text{a.s.} \tag{7}$$

for some X with $E|X|^p < \infty$. Relations (4)–(7) are the analogues of the central limit theorem, law of the iterated logarithm, strong law of large numbers, and Marczinkiewicz's strong law, respectively, in a slightly modified form. Note the randomization in all these examples: the role of the mean and variance of the subsequence (X_{n_k}) is played by random variables X, Y. On the basis of these and several other examples, Chatterji [29] formulated the following heuristic principle:

Subsequence Principle *Let T be a probability limit theorem valid for all sequences of i.i.d. random variables belonging to an integrability class L defined by the finiteness of a norm $\| \cdot \|_L$. Then if (X_n) is an arbitrary (dependent) sequence of random variables satisfying $\sup_n \|X_n\|_L < +\infty$, then there exists a subsequence (X_{n_k}) satisfying T in a randomized form.*

In a profound paper, Aldous [11] proved the validity of this principle for all distributional and almost sure limit theorems for i.i.d. random variables, subject to mild technical conditions. For a precise formulation, we refer to [11]. This result closes a long series of investigations and, together with a wide asymptotic theory of "concrete" (e.g., Hadamard lacunary) subsequences of classical function systems such as the trigonometric and Walsh system, orthogonal polynomials, dilated series $\sum c_k f(kx)$ (for a survey see Gaposhkin [41]), gives a quite satisfactory picture of the asymptotic behavior of lacunary series. Despite this fact, some important questions in the field remain open. By a classical result of Menshov [52], every orthonormal system (φ_n) contains a subsequence (φ_{n_k}) which is a convergence system, i.e., under $\sum_{k=1}^{\infty} c_k^2 < \infty$ the series $\sum_{k=1}^{\infty} c_k \varphi_{n_k}$ converges almost everywhere. Since every sequence of independent random variables with mean 0 and variance 1 has this property, Aldous' theorem guarantees, for every fixed (c_k) with $\sum_{k=1}^{\infty} c_k^2 < \infty$, the existence of a subsequence (φ_{n_k}) such that $\sum_{k=1}^{\infty} c_k \varphi_{n_k}$ converges a.e., but the subsequence (n_k) depends on the coefficients c_k and the existence of a subsequence having the above convergence property simultaneously for all (c_k) does not follow. A related famous problem of the theory of orthogonal sequences (see Uljanov [65], p. 48) was if every orthonormal sequence (φ_n) has a subsequence which is a convergence system after any rearrangement of its terms. For a specific permutation of its terms, this follows again from Menshov's theorem, but whether there exists a subsequence (φ_{n_k}) having this property simultaneously after all rearrangements remained open until settled by Komlós [51]. By a classical result of Kadec and Pełczyński [49], every normalized weakly null sequence in L^p, $p > 2$ contains a subsequence spanning a subspace isomorphic to ℓ^2 or ℓ^p and another profound result of Aldous [12] states that every closed infinite dimensional subspace H of L^1 contains a subspace isomorphic to ℓ^p for some $1 \leq p \leq 2$. Equivalently, these results concern the behavior of norms $\| \sum_{k=1}^{n} c_k f_{n_k} \|$ for lacunary sequences (f_{n_k}) in $L^p, p > 2$ or in H with the additional requirement that the estimates must be uniform for all $(c_1, \ldots, c_k) \in \mathbb{R}^k$, $k = 1, 2, \ldots$. Since an i.i.d. sequence is a symmetric structure, from the subsequence principle one would expect that limit theorems for lacunary series are permutation-invariant but, as we will see, this is not the case and establishing permutation-invariance of series $\sum c_k f_{n_k}$ requires considerable effort. Note that the permutation-invariance of lacunary series is also a uniformity problem, requiring the study of series $\sum c_k f_{\sigma(k)}$ simultaneously for all permutations σ of $\mathbb{N}$.

Studying trigonometric series

$$\sum_{k=1}^{\infty} (a_k \cos n_k x + b_k \sin n_k x)$$

with random frequencies n_k (such series provide an important tool, e.g., for constructing counterexamples in analysis) also requires studying a class of trigonometric series simultaneously, namely for all realizations $(n_k(\omega))$, $\omega \in \Omega$ in the underlying probability space $(\Omega, \mathcal{F}, P)$. Finally, the metric discrepancy theory of lacunary series, a field where considerable progress has taken place in recent years, also involves uniform limit theorems for lacunary sequences. The purpose of the present paper is to give a general study of such uniformity problems and to formulate recent results in the field.

2 Limit Random Measure and a Structure Theorem

Using the terminology of [20], we call a sequence (X_n) of random variables *determining* if it has a limit distribution relative to any set A in the probability space with $P(A) > 0$, i.e., for any $A \subset \Omega$ with $P(A) > 0$ there exists a distribution function F_A such that

$$\lim_{n \to \infty} P(X_n \le t \mid A) = F_A(t)$$

for all continuity points t of F_A. By an extension of the Helly–Bray theorem (see [20]), every tight sequence of r.v.'s contains a determining subsequence. As is shown in [11, 20], for any determining sequence (X_n) there exists a random measure μ (i.e., a measurable map from the underlying probability space $(\Omega, \mathcal{F}, \mathcal{P})$ to the class $\mathcal{M}$ of probability measures on $\mathbb{R}$ equipped with the Lévy metric) such that for any A with $P(A) > 0$ and any continuity point t of F_A we have

$$F_A(t) = \mathbb{E}_A(\mu(-\infty, t]) \tag{8}$$

where $\mathbb{E}_A$ denotes conditional expectation given A. We call μ the *limit random measure* of (X_n). Let (Y_n) be a sequence of random variables, defined on the same probability space, conditionally i.i.d. with respect to their tail σ-field $\mathcal{F}_\infty$ with conditional distribution μ. (For the construction of such an (Y_n) one may need to enlarge the probability space.) Clearly, (Y_n) is exchangeable (i.e., for any permutation $\sigma : \mathbb{N} \to \mathbb{N}$ of the positive integers, the sequence $(Y_{\sigma(n)})$ has the same distribution as (Y_n)) and satisfies limit theorems of i.i.d. random variables in a mixed form. For example, if $EY_1^2 < \infty$ and $Y = E(Y_1 \mid \mathcal{F}_\infty), Z = \mathrm{Var}\,(Y_1 \mid \mathcal{F}_\infty)$, then

$$N^{-1/2} \sum_{k=1}^{N} (Y_k - Y) \xrightarrow{d} N(0, Z)$$

and

$$\limsup_{N \to \infty} (2N \log \log N)^{-1/2} \sum_{k=1}^{N} (Y_k - Y) = Z^{1/2} \quad \text{a.s.}$$

If $E|Y_1| < \infty$ and $Y = E(Y_1 \mid \mathcal{F}_\infty)$, then

$$\lim_{N \to \infty} N^{-1} \sum_{k=1}^{N} Y_k = Y \quad \text{a.s.}$$

We call (Y_n) the *limit exchangeable sequence* of (X_n). Aldous' subsequence theorem, mentioned in Sect. 1, states that if (X_n) is a tight sequence of random variables with limit random measure μ, then sufficiently thin subsequences of (X_n) satisfy all limit theorems valid for (Y_n). To make this precise, one has to formalize the concept "limit theorem," which is rather technical, for the details we refer to Aldous [11].

The simplest exchangeable sequences are finite mixtures of i.i.d. sequences, i.e., sequences (X_n) for which there exists a finite partition $\Omega = \cup_{j=1}^{r} B_j$ of the underlying probability space such that, restricted to any (B_j), (X_n) is an i.i.d. sequence. The following theorem, proved by Berkes and Péter [17], shows that given a tight sequence (X_n) of random variables, sufficiently thin subsequences (X_{n_k}) can be approximated arbitrary closely by finite mixtures of i.i.d. sequences. This theorem yields an immediate, simple way to prove the subsequence principle and it will be the main tool for the proof of our results discussed in the later chapters.

Theorem 2.1 *Let (X_n) be a determining sequence of r.v.'s and (ε_n) a positive numerical sequence. Then there exists a subsequence (X_{m_k}) and a sequence (Y_k) of discrete r.v.'s such that*

$$P\big(|X_{m_k} - Y_k| \geq \varepsilon_k\big) \leq \varepsilon_k \quad k = 1, 2 \ldots \tag{9}$$

and for each $k > 1$ the atoms of the finite σ-field $\sigma\{Y_1, \ldots, Y_{k-1}\}$ can be divided into two classes Γ_1 and Γ_2 such that

(i) $\displaystyle \sum_{B \in \Gamma_1} P(B) \leq \varepsilon_k$

(ii) *For any $B \in \Gamma_2$ there exist P_B-independent r.v.'s $\{Z_j^{(B)}, j = k, k+1, \ldots\}$ defined on B with common distribution function F_B such that*

$$P_B\left(|Y_j - Z_j^{(B)}| \geq \varepsilon_k\right) \leq \varepsilon_k \quad j = k, k+1, \ldots \tag{10}$$

Here F_B denotes the limit distribution of (X_n) relative to B and P_B denotes conditional probability given B.

3 Permutation-Invariance

Studying rearrangements of function series $\sum_{k=1}^{\infty} f_k$ is a much investigated problem of analysis. By the Riemann rearrangement theorem, a conditionally convergent numerical series $\sum_{k=1}^{\infty} c_k$ can be rearranged to converge to any prescribed real number and to $+\infty$ and $-\infty$. The corresponding problem for function series, called Banach's problem, is much harder and is still unsolved. (See Nikishin [55], Theorem 22 for the strongest existing result.) Another famous conjecture of analysis is that any orthonormal system (φ_n) can be rearranged to become a convergence system. (See Garsia [44] for a partial result.) Unconditional a.e. convergence (i.e., almost everywhere convergence after every rearrangement) of function series is also an important topic in functional analysis. Lacunary series play a special role in this field. An exchangeable sequence (Y_n) is a completely symmetric structure in the sense that the distribution of any functional of (Y_n) is the same as that of $(Y_{\sigma(n)})$ where $\sigma : \mathbb{N} \to \mathbb{N}$ is a permutation of the positive integers. Consequently, limit theorems valid for (Y_n) remain valid also for the permuted sequence $(Y_{\sigma(n)})$. Hence on the basis of Aldous' theorem it is natural to expect that limit theorems valid for thin subsequences (X_{n_k}) of general tight sequences (X_n) remain valid after any permutation of their terms. This is indeed the case, as proved by Berkes and Tichy [21], but it is far from obvious: the proof of Aldous' theorem is not permutation-invariant and a further thinning of the subsequence constructed in [11] is required to get a subsequence satisfying i.i.d. limit theorems after permutations. As far as limit theorems for lacunary trigonometric and related sums under the Hadamard gap condition are concerned, their proofs depend on number-theoretic properties of (n_k) which are, except for a few cases, not permutation-invariant and the CLT and LIL for such series can indeed fail after permutations. The purpose of the present chapter is to provide a complete solution of the permutation-invariance of such results. For the proofs we refer to Aistleitner et al. [6–8, 10].

Theorem 3.1 *Let (n_k) be a sequence of positive integers satisfying (1) and let $\sigma : \mathbb{N} \to \mathbb{N}$ be a permutation of the set of positive integers. Then we have*

$$N^{-1/2} \sum_{k=1}^{N} \cos 2\pi n_{\sigma(k)} x \xrightarrow{d} N(0, 1/2) \tag{11}$$

and

$$\limsup_{N\to\infty} (N \log \log N)^{-1/2} \sum_{k=1}^{N} \cos 2\pi n_{\sigma(k)} x = 1 \qquad \text{a.e.} \tag{12}$$

where $\xrightarrow{d}$ denotes convergence in distribution over the probability space $(0,1)$ equipped with the Borel σ-algebra and the Lebesgue measure.

Note that for the unpermuted CLT and LIL we need much weaker gap conditions than (1). In fact, Erdős [32] and Takahashi [61–63] showed that if a sequence (n_k) of integers satisfies

$$n_{k+1}/n_k \geq 1 + k^{-\alpha}, \qquad 0 \leq \alpha < 1/2 \tag{13}$$

then we have the CLT (2) and LIL (3). Note, however, that (13) does not imply permutation-invariance and in fact Aistleitner et al. [8] showed that permutation-invariance fails under any gap condition weaker than (1).

We pass now to the permutation-invariance of the more general sums $\sum_{k=1}^{n} f(n_k x)$, where f is a measurable function satisfying

$$f(x+1) = f(x), \qquad \int_0^1 f(x)\,dx = 0, \qquad \int_0^1 f^2(x)\,dx < \infty \tag{14}$$

and (n_k) is a sequence of integers satisfying the Hadamard gap condition (1). The central limit theorem for $f(n_k x)$ has a long history. Kac [47] proved that if f is Lipschitz continuous, then $f(n_k x)$ satisfies the CLT for $n_k = 2^k$ and not much later Erdős and Fortet (see [48], p. 655) showed that this is not any more valid if $n_k = 2^k - 1$. Gaposhkin [43] proved that $f(n_k x)$ obeys the CLT if $n_{k+1}/n_k \to \alpha$ where α^r is irrational for $r = 1, 2 \ldots$ and the same holds if all the fractions n_{k+1}/n_k are integers. To formulate our results on permutation-invariance, we introduce a series of number-theoretic conditions playing an important role in the circle of problems studied. We say that a sequence (n_k) of positive integers satisfies
Condition D_2, if for any fixed nonzero integers a, b, c the number of solutions of the Diophantine equation

$$an_k + bn_l = c \tag{15}$$

is bounded by a constant $K(a,b)$, independent of c.
Condition $D_2^{(s)}$ (strong D_2), if for any fixed integers $a \neq 0$, $b \neq 0$, c the number of solutions of the Diophantine equation (15) is bounded by a constant $K(a,b)$, independent of c, where for $c = 0$ we require also $k \neq l$.

Condition $D_2^{(w)}$ (weak D_2), if for any fixed nonzero integers a, b, c the number of solutions of the Diophantine equation

$$an_k + bn_l = c, \qquad 1 \le k, l \le N \tag{16}$$

is $o(N)$, uniformly in c.

Condition $D_2^{(0)}$, if for any fixed nonzero integers a, b the number of solutions of the Diophantine equation

$$an_k + bn_l = 0, \qquad 1 \le k, l \le N \tag{17}$$

is $o(N)$.

Condition D_2 was introduced by Gaposhkin [43], who proved that under minor smoothness assumptions on f, it implies the CLT for $f(n_k x)$. Aistleitner and Berkes [3] proved that the CLT holds for $f(n_k x)$ also under $D_2^{(w)}$ and this condition is necessary. Below we give a precise description of the CLT and LIL behavior of permuted sums $\sum_{k=1}^{N} f(n_{\sigma(k)} x)$ and, in particular, we obtain characterizations of permutation-invariance.

Our first result shows that if we assume the slightly stronger gap condition

$$n_{k+1}/n_k \to \infty \tag{18}$$

then the behavior of $f(n_k x)$ is permutation-invariant regardless of the number-theoretic structure of (n_k). In what follows, let $\| \cdot \|$ and $\| \cdot \|_p$ denote the $L^2(0, 1)$, resp. $L^p(0, 1)$ norm.

Theorem 3.2 *Let (n_k) be a sequence of positive integers satisfying the gap condition (18). Then for any permutation $\sigma : \mathbb{N} \to \mathbb{N}$ of the integers and for any measurable function $f : \mathbb{R} \to \mathbb{R}$ satisfying*

$$f(x + 1) = f(x), \qquad \int_0^1 f(x)\, dx = 0, \qquad \mathrm{Var}_{[0,1]} f < +\infty \tag{19}$$

we have

$$\frac{1}{\sqrt{N}} \sum_{k=1}^{N} f(n_{\sigma(k)} x) \xrightarrow{d} \mathcal{N}(0, \|f\|^2) \tag{20}$$

and

$$\limsup_{N \to \infty} \frac{1}{\sqrt{2N \log \log N}} \sum_{k=1}^{N} f(n_{\sigma(k)} x) = \|f\| \quad \text{a.e.} \tag{21}$$

Assuming only the Hadamard gap condition (1), the situation becomes more complex and the number-theoretic structure of (n_k) comes into play. Our next result gives a necessary and sufficient condition for permuted partial sums $\sum_{k=1}^{N} f(n_{\sigma(k)}x)$ to have only Gaussian limit distributions and gives precise criteria for this to happen for a specific permutation σ.

Theorem 3.3 *Let (n_k) be a sequence of positive integers satisfying the Hadamard gap condition (1) and condition* $\mathbf{D_2}$. *Let f satisfy (19) and let σ be a permutation of* $\mathbb{N}$. *Then $N^{-1/2} \sum_{k=1}^{N} f(n_{\sigma(k)}x)$ has a limit distribution iff*

$$\gamma = \lim_{N \to \infty} N^{-1} \int_0^1 \left(\sum_{k=1}^{n} f(n_{\sigma(k)}x) \right)^2 dx \tag{22}$$

exists, and then

$$N^{-1/2} \sum_{k=1}^{N} f(n_{\sigma(k)}x) \xrightarrow{d} N(0, \gamma). \tag{23}$$

Theorem 3.3 is best possible in the following sense:

Theorem 3.4 *If condition* $\mathbf{D_2}$ *fails, there exists a permutation σ such that the normed partial sums in (23) have a non-Gaussian limit distribution.*

In other words, under the Hadamard gap condition and condition $\mathbf{D_2}$, the limit distribution of $N^{-1/2} \sum_{k=1}^{N} f(n_{\sigma(k)}x)$ can only be Gaussian, but the variance of the limit distribution depends on the constant γ in (22) which, as simple examples show, is not permutation-invariant. For example, if $n_k = 2^k$ and σ is the identity permutation, then (22) holds with

$$\gamma = \gamma_f = \int_0^1 f^2(x)dx + 2 \sum_{k=1}^{\infty} \int_0^1 f(x)f(2^k x)dx \tag{24}$$

(see Kac [47]). Using an idea of Fukuyama [36], one can construct permutations σ of $\mathbb{N}$ such that

$$\lim_{N \to \infty} \frac{1}{N} \int_0^1 \left(\sum_{k=1}^{N} f(n_{\sigma(k)}x) \right)^2 dx = \gamma_{\sigma f} \tag{25}$$

with $\gamma_{\sigma f} \neq \gamma_f$. Actually, the set of possible values $\gamma_{\sigma f}$ belonging to all permutations σ contains the interval $I_f = [\gamma_f, \|f\|^2]$ and it is equal to this interval provided the Fourier coefficients of f are nonnegative.

Under the slightly stronger condition $\mathbf{D_2^{(s)}}$ we get

Theorem 3.5 *Let (n_k) be a sequence of positive integers satisfying the Hadamard gap condition (1) and condition $\mathbf{D}_2^{(s)}$. Let f satisfy (19) and let σ be a permutation of $\mathbb{N}$. Then the central limit theorem (23) holds with $\gamma = \|f\|^2$.*

We now pass to the problem of the LIL.

Theorem 3.6 *Let (n_k) be a sequence of positive integers satisfying the Hadamard gap condition (1) and condition $\mathbf{D}_2$. Let f be a measurable function satisfying (19), let σ be a permutation of $\mathbb{N}$ and assume that*

$$\gamma = \lim_{N \to \infty} N^{-1} \int_0^1 \left(\sum_{k=1}^N f(n_{\sigma(k)}x) \right)^2 dx \tag{26}$$

for some $\gamma \geq 0$. Then we have

$$\limsup_{N \to \infty} \frac{\sum_{k=1}^N f(n_{\sigma(k)}x)}{\sqrt{2N \log \log N}} = \gamma^{1/2} \quad \text{a.e.} \tag{27}$$

Theorem 3.7 *Let (n_k) be a lacunary sequence of positive integers satisfying condition $\mathbf{D}_2^{(s)}$, and let f be a function satisfying (19). Then for any permutation $\sigma : \mathbb{N} \to \mathbb{N}$ we have*

$$\limsup_{N \to \infty} \frac{\sum_{k=1}^N f(n_{\sigma(k)}x)}{\sqrt{2N \log \log N}} = \|f\| \quad \text{a.e.}$$

Remark If f is a trigonometric polynomial of degree d, then in conditions $\mathbf{D}_2$ resp. $\mathbf{D}_2^{(s)}$ it suffices to have the bound for the number of solutions of (15) for coefficients a, b satisfying $|a| \leq d, |b| \leq d$. Applying this with $d = 1$ and using the fact that for a Hadamard lacunary sequence (n_k) and $c \in \mathbb{Z}$ the number of solutions $(k, l), k \neq l$ of

$$n_k \pm n_l = c$$

is bounded by a constant which is independent of c (see Zygmund [68, p. 203]), we get Theorem 3.1.

All the results formulated so far assume the Hadamard gap condition (1) or the stronger condition (18). If we drop (1), i.e., we allow subexponential sequences (n_k), we need much stronger Diophantine conditions even for the unpermuted CLT and LIL for $f(n_k x)$. Specifically, we need uniform bounds for the number of solutions of Diophantine equations of the type

$$a_1 n_{k_1} + \cdots + a_p n_{k_p} = b. \tag{28}$$

Call a solution of (28) *nondegenerate* if no subsum of the left-hand side equals 0. Let us say that a sequence (n_k) of positive integers satisfies
Condition $\mathbf{A}_p$, if there exists a constant $C_p \geq 1$ such that for any integer $b \neq 0$ and any nonzero integers $a_1, \ldots, a_p$ the number of nondegenerate solutions of the Diophantine equation (28) is at most C_p.

The following results are the analogues of the previous results without growth conditions on (n_k).

Theorem 3.8 *Let (n_k) be an increasing sequence of positive integers satisfying condition $\mathbf{A}_p$ for all $p \geq 2$. Let f satisfy (19), let σ be a permutation of $\mathbb{N}$ and assume that the limit (22) exists. Then the permuted CLT (23) is valid.*

Theorem 3.9 *Let (n_k) be an increasing sequence of positive integers satisfying condition $\mathbf{A}_p$ for all $p \geq 2$ with $C_p \leq \exp(Cp^\alpha)$ for some $\alpha > 0$. Assume also that f satisfies (19). Then for any permutation σ of $\mathbb{N}$ we have*

$$\limsup_{N \to \infty} \frac{1}{\sqrt{2N \log \log N}} \sum_{k=1}^{N} f(n_{\sigma(k)}x) = \gamma^{1/2} \quad \text{a.e.} \tag{29}$$

Note that for the validity of the LIL we require a specific bound for the constants C_p in condition $\mathbf{A}_p$. For subexponentially growing (n_k), verifying property $\mathbf{A}_p$ is a difficult number-theoretic problem. Classical examples of such sequences are the Hardy–Littlewood–Pólya sequences, i.e., increasing sequences (n_k) consisting of all positive integers of the form $q_1^{\alpha_1} \cdots q_\tau^{\alpha_\tau}$ $(\alpha_1, \ldots \alpha_\tau \geq 0)$, where $q_1, \ldots, q_\tau$ is a fixed set of coprime integers. Clearly, such sequences grow subexponentially; Tijdeman [64] proved that

$$n_{k+1} - n_k \geq \frac{n_k}{(\log n_k)^\alpha} \tag{30}$$

for some $\alpha > 0$, i.e., the growth of (n_k) is almost exponential. Hardy–Littlewood–Pólya sequences have remarkable probabilistic and ergodic properties. Nair [54] proved that if f is 1-periodic and integrable in $(0, 1)$, then

$$\lim_{N \to \infty} \frac{1}{N} \sum_{k=1}^{N} f(n_k x) = \int_0^1 f(t)dt \quad \text{a.e.}$$

Fukuyama and Petit [38] also showed that the central limit theorem

$$N^{-1/2} \sum_{k=1}^{N} f(n_k x) \xrightarrow{d} N(0, \gamma_f^*)$$

holds with

$$\gamma_f^* = \sum_{k,l:(n_k,n_l)=1} \int_0^1 f(n_k x)f(n_l x)dx. \qquad (31)$$

The Diophantine properties of (n_k) have been studied in great detail in recent years. Amoroso and Viada [13] showed recently that Hardy–Littlewood–Pólya sequences satisfy condition $\mathbf{A}_p$ for any $p \geq 2$ with $C_p = \exp(p^6)$. This is a deep result, involving a substantial sharpening of the subspace theorem of Evertse, Schlickewei and Schmidt (see [34]). Again, the limit γ in (22) depends on the permutation σ; one can show that if the Fourier coefficients of f are nonnegative, the class of numbers γ contains the interval $[\|f\|^2, \gamma_f^*]$, with γ_f^* in (31).

Since verifying condition $\mathbf{A}_p$ for a concrete subexponential sequence (n_k) is difficult, it is worth looking for Diophantine conditions which are strong enough to imply the permutation-invariant CLT and LIL, but which hold for a sufficiently large class of subexponential sequences. In the sequel we will give such a Diophantine condition $\mathbf{A}_\omega$. Actually, one can show that this $\mathbf{A}_\omega$ is satisfied, in a certain statistical sense, for "almost all" sequences (n_k), and thus the permutation-invariant CLT and LIL are the "typical" behavior of sequences $f(n_k x)$. Given a nondecreasing sequence $\omega = (\omega_1, \omega_2, \ldots)$ of positive integers tending to $+\infty$, let us say that an increasing sequence (n_k) of positive integers satisfies
Condition $\mathbf{A}_\omega$, if the Diophantine equation

$$a_1 n_{k_1} + \cdots + a_p n_{k_p} = 0, \qquad k_1 < \cdots < k_p, \ p = \omega_N, \ 0 < |a_1|, \ldots, |a_p| \leq N^{\omega_N}$$

has no solutions with $k_p > N$.

This condition is more technical than $\mathbf{A}_p$ and for each $N \geq 1$ it requires a bound for the number of solutions of the p-term Diophantine equation $a_1 n_{k_1} + \cdots + a_p n_{k_p} = 0$, where $p = \omega_N$ and the coefficients are bounded by N^{ω_N}, where we usually choose (ω_N) to grow slowly.

Theorem 3.10 *Let $\omega = (\omega_1, \omega_2, \ldots)$ be a nondecreasing sequence of positive integers tending to $+\infty$ and let (n_k) be a sequence of positive integers satisfying condition $\mathbf{A}_\omega$. Then, for any f satisfying (19) and any permutation σ of $\mathbb{N}$, relations (20) and (21) are valid.*

Fix $\omega_N \to \infty$. One can show that, in a certain statistical sense, "almost all" sequences $n_k \leq k^{\omega_k}$ satisfy condition $\mathbf{A}_\omega$. To make this precise, we need to define a probability measure over the set of such sequences, or, equivalently, a natural random procedure to generate such sequences. Clearly, the simplest procedure is to choose n_k independently and uniformly from the integers in the interval $I_k = [1, k^{\omega_k}]$ $(k = 1, 2, \ldots)$. Denote the so obtained measure by μ.

Theorem 3.11 *With probability one with respect to μ, the sequence (n_k) satisfies (20), (21).*

Clearly, for slowly increasing (ω_k) the so obtained sequence grows almost polynomially (as a comparison, Hardy–Littlewood–Pólya sequences grow almost exponentially by (30)). We do not know if there exist polynomially growing sequences (n_k) satisfying any of (20), (21). As a simple combinatorial argument shows, sequences (n_k) satisfying $\mathbf{A}_p$ for all $p \geq 2$ cannot grow polynomially.

4 Lacunary Sequences Outside L^2

For lacunary series $\sum c_k \varphi_{n_k}$ in L^2, a number of interesting uniform limit theorems have been obtained by ad hoc methods, such as the selection theorem of Menshov [52] or its permutation-invariant version by Komlós [50], already mentioned in Sect. 1. A further example is the result of Gaposhkin [41] stating that for every uniformly bounded sequence (X_n) of random variables there exists a subsequence (X_{n_k}) and random variables X and $Y \geq 0$ such that if (a_k) is a sequence of real numbers satisfying

$$a_N = o(A_N) \quad \text{with} \quad A_N = \left(\sum_{k=1}^{N} a_k^2 \right)^{1/2} \longrightarrow \infty \tag{32}$$

then

$$A_N^{-1} \sum_{k=1}^{N} a_k (X_{n_k} - X) \xrightarrow{d} N(0, Y) \tag{33}$$

and if

$$a_N = o(A_N / (\log \log A_N)^{1/2}), \quad A_N \to \infty, \tag{34}$$

then

$$\limsup_{N \to \infty} (2A_N^2 \log \log A_N)^{-1/2} \sum_{k=1}^{N} a_k (X_{n_k} - X) = Y^{1/2} \quad \text{a.s.} \tag{35}$$

Note that the conditions (32) and (34) are the classical (and optimal) conditions for a weighted sum $\sum_{k=1}^{N} a_k X_k$ of bounded independent random variables with mean 0 to satisfy the central limit theorem and the law of the iterated logarithm. In contrast, no uniform results seem to be known for sequences (X_n) which are not in L^2 and the purpose of this section is to formulate such results.

Our first result is the analogue of the uniform CLT (32)–(33) in the case when the limit distribution is, instead of the Gaussian law, a symmetric stable distribution $G_{\alpha,c}$ with parameter $0 < \alpha < 2$, having characteristic function (Fourier transform) $\exp(-c|t|^\alpha)$. By a classical result in probability theory, if $0 < \alpha < 2$, (X_n) is a sequence of i.i.d. symmetric random variables with characteristic function φ satisfying

$$\varphi(t) = 1 - c|t|^\alpha + o(|t|^\alpha) \qquad \text{as } t \to 0 \tag{36}$$

and (a_k) is a numerical sequence with

$$a_N = o(A_N), \quad A_N = \left(\sum_{k=1}^{N} |a_k|^\alpha \right)^{1/\alpha} \longrightarrow \infty, \tag{37}$$

then

$$A_N^{-1/\alpha} \sum_{k=1}^{N} a_k X_k \xrightarrow{d} G_{\alpha,c}.$$

Using Theorem 2.1, Berkes and Tichy [22] proved the corresponding uniform limit theorem for lacunary series.

Theorem 4.1 *Let $0 < \alpha < 2$, $c > 0$ and let (X_n) be a determining sequence of r.v.'s with limit random measure μ. Assume that the characteristic function φ of μ satisfies (36) with probability 1. Then there exists a subsequence (X_{n_k}) such that for any real sequence (a_k) satisfying (37) we have*

$$A_N^{-1} \sum_{k=1}^{N} a_k X_{n_k} \xrightarrow{d} G_{\alpha,c}.$$

Using the same method, one can prove analogous uniform limit theorems for partial maxima, sample extrema, and other nonlinear functionals of lacunary series without assuming finite second moments, but instead of elaborating on this, we investigate lacunary subsequences $f(n_k x)$ of the dilated system $f(nx)$, already studied in the previous section, and formulate very general uniform results for their lacunary subsequences $f(n_k x)$ without assuming the square integrability of (or, for that matter, any integrability condition for) the periodic function f. Our starting point is Gaposhkin's observation [42] that sufficiently thin subsequences of $f(nx)$ are almost independent in a very strong sense.

Theorem 4.2 *Let $f : \mathbb{R} \to \mathbb{R}$ be a measurable function with period 1. Then there exists an increasing sequence (n_k) of positive integers and measurable functions $g_k(x)$, $k = 1, 2, \ldots$ on $(0, 1)$ such that the g_k are stochastically independent and*

$$f(n_k x) = g_k(x) + \psi_k(x) + \eta_k(x) \tag{38}$$

where

$$\sum_{k=1}^{\infty} \|\psi_k\|_M < \infty \quad \text{and} \quad \sum_{k=1}^{\infty} \mu\{x : \eta_k(x) \neq 0\} < \infty. \tag{39}$$

Here $\| \cdot \|_M$ denotes the norm in the space $M(0, 1)$ of measurable functions on $(0, 1)$ defined by $\|\psi\|_M = \inf\{\epsilon > 0 : \mu(x : |\psi(x)| \geq \epsilon) \leq \epsilon\}$. If $f \in L^p(0, 1)$ $(p \geq 1)$ or $f \in C(0, 1)$, then the conclusion remains valid with the g_k belonging to the corresponding spaces and $\| \cdot \|_M$ replaced by $\| \cdot \|_p$ or $\| \cdot \|_C$, respectively.

As an immediate consequence, we get

$$\sum_{k=1}^{\infty} |f(n_k x) - g_k(x)| < \infty \qquad \text{a.e.} \tag{40}$$

in all cases covered by the theorem. Clearly (40) extends any limit theorem T with the property

$$(X_k) \text{ satisfies } T \text{ and } \sum_{k=1}^{\infty} |X_k - X_k'| < \infty \text{ a.s.} \implies (X_k') \text{ satisfies } T \tag{41}$$

from (g_k) to $(f(n_k x))$. Moreover, relation (40) remains valid after any permutation of the indices and thus limit theorems for $f(n_k x)$ provided by Gaposhkin's theorem are automatically permutation-invariant. We observe further that by (40) we have $f(n_k x) - g_k(x) \to 0$ a.s. and thus given any sequence $\Psi(k) \to \infty$ one can select a subsequence (m_k) such that

$$\sum_{k=1}^{\infty} \Psi(k) |f(n_{m_k} x) - g_{m_k}(x)| < \infty \quad \text{a.e.}$$

Thus, if (a_k) is a weight sequence such that $|a_k| \leq \Psi(k)$, then

$$\left| \sum_{k=1}^{N} a_k f(n_{m_k} x) - \sum_{k=1}^{N} a_k g_{m_k}(x) \right| = O(1) \quad \text{a.e.}$$

a relation leading to uniform limit theorems of weighted sums of $f(n_{m_k}x)$ by using the corresponding limit theorem for weighted sums of independent random variables. The only shortcoming of Gaposhkin's result is that it gives no information on the growth speed of (n_k): the construction of (n_k), although theoretically explicit, is very complicated and yields an extremely rapidly growing (n_k). Since all classical lacunarity results for "concrete" function systems involve concrete lacunarity conditions (such as the Hadamard gap condition), it would be desirable to give the dependence of the growth of (n_k) in Gaposhkin's theorem on the function f explicitly. In the sequel, we formulate such explicit results; for the proofs, see Aistleitner et al. [9].

Theorem 4.3 *Let (n_k) be an increasing sequence of positive integers. Then there exists a sequence (g_k) of measurable functions on $(0, 1)$, independent and uniformly distributed over $(0, 1)$ in the sense of probability theory, such that*

$$\lambda\{x \in (0, 1) : |\langle n_k x \rangle - g_k(x)| \geq 2n_k/n_{k+1}\} \leq 2n_k/n_{k+1} \qquad k = 1, 2, \ldots. \qquad (42)$$

where $\langle \cdot \rangle$ denotes fractional part.

Theorem 4.3 improves a weaker result of Hawkes [46] which gives a similar approximation on large subsets of $(0, 1)$ and after a suitable transformation of $(0, 1)$. It concerns the specific sequence $\langle n_k x \rangle$, but depending on the properties of the function f, it leads automatically to a corresponding approximation theorem for general sequences $(f(n_k x))$. For example, if f is continuous with continuity modulus $\omega(f, \delta)$, then (42) implies

$$\sum_{k=1}^{\infty} |f(n_k x) - g_k^*(x)| < \infty \qquad \text{a.e.} \qquad (43)$$

with $g_k^*(x) = f(g_k(x))$, provided

$$\sum_{k=1}^{\infty} n_k/n_{k+1} < \infty, \qquad \sum_{k=1}^{\infty} \omega(f, n_k/n_{k+1}) < \infty.$$

A much more general consequence of Theorem 4.3 is the following:

Corollary 4.1 *Let $f : \mathbb{R} \to \mathbb{R}$ be a measurable function with period 1 and let (n_k) be an increasing sequence of positive integers. Let T_k be positive numbers such that*

$$\lambda\{f \in (0, 1) : |f| \geq T_k\} \leq k^{-2} \qquad (k = 1, 2, \ldots) \qquad (44)$$

and assume that

$$\sum_{k=1}^{\infty} \left[(n_k/n_{k+1})^{1/4} T_k + \omega_2(f_{T_k}, (n_k/n_{k+1})^{1/4}) \right] < \infty. \qquad (45)$$

Then there exists a sequence (g_k^*) of measurable functions on $(0, 1)$, independent and having the same distribution over $(0, 1)$ as f, such that (43) holds.

Here

$$\omega_2(f, \delta) = \sup_{0 \leq h \leq \delta} \left(\int_0^1 |f(x + h) - f(x)|^2 dx \right)^{1/2}$$

is the L^2 modulus of continuity of f and f_T denotes the truncated function $f \cdot I\{|f| \leq T\}$.

Given any periodic measurable function f, one can choose T_k so that (44) holds, and then (45) becomes a "concrete" gap condition. Clearly, the larger the function f is, the faster n_k tends to ∞. In particular for functions f not belonging to any L^p, $p > 0$, (n_k) is growing very fast. We illustrate the procedure on two classical limit theorems mentioned above.

Corollary 4.2 *Let $f : \mathbb{R} \to \mathbb{R}$ be a measurable function with period 1 such that the distribution function*

$$F(x) = \mu\{t \in (0, 1) : f(t) \leq x\} \tag{46}$$

satisfies

$$1 - F(x) \sim px^{-\alpha}L(x), \quad F(-x) \sim qx^{-\alpha}L(x) \quad \text{as } x \to \infty \tag{47}$$

for some constants $p, q \geq 0$, $p + q = 1$, $0 < \alpha < 2$ and a slowly varying function L. Let (n_k) be an increasing sequence of positive integers satisfying (45). Then letting $S_n = \sum_{k=1}^n f(n_k x)$, we have

$$(S_n - a_n)/b_n \xrightarrow{d} G_\alpha \tag{48}$$

for some numerical sequences (a_n), (b_n) and an α-stable distribution G_α.

Corollary 4.3 *Let $f : \mathbb{R} \to \mathbb{R}$ be a measurable function with period 1 such that the distribution function F of f in (46) satisfies $F(x) < 1$ for all x and $1 - F$ is regularly varying at $+\infty$ in the Karamata sense with a negative exponent $-\alpha$. Let (n_k) be an increasing sequence of positive integers satisfying (45). Then letting $M_n = \max_{1 \leq k \leq n} f(n_k x)$, we have*

$$(M_n - a_n)/b_n \xrightarrow{d} H_\alpha \tag{49}$$

where $H_\alpha(x) = \exp(-x^{-\alpha})I_{(0,\infty)}(x)$.

Note that (47) is the classical necessary and sufficient condition for the partial sums S_n of an i.i.d. sequence with distribution function F to satisfy the limit theorem (48) with suitable norming and centering sequences (a_n), (b_n). Corollary 4.2

shows that if the distribution function F of the periodic function f satisfies (47), then the partial sums of $f(n_k x)$ for any (n_k) satisfying (45) obey the stable limit theorem (48). Similarly, the assumption on F in Corollary 4.3 is the well-known necessary and sufficient condition for the centered and normed maxima of an i.i.d. sequence with distribution F to converge weakly to the distribution $H(x) = \exp(-x^{-\alpha})I_{(0,\infty)}(x)$, the so-called Fréchet distribution. As we know (see, e.g., [40]), the limit distribution in (49) for any i.i.d. sequence can be only one of the Fréchet, Gumbel, and Weibull distributions with respective distribution functions $\exp(-x^{-\alpha})I_{(0,\infty)}(x)$, $\exp(-(-x)^{\alpha})(1 - I_{(0,\infty)}(x))$, and $\exp(-e^{-x})$. The analogue of Corollary 4.3 holds for the other two limiting classes, too.

5 Applications to Banach Space Theory

An old (and only partially solved) problem of Banach space theory is to characterize Banach spaces containing an isomorphic copy of classical spaces like $L^p(0, 1)$, $C(0, 1)$, etc. A modern approach to this problem via lacunary series is due to Kadec and Pełczyński [49]. Call two sequences (x_n) and (y_n) in a Banach space $(B, \|\cdot\|)$ equivalent if there exists a constant $K > 0$ such that

$$K^{-1}\left\|\sum_{i=1}^{n} a_i x_i\right\| \leq \left\|\sum_{i=1}^{n} a_i y_i\right\| \leq K\left\|\sum_{i=1}^{n} a_i x_i\right\|$$

for every $n \geq 1$ and every $(a_1, \ldots, a_n) \in \mathbb{R}^n$. Clearly, in this case the subspaces of B spanned by (x_n) and (y_n) are isomorphic. In particular, if (x_n) is a sequence in $L^p(0, 1)$, $p \geq 1$ such that for some constant $K > 0$

$$K^{-1}\left(\sum_{k=1}^{N} a_k^2\right)^{1/2} \leq \left\|\sum_{k=1}^{N} a_k x_k\right\|_p \leq K\left(\sum_{k=1}^{N} a_k^2\right)^{1/2} \tag{50}$$

or

$$K^{-1}\left(\sum_{k=1}^{N} |a_k|^p\right)^{1/p} \leq \left\|\sum_{k=1}^{N} a_k x_k\right\|_p \leq K\left(\sum_{k=1}^{N} |a_k|^p\right)^{1/p} \tag{51}$$

then the subspace spanned by (x_n) is isomorphic to ℓ^2 and ℓ^p, respectively. Kadec and Pełczyński [49] proved that for any normalized weakly null sequence (x_n) in $L^p(0, 1)$, $p > 2$ there exists a subsequence (x_{n_k}) which is equivalent to the unit vector basis of ℓ^2 or ℓ^p. Thus every infinite dimensional closed subspace H of $L^p(0, 1)$, $p > 2$ contains a subspace isomorphic to ℓ^2 or ℓ^p. In the case when (x_n) is uniformly integrable in $L^p(0, 1)$ then the first alternative holds, while if the functions (x_n) have disjoint support, the second alternative holds trivially with $K = 1$. The general case follows via a subsequence splitting argument as in [49].

In the case of a sequence $(x_n) \in L^p$, $1 \le p < 2$ the problem is considerably harder. Sufficient conditions for the existence of a subsequence equivalent to the unit vector basis of ℓ^2 were given by Berkes [15] and Guerre [45]. The following theorem in Berkes and Tichy [23] gives a complete solution of the problem.

Theorem 5.1 *Let* $1 \le p < 2$ *and let* (X_n) *be a determining sequence of random variables such that* $\|X_n\|_p = 1$ $(n = 1, 2, \ldots)$, $\{|X_n|^p,\ n \ge 1\}$ *is uniformly integrable and* $X_n \to 0$ *weakly*[1] *in* L^p. *Let* μ *be the limit random measure of* (X_n). *Then there exists a subsequence* (X_{n_k}) *equivalent to the unit vector basis of* ℓ^2 *if and only if*

$$\int_{-\infty}^{\infty} x^2 d\mu(x) \in L^{p/2}. \tag{52}$$

Note that (50), (51) are required for all $(a_1, \ldots, a_n) \in \mathbb{R}^n$ and all $n \ge 1$ with the same constant K, i.e., they are uniform statements. Using an observation of Aldous [11], such a uniformity can be obtained from an equicontinuity statement. Let (X_n) be a determining sequence in $L^p(0, 1)$ with limit random measure μ and limit exchangeable sequence (Y_n). Our purpose is to construct, given $\varepsilon > 0$, a sequence $n_1 < n_2 < \cdots$ of integers such that

$$(1 - \varepsilon)\psi(a_1, \ldots, a_k) \le \left\| \sum_{i=1}^{k} a_i X_{n_i} \right\|_p \le (1 + \varepsilon)\psi(a_1, \ldots, a_k)$$

for every $k \ge 1$ and $(a_1, \ldots, a_k) \in \mathbb{R}^k$ where

$$\psi(a_1, \ldots, a_n) = \left\| \sum_{i=1}^{n} a_i Y_i \right\|_p.$$

To construct n_1 we set

$$Q(\mathbf{a}, n, \ell) = |a_1 X_n + a_2 Y_2 + \cdots + a_\ell Y_\ell|^p$$

$$R(\mathbf{a}, \ell) = |a_1 Y_1 + a_2 Y_2 + \cdots + a_\ell Y_\ell|^p$$

for every $n \ge 1$, $\ell \ge 2$ and $\mathbf{a} = (a_1, \ldots, a_\ell) \in \mathbb{R}^\ell$. We claim that

$$E\left\{ \frac{Q(\mathbf{a}, n, \ell)}{\psi(\mathbf{a})^p} \right\} \longrightarrow E\left\{ \frac{R(\mathbf{a}, \ell)}{\psi(\mathbf{a})^p} \right\} \quad \text{as } n \to \infty \quad \text{uniformly in } \mathbf{a}, \ell \tag{53}$$

---

[1]This is meant as $\lim_{n \to \infty} \mathbb{E}(X_n Y) = 0$ for all $Y \in L_q$ where $1/p + 1/q = 1$. This convergence should not be confused with weak convergence of probability distributions, also called convergence in distribution.

(The right side of (53) equals 1.) To do this we note that by the properties of the limit exchangeable sequence, relation (53) holds for every fixed vector $\mathbf{a} = (a_1, \ldots, a_\ell)$ and by a well-known result of Ranga Rao [58], for uniformity we have to verify a certain equicontinuity property of the functions $Q(\mathbf{a}, n, \ell)/\psi(\mathbf{a})^p$ in (53). If $n_1 < n_2 < \cdots < n_{k-1}$ are already constructed, an analogous equicontinuity argument implies that for $\ell > k$

$$\psi(\mathbf{a})^{-1} \| a_1 X_{n_1} + \cdots + a_{k-1} X_{n_{k-1}} + a_k X_n + a_{k+1} Y_{k+1} + \cdots + a_\ell Y_\ell \|_p$$
$$\longrightarrow \psi(\mathbf{a})^{-1} \| a_1 X_{n_1} + \cdots + a_{k-1} X_{n_{k-1}} + a_k Y_k + \cdots + a_\ell Y_\ell \|_p \quad \text{as } n \to \infty$$

uniformly in $\mathbf{a}$ and ℓ, a relation that can be used to choose n_k. Carrying out this argument, we get the following general theorem, proved in Berkes [15]. Let $\mathcal{M}$ be the set of all probability measures on $\mathbb{R}^1$ and let π be the Prohorov metric on $\mathcal{M}$ defined by

$$\pi(\nu, \lambda) = \inf \{ \varepsilon > 0 : \nu(A) \leq \lambda(A^\varepsilon) + \varepsilon \text{ and}$$
$$\lambda(A) \leq \nu(A^\varepsilon) + \varepsilon \text{ for all Borel sets } A \subset \mathbb{R}^1 \}.$$

Here A^ε denotes the open ε-neighborhood of A.

Theorem 5.2 *Let $p \geq 1$ and let (X_n) be a sequence of r.v.'s so that $\{|X_n|^p, \, n \geq 1\}$ is uniformly integrable. Let μ and (Y_n) denote the limit random measure and limit exchangeable sequence of (X_n), respectively. Let S be a Borel subset of $(\mathcal{M}, \pi)$ such that μ is concentrated on S with probability 1. Assume that there exists a separable metric d on S, Borel-equivalent to the Prohorov metric π such that $\mathbb{E}d(\mu, 0)^p < +\infty$ (0 denotes the zero distribution) and*

$$\left| \frac{\left\| t + \sum_{k=1}^{n} a_k \xi_k^{(\nu)} \right\|_p - \left\| t + \sum_{k=1}^{n} a_k \xi_k^{(\lambda)} \right\|_p}{\left\| \sum_{k=1}^{n} a_k Y_k \right\|_p} \right| \leq d(\nu, \lambda) \tag{54}$$

for every $n \geq 1$, $\nu, \lambda \in S$, real numbers $t, a_1, \ldots, a_n$ and i.i.d. sequences $(\xi_n^{(\nu)}), (\xi_n^{(\lambda)})$ with respective distributions ν and λ. Then for every $\varepsilon > 0$ there exists an increasing sequence (n_k) of positive integers such that

$$(1 - \varepsilon)\psi(a_1, \ldots, a_k) \leq \left\| \sum_{i=1}^{k} a_i X_{n_i} \right\|_p \leq (1 + \varepsilon)\psi(a_1, \ldots, a_k) \tag{55}$$

for every $k \geq 1$ and $(a_1, \ldots, a_k) \in \mathbb{R}^k$.

Condition (54) is the crucial equicontinuity assumption assuring uniformity in (55). The proof of Theorem 5.1 uses Theorem 5.2, Theorem 2.1, and concentration arguments.

Theorem 5.2 remains valid if we replace the linear functional $\| \sum_{k=1}^{n} a_k x_k \|$ by a general functional $f_n(a_1 x_1, \ldots, a_n x_n)$; the only difference is that the equicontinuity condition (54) should be replaced by the corresponding assumption for the functional f_n:

Theorem 5.3 *Let (X_n) be a tight sequence of r.v.'s and $f_k : \mathbb{R}^k \to \mathbb{R}$ ($k = 1, 2, \ldots$) be measurable functions. Let μ and (Y_n) denote the limit random measure and limit exchangeable sequence of (X_n), respectively. Put*

$$\psi^*(a_1, \ldots, a_k) = \mathbb{E} f_k(a_1 Y_1, \ldots, a_k Y_k).$$

Let S be a Borel subset of $(\mathcal{M}, \pi)$ such that $P(\mu \in S) = 1$. Assume that the following conditions are satisfied:

(a) $|\psi^*(a_1, \ldots, a_k)| \geq |\psi^*(a_1, 0, \ldots, 0)|,$

(b)   *There exists a separable metric d on S, Borel-equivalent to the Prohorov metric π such that*

$$|\mathbb{E} f_{k+1}(t, a_1 \xi_1^{(\nu)}, \ldots, a_k \xi_k^{(\nu)}) - \mathbb{E} f_{k+1}(t', a_1 \xi_1^{(\lambda)}, \ldots, a_k \xi_k^{(\lambda)})|$$

$$\leq |t - t'| + \psi^*(a_1, \ldots, a_k) \, d(\nu, \lambda) \tag{56}$$

for every $k \geq 1$, $\nu, \lambda \in S$, real numbers $t, a_1, \ldots, a_k$ and i.i.d. sequences $(\xi_n^{(\nu)}), (\xi_n^{(\lambda)})$ with respective distributions ν and λ.

Then for every $\varepsilon > 0$ there exists a subsequence (X_{n_k}) such that

$$(1 - \varepsilon)\psi^*(a_1, \ldots, a_k) \leq \mathbb{E} f_k(a_1 X_{n_1}, \ldots, a_k X_{n_k}) \leq (1 + \varepsilon)\psi^*(a_1, \ldots, a_k)$$

for any $k \geq 1$ and any $(a_1, \ldots, a_k) \in \mathbb{R}^k$.

Theorem 5.3 states the uniform asymptotic behavior of $\mathbb{E} f_k(a_1 X_{n_1}, \ldots, a_k X_{n_k})$ for lacunary sequences (n_k) and general functionals f_k. The equicontinuity condition (56) can be readily verified for various functionals f_k corresponding to actual limit theorems, leading to a widely applicable uniform version of the subsequence principle. For a detailed discussion and for the proof of the theorem we refer to Berkes and Tichy [25].

6   Resonance Theorems

Call a sequence (f_n) of measurable functions on $(0, 1)$ a *convergence system in measure* if for any real sequence (c_n) with $\sum_{n=1}^{\infty} c_n^2 < \infty$ the series $\sum_{n=1}^{\infty} c_n f_n$ converges in measure. The following interesting result was proved by Nikishin [55]:

Theorem *A function system (f_n) over $(0, 1)$ is a convergence system in measure if and only if for any $\varepsilon > 0$ there exists a measurable set $A_\varepsilon \subset (0, 1)$ with measure exceeding $1 - \varepsilon$ and a constant $K_\varepsilon > 0$ such that for all $N \geq 1$, $(a_1, \ldots, a_N) \in \mathbb{R}^N$ we have*

$$\int_{A_\varepsilon} \left(\sum_{k=1}^{N} a_k f_k \right)^2 dx \leq K_\varepsilon \sum_{k=1}^{N} a_k^2. \tag{57}$$

The sufficiency of (57) is obvious from Cauchy's criterion, the crucial statement is the converse: if a sequence (f_n) is a convergence system in measure then, except a subset of $(0, 1)$ with arbitrary small measure, (f_n) behaves like an orthonormal sequence. For reasons explained in [55], p. 128 such a theorem, and its analogue for function series $\sum_{n=1}^{\infty} c_n f_n$ with $(c_n) \in \ell^p$, $1 \leq p \leq \infty$ are called *resonance* theorems. The purpose of the present section is to give analogues of Nikishin's resonance theorem for the central limit theorem. Our first result is

Theorem 6.1 *Let (X_n) be a sequence of random variables over a probability space $(\Omega, \mathcal{F}, P)$ such that for any bounded real sequence (a_n) satisfying*

$$A_N^2 := \sum_{k=1}^{N} a_k^2 \to \infty \tag{58}$$

we have

$$\frac{1}{A_N} \sum_{k=1}^{N} a_k X_k \xrightarrow{d} N(0, 1). \tag{59}$$

Then for any $\varepsilon > 0$ there exists a set $A \subset \Omega$ with $P(A) \geq 1 - \varepsilon$ such that

$$\sup_n \int_A X_n^2 dP < \infty. \tag{60}$$

Note that we do not assume here the independence (or anything about the joint distribution) of the X_n. Because of that, the converse of the theorem is obviously false. As we will see, however, a necessary and sufficient "almost L^2 type" characterization of the weighted CLT can be given in the lacunary case. Call a sequence (X_n) of r.v.'s *nontrivial* if it has no subsequence converging with positive probability.

Theorem 6.2 *Let (X_n) be a nontrivial sequence of r.v.'s. Then the following statements are equivalent:*

(A) There exists a subsequence (X_{n_k}) and r.v.'s X, Y with $Y > 0$ such that for all further subsequences (X_{m_k}) of (X_{n_k}) we have

$$\frac{\sum_{k=1}^{N}(X_{m_k} - X)}{Y\sqrt{N}} \xrightarrow{d} N(0, 1) \tag{61}$$

relatively to any set $A \subset \Omega$ with $P(A) > 0$.

(B) For every $\varepsilon > 0$ there is a subsequence (X_{n_k}) and a set $A \subset \Omega$ with $P(A) \geq 1-\varepsilon$ such that

$$\sup_{k} \int_{A} X_{n_k}^2 dP < +\infty. \tag{62}$$

If (X_n) is determining with limit random measure μ, a further equivalent statement is

(C) We have

$$\int_{-\infty}^{+\infty} x^2 d\mu(x) < +\infty \quad a.s. \tag{63}$$

For the proof we refer to Berkes and Tichy [24].

7 Series with Random Gaps

In this chapter we investigate the behavior of trigonometric sums $S_N = \sum_{k=1}^{N} \sin n_k x$, where (n_k) is an increasing random sequence of integers. There are many different types of such random sequences and we investigate the simplest case when $n_1, n_2, \ldots$ are independent random variables having discrete uniform distribution on disjoint blocks $I_1, I_2, \ldots$ of integers. The case when $\cup_{k=1}^{\infty} I_k = \mathbb{N}$ and $|I_k|$ is constant, or tend to $+\infty$ was settled by Berkes [14] and Bobkov and Götze [27] and in the present chapter we investigate the case of general I_k, exhibiting a number of interesting new phenomena. We show that S_N has a decomposition $S_N^{(1)} + S_N^{(2)}$, where $S_N^{(1)}$ satisfies, with probability 1, a self-normalized central limit theorem and under mild regularity conditions on the sizes $|I_k|$ of the blocks I_k, $S_N^{(1)}/\sqrt{N}$ has a pure or mixed Gaussian limit distribution. Moreover, $S_N^{(2)}$ is a nonrandom trigonometric sum, asymptotically independent of $S_N^{(1)}$, whose asymptotic distribution depends sensitively of the gaps Δ_k between the blocks I_k and which can be non-Gaussian.

Theorem 7.1 *Let $I_1, I_2, \ldots$ be disjoint intervals of positive integers with cardinalities $|I_k| \geq 2$ and let $n_1, n_2, \ldots$ be independent random variables defined on a probability space $(\Omega, \mathcal{A}, \mathbb{P})$ such that n_k is uniformly distributed on I_k. Let*

$$\lambda_k(x) = \mathbb{E}(\sin n_k x), \quad \gamma_N^2(x) = \sum_{k=1}^{N}(\sin n_k x - \lambda_k(x))^2. \tag{64}$$

Then $\mathbb{P}$-almost surely

$$\frac{1}{\gamma_N(x)}\sum_{k=1}^{N}(\sin n_k x - \lambda_k(x)) \xrightarrow{d} N(0, 1) \tag{65}$$

with respect to the probability space $((0, 2\pi), \mathcal{B}, \lambda)$, where $\mathcal{B}$ is the Borel σ-algebra and λ is normalized Lebesgue measure on $(0, 2\pi)$. If the asymptotic densities μ_d of the sets $\{k \in \mathbb{N} : |I_k| = d\}$, $d = 1, 2, \ldots$ exist, then for every $x \in \mathbb{R}$ we have

$$\lim_{N\to\infty} \frac{\gamma_N^2(x)}{N} = g(x) \quad \mathbb{P} - a.s. \tag{66}$$

where

$$g(x) = \frac{1}{2} - \sum_{d=1}^{\infty}\frac{\sin^2(dx/2)}{d^2 \sin^2(x/2)}\mu_d \tag{67}$$

and

$$\frac{1}{\sqrt{N}}\sum_{k=1}^{N}(\sin n_k x - \lambda_k(x)) \xrightarrow{d} N(0, g). \tag{68}$$

Note that the self-normalized CLT (65) holds for $\sin n_k x - \lambda_k(x)$ without any regularity condition on the sequence $|I_k|$; in particular, the existence of the asymptotic densities μ_d is not required for (68). Without the existence of μ_d, however, the sequence $\gamma_N(x)^2/N$ in (66) can converge to different functions g along different subsequences and thus the limit distribution of $N^{-1/2}\sum_{k=1}^{N}(\sin n_k x - \lambda_k(x))$ may not exist.

If there is no gap between the blocks I_k, i.e., $\cup_{k=1}^{\infty}I_k = \mathbb{N}$ and $|I_k| \uparrow \infty$, then Abel rearrangement shows that $\sum_{k=1}^{N}\lambda_k(x) = O(1)$, and thus in (65) the centering factor $\lambda_k(x)$ can be omitted. For a motivation of (65) and the centering factors $\lambda_k(x)$, let us note that for any fixed $x \in \mathbb{R}$ the law of the iterated logarithm for independent bounded r.v.'s implies

$$\limsup_{N\to\infty} \frac{\sum_{k=1}^{N}(\sin n_k x - \lambda_k(x))}{(2\gamma_N^{*2}(x) \log\log \gamma_N^{*2}(x))^{1/2}} = 1 \quad \mathbb{P} - a.s., \tag{69}$$

where

$$\gamma_N^{*2}(x) = \sum_{k=1}^{N} \mathbb{E}(\sin n_k x - \lambda_k(x))^2 \sim \gamma_N^2(x) \tag{70}$$

and the last relation follows from the strong law of large numbers. Clearly, (69) and (70) yield for any fixed $x \in \mathbb{R}$

$$\limsup_{N \to \infty} \frac{\sum_{k=1}^{N}(\sin n_k x - \lambda_k(x))}{(2\gamma_N^2(x) \log \log \gamma_N^2(x))^{1/2}} = g(x) \qquad \mathbb{P} - \text{a.s.} \tag{71}$$

By Fubini's theorem, with $\mathbb{P}$-probability 1, (71) holds for almost all $x \in (0, 2\pi)$ providing the LIL corresponding to (65). Of course, Fubini's theorem cannot be applied for distributional limit theorems like the CLT (and in case of the CLT, the factor $g(x)$ in the denominator cannot be brought to the right-hand side) and the proof of (68) requires an elaborate argument.

We pass now to the study of non-centered partial sums $S_N = \sum_{k=1}^{N} \sin n_k x$. In contrast to $S_N - \mathbb{E}S_N$, the behavior of S_N depends on the size of the gaps Δ_k between I_k and I_{k+1}: for $\Delta_k = 0$ the limit distribution of $S_N / \sqrt{N}$ is mixed Gaussian with $\mathbb{P}$-probability 1 by a result of Bobkov and Götze [27] and for exponentially growing Δ_k the Salem–Zygmund CLT implies that $S_N / \sqrt{N} \xrightarrow{d} N(0, 1/2)$. Our next result shows that in regular cases $(S_N - \mathbb{E}S_N) / \sqrt{N}$ and $\mathbb{E}S_N / \sqrt{N}$ are asymptotically independent, reducing the behavior of $S_N / \sqrt{N}$ to that of the nonrandom trigonometric sum $\mathbb{E}S_N / \sqrt{N}$.

Theorem 7.2 *Let $I_1, I_2, \ldots$ be disjoint intervals of positive integers such that the sets $\{k \in \mathbb{N} : |I_k| = d\}, d = 1, 2, \ldots$ have asymptotic densities μ_d. Let $n_1, n_2, \ldots$ be independent random variables defined on a probability space $(\Omega, \mathcal{A}, \mathbb{P})$ such that n_k is uniformly distributed on I_k. Let $\lambda_k(x)$ and $g(x)$ be defined by (64) and (67). Then $\mathbb{P}$-almost surely*

$$\frac{1}{\sqrt{Ng(x)}} \sum_{k=1}^{N}(\sin n_k x - \lambda_k(x)) \xrightarrow{d} N(0, 1/2) \tag{72}$$

and

$$\frac{1}{\sqrt{N}} \sum_{k=1}^{N}(\sin n_k x - \lambda_k(x)) \xrightarrow{d} F \tag{73}$$

with respect to the probability space $((0, 2\pi), \mathcal{B}, \lambda)$, where F is the mixed Gaussian distribution with characteristic function

$$\phi(\lambda) = \int_0^{2\pi} \exp\left(-\frac{\lambda^2}{2}g(x)\right) dx. \tag{74}$$

If in addition we have

$$\frac{1}{\sqrt{N}} \sum_{k=1}^{N} \lambda_k(x) \xrightarrow{d} G \tag{75}$$

with respect to any interval $E \subset (0, 2\pi)$ with positive measure, then $\mathbb{P}$-almost surely

$$\left(\frac{1}{\sqrt{N}} \sum_{k=1}^{N} (\sin n_k x - \lambda_k(x)), \; \frac{1}{\sqrt{N}} \sum_{k=1}^{N} \lambda_k(x) \right) \xrightarrow{d} (F, G) \tag{76}$$

where the components of the limit vector are independent.

Theorem 7.2 shows that $S_N^{(1)} = S_N - \mathbb{E}S_N$ and $S_N^{(2)} = \mathbb{E}S_N$ are asymptotically independent and thus we can study their contributions separately. Note that $S_N^{(2)}$, the averaged version of S_N, is a nonrandom trigonometric sum. In the case when $|I_k| = d$ for all k, we have

$$\lambda_k(x) = \mathbb{E}(\sin n_k x) = d^{-1} \sum_{j \in I_k} \sin jx = \frac{\sin(dx/2)}{d \sin(x/2)} \sin(A_k + d/2 + 1/2)x \tag{77}$$

where A_k is the smallest integer of I_k and thus

$$S_N^{(2)} = \frac{\sin(dx/2)}{d \sin(x/2)} \sum_{k=1}^{N} \sin(A_k + d/2 + 1/2)x.$$

As one can show, if the gaps $\Delta_k = A_{k+1} - A_k - d$ between the intervals remain constant or if the A_k are integers and $\Delta_k \uparrow \infty$, $\Delta_k = O(k^\gamma)$ with $\gamma < 1/4$ (small gaps), then (65) holds with $\lambda_k = 0$, i.e., without a centering factor. At the other end of the spectrum, i.e., for rapidly increasing A_k, the centering factors themselves contribute to the limit distribution, i.e.,

$$\frac{1}{\sqrt{N}} \sum_{k=1}^{N} \lambda_k(x) \tag{78}$$

has a nondegenerate limit distribution. More precisely, if A_k satisfies the Erdős gap condition

$$A_{k+1}/A_k \geq 1 + c_k/\sqrt{k}, \qquad c_k \to \infty \tag{79}$$

then (78) has the limit distribution with characteristic function

$$\phi(\lambda) = \frac{1}{2\pi} \int_0^{2\pi} \exp\left(-\frac{\lambda^2}{4} \cdot \frac{\sin^2(dx/2)}{d^2 \sin^2(x/2)}\right) dx \tag{80}$$

and thus by the asymptotic independence of the components of (76) it follows that

$$N^{-1/2} \sum_{k=1}^{N} \sin n_k x \tag{81}$$

has a pure Gaussian limit distribution $N(0, 1/2)$. Since in this case (n_k) also satisfies the analogue of (79), the asymptotic normality of (81) follows from Erdős' central limit theorem [32] even for nonrandom (n_k), i.e., in this case Theorem 7.2 reduces to a result in classical lacunarity theory. It is interesting to note that in this case the pure Gaussian limit distribution of $S_N/\sqrt{N}$ is obtained as the convolution of two mixed Gaussian distributions. In the intermediate case between slowly and rapidly increasing (n_k), the centering factors $\lambda_k(x)$ in (64) may or may not contribute to the limit distribution F and F may be non-Gaussian. In view of (77), from the results of Berkes [16] it follows that there exist sequences (A_k) satisfying (79) with $c_k \to \infty$ replaced by $c_k = c > 0$ such that (78) has a non-Gaussian limit distribution and for any positive sequence $c_k \to 0$ there exist sequences (A_k) satisfying (79) such that (78) tends to 0 in probability. This shows that non-Gaussian limits of (78) can occur arbitrary close to the gap condition (79), i.e., (79) is critical in the theory. Theorem 7.2 also shows that the limit distribution of (81), if it exists, is the convolution of a mixed normal distribution and the limit distribution of a normed trigonometric sum with nonrandom frequencies $A_k + (d + 1)/2$. The asymptotic behavior of such nonrandom sums is an arithmetic rather than a probabilistic problem and we do not discuss it here.

For the proof of the results in this chapter we refer to Berkes and Raseta [19].

8 Discrepancy of Lacunary Series

To conclude our paper, we mention here briefly another important field of uniform limit theorems for lacunary series, namely metric results for the discrepancy of lacunary series $\{n_k x\}$. For a survey of the field until 2011, see Aistleitner and Berkes [4]. Given a sequence $(x_k)_{k \geq 1}$ of real numbers, the discrepancy $D_N(\{x_k\})$ is defined by

$$D_N(\{x_k\}) = \sup_{0 \leq a < b < 1} \left| \frac{1}{N} \sum_{k=1}^{N} I_{(a,b)}(x_k) - (b - a) \right|, \tag{82}$$

where $I_{(a,b)}$ denotes the indicator function of (a,b), extended with period 1. Verifying a conjecture of Erdős and Gál, Philipp [56, 57] proved the following result:

Theorem 8.1 *Let (n_k) be a sequence of positive integers satisfying the Hadamard gap condition (1). Then we have for almost all x*

$$\frac{1}{4\sqrt{2}} < \limsup_{N \to \infty} \frac{N D_N(\{n_k x\})}{\sqrt{2N \log \log N}} \le C \tag{83}$$

where $C = 166 + 664(q^{1/2} - 1)^{-1}$.

Note that by the Chung–Smirnov law of the iterated logarithm for empirical distribution functions (see, e.g., Shorack and Wellner [60], p. 504), the discrepancy $D_N(\{\xi_k\})$ of an i.i.d. sequence of random variables with uniform distribution in $(0, 1)$ satisfies the LIL (83) with the limsup equal to $1/2$ a.s. In contrast, Philipp's theorem does not state that the limsup equals $1/2$ or even if the limsup is constant almost everywhere, and these statements are generally not true. The first examples for a nonconstant limsup in (83) were given by Aistleitner [1] and Fukuyama [37]. Fukuyama [35] proved that if $n_k = \theta^k$, $\theta > 1$ and θ is not the n-th root of a rational number for any $n \ge 1$, then the limsup in (83) equals $1/2$. He also computed the limsup for $\theta = \sqrt[n]{p/q}$, $(p, q) = 1$ when either $q = 1$, or both p and p are odd. The case when one of the p and q is odd and the other one is even was settled by Fukuyama and Yamashita [39] for $\theta \ge \theta_0$; the remaining cases present extreme technical difficulties and remain open. Similar results hold for $\theta < -1$, when the sequence (n_k) has alternating signs. Aistleitner [2] proved that the limsup is $1/2$ provided (n_k) satisfies a Diophantine condition slightly stronger than $\mathbf{D}_2^{(s)}$ in Sect. 3. Under $\mathbf{D}_2^{(s)}$ the limit distribution of $\sqrt{N} D_N(\{n_k x\})$ was determined by Aistleitner and Berkes [5]. For the asymptotic behavior of $D_N(\{n_k x\})$ for random sequences (n_k) see Weber [66], Berkes and Weber [26], and Berkes and Raseta [18].

Acknowledgement The research is supported by FWF grant P24302-N18 and NKFIH grant K 108615.

References

1. C. Aistleitner, Irregular discrepancy behavior of lacunary series II. Monatsh. Math. **161**, 255–270 (2010)
2. C. Aistleitner, On the law of the iterated logarithm for the discrepancy of lacunary sequences. Trans. Am. Math. Soc. **362**, 5967–5982 (2010)
3. C. Aistleitner, I. Berkes, On the central limit theorem for $f(n_k x)$. Prob. Theory Rel. Fields **146**, 267–289 (2010)
4. C. Aistleitner, I. Berkes, Probability and metric discrepancy theory. Stochastics Dyn. **11**, 183–207 (2011)
5. C. Aistleitner, I. Berkes, Limit distributions in metric discrepancy theory. Monatsh. Math. **169**, 253–265 (2013)

6. C. Aistleitner, I. Berkes, R. Tichy, On permutations of Hardy-Littlewood-Pólya sequences. Trans. Am. Math. Soc. **363**, 6219–6244 (2011)
7. C. Aistleitner, I. Berkes, R. Tichy, On the asymptotic behavior of weakly lacunary sequences. Proc. Am. Math. Soc. **139**, 2505–2517 (2011)
8. C. Aistleitner, I. Berkes, R. Tichy, On permutations of lacunary series. RIMS Kokyuroku Bessatsu B **34**, 1–25 (2012)
9. C. Aistleitner, I. Berkes, R. Tichy, On the system $f(nx)$ and probabilistic number theory, in *Anal. Probab. Methods Number Theory*, Vilnius, 2012, ed. by E. Manstavicius et al., pp. 1–18
10. C. Aistleitner, I. Berkes, R. Tichy, On the law of the iterated logarithm for permuted lacunary sequences. Proc. Steklov Inst. Math. **276**, 3–20 (2012)
11. D.J. Aldous, Limit theorems for subsequences of arbitrarily-dependent sequences of random variables. Z. Wahrsch. verw. Gebiete **40**, 59–82 (1977)
12. D.J. Aldous, Subspaces of L_1 via random measures. Trans. Am. Math. Soc. **267**, 445–463 (1981)
13. F. Amoroso, E. Viada, Small points on subvarieties of a torus. Duke Math. J. **150**, 407–442 (2009)
14. I. Berkes, A central limit theorem for trigonometric series with small gaps. Z. Wahrsch. verw. Gebiete **47**, 157–161 (1979)
15. I. Berkes, On almost symmetric sequences in L_p. Acta Math. Hung. **54**, 269–278 (1989)
16. I. Berkes, Nongaussian limit distributions of lacunary trigonometric series. Can. J. Math. **43**, 948–959 (1991)
17. I. Berkes, E. Péter, Exchangeable random variables and the subsequence principle. Prob. Theory Rel. Fields **73**, 395–413 (1986)
18. I. Berkes, M. Raseta, On the discrepancy and empirical distribution function of $\{n_k\alpha\}$. Unif. Distr. Theory **10**, 1–17 (2015)
19. I. Berkes, M. Raseta, On trigonometric sums with random frequencies, Preprint 2016
20. I. Berkes, H.P. Rosenthal, Almost exchangeable sequences of random variables. Z. Wahrsch. verw. Gebiete **70**, 473–507 (1985)
21. I. Berkes, R. Tichy, On permutation-invariance of limit theorems. J. Complexity **31**, 372–379 (2015)
22. I. Berkes, R. Tichy, Lacunary series and stable distributions, in *Mathematical Statistics and Limit Theorems. Festschrift for P. Deheuvels*, ed. by M. Hallin, D.M. Mason, D. Pfeifer, J. Steinebach (Springer, Berlin, 2015), pp. 7–19
23. I. Berkes, R. Tichy, The Kadec-Pełczynski theorem in L^p, $1 \leq p < 2$. Proc. Am. Math. Soc. **144**, 2053–2066 (2016)
24. I. Berkes, R. Tichy, Resonance theorems and the central limit theorem for lacunary series, Preprint 2016
25. I. Berkes, R. Tichy, A uniform version of the subsequence principle, Preprint 2016
26. I. Berkes, M. Weber, On the convergence of $\sum c_k f(n_k x)$. Mem. Am. Math. Soc. **201**(943), viii+72 pp. (2009)
27. S. Bobkov, F. Götze, Concentration inequalities and limit theorems for randomized sums. Probab. Theory Rel. Fields **137**, 49–81 (2007)
28. S.D. Chatterji, A general strong law. Invent. Math. **9**, 235–245 (1970)
29. S.D. Chatterji, Un principe de sous-suites dans la théorie des probabilités, in *Séminaire des probabilités VI, Strasbourg*. Lecture Notes in Mathematics, vol. 258 (Springer, Berlin, 1972), pp. 72–89
30. S.D. Chatterji, A principle of subsequences in probability theory: the central limit theorem. Adv. Math. **13**, 31–54 (1974)
31. S.D. Chatterji, A subsequence principle in probability theory II. The law of the iterated logarithm. Invent. Math. **25**, 241–251 (1974)
32. P. Erdős, On trigonometric sums with gaps. Magyar Tud. Akad. Mat. Kut. Int. Közl. **7**, 37–42 (1962)
33. P. Erdős, I.S. Gál, On the law of the iterated logarithm. Proc. Nederl. Akad. Wetensch. Ser. A **58**, 65–84 (1955)

34. J.-H. Evertse, R.H.-P. Schlickewei, W.M. Schmidt, Linear equations in variables which lie in a multiplicative group. Ann. Math. **155**, 807–836 (2002)
35. K. Fukuyama, The law of the iterated logarithm for discrepancies of $\{\theta^n x\}$. Acta Math. Hungar. **118**, 155–170 (2008)
36. K. Fukuyama, The law of the iterated logarithm for the discrepancies of a permutation of $\{n_k x\}$. Acta Math. Acad. Sci. Hung. **123**, 121–125 (2009)
37. K. Fukuyama, A law of the iterated logarithm for discrepancies: non-constant limsup. Monatsh. Math. **160**, 143–149 (2010)
38. K. Fukuyama, B. Petit, Le théorème limite central pour les suites de R. C. Baker. Ergodic Theory Dyn. Syst. **21**, 479–492 (2001)
39. F.K. Fukuyama, M. Yamashita, Metric discrepancy results for geometric progressions with large ratios. Monatsh. Math. **180**, 731–742 (2016)
40. J. Galambos, *The Asymptotic Theory of Extreme Order Statistics*, 2nd ed. (Robert E. Krieger Publishing Co., Melbourne, FL, 1987)
41. V.F. Gaposhkin, Lacunary series and independent functions. Russian Math. Surv. **21**, 1–82 (1966)
42. V.F. Gaposhkin, On some systems of almost independent functions. Siberian Math. J. **9**, 198–210 (1968)
43. V.F. Gaposhkin, The central limit theorem for some weakly dependent sequences. Theory Probab. Appl. **15**, 649–666 (1970)
44. A. Garsia, Existence of almost everywhere convergent rearrangements for Fourier series of L_2 functions. Ann. Math. **79**, 623–629 (1964)
45. S. Guerre, Types and suites symétriques dans L^p, $1 \leq p < +\infty$. Israel J. Math. **53**, 191–208 (1986)
46. J. Hawkes, Probabilistic behaviour of some lacunary series. Z. Wahrsch. verw. Gebiete **53**, 21–33 (1980)
47. M. Kac, On the distribution of values of sums of the type $\sum f(2^k t)$. Ann. Math. **47**, 33–49 (1946)
48. M. Kac, Probability methods in some problems of analysis and number theory. Bull. Am. Math. Soc. **55**, 641–665 (1949)
49. M.I. Kadec, W. Pełczyński, Bases, lacunary sequences and complemented subspaces in the spaces L_p. Studia Math. **21**, 161–176 (1961/1962)
50. J. Komlós, A generalization of a problem of Steinhaus. Acta Math. Acad. Sci. Hungar. **18**, 217–229 (1967)
51. J. Komlós, Every sequence converging to 0 weakly in L_2 contains an unconditional convergence sequence. Ark. Math. **12**, 41–49 (1974)
52. D.E. Menshov, Sur la convergence et la sommation des séries de fonctions orthogonales. Bull. Soc. Math. France **64**, 147–170 (1936)
53. W. Morgenthaler, A central limit theorem for uniformly bounded orthonormal systems. Trans. Am. Math. Soc. **79**, 281–311 (1955)
54. R. Nair, On strong uniform distribution. Acta Arith. **56**, 183–193 (1990)
55. E.M. Nikishin, Resonance theorems and superlinear operators. Russian Math. Surv. **25/6**, 125–187 (1970)
56. W. Philipp, Limit theorems for lacunary series and uniform distribution mod 1. Acta Arith. **26**, 241–251 (1974/1975)
57. W. Philipp, The functional law of the iterated logarithm for empirical distribution functions of weakly dependent random variables. Ann. Probab. **5**, 319–350 (1977)
58. R. Ranga Rao, Relations between weak and uniform convergence of measures with applications. Ann. Math. Stat. **33**, 659–680 (1962)
59. R. Salem, A. Zygmund, On lacunary trigonometric series. Proc. Natl. Acad. Sci. USA **33**, 333–338 (1947)
60. G. Shorack, J. Wellner, *Empirical Processes with Applications in Statistics* (Wiley, New York, 1986)
61. S. Takahashi, On lacunary trigonometric series. Proc. Jpn. Acad. **41**, 503–506 (1965)

62. S. Takahashi, On the law of the iterated logarithm for lacunary trigonometric series. Tohoku Math. J. **24**, 319–329 (1972)
63. S. Takahashi, On the law of the iterated logarithm for lacunary trigonometric series. II. Tohoku Math. J. **27**, 391–403 (1975)
64. R. Tijdemann, On integers with many small prime factors. Compositio Math. **26**, 319–330 (1973)
65. P. Uljanov, Solved and unsolved problems in the theory of trigonometric and orthogonal series (Russian). Uspehi Mat. Nauk **19/1**, 1–69 (1964)
66. M. Weber, Discrepancy of randomly sampled sequences of reals. Math. Nachr. **271**, 105–110 (2004)
67. M. Weiss, On the law of the iterated logarithm for uniformly bounded orthonormal systems. Trans. Am. Math. Soc. **92**, 531–553 (1959)
68. A. Zygmund, *Trigonometric Series*, vols. I, II, 3rd ed. (Cambridge University Press, Cambridge, 2002)

Diversity in Parametric Families of Number Fields

Yuri Bilu and Florian Luca

To Robert Tichy, a colleague and a friend

Abstract Let X be a projective curve defined over $\mathbb{Q}$ and $t \in \mathbb{Q}(X)$ a non-constant rational function of degree $\nu \geq 2$. For every $n \in \mathbb{Z}$ pick $P_n \in X(\bar{\mathbb{Q}})$ such that $t(P_n) = n$. A result of Dvornicich and Zannier implies that, for large N, among the number fields $\mathbb{Q}(P_1), \ldots, \mathbb{Q}(P_N)$ there are at least $cN/\log N$ distinct; here, $c > 0$ depends only on the degree ν and the genus $\mathbf{g} = \mathbf{g}(X)$. We prove that there are at least $N/(\log N)^{1-\eta}$ distinct fields, where $\eta > 0$ depends only on ν and $\mathbf{g}$.

1 Introduction

Everywhere in this paper "curve" means "smooth geometrically irreducible projective algebraic curve."

Let X be a curve over $\mathbb{Q}$ of genus $\mathbf{g}$ and $t \in \mathbb{Q}(X)$ a non-constant rational function of degree $\nu \geq 2$. We fix, once and for all, an algebraic closure $\bar{\mathbb{Q}}$. Our starting point is the celebrated Hilbert Irreducibility Theorem.

Theorem 1.1 (Hilbert) *In the above set-up, for infinitely many $n \in \mathbb{Z}$ the fiber $t^{-1}(n) \subset X(\bar{\mathbb{Q}})$ is $\mathbb{Q}$-irreducible; that is, the Galois group $G_{\bar{\mathbb{Q}}/\mathbb{Q}}$ acts on $t^{-1}(n)$ transitively.*

This can also be re-phrased as follows: for every $n \in \mathbb{Z}$ pick $P_n \in t^{-1}(n)$; then for infinitely many $n \in \mathbb{Z}$ we have $[\mathbb{Q}(P_n) : \mathbb{Q}] = \nu$.

Y. Bilu (✉)
Institut de Mathématiques de Bordeaux, Université de Bordeaux & CNRS, Talence, France
e-mail: yuri@math.u-bordeaux.fr

F. Luca
School of Mathematics, Wits University, Johannesburg, South Africa
e-mail: Florian.Luca@wits.ac.za

C. Elsholtz, P. Grabner (eds.), *Number Theory – Diophantine Problems,
Uniform Distribution and Applications*, DOI 10.1007/978-3-319-55357-3_7

"Infinitely many" in the Hilbert Irreducibility Theorem means, in fact, "overwhelmingly many": for sufficiently large positive N we have

$$\left|\{n \in [1, N] \cap \mathbb{Z} : t^{-1}(n) \text{ is reducible}\}\right| \leq c(v)N^{1/2}. \tag{1}$$

Everywhere in the introduction "sufficiently large" means "exceeding a certain positive number depending on X and t."

For the proof of (1) we invite the reader to consult Chap. 9 of Serre's book [8]. See, in particular, Sect. 9.2 and the theorem on page 134 of [8], where (1) is proved with $\mathbb{Q}$ replaced by an arbitrary number field and $\mathbb{Z}$ by its ring of integers.

Hilbert's Irreducibility Theorem, however, does not answer the following natural question: among the field $\mathbb{Q}(P_n)$, are there "many" distinct (in the fixed algebraic closure $\bar{\mathbb{Q}}$)? This question is addressed in the article of Dvornicich and Zannier [6], where the following theorem is proved (see [6, Theorem 2(a)]).

Theorem 1.2 (Dvornicich, Zannier) *In the above set-up, there exists a real number $c = c(\mathbf{g}, v) > 0$ such that for sufficiently large integer N the number field $\mathbb{Q}(P_1, \ldots, P_N)$ is of degree at least $e^{cN/\log N}$ over $\mathbb{Q}$.*

One may note that the statement holds true independently of the choice of the points P_n.

An immediate consequence is the following result.

Corollary 1.3 *In the above set-up, there exists a real number $c = c(\mathbf{g}, v) > 0$ such that for every sufficiently large integer N, there are at least $cN/\log N$ distinct fields among the number fields $\mathbb{Q}(P_1), \ldots, \mathbb{Q}(P_N)$.*

Theorem 1.2 is best possible, as obvious examples show. Say, if X is (the projectivization of) the plane curve $t = u^2$ and t is the coordinate function, then the field

$$\mathbb{Q}(P_1, \ldots, P_N) = \mathbb{Q}(\sqrt{1}, \sqrt{2}, \ldots, \sqrt{N}) = \mathbb{Q}(\sqrt{p} : p \leq N)$$

is of degree $2^{\pi(N)} \leq e^{cN/\log N}$.

On the contrary, Corollary 1.3 does not seem to be best possible. For instance, in the same example, if n runs the square-free numbers among $1, \ldots, N$, then the fields $Q(P_n) = \mathbb{Q}(\sqrt{n})$ are pairwise distinct. It is well known that among $1, \ldots, N$ there are, asymptotically, $\zeta(2)^{-1}N$ square-free numbers as $N \to \infty$.

We suggest the following conjecture.

Conjecture 1.4 (Weak Diversity Conjecture) Let X be a curve over $\mathbb{Q}$ and $t \in \mathbb{Q}(X)$ a non-constant $\mathbb{Q}$-rational function of degree at least 2. Then there exists a real number $c > 0$ such that for every sufficiently large integer N, among the number fields $\mathbb{Q}(P_1), \ldots, \mathbb{Q}(P_N)$ there are at least cN distinct.

There is also a stronger conjecture, attributed in [6, 7] to Schinzel, which relates to Theorem 1.2 in the same way as Conjecture 1.4 relates to Corollary 1.3. To state it, we need to recall the notion of *critical value*.

We call $\alpha \in \bar{\mathbb{Q}} \cup \{\infty\}$ a *critical value* (or a *branch point*) of t if the rational function[1] $t - \alpha$ has at least one multiple zero in $X(\bar{\mathbb{Q}})$. It is well known that any rational function $t \in \bar{\mathbb{Q}}(X)$ has at most finitely many critical values, and that t has at least 2 distinct critical values if it is of degree $\nu \geq 2$ (a consequence of the Riemann–Hurwitz formula). In particular, in this case t admits at least one *finite* critical value.

Conjecture 1.5 (Strong Diversity Conjecture (Schinzel)) In the set-up of Conjecture 1.4, assume that either t has at least one finite critical value not belonging to $\mathbb{Q}$ or the field extension $\bar{\mathbb{Q}}(X)/\bar{\mathbb{Q}}(t)$ is not abelian. Then there exists a real number $c > 0$ such that for every sufficiently large integer N the number field $\mathbb{Q}(P_1, \ldots, P_N)$ is of degree at least e^{cN} over $\mathbb{Q}$.

As Dvornicich and Zannier remark, the hypothesis in the Strong Diversity Conjecture is necessary. Indeed, when all critical values belong to $\mathbb{Q}$ and the field extension $\bar{\mathbb{Q}}(X)/\bar{\mathbb{Q}}(t)$ is abelian, it follows from Kummer's Theory that $\mathbb{Q}(X)$ is contained in a field of the form $L(t, (t - \alpha_1)^{1/e_1}, \ldots, (t - \alpha_s)^{1/e_s})$, where L is a number field, $\alpha_1, \ldots, \alpha_s$ are rational numbers and $e_1, \ldots, e_s$ are positive integers. Clearly, in this case the degree of the number field generated by $P_1, \ldots, P_N$ cannot exceed $e^{cN/\log N}$ for some $c > 0$.

On the other hand, Conjecture 1.4 does hold [2] in the case excluded in Conjecture 1.5, when the finite critical values of t are all in $\mathbb{Q}$, and the field extension $\bar{\mathbb{Q}}(X)/\bar{\mathbb{Q}}(t)$ is abelian. Hence, *the Strong Conjecture implies the Weak Conjecture.*

Dvornicich and Zannier [6, 7] obtain several results in favor of Schinzel's Conjecture. In particular, they show that Conjecture 1.5 holds true in the following cases:

- when t admits a critical value of degree 2 or 3 over $\mathbb{Q}$, see [6, Theorem 2(b)];
- when all finite critical values are in $\mathbb{Q}$ and the Galois group of the normal closure of $\bar{\mathbb{Q}}(X)$ over $\bar{\mathbb{Q}}(t)$ is "sufficiently large" (for instance, symmetric or alternating), see [7].

A result of Corvaja and Zannier [3, Corollary 1] implies that, in the case when t has at least three zeros in $X(\bar{\mathbb{Q}})$, a number field K of degree ν or less may appear as $\mathbb{Q}(P_n)$ for at most $c(X, t, \nu)$ possible n. In particular, the Weak Conjecture holds in this case (but the Strong Conjecture remains open).

We mention also the work of Zannier [9], who studies the following problem: given a number field K, how many fields among $\mathbb{Q}(P_1), \ldots, \mathbb{Q}(P_N)$ contain K? He proves that, under suitable assumptions, the number of such fields is $o(N^\varepsilon)$ as $N \to \infty$ for any $\varepsilon > 0$.

In the present article we go a different way: instead of imposing additional restrictions on X and t, we work in full generality, improving on Corollary 1.3 quantitatively in the direction of Conjecture 1.4. Here is our principal result.

---

[1] Here and everywhere below we use the standard convention $t - \infty = t^{-1}$.

Theorem 1.6 *In the set-up of Conjecture 1.4, there exists a positive real number* $\eta = \eta(\mathbf{g}, v)$ *such that for every sufficiently large integer N, among the number fields* $\mathbb{Q}(P_1), \ldots, \mathbb{Q}(P_N)$ *there are at least* $N/(\log N)^{1-\eta}$ *distinct.*

The proof shows that $\eta = 10^{-6}\big((\mathbf{g} + v)\log(\mathbf{g} + v)\big)^{-1}$ would do.

1.1 Plan of the Article

In Sect. 2 we introduce the notation and recall basic facts, to be used throughout the article.

In Sect. 3 we review the argument of Dvornicich and Zannier, and explain how it should be modified for our purposes.

Sections 4–6 are the technical heart of the article. In Sects. 4 and 5 we introduce a certain set of square-free numbers and study its properties. A key lemma used in Sect. 5 is proved in Sect. 6.

After all this preparatory work, the proof of Theorem 1.6 becomes quite transparent, see Sect. 7.

2 Notation and Conventions

Unless the contrary is stated explicitly, everywhere in the article:

- n (with or without indexes) denotes a positive integer;
- m (with or without indexes) denotes a square-free positive integer;
- p (with or without indexes) denotes a prime number;
- x, y, z denote positive real numbers.

We use the notation

$$p_{\max}(n) = \max\{p : p \mid n\}, \quad p_{\min}(n) = \min\{p : p \mid n\}.$$

As usual, we denote by $\omega(n)$ (respectively, $\Omega(n)$) the number of prime divisors of n counted without (respectively, with) multiplicities.

For a separable polynomial $F(T) \in \mathbb{Z}[T]$ we denote:

- Δ_F the discriminant of F;
- $\mathscr{P}_F$ the set of p for which $F(T)$ has a root mod p, and which do not divide Δ_F.
- $\mathscr{M}_F$ the set of square-free integers composed of primes from $\mathscr{P}_F$.

By the Chebotarev Density Theorem, the set $\mathscr{P}_F$ is of positive density among all the primes. We call it the *Chebotarev density* of F and denote it by δ_F. Note that

$$\delta_F \geq \frac{1}{d}, \tag{2}$$

where $d = \deg F$.

3 The Argument of Dvornicich-Zannier

In this section we briefly review the beautiful ramification argument of Dvornicich and Zannier[2] and explain which changes are to be made therein to adapt it for proving Theorem 1.6.

Like in introduction, in this section "sufficiently large" means "exceeding some quantity depending on X and t."

Let $F(T) \in \mathbb{Z}[T]$ be the primitive separable polynomial whose roots are exactly the finite critical values of t, and let $d = \deg F$. Using the Riemann–Hurwitz formula, one bounds the total number of critical values by $2\mathbf{g} - 2 + 2\nu$, where $\mathbf{g} = \mathbf{g}(X)$ is the genus of the curve X. Hence

$$d \leq 2\mathbf{g} - 2 + 2\nu. \tag{3}$$

The basic properties of the polynomial $F(T)$ are summarized below.

A *For sufficiently large p, if p ramifies in $\mathbb{Q}(P)$ for some $P \in t^{-1}(n)$, then $p \mid F(n)$.*
B *For sufficiently large p, if $p \parallel F(n)$, then p ramifies in $\mathbb{Q}(P)$ for some $P \in t^{-1}(n)$.*
C *For all p not dividing the discriminant Δ_F (which is non-zero because F is a separable polynomial) the following holds: if for some n we have $p^2 \mid F(n)$, then $p \parallel F(n + p)$.*
D *For every $p \in \mathscr{P}_F$ there exists $n \leq 2p$ such that $p \parallel F(n)$.*
E *When n is sufficiently large, $F(n)$ has at most d prime divisors $p \geq n/4$.*

Here properties **A** and **B** are rather standard statements linking geometric and arithmetical ramification, see [1, Theorem 7.8].

Property **C** is very easy: write

$$F(n + p) \equiv F(n) + F'(n)p \bmod p^2.$$

If p^2 divides both $F(n)$ and $F(n + p)$, then $p \mid F'(n)$, which means that p must divide the discriminant Δ_F, a contradiction.

---

[2]In [6] they trace it back to the work of Davenport et al. [4] from sixties.

Property **D** follows from **C**, and property **E** is obvious: if there are $d + 1$ such primes, then $(n/4)^{d+1} \leq |F(n)|$, which is impossible for large n.

One may also note that our definition of the polynomial $F(T)$ is relevant only for properties **A** and **B**; the other properties hold for any separable polynomial $F(T) \in \mathbb{Z}[T]$.

Now we are ready to sketch the proof of Theorem 1.2. Denote by K_n the number field $\mathbb{Q}(t^{-1}(n))$, generated by all the points in the fiber of n, and by L_n the compositum of the fields $K_1, \ldots, K_n$. Then K_n is a Galois extension of $\mathbb{Q}$ containing $\mathbb{Q}(P_n)$, and L_n is a Galois extension of $\mathbb{Q}$ containing $\mathbb{Q}(P_1, \ldots, P_n)$.

We call p *primitive* for some n if p ramifies in K_n, but not in L_{n-1}. The observations above have the following two consequences.

F *Every sufficiently large $p \in \mathscr{P}_F$ is primitive for some $n \leq 2p$.*
G *Every sufficiently large n has at most d primitive $p \in [n/4, n]$.*

Here **F** follows from **B** and **D**, and **G** follows from **A** and **E**.
For a given N let S_N be the set of n with the property

$$n \text{ has a primitive } p \in [N/4, N/2].$$

It follows from **F** that $S_N \subset [1, N]$, and from **G**, the Chebotarev Theorem and the Prime Number Theorem that, for sufficiently large N

$$|S_N| \geq \frac{1}{d} \left| \mathscr{P}_F \cap [N/4, N/2] \right| \geq \frac{\delta_F}{5d} \frac{N}{\log N}.$$

Furthermore, let S_N' be the subset of S_N consisting of n such that the fiber $t^{-1}(n)$ is irreducible. The quantitative Hilbert Irreducibility Theorem 1 implies that, for large N we have $|S_N \smallsetminus S_N'| \leq c(v)N^{1/2}$, which means that, for large N,

$$|S_N'| \geq \frac{\delta_F}{6d} \frac{N}{\log N}.$$

It is clear that if n admits a primitive p then K_n is not contained in L_{n-1}. If, in addition to this, the fiber $t^{-1}(n)$ is irreducible, then $\mathbb{Q}(P_n)$ is not contained in $\mathbb{Q}(P_1, \ldots, P_{n-1})$, because in this case K_n is the Galois closure (over $\mathbb{Q}$) of $\mathbb{Q}(P_n)$. It follows that

$$[\mathbb{Q}(P_1, \ldots, P_N) : \mathbb{Q}] \geq 2^{|S_N'|},$$

which, in view of (2) and (3), proves Theorem 1.2.

The (already mentioned in Sect. 1) example of the curve $u = t^2$ suggests that we can make progress towards Conjecture 1.4 replacing prime numbers in the argument above by (suitably chosen) square-free numbers. This means that we have to obtain analogues of properties **F** and **G** above with primes replaced by square-free numbers.

Let m be a square-free integer, and n an arbitrary integer. We say that $m \| n$ if $m \mid n$ and $\gcd(m, n/m) = 1$.

A "square-free analogue" of $\mathbf{F}$ is relatively easy: one uses the following lemma, which generalizes property $\mathbf{C}$.

Lemma 3.1 *Let m be a square free positive integer, coprime with Δ_F and such that $p_{\min}(m) > \omega(m)$. Assume that for some n we have $m \mid F(n)$. Then there exists $\ell \in \{0, 1, \dots, \omega(m)\}$ such that $m \| F(n + \ell m)$.*

Proof Assume the contrary: for every $\ell \in \{0, 1, \dots, \omega(m)\}$ there exists $p \mid m$ such that $p^2 \mid f(n + \ell m)$. By the box principle some p would occur for two distinct values ℓ_1 and ℓ_2; we will assume that $0 \le \ell_1 < \ell_2 \le \omega(m)$. We obtain

$$
\begin{aligned}
0 &\equiv F(n + \ell_2 m) & \mod p^2 \\
&\equiv F(n + \ell_1 m) + F'(n + \ell_1 m)(\ell_2 - \ell_1)m & \mod p^2 \\
&\equiv F'(n + \ell_1 m)(\ell_2 - \ell_1)m & \mod p^2.
\end{aligned}
$$

We have $p \| m$ and, since

$$
0 < \ell_2 - \ell_1 \le \omega(m) < p_{\min}(m) \le p,
$$

we have $p \nmid (\ell_2 - \ell_1)$. Hence $p \mid F'(n + \ell_1 m)$, which implies that $p \mid \Delta_F$, a contradiction. $\qquad\square$

Recall that the set $\mathscr{P}_F$ consists of primes p not dividing the discriminant Δ_F and such that F has a root $\mod p$, and that $\mathscr{M}_F$ is the set of square-free numbers composed of primes from $\mathscr{P}_F$. The following consequence is immediate.

Corollary 3.2 *Let $m \in \mathscr{M}_F$ have the property $p_{\min}(m) > \omega(m)$. Then there exists $n \le m(\omega(m) + 1)$ such that $m \| f(n)$.*

Proof The Chinese Remainder Theorem implies that for any $m \in \mathscr{M}_F$ there exists $n \le m$ such that $m \mid F(n)$. Now use Lemma 3.1. $\qquad\square$

Call $m \in \mathscr{M}_F$ *primitive* for n if every $p \mid m$ ramifies in K_n, and for every $n' < n$ some $p \mid m$ does not ramify in $K_{n'}$. Combining Corollary 3.2 with property $\mathbf{A}$, we obtain a quite satisfactory generalization of property $\mathbf{F}$ to square-free numbers.

Corollary 3.3 *Let m be like in Corollary 3.2. Then m is primitive for some $n \le m(\omega(m) + 1)$.*

Another task to accomplish is extending to square-free numbers property $\mathbf{G}$. This is much more intricate, see Sects. 4–6.

4   A Special Set of Square-Free Numbers

In this section we fix a separable polynomial $F(T) \in \mathbb{Z}[T]$ of degree d and a real number ε satisfying $0 < \varepsilon \le 1/2$. "Sufficiently large" will always mean "exceeding a certain quantity depending on F and ε", and the constants implied by the " $O(\cdot)$ " and " $\ll$ " notation depend on F and ε unless the contrary is stated explicitly.

Recall that $\mathscr{P}_F$ denotes the set of primes p not dividing the discriminant Δ_F and such that F has a root $\bmod\, p$, and $\mathscr{M}_F$ denotes the set of the square-free numbers composed of primes from $\mathscr{P}_F$. Recall also that we denote by $\delta = \delta_F$ the density of $\mathscr{P}_F$. We have, as $x \to \infty$,

$$\left| \mathscr{P}_F \cap [0, x] \right| \sim \delta \frac{x}{\log x}, \qquad \left| \mathscr{M}_F \cap [0, x] \right| \sim \gamma \frac{x}{(\log x)^{1-\delta}}$$

where $\gamma = \gamma(F)$ is a certain positive real number.

Recall that, unless the contrary is stated explicitly, the letter n always denotes a positive integer, m a square-free positive integer, and p a prime number.

We fix a big positive real number x and set

$$\kappa = \log\log x, \qquad k = \lfloor \varepsilon\delta \log\log x \rfloor + 1, \qquad y = e^{(\log x)^{1-\varepsilon}}.$$

Furthermore, we denote by $\mathscr{M}_F(x)$ the set of $m \in \mathscr{M}_F$ satisfying

$$\frac{x}{2\kappa} \le m \le \frac{x}{\kappa}, \quad p_{\max}(m) \ge x^{9/10}, \quad p_{\min}(m) \ge y, \quad \omega(m) = k + 1.$$

Proposition 4.1 *We have* $|\mathscr{M}_F(x)| = x(\log x)^{-1+\varepsilon\delta+o(1)}$ *as* $x \to \infty$.

Proof If $m \in \mathscr{M}_F(x)$, then $m = P m_1$, where $P = p_{\max}(m) \ge x^{9/10}$. We denote by $\mathscr{M}_F'(x)$ be the set of such m_1's. Then $\mathscr{M}_F'(x) \subset \mathscr{M}_F$ and for every $m_1 \in \mathscr{M}_F'(x)$ we have

$$m_1 \le x^{1/10}, \quad p_{\min}(m_1) \ge y, \quad \omega(m_1) = k. \tag{4}$$

Let us count suitable P for a fixed m_1. These are exactly the primes $P \in \mathscr{P}_F$ from the interval $[x/(2\kappa m_1), x/(\kappa m_1)]$ satisfying $P \ge x^{9/10}$. The following observations are crucial.

- Since $m_1 \le x^{1/10}$, we have $x/(\kappa m_1) > x^{4/5}$ for sufficiently large x. Hence, for a fixed m_1, the number of suitable P is bounded from above by

$$\pi\left(\frac{x}{\kappa m_1}\right) \ll \frac{x}{\kappa m_1 \log x}.$$

- If $m_1 \le x^{1/10}/2\kappa$, then every prime $P \in \mathscr{P}_F \cap [x/(2\kappa m_1), x/(\kappa m_1)]$ is suitable. Hence, for a fixed $m_1 \le x^{1/10}/2\kappa$, the number of suitable P is bounded from below by

$$\pi_F\left(\frac{x}{\kappa m_1}\right) - \pi_F\left(\frac{x}{2\kappa m_1}\right) = \left(\frac{\delta}{2} + o(1)\right) \frac{x}{\kappa m_1 \log(x/(\kappa m_1))} \gg \frac{x}{\kappa m_1 \log x}.$$

Here, $\pi_F(T)$ counts the number of primes in $\mathscr{P}_F \cap [0, T]$.

Summing up over $m_1 \in \mathscr{M}'_F(x)$, we obtain

$$\frac{x}{\kappa \log x} \sum_{\substack{m_1 \in \mathscr{M}'_F(x) \\ m_1 \le x^{1/10}/2\kappa}} \frac{1}{m_1} \ll |\mathscr{M}_F(x)| \ll \frac{x}{\kappa \log x} \sum_{m_1 \in \mathscr{M}'_F(x)} \frac{1}{m_1}. \tag{5}$$

We will show that the right-hand side of (5) is bounded by $x(\log x)^{-1+\varepsilon\delta+o(1)}$ from above, and the left-hand side from below.

The **upper bound** is easy:

$$\sum_{m_1 \in \mathscr{M}'_F(x)} \frac{1}{m_1} \le \frac{1}{k!} \left(\sum_{\substack{y \le p \le x \\ p \in \mathscr{P}_F}} \frac{1}{p}\right)^k$$

$$\ll \frac{1}{(k/e)^k} \left((\delta + o(1)) \log \log x - (\delta + o(1)) \log \log y\right)^k$$

$$\ll \left(\frac{(e + o(1))\varepsilon\delta \log \log x}{k}\right)^k$$

$$= (\log x)^{\varepsilon\delta+o(1)} \tag{6}$$

as $x \to \infty$. Hence, $|\mathscr{M}_F(x)| \le x(\log x)^{-1+\varepsilon\delta+o(1)}$ as $x \to \infty$.

For the **lower bound**, set $z = x^{(1/11 \log \log x)}$ and $\mathscr{I} = [y, z]$ and consider the following two sets:

- the set $\mathscr{M}''_F(x)$ of square-free numbers m_1 with prime divisors in $\mathscr{P}_F \cap \mathscr{I}$ and with $\omega(m_1) = k$;
- the set $\mathscr{N}''_F(x)$ of *non-square-free* numbers n_1 with prime divisors in $\mathscr{P}_F \cap \mathscr{I}$ and with $\Omega(n_1) = k$.

Clearly, every $m_1 \in \mathscr{M}''_F(x)$ satisfies

$$m_1 \le x^{k/(11 \log \log x)} < x^{1/11} \le \frac{x^{1/10}}{2\kappa}$$

for large x. Hence the sum in the left-hand side of (5) can be bounded as follows:

$$\sum_{\substack{m_1 \in \mathcal{M}_F'(x) \\ m_1 \le x^{1/10}/2\kappa}} \frac{1}{m_1} \ge \sum_{m_1 \in \mathcal{M}_F''(x)} \frac{1}{m_1}$$

$$\ge \frac{1}{k!} \left(\sum_{p \in \mathcal{P}_F \cap [y,z]} \frac{1}{p} \right)^k - \sum_{n_1 \in \mathcal{N}_F''(x)} \frac{1}{n_1}. \tag{7}$$

We need to estimate the first sum in (7) from below and the second sum from above. For the first sum we use the same argument as before and get

$$\frac{1}{k!} \left(\sum_{p \in \mathcal{P}_F \cap [y,z]} \frac{1}{p} \right)^k \gg \frac{1}{\sqrt{k}} \frac{1}{(k/e)^k} \left((\delta + o(1)) \log \log z - (\delta + o(1)) \log \log y \right)^k$$

$$\gg \left(\frac{(e + o(1))\varepsilon\delta \log \log x}{k} \right)^k$$

$$= (\log x)^{\varepsilon\delta + o(1)}.$$

Now let us estimate the second sum in (7). Note that every $n_1 \in \mathcal{N}_F''(x)$ satisfies $n_1 \le z^k < x$ and is divisible by the square of a prime $p \ge y$. Hence, $n_1 = p^2 n_2$ for some $n_2 \le x$. It follows that

$$\sum_{n_1 \in \mathcal{N}_F''(x)} \frac{1}{n_1} \le \left(\sum_{p \ge y} \frac{1}{p^2} \right) \left(\sum_{n_2 \le x} \frac{1}{n_2} \right) \ll \frac{\log x}{y} = o(1)$$

as $x \to \infty$.

Putting all the estimates together, we conclude that

$$|\mathcal{M}_F(x)| \gg \frac{x(\log x)^{\varepsilon\delta + o(1)}}{\log x \log \log x} = \frac{x}{(\log x)^{1 - \varepsilon\delta + o(1)}}$$

as $x \to \infty$, which is what we wanted. $\qquad\square$

5   Greedy and Generous Square-Free Numbers

We retain the notation and set-up of Sect. 4.

As we have already remarked in Sect. 3, the Chinese Remainder Theorem implies that for any $m \in \mathcal{M}_F$ there exists a positive integer n such that $m \mid F(n)$. Moreover, if $m \in \mathcal{M}_F(x)$, then we can choose such n satisfying $n \leq x$. Of course, there can be several n with this property; pick one of them and call it n_m.

Thus, for every $m \in \mathcal{M}_F(x)$ we pick $n_m \leq x$ such that $m \mid f(n_m)$; we fix this choice of the numbers n_m until the end of this section.

It might happen that $n_m = n_{m'}$ for distinct $m, m' \in \mathcal{M}_F(x)$. It turns out, however, that, with a suitable choice of our parameter ε, the repetitions are "not too frequent."

Call $m \in \mathcal{M}_F(x)$ *generous* if it shares its n_m with at least $6d$ other elements of $\mathcal{M}_F(x)$, and *greedy* otherwise.

Proposition 5.1 *Specify*

$$\varepsilon = \frac{1}{10^3 \log(2d)}. \tag{8}$$

Then for sufficiently large x at least half of the elements of the set $\mathcal{M}_F(x)$ are greedy. In particular,

$$\left| \{n_m : m \in \mathcal{M}_F(x)\} \right| \geq \frac{1}{12d} |\mathcal{M}_F(x)|.$$

The crucial tool in the proof of this proposition is the following lemma, which might be viewed as a partial "square-free" version of Property **E** from Sect. 3. We cannot affirm that $F(n)$ has "few" divisors in $\mathcal{M}_F$ for all n; but we can affirm that, with "few" exceptions, $F(n)$ has "few" divisors in $\mathcal{M}_F(x)$.

Lemma 5.2 *For sufficiently large x, the set of $n \leq x$ such that $F(n)$ has more than $6d$ divisors in $\mathcal{M}_F(x)$ is of cardinality at most $x(\log x)^{-2+30\varepsilon \log(2d)}$.*

We postpone the proof of this lemma until Sect. 6.

5.1 Initializing the Proof of Proposition 5.1

Starting from this subsection we work on the proof of Proposition 5.1.

We set $\mathcal{J} = [y, x]$ and we try to understand the function $\omega_{\mathcal{J}}(F(n))$, where $\omega_{\mathcal{J}}(\cdot)$ is the number of prime factors of the argument in the interval $\mathcal{J}$. We split n into three sets as follows:

(i) $E(x)$ (enormous), which is the set of $n \leq x$ for which

$$\omega_{\mathcal{J}}(F(n)) \geq 3d(\log \log x)^2.$$

(ii) $L(x)$ (large), which is the set of $n \leq x$ for which

$$\omega_{\mathcal{J}}(F(n)) \in [10^5 d^2 \log \log x, 3d(\log \log x)^2].$$

(iii) $R(x)$ (reasonable), which is the set of $n \le x$ such that

$$\omega_{\mathscr{J}}(F(n)) \le 10^5 d^2 \log \log x.$$

For the purpose of this argument, if $s = \omega_{\mathscr{J}}(F(n))$ then we denote all the prime factors of $F(n)$ in $\mathscr{J}$ by $p_1 < p_2 < \cdots < p_s$.

We will use the multiplicative function ρ_F, defined for a positive integer u by

$$\rho_F(u) = |\{0 \le n \le u - 1 : F(n) \equiv 0 \bmod u\}|. \tag{9}$$

Clearly, $\rho_F(m) \le d^{\omega(m)}$ holds for all square-free positive integers m.

5.2   *Counting m with $n_m \in E(x)$*

Since $|F(n)| \ll n^d \ll x^d$ it follows that in case (i), if we put $U = \lfloor (\log \log x)^2 \rfloor$, then $p_1 \cdots p_U \le x^{1/2}$ for large x.

To count $E(x)$, fix $p_1 < p_2 < \cdots < p_U$ all in $\mathscr{J}$ and let us count the number of $n \le x$ such that $m_1 \mid f(n)$, where $m_1 = p_1 \cdots p_U$. The number of such n is

$$\frac{\rho_F(m_1)}{m_1} x + O(\rho_F(m_1)) \ll \frac{d^{\omega(m_1)}}{m_1} x + d^{\omega(m_1)} \ll \frac{d^{\omega(m_1)}}{m_1} x. \tag{10}$$

In the middle of (10), the first term $d^{\omega(m_1)} x / m_1$ dominates because $m_1 \le x^{1/2}$.

We sum up over the possible m_1 getting

$$|E(x)| \ll x d^U \sum \frac{1}{m_1}, \tag{11}$$

where the sum runs over all square-free m_1 satisfying $\omega(m_1) = U$ and having all prime divisors in $\mathscr{J}$. We estimate this sum by the multinomial coefficient trick, already used in the proof of Proposition 4.1:

$$\sum \frac{1}{m_1} \ll \frac{1}{U!} \left(\sum_{y \le p \le x} \frac{1}{p} \right)^U \ll \left(\frac{3 \log \log x}{U} \right)^U.$$

This gives us the estimate

$$|E(x)| \ll x \left(\frac{3d \log \log x}{U} \right)^U,$$

which, with our definition $U = \lfloor (\log\log x)^2 \rfloor$, implies that

$$|E(x)| \le xe^{-(1+o(1))(\log\log x)^2 \log\log\log x}$$

as $x \to \infty$.

Having bounded $|E(x)|$, we may now estimate the number of m such that $n_m \in E(x)$. For each $n \le x$ we have $|F(n)| \ll n^d \le x^d$ which implies that, for large x, we have $\omega_{\mathscr{J}}(F(n)) \le \log x$. Thus, for large x, the divisor $m \mid F(n)$ with $\omega(m) = k$ can be chosen in at most

$$\binom{\lfloor \log x \rfloor}{k+1} \le (\log x)^{k+1} \ll e^{2(\log\log x)^2}$$

ways. This implies that, as $x \to \infty$,

$$\left| \{ m \in \mathscr{M}_F(x) : n_m \in E(x) \} \right| \le |E(x)| e^{2(\log\log x)^2}$$

$$\le xe^{-(1+o(1))(\log\log x)^2 \log\log\log x}.$$

Proposition 4.1 implies that this is $o\big(|\mathscr{M}_F(x)|\big)$ as $x \to \infty$.

5.3 Counting m with $n_m \in L(x)$

Let us deal with (ii) now. We let i_0 and i_1 be the maximal and the minimal positive integers such that $2^{i_0} \le 10^5 d$ and $2^{i_1} \ge 3(\log\log x)$, respectively. Clearly, $i_1 - i_0 = O(\log\log\log x)$. Consider an integer $j \in [i_0, i_1 - 1]$ and denote by $L_j(x)$ the subset of $L(x)$ consisting of n such that

$$\omega_{\mathscr{J}}(F(n)) \in [2^j d \log\log x, 2^{j+1} d \log\log x].$$

We revisit the previous argument. We now take $U = \lfloor 2^{j-1} \log\log x \rfloor$, and let $m_1 = p_1 \cdots p_U$. Then $m_1^{2d} \le |F(n)| \ll x^d$, therefore $m_1 \ll x^{1/2}$. Now exactly as before we prove that

$$|L_j(x)| \ll x \left(\frac{3d \log\log x}{U} \right)^U,$$

which, with our definition $U = \lfloor 2^{j-1} \log\log x \rfloor$, implies that

$$|L_j(x)| \ll \frac{x}{(\log x)^{2^{j-2} \log(2^{j-2}/3d)}}.$$

Since

$$\log \frac{2^{j-2}}{3d} \geq \log \frac{2^{i_0-2}}{3d} \geq \log \frac{10^5 d}{24d} \geq 8,$$

we have

$$|L_j(x)| \ll \frac{x}{(\log x)^{2^{j+1}}}.$$

On the other hand, for $n \in L_j(x)$ we have $\omega_{\mathscr{J}}(F(n)) \leq 2^{j+1} d \log \log x$. It follows that, for large x, the number of choices for m for a given $n \in L_j(x)$ is at most

$$\binom{\lfloor 2^{j+1} d \log \log x \rfloor}{k+1} \leq \frac{\left(2^{j+1} d \log \log x\right)^{k+1}}{(k+1)!}$$

$$\leq \left(\frac{2^{j+3} d}{\delta \varepsilon}\right)^{2\delta\varepsilon \log \log x}$$

$$= (\log x)^{2\delta\varepsilon \log\left(2^{j+3} d/\delta\varepsilon\right)}. \tag{12}$$

Since

$$\frac{2^{j-1}}{\delta \varepsilon} \geq 2^{i_0-1} \geq \frac{10^5 d}{4} \geq 10^4 d,$$

we have

$$\frac{2^{j-1}}{\delta \varepsilon} \geq 2 \log \frac{2^{j-1}}{\delta \varepsilon} \geq \log \left(\frac{2^{j-1}}{\delta \varepsilon} \cdot 10^4 d\right) \geq \log \frac{2^{j+3} d}{\delta \varepsilon},$$

which shows that the exponent in (12) does not exceed 2^j.

Thus, for large x

$$\left|\{m \in \mathscr{M}_F(x) : n_m \in L_j(x)\}\right| \leq |L_j(x)|(\log x)^{2^j} \ll \frac{x}{(\log x)^{2^j}} \leq \frac{x}{(\log x)^2},$$

because $2^j \geq 2^{i_0} \geq 10^5 d/2 \geq 2$. Since there are $O(\log \log \log x)$ possible j, we conclude that

$$\left|\{m \in \mathscr{M}_F(x) : n_m \in L(x)\} \ll \frac{x \log \log \log x}{(\log x)^2},\right.$$

which is again $o\left(|\mathscr{M}_F(x)|\right)$ as $x \to \infty$.

Thus, we have proved that

$$\left|\{m : n_m \in E(x) \cup L(x)\}\right| = o\left(|\mathcal{M}_F(x)|\right) \tag{13}$$

as $x \to \infty$.

5.4 Completing the Proof

We are ready now to complete the proof of Proposition 5.1. It remains to deal with $n \in R(x)$. If $n \in R(x)$, then $\omega_{\mathscr{G}}(F(n)) \le 10^5 d^2 \log\log x$. Thus, for fixed $n \in R(x)$ we have

$$
\begin{aligned}
\left|\{m \in \mathcal{M}_F(x) : n_m = n\}\right| &\le \binom{\lfloor 10^5 d^2 \log\log x \rfloor}{k+1} \\[2mm]
&\le \frac{(10^5 d^2 \log\log x)^{k+1}}{(k+1)!} \\[2mm]
&\le \left(\frac{10^6 d^2}{\varepsilon\delta}\right)^{2\varepsilon\delta \log\log x} \\[2mm]
&= (\log x)^{2\varepsilon\delta \log(10^6 d^2/\varepsilon\delta)}.
\end{aligned}
\tag{14}
$$

Now we are done: Lemma 5.2 combined with estimate (14) implies that there exists at most

$$\frac{x}{(\log x)^{2-30\varepsilon \log(2d)-2\varepsilon\delta \log(10^6 d^2/\varepsilon\delta)}} \tag{15}$$

generous $m \in \mathcal{M}_F(x)$ with the property $n_m \in R(x)$. When ε is chosen as in (8), a quick calculation shows that

$$30\varepsilon \log(2d) + 2\varepsilon\delta \log\left(\frac{10^6 d^2}{\varepsilon\delta}\right) < \frac{1}{2}.$$

Hence (15) is $o(|\mathcal{M}_F(x)|)$ as $x \to \infty$. In particular, when x is sufficiently large, at least half of elements of $\mathcal{M}_F(x)$ are greedy. $\square$

It remains to prove Lemma 5.2.

6   Proof of Lemma 5.2

We keep the notation of Sect. 4, especially $y = \exp((\log x)^{1-\varepsilon})$.

6.1 Two Simple Lemmas

Let A be the subset of $\mathcal{M}_F$ consisting of m with $p_{\min}(m) \geq y$. We study the set $A(z) = A \cap [y, z]$ for $z \in [y, x]$.

Lemma 6.1 *When x is sufficiently large we have $|A(z)| \leq z(\log x)^{-1+3\varepsilon}$ for all $z \in [y, x]$.*

Proof Let $g(n)$ be the characteristic function of A. Then for any $z > 1$ we have

$$\sum_{p \leq z} g(p) \log p \leq 2z,$$

and $g(p^n) = 0$ for $n \geq 2$. Using Lemma 9.6 on page 138 in [5], we obtain

$$|A(z)| = \sum_{n \leq z} g(n) \leq 3 \frac{z}{\log z} \sum_{n \in A(z)} \frac{1}{n}. \tag{16}$$

Clearly, $\log z \geq (\log x)^{1-\varepsilon}$ for $z \in [y, x]$. As for the sum above, we have

$$\sum_{n \in A(z)} \frac{1}{n} \leq \prod_{y \leq p \leq z} \left(1 + \frac{1}{p}\right) \leq (\log x)^{\varepsilon + o(1)}$$

as $x \to \infty$. Together with (16) this finishes the proof. $\quad\square$

Lemma 6.2 *Assuming x sufficiently large, for $y \leq a \leq b \leq x$ we have*

$$\sum_{\substack{a \leq n \leq b \\ n \in A}} \frac{1}{n} \leq \frac{\log b - \log a + 1}{(\log x)^{1-3\varepsilon}}.$$

Proof Using Abel summation and Lemma 6.1, we obtain

$$\sum_{\substack{a \leq n \leq b \\ n \in A}} \frac{1}{n} = \int_a^b \frac{d|A(z)|}{z}$$

$$= \frac{|A(b)|}{b} - \frac{|A(a)|}{a} + \int_a^b \frac{|A(z)|}{z^2} dz$$

$$\leq \frac{|A(b)|}{b} + \frac{1}{(\log x)^{1-3\varepsilon}} \int_a^b \frac{dz}{z}$$

$$\leq \frac{1}{(\log x)^{1-3\varepsilon}} + \frac{\log b - \log a}{(\log x)^{1-3\varepsilon}},$$

as wanted. $\quad\square$

6.2 Cliques

Starting from this subsection we begin the proof of Lemma 5.2. Recall that every $m \in \mathcal{M}_F(x)$ writes as $m = m_1 P$, where $P = p_{\max}(m) \geq x^{9/10}$. As in Sect. 4 we denote by $\mathcal{M}'_F(x)$ the set of all m_1 obtained this way. They satisfy (4), which will be used in the sequel without special reference.

Let $n \leq x$ be such that $F(n)$ has at least $6d$ distinct divisors in $\mathcal{M}_F(x)$. Write each of them $m_1 P$ as above and let s be the number of such P. Then $x^{9s/10} \leq |f(n)| \ll x^d$, so $s \leq 10d/9 + o(1)$ as $x \to \infty$. In particular, $s < 2d$ for large x. Hence among the $6d$ divisors there are three with the same P; write them $m_1 P$, $m_2 P$, and $m_3 P$.

Let us call an (unordered) triple of pairwise distinct $m_1, m_2, m_3 \in \mathcal{M}'_F(x)$ a *clique* if there exists a prime $P \geq x^{9/10}$ such that $m_1 P, m_2 P, m_3 P \in \mathcal{M}_F(x)$. If $\{m_1, m_2, m_3\}$ is a clique, then $m_1 P, m_2 P, m_3 P \in [x/(2\kappa), x/\kappa]$. This implies that in a clique we have

$$\frac{m_j}{2} \leq m_i \leq 2m_j \tag{17}$$

for any i, j. In addition to this, since m_1, m_2, m_3 in a clique are square-free with the same number of prime factors, we have

$$\gcd(m_i, m_j) < m_i < [m_i, m_j], \qquad (i \neq j). \tag{18}$$

where $[\cdots]$ denotes the least common multiple. We will repeatedly use these properties.

6.3 The Sum over Cliques

To prove the lemma, it suffices to estimate the number of n such that $F(n)$ has three distinct divisors forming a clique. When a clique $\{m_1, m_2, m_3\}$ is fixed, the number of such n is at most

$$\frac{\rho_F([m_1, m_2, m_3])}{[m_1, m_2, m_3]} x + O(\rho_F([m_1, m_2, m_3])), \tag{19}$$

where $\rho_F(\cdot)$ is defined in (9). When x is large, we have

$$\omega([m_1, m_2, m_3]) \leq 3k \leq 4\varepsilon \log\log x,$$

which implies

$$\rho_F([m_1, m_2, m_3]) \leq d^{\omega([m_1, m_2 m_3])} \leq (\log x)^{4\varepsilon \log d}.$$

Further, since $m_i \leq x^{1/10}$, we have $[m_1 m_2, m_3] \leq x^{3/10} \leq x^{1/2}$. It follows that in (19) the first term dominates over the second one, and the number of our n (for the fixed m_1, m_2, m_3) is bounded, for large x, by

$$x(\log x)^{5\varepsilon \log d} \frac{1}{[m_1, m_2, m_3]}.$$

Hence the total number of n (for all possible choices of m_1, m_2, m_3) is bounded by $x(\log x)^{5\varepsilon \log d} S$, where

$$S = \sum_{\{m_1, m_2, m_3\}} \frac{1}{[m_1, m_2, m_3]},$$

the summation being over all cliques. The rest of the argument is estimating this sum S.

We write $S = S' + S''$, where S' is the sum over the cliques with the property

$$\text{there is a relabeling of the indices such that } [m_1, m_2] < [m_1, m_2, m_3], \qquad (20)$$

and S'' is over the cliques such that

$$[m_1, m_2] = [m_1, m_3] = [m_2, m_3] = [m_1, m_2, m_3]. \qquad (21)$$

6.4 Estimating S'

We are starting now to estimate S'. All cliques appearing in this subsection satisfy (20).

6.4.1 The Estimate with m_1 and m_2 Fixed

Fix m_1 and m_2. Then $m_3 \nmid [m_1, m_2]$ by (20). Set $u = \gcd(m_3, [m_1, m_2])$. With m_1 and m_2 being fixed, there are at most

$$2^{2k} \ll (\log x)^{3\varepsilon \delta}$$

choices for u as a divisor of $[m_1, m_2]$.

Writing $m_3 = uv$. Clearly, $v \in A$, where A is the set from Sect. 6.1. Using (17), we obtain $m_1/(2u) \leq v \leq 2m_1/u$. Since u is a proper divisor of m_3, we also have $v > 1$, which implies $v \geq y$, because $v \in A$. Also, clearly $v \leq m_3 \leq x$. This shows that

$$\max\left\{y, \frac{m_1}{2u}\right\} \le v \le \min\left\{x, 2\frac{m_1}{u}\right\}. \tag{22}$$

We have $[m_1, m_2, m_3] = [m_1, m_2]v$. Thus, assuming m_1 and m_2 fixed, and summing up over all possible m_3, we get

$$\sum \frac{1}{[m_1, m_2, m_3]} \le \frac{1}{[m_1, m_2]} \sum_{u \mid [m_1, m_2]} \sum_{v \in A \text{ satisfying (22)}} \frac{1}{v} \tag{23}$$

$$\ll \frac{1}{[m_1, m_2](\log x)^{1-4\varepsilon}} \sum_{u \mid [m_1, m_2]} 1$$

$$\ll \frac{1}{(\log x)^{1-8\varepsilon}[m_1, m_2]}.$$

Here, in the inner sum in (23), we applied Lemma 6.2 with the choices

$$b = \min\left\{x, \frac{2m_1}{u}\right\}, \quad a = \max\left\{y, \frac{m_1}{2u}\right\},$$

and we used the fact that $\log b - \log a \ll 1$.

6.4.2 The Estimate with m_1 Fixed

We now fix m_1 and vary m_2. This time we set $u = \gcd(m_1, m_2)$ and again write $m_2 = uv$. There are at most $2^k \ll (\log x)^{2\varepsilon\delta}$ choices for u. Furthermore, it follows from (18) that u is a proper divisor of m_2, which implies $v > 1$. Thus, our v again belongs to the set A and satisfies (22).

Keeping m_1 fixed, we argue as above:

$$\sum \frac{1}{[m_1, m_2]} = \frac{1}{m_1} \sum_{u \mid m_1} \sum_{v \in A \text{ satisfying (22)}} \frac{1}{v}$$

$$\ll \frac{1}{m_1 (\log x)^{1-4\varepsilon}} \sum_{u \mid m_1} 1$$

$$\ll \frac{1}{m_1 (\log x)^{1-7\varepsilon}}.$$

6.4.3 Estimating S'

Now we are ready to estimate S':

$$S' \ll \frac{1}{(\log x)^{2-15\varepsilon}} \sum_{m_1 \in \mathcal{M}'_F} \frac{1}{m_1} \ll \frac{1}{(\log x)^{2-17\varepsilon}},$$

where for the last estimate we used (6).

6.5 Estimating S''

Now let $\{m_1, m_2, m_3\}$ be a clique satisfying (21). Setting $u = \gcd(m_1, m_2, m_3)$ and $v_i = [m_1, m_2, m_3]/m_i$, we obtain

$$m_1 = uv_2v_3, \quad m_2 = uv_1v_3, \quad m_3 = uv_1v_2,$$

$$[m_1, m_2] = [m_1, m_3] = [m_2, m_3] = [m_1, m_2, m_3] = uv_1v_2v_3.$$

We again use (18) to obtain $v_i > 1$, which implies $v_i \geq y$ because $v_i \in A$. Also, $v_i \leq x$. Together with (17) this gives

$$\max\left\{y, \frac{v_1}{2}\right\} \leq v_i \leq \min\{x, 2v_1\} \qquad (i = 2, 3). \tag{24}$$

It follows that

$$S'' \leq \sum_{\substack{u,v_1,v_2,v_3 \in A \\ \text{satisfying (24)}}} \frac{1}{uv_1v_2v_3}.$$

When u and v_1 are fixed, we have

$$\sum_{\substack{v_2,v_3 \in A \\ \text{satisfying (24)}}} \frac{1}{uv_1v_2v_3} \leq \frac{1}{uv_1}\left(\sum_{\substack{v \in A \\ \max\{y,v_1/2\}\leq v \leq \min\{x,2v_1\}}} \frac{1}{v}\right)^2,$$

and the squared sum can be estimated, using Lemma 6.2, as $O\big((\log x)^{-1+4\varepsilon}\big)$. Hence

$$S'' \ll \frac{1}{(\log x)^{2-8\varepsilon}} \sum \frac{1}{uv_1}, \tag{25}$$

the latter sum being over all possible values of u and v_1.

To estimate the latter, we make the following observations.

- The number uv_1 belongs to A, satisfies $y \le uv_1 \le x$ and $\omega(yv_1) \le k$.
- Given $m \in A$ with $\omega(m) \le k$, it can be written as $m = uv_1$ in at most $2^k \ll (\log x)^{2\varepsilon}$ ways.

It follows that

$$\sum \frac{1}{uv_1} \ll (\log x)^{2\varepsilon} \sum_{m \in A \cap [y,x]} \frac{1}{m} \ll (\log x)^{6\varepsilon}, \qquad (26)$$

the latter sum being $O\big((\log x)^{4\varepsilon}\big)$ by Lemma 6.2 with $b = x$ and $a = y$.

Combining (25) and (26), we conclude that

$$S'' \ll \frac{1}{(\log x)^{2-14\varepsilon}}.$$

6.6 Proof of Lemma 5.2

Thus, for large x, the total number of n such that $F(n)$ has at least $6d$ distinct divisors in $\mathcal{M}_F(x)$ is bounded by

$$x(\log x)^{5\varepsilon \log d}(S' + S'') \ll \frac{x}{(\log x)^{2-5\varepsilon \log d - 17\varepsilon}},$$

which proves Lemma 5.2.

7 Proof of Theorem 1.6

We are ready now to prove Theorem 1.6. Thus, let X and t be as in Theorem 1.6, and, as in Sect. 3, let $F(T) \in \mathbb{Z}[T]$ be the primitive separable polynomial whose roots are exactly the finite critical values of t, with $d = \deg F$. We use all notation and conventions from Sect. 4. In particular, we fix ε satisfying $0 < \varepsilon \le 1/2$ (which will be specified later) and for sufficiently large x we consider the set $\mathcal{M}_F(x)$.

Recall (see Sect. 3) that we denote by K_n the field $\mathbb{Q}(t^{-1}(n))$. We call $m \in \mathcal{M}_F$ primitive for n if every $p \mid m$ ramifies in K_n, but for every $n' < n$ some $p \mid m$ does not ramify in $K_{n'}$. Clearly, if n admits a primitive $m \in \mathcal{M}_F$, then the field K_n is distinct from $K_1, \ldots, K_{n-1}$.

Our starting point is Corollary 3.3, which asserts that every $m \in \mathcal{M}_F$ with the property $p_{\min}(m) > \omega(m)$ serves as a primitive for some $n_m \le m(\omega(m) + 1)$. If $m \in \mathcal{M}_F(x)$, then this property is trivially satisfied when x is large enough; hence every $m \in \mathcal{M}_F(x)$ serves as primitive for some $n_m \le m(k+2)$, and we have

$$n_m \le m(k+2) \le \frac{x}{\log\log x}(\varepsilon\delta \log\log x + 3) \le x, \tag{27}$$

again provided x is sufficiently large.

Set

$$\mathcal{N}(x) = \{n_m : m \in \mathcal{M}_F(x)\},$$

$$\mathcal{N}'(x) = \{n \in \mathcal{N}(x) : \text{the fiber } t^{-1}(n) \text{ is } \mathbb{Q}\text{-irreducible}\}.$$

It follows from (27) that

$$\mathcal{N}'(x) \subset \mathcal{N}(x) \subset [1, x],$$

and Hilbert's Irreducibility Theorem implies that

$$|\mathcal{N}'(x)| \ge |\mathcal{N}(x)| - O(x^{1/2}). \tag{28}$$

The fields

$$K_n \qquad (n \in \mathcal{N}(x))$$

are pairwise distinct, and, since for $n \in \mathcal{N}'(x)$ the field K_n is the Galois closure of $\mathbb{Q}(P_n)$, the fields

$$\mathbb{Q}(P_n) \qquad (n \in \mathcal{N}'(x)) \tag{29}$$

are pairwise distinct as well.

Thus, to prove Theorem 1.6, we only have to show that, with suitable choice of ε, the lower estimate

$$|\mathcal{N}'(x)| \ge \frac{x}{(\log x)^{1-\eta}} \tag{30}$$

holds for sufficiently large x. Here η is a positive number depending only on d (which, through (3), translates into dependence in ν and $\mathbf{g}$).

This can be accomplished using the results of Sects. 4 and 5. Since every $p \mid m$ ramifies in K_{n_m}, we have $m \mid F(n_m)$ (see Property **A** in Sect. 3). Hence Proposition 5.1 applies to our definition of n_m. Thus, setting ε as in (8), Proposition 5.1 implies that, for sufficiently large x, we have $|\mathcal{N}(x)| \ge (12d)^{-1}|\mathcal{M}_F(x)|$. Together with Proposition 4.1 this implies that $|\mathcal{N}(x)| \ge x(\log x)^{-1+\delta\varepsilon+o(1)}$ as $x \to \infty$, which, combined with (28), implies the same lower estimate for $|\mathcal{N}'(x)|$. In particular, for sufficiently large x we have (30) with $\eta = \delta\varepsilon/2$.

In view of (2) and (8) we have $\eta \geq 10^{-4}(d\log(2d))^{-1}$. Using (3) we deduce that $\eta \geq 10^{-6}\big((\mathbf{g}+v)\log(\mathbf{g}+v)\big)^{-1}$. $\qquad\qquad\square$

Acknowledgements During the work on this article Yuri Bilu was partially supported by the University of Xiamen, and by the binational research project MuDeRa, funded jointly by the French ANR and the Austrian FWF. Florian Luca was supported by an A-rated researcher award of the NRF of South Africa.

We thank Jean Gillibert and Felipe Voloch for useful discussions. We also thank the referees who carefully read the manuscript and detected several inaccuracies.

References

1. Yu. Bilu, Counting Number Fields in Fibers (With An Appendix by Jean Gillibert). Math. Z. (2017, to appear). arXiv:1606.02341[math.NT]
2. Yu. Bilu, F. Luca, Number Fields in Fibers: The Geometrically Abelian Case with Rational Critical Values. Periodica Math. Hung. (to appear). arXiv:1606.09164[math.NT]
3. P. Corvaja, U. Zannier, On the number of integral points on algebraic curves. J. Reine Angew. Math. **565**, 27–42 (2003)
4. H. Davenport, D. Lewis, A. Schinzel, Polynomials of certain special types. Acta Arith. **9**, 107–116 (1964)
5. J.-M. De Koninck, F. Luca, *Analytic Number Theory: Exploring the Anatomy of Integers*. Graduate Studies in Mathematics, vol. 134 (AMS, Providence, RI, 2012)
6. R. Dvornicich, U. Zannier, Fields containing values of algebraic functions. Ann. Scuola Norm. Sup. Pisa Cl. Sci. (4) **21**, 421–443 (1994)
7. R. Dvornicich, U. Zannier, Fields containing values of algebraic functions II (On a conjecture of Schinzel). Acta Arith. **72**, 201–210 (1995)
8. J.-P. Serre, *Lectures on the Mordell-Weil Theorem*, 3rd edn. (Vieweg & Sohn, Braunschweig, 1997)
9. U Zannier, On the number of times a root of $f(n,x) = 0$ generates a field containing a given number field. J. Number Theory **72**, 1–12 (1998)

Local Oscillations in Moderately Dense Sequences of Primes

Jörg Brüdern and Christian Elsholtz

To Robert Tichy, on the occasion of his 60th birthday

Abstract The distribution of differences of consecutive members of sequences of primes is investigated. A quantitative measure for oscillations among these differences is the curvature of the sequence. If the sequence is not too sparse, then sharp estimates for its curvature are provided.

2010 *Mathematics Subject Classification* 11N05

1 Introduction

In an influential paper, Erdős and Turán [2] showed that when (p_n) denotes the sequence of all prime numbers arranged in increasing order, then there are infinitely many sign changes among the numbers

$$p_{n+1}^2 - p_n p_{n+2}. \tag{1}$$

Motivated by quantitative versions of this result due to Rényi [8] and Erdős and Rényi [1], we develop this theme further in the context of sequences that are not too sparse.

J. Brüdern
Mathematisches Institut, Bunsenstrasse 3–5, 37073 Göttingen, Germany
e-mail: Joerg.Bruedern@mathematik.uni-goettingen.de

C. Elsholtz (✉)
Institut für Analysis und Zahlentheorie, Technische Universität Graz, Kopernikusgasse 24, A-8010 Graz, Austria
e-mail: elsholtz@math.tugraz.at

© Springer International Publishing AG 2017

C. Elsholtz, P. Grabner (eds.), *Number Theory – Diophantine Problems, Uniform Distribution and Applications*, DOI 10.1007/978-3-319-55357-3_8

Theorem 1 *Let $\mathscr{P}$ be a set of primes with the property that*

$$\frac{(\log x)^{4/3}}{x}\#\{p \in \mathscr{P} : p \le x\} \tag{2}$$

tends to infinity with x. If p_n denotes the enumeration of the set $\mathscr{P}$ in increasing order, then the sequence (1) *changes sign infinitely often.*

Our main object of study is the *curvature* of sequences. The idea is due to Rényi [8]. Consider at least three distinct points $z_1, \ldots, z_N$ in the complex plane. With the argument of a complex number chosen in the interval $(-\pi, \pi]$, the sum

$$\sum_{n=1}^{N-2} \left| \arg \frac{z_{n+2} - z_{n+1}}{z_{n+1} - z_n} \right| \tag{3}$$

is referred to as the total curvature of the polygonal line connecting z_{n-1} with z_n for $2 \le n \le N$, because this adds up the (non-negative) angles between the line segments from z_n to z_{n+1}, and on to z_{n+2}. For a set of primes $\mathscr{P}$, again enumerated in increasing order as p_n, we take $z_n = n + i \log p_n$, and then let $K_N(\mathscr{P})$ denote the sum in (3) with this special choice of z_n. This is the curvature of $\mathscr{P}$, truncated at N.

Now suppose that we knew that $K_N(\mathscr{P})$ were unbounded. Then, if the segment $(\log p_n)_{n_0 \le n \le N}$ is either concave or convex, then $K_N(\mathscr{P}) - K_{n_0}(\mathscr{P}) \le \frac{1}{2}\pi$ which is impossible for large N. We conclude that the sequence

$$\log p_{n+2} - 2 \log p_{n+1} + \log p_n$$

changes sign infinitely often, and on taking exponentials this is the same as exhibiting sign changes in the sequence (1). In particular, Theorem 1 will follow once we have established that $K_N(\mathscr{P})$ is unbounded for the sequences of primes satisfying (2). Further, we see that the growth rate of $K_N(\mathscr{P})$ is a rough measure for the oscillations in the sequence (1).

Rényi [8] in 1950 considered the sequence of all primes and bounded their curvature, hereafter denoted by K_N, from below by

$$K_N \gg \log \log \log N.$$

Shortly afterwards, in collaboration with Erdős [1] (see also [7]) he determined the order of magnitude of K_N, now showing that

$$\log N \ll K_N \ll \log N. \tag{4}$$

Their methods rely on the prime number theorem. Our concern in this paper is with estimates for the curvature that are based solely on lower bounds for the number of primes in a given sequence, such as in (2). Before we can formulate our principal estimate, we have to set up some notation.

We work relative to an arithmetic progression. When $a, q \in \mathbb{N}$ with $1 \le a \le q$ and $(a, q) = 1$, let $\mathscr{P}_{q,a}$ denote the set of all primes $p \equiv a \bmod q$. We refer to a subset $\mathscr{P} \subset \mathscr{P}_{q,a}$ as *dense* if there are positive numbers δ and x_0 with the property that whenever $x \ge x_0$, then

$$\#\{p \in \mathscr{P} : p \le x\} \ge \delta \pi(x; q, a) \tag{5}$$

where as usual $\pi(x; q, a)$ is the number of primes not exceeding x in $\mathscr{P}_{q,a}$. More generally, if $\delta : [3, \infty) \to (0, 1]$ is monotonically decreasing with $\delta(x) \ge (\log x)^{-1}$, and (5) is satisfied with $\delta = \delta(x)$ for all $x \ge x_0$, then[1] the set $\mathscr{P}$ is called δ-dense (relative to x_0 and $\mathscr{P}_{q,a}$). The lower bound on δ ensures that $\mathscr{P}$ is an infinite set, enumerated in ascending order by p_n, as before. Then $K_N(\mathscr{P})$ is defined for all $N \ge 3$. We also put $\delta_N = \delta(p_N)$.

Theorem 2 *Fix a number $x_0 \ge 3$ and a decreasing function $\delta : [3, \infty) \to (0, 1]$ with $\delta(x) \ge (\log x)^{-1}$ for all $x \ge 3$. Then there is a sequence of natural numbers $N_0(q)$ with the property that for all $N \ge N_0(q)$ and for all sets of primes $\mathscr{P}$ that are δ-dense relative to x_0 and some $\mathscr{P}_{q,a}$, one has*

$$K_N(\mathscr{P}) \le 500 \delta_N^{-1} \log N. \tag{6}$$

If $\delta(x)^2 \log x$ tends to infinity with x, then one also has

$$K_N(\mathscr{P}) \ge 10^{-8} \delta_N^3 \log N.$$

Theorem 2 may be applied to the arithmetic progression $\mathscr{P}_{q,a}$ itself, with $\delta = 1$. We then conclude as follows.

Corollary *With $N_0(q)$ as in the preceding Theorem, for $N \ge N_0(q)$ one has*

$$10^{-8} \log N \le K_N(\mathscr{P}_{q,a}) \le 500 \log N.$$

This contains (4) as a very special case. Note that here as well as in Theorem 2 no effort has been made to optimize the numerical constants.

When δ decays it is important to have at hand a lower bound for δ_N. One has

$$\delta_N \ge \delta(4\varphi(q)N(\log N)^2) \tag{7}$$

for all large N. We show this in passing, in Sect. 3 below. In particular, if δ is a decreasing function such that $\delta(x)^3 \log x$ tends to infinity with x and $\mathscr{P}$ is δ-dense,

---

[1]It may seem unnatural to include the lower bound on δ in this definition, but more rapidly decaying functions will play no role in this paper, and it simplifies the exposition later that δ is not too small, *a fortiori*.

then by (7) and Theorem 2 we see that $K_N(\mathscr{P})$ does not remain bounded. Hence, Theorem 1 is merely a corollary of Theorem 2.

We are not aware of earlier results of the type considered in Theorem 1 or Theorem 2 for sequences that are not quite dense. For other developments of the ideas deriving from [1, 2, 8], see Pomerance [6].

With the sequence of primes comprising $\mathscr{P}$ we associate their *second differences*

$$\Delta_n = p_{n+2} - 2p_{n+1} + p_n. \tag{8}$$

Following Rényi in spirit, our approach to Theorem 2 rests on the observation that Δ_n is not too small for many values of n. Our next theorem is a strong quantitative version of this principle.

Theorem 3 *Fix x_0 and δ as in Theorem 2. Then there is a sequence of natural numbers $N_0(q)$ with the property that for all $N \geq N_0(q)$ and for all sets of primes $\mathscr{P}$ that are δ-dense relative to x_0 and some $\mathscr{P}_{q,a}$, one has*

$$\sum_{N < n \leq 2N} \frac{|\Delta_n|}{p_n} \leq \frac{11}{\delta_{2N+2}}.$$

If $\delta(x)^2 \log x$ tends to infinity with x, then one also has

$$\sum_{N < n \leq 2N} \frac{|\Delta_n|}{p_n} \geq 10^{-7}\delta_{2N}^3.$$

Perhaps it is worth remarking that the upper bound recorded in Theorem 3 is nearly the best possible. We demonstrate this with a scattered sequence that we briefly discuss at the end of the paper.

The proof of Theorem 3 invokes upper bounds for the number of triplets of primes that come close to an arithmetic 3-progression. In Sect. 2 we use Selberg's sieve and a method of Gallagher [3] to manufacture a suitable estimate, but it is worth pointing out that the older Brun's sieve and a technique of Hardy and Littlewood [5] would yield results of comparable strength. Equipped with the sieve estimates, the transition to the lower bound announced in Theorem 3 is elementary, and is performed in Sect. 3. For Theorem 2, we need a more explicit version of Theorem 3 (see Lemma 4 below) and the method of Rényi [8]. The latter depends, in its original form, on the prime number theorem and is therefore not directly applicable to subsets of the primes. In Sect. 4, we reconfigure Rényi's approach and establish Theorem 2. Thus our arguments that have elements in common with the work of Erdős and Rényi [1] rely on methods that have been familiar for decades and yet, are of strength sufficient to address sequences of primes that are not quite dense.

If more is known about the distribution of the sequence (p_n), then our arguments sometimes produce estimates that are superior to those recorded in the theorems. For example, this is the case when the δ-dense set $\mathscr{P}$ has the additional property that the numbers p_{2n}/p_n remain bounded. With this extra assumption, the factor δ^{-1} can be deleted from the upper bounds in Theorems 2 and 3. For more details on this refinement, the reader is referred to Sects. 3 and 4 below.

2   A Sieve Estimate

In this section we establish an auxiliary estimate concerned with triplets of primes. The main result is Lemma 2 below, and this depends on a certain singular series average that we now describe.

Throughout this section, let $\mathbf{h} = (h, h') \in \mathbb{N}^2$ and suppose that $h' < h$. Then

$$D = hh'(h - h') \tag{9}$$

is an even natural number. For a prime p, let $\nu_{\mathbf{h}}(p)$ denote the number of distinct residue classes, modulo p, in which the numbers $0, h, h'$ lie. Then $1 \leq \nu_{\mathbf{h}}(2) \leq 2$ and $1 \leq \nu_{\mathbf{h}}(p) \leq 3$ for all odd primes p. Further, it is immediate that one has $\nu_{\mathbf{h}}(p) = 3$ if and only if $p \nmid D$. For a given $q \in \mathbb{N}$, we now define the *singular product*

$$\mathfrak{S}_{q,\mathbf{h}} = \prod_{p \nmid q} \left(1 - \frac{\nu_{\mathbf{h}}(p)}{p}\right)\left(1 - \frac{1}{p}\right)^{-3}. \tag{10}$$

Note that

$$\left(1 - \frac{\nu_{\mathbf{h}}(p)}{p}\right)\left(1 - \frac{1}{p}\right)^{-3} = 1 + a(p, \nu_{\mathbf{h}}(p)) \tag{11}$$

where

$$a(p, \nu) = \frac{p^3 - (p - 1)^3 - \nu p^2}{(p - 1)^3}.$$

One readily checks that for odd primes one has

$$|a(p, \nu)| \leq 3/(p - 1) \quad (\nu = 1, 2), \quad |a(p, 3)| \leq 4/(p - 1)^2. \tag{12}$$

Hence, recalling that $\nu_{\mathbf{h}}(p) = 3$ holds for all $p \nmid D$, one finds that the product (10) converges absolutely to a non-negative limit.

Lemma 1 *Let $\varepsilon > 0$. Then, uniformly for $q \in \mathbb{N}$ and $0 < \alpha \leq 1$ one has*

$$\sum_{1 \leq h \leq H} \sum_{|2h'-h| < \alpha h} \mathfrak{S}_{q,\mathbf{h}} = \frac{1}{2}\alpha H^2 + O(H^{1+\varepsilon}).$$

When $q = 1$ a similar estimate occurs in Gallagher [3], but the average there is over more parameters, and is more symmetric. We therefore give a complete proof, although we shall follow [3] quite closely. For convenience, it is appropriate to put

$$\mathcal{H} = \{(h, h') \in \mathbb{N}^2 : h \leq H, \ |2h' - h| < \alpha h\}.$$

Note that $\alpha \leq 1$ ensures that for any pair $(h, h') \in \mathcal{H}$ one has $h' < h$. In particular, $\mathfrak{S}_{q,\mathbf{h}}$ is defined. Now put

$$a_{\mathbf{h}}(r) = \prod_{p \mid r} a(p, v_{\mathbf{h}}(p)). \tag{13}$$

Then, by (11), the absolutely convergent product in (10) can be rewritten as

$$\mathfrak{S}_{q,\mathbf{h}} = \sum_{\substack{r=1 \\ (r,q)=1}}^{\infty} \mu(r)^2 a_{\mathbf{h}}(r). \tag{14}$$

From (12), (13) and a familiar divisor estimate, we infer that $a_{\mathbf{h}}(r) \ll r^{\varepsilon-2}(r, D)$. Consequently, since $\mathbf{h} \in \mathcal{H}$ implies $0 < D \leq H^3$, we conclude that

$$\sum_{r>H} \mu(r)^2 |a_{\mathbf{h}}(r)| \ll H^{\varepsilon-1}$$

holds uniformly in $\mathbf{h} \in \mathcal{H}$. Then, the crude bound $\#\mathcal{H} \leq H^2$ and (14) suffice to deduce that

$$\sum_{\mathbf{h} \in \mathcal{H}} \mathfrak{S}_{q,\mathbf{h}} = \sum_{\substack{r \leq H \\ (r,q)=1}} \mu(r)^2 \sum_{\mathbf{h} \in \mathcal{H}} a_{\mathbf{h}}(r) + O(H^{1+\varepsilon}). \tag{15}$$

Note that this estimate is uniform with respect to α and q.

Consider the inner sum over $\mathbf{h}$ in (15) for a given square-free number $r \leq H$. Let $r = p_1 \cdots p_\omega$ be the prime factorization. We apply (13) and sort the $\mathbf{h} \in \mathcal{H}$ according to given values of $v_{\mathbf{h}}(p_j)$ $(1 \leq j \leq \omega)$ to conclude that

$$\sum_{\mathbf{h} \in \mathcal{H}} a_{\mathbf{h}}(r) = \sum_{\substack{1 \leq v_j \leq 3 \\ 1 \leq j \leq \omega}} a(p_1, v_1) \cdots a(p_\omega, v_\omega) S(r, \mathbf{v}) \tag{16}$$

where $S(r, \boldsymbol{v})$ is the number of $\mathbf{h} \in \mathscr{H}$ with $v_\mathbf{h}(p_j) = v_j$ for all $1 \le j \le \omega$. Note that the condition $v_\mathbf{h}(p_j) = v_j$ depends only on the residue classes of h and h', modulo p_j. Hence, we may arrange h and h' into residue classes, modulo r, and then apply the Chinese Remainder Theorem to see that

$$S(r, \boldsymbol{v}) = \sum_{\substack{1 \le a, a' \le r \\ v_a(p_j) = v_j \\ 1 \le j \le \omega}} \sum_{\substack{\mathbf{h} \in \mathscr{H} \\ h \equiv a \bmod r \\ h' \equiv a' \bmod r}} 1.$$

For $r \le H$, we also have

$$\sum_{\substack{\mathbf{h} \in \mathscr{H} \\ h \equiv a \bmod r \\ h' \equiv a' \bmod r}} 1 = \sum_{\substack{1 \le h \le H \\ h \equiv a \bmod r}} \left(\frac{\alpha h}{r} + O(1) \right) = \frac{1}{2} \alpha \left(\frac{H}{r} \right)^2 + O\left(\frac{H}{r} \right).$$

Now let $t(p, v)$ denote the number of choices for a, a' with $1 \le a, a' \le p$ such that the numbers $0, a, a'$ lie in exactly v residue classes, modulo p. Then, again by the Chinese Remainder Theorem,

$$\sum_{\substack{1 \le a, a' \le r \\ v_a(p_j) = v_j \\ 1 \le j \le \omega}} 1 = \prod_{j=1}^{\omega} t(p_j, v_j),$$

and on collecting together we infer that

$$S(r, \boldsymbol{v}) = \left(\frac{1}{2} \alpha \left(\frac{H}{r} \right)^2 + O\left(\frac{H}{r} \right) \right) \prod_{j=1}^{\omega} t(p_j, v_j).$$

Now (16) delivers

$$\sum_{\mathbf{h} \in \mathscr{H}} a_\mathbf{h}(r) = \frac{1}{2} \alpha \left(\frac{H}{r} \right)^2 \prod_{p|r} \sum_{v=1}^{3} a(p, v) t(p, v) + O\left(\frac{H}{r} \prod_{p|r} \sum_{v=1}^{3} |a(p, v)| t(p, v) \right).$$

$$(17)$$

An inspection of the definition of $t(p, v)$ readily shows that

$$t(p, 1) = 1, \quad t(p, 2) = 3(p - 1), \quad t(p, 3) = (p - 1)(p - 2).$$

A short calculation leads to the identity

$$\sum_{v=1}^{3} a(p, v) t(p, v) = 0$$

for all primes p, and for odd primes, by (12) we also have

$$\sum_{v=1}^{3} |a(p,v)| t(p,v) \le 15.$$

It follows that the leading term in (17) vanishes except when $r = 1$. Moreover, again using a divisor estimate, we see that the error term in (17) does not exceed $O(Hr^{\varepsilon-1})$. Hence, by (15),

$$\sum_{\mathbf{h}\in\mathcal{H}} \mathfrak{S}_{q,\mathbf{h}} = \frac{1}{2}\alpha H^2 + O(H^{1+\varepsilon}) + O\left(H\sum_{r\le H} r^{\varepsilon-1}\right),$$

and the conclusion of Lemma 1 follows.

Lemma 2 *Suppose that $0 < \alpha \le 1 \le H \le x$ and that $a, q \in \mathbb{N}$ are coprime with $1 \le a \le q$. Let $U = U_{\alpha,q,a}(x, H)$ denote the number of primes p, p', p'' with $p \equiv p' \equiv p'' \equiv a \bmod q$ that satisfy the inequalities*

$$5 \le p \le x, \quad p < p'' \le p + qH, \quad |p'' - 2p' + p| < \alpha(p'' - p). \tag{18}$$

Further let $\varepsilon > 0$. Then there are a number $x_2 = x_2(q)$ depending only on q and a number $E = E_\varepsilon$ depending only on ε such that whenever $x \ge x_2$ one has

$$U \le (25\alpha H^2 + EH^{1+\varepsilon})\frac{x}{\varphi(q)(\log x)^3}. \tag{19}$$

Proof Suppose that p, p', p'' is a triple counted by U. We write

$$p = a + ql, \quad p'' - p = qh, \quad p' - p = qh'. \tag{20}$$

Then $l \in \mathbb{N}_0$, $(h, h') \in \mathbb{N}^2$, and the conditions (18) imply that

$$0 \le l \le x/q, \quad h \le H, \quad |h - 2h'| \le \alpha h. \tag{21}$$

By (20), it follows that U does not exceed the number of $l \in \mathbb{N}_0$, $(h, h') \in \mathbb{N}^2$ satisfying (21) and $a + ql \ge 5$ for which the three numbers

$$a + ql, \quad a + q(l + h), \quad a + q(l + h') \tag{22}$$

are all prime.

Let $V(h, h') = V(\mathbf{h})$ denote the number of integers l with $0 \le l \le x/q$ and $a + ql \ge 5$ for which the numbers (22) are simultaneously prime. Then, in the notation of the proof of Lemma 1, the above argument shows that

$$U \le \sum_{\mathbf{h} \in \mathscr{H}} V(\mathbf{h}).$$

Further, the quantity $V(\mathbf{h})$ is readily estimated by an upper bound sieve. We wish to apply [4, Theorem 5.7], and with this end in view we consider, for a prime p, the number $\varrho_\mathbf{h}(p)$ of incongruent solutions in z of the congruence

$$(a + qz)(a + q(z + h))(a + q(z + h')) \equiv 0 \bmod p.$$

Then, whenever $p \mid q$, one has $\varrho_\mathbf{h}(p) = 0$ while in the contrary case $p \nmid q$ it is immediate that $\varrho_\mathbf{h}(p) = \nu_\mathbf{h}(p)$. If $\mathbf{h} \in \mathscr{H}$ is such that $\varrho_\mathbf{h}(p) < p$ holds for all primes p, then [4, Theorem 5.7] is applicable and delivers the inequality

$$V(\mathbf{h}) \le 50\mathfrak{S}_{q,\mathbf{h}} \frac{x}{\varphi(q)} (\log x)^{-3} \tag{23}$$

for all x that are sufficiently large in terms of q, as one readily confirms by inspecting (10) and the Euler product in [4, (5.8.3)].

It remains to evaluate $V(\mathbf{h})$ in those cases where $\varrho_\mathbf{h}(p) < p$ fails for some prime p. The trivial upper bound $\varrho_\mathbf{h}(p) \le \min(3, p)$ shows that this is possible only when $p = 2$ or 3. Further, the hypothesis that $\varrho_\mathbf{h}(2) = 2$ implies that $2 \nmid q$, and that at least one of h, h' is odd. By (21) we then find that one of the differences $p' - p, p'' - p$ is odd which is impossible for $p \ge 5$. This shows that $\varrho_\mathbf{h}(2) = 2$ implies $V(\mathbf{h}) = 0$, and a similar argument confirms that the same is true when $\varrho_\mathbf{h}(3) = 3$. In particular, we now see that (23) holds for all $\mathbf{h} \in \mathscr{H}$. Summing (23) over these $\mathbf{h}$ with the aid of Lemma 1 yields Lemma 2.

3 Second Differences: Proof of Theorem 3

We launch an attack toward the estimates claimed in Theorem 3 with a preliminary remark. Throughout, suppose that x_0 and δ are fixed, as in Theorem 2. Let $\mathscr{P}$ be a set of primes, choose a, q with $\mathscr{P} \subset \mathscr{P}_{q,a}$, and assume that (5) holds for all $x \ge x_0$. Suppose it were the case that $p_n > n^2$ holds for some n with $n^2 \ge x_0$. Then, in (5) we take $x = n^2$ and use the lower bound for $\delta(x)$ to infer that

$$n \ge \delta(n^2)\pi(n^2; q, a) \ge (2 \log n)^{-1}\pi(n^2; q, a).$$

The prime number theorem in arithmetic progressions supplies a number $x_1(q)$ such that whenever $x \ge x_1(q)$ then one has $\pi(x; q, a) \ge x/(2\varphi(q) \log x)$. Hence, for

$n^2 \geq \max(x_0, x_1(q))$, we conclude that

$$n \geq \frac{n^2}{4\varphi(q)(\log n)^2}.$$

This is absurd for n sufficiently large in terms of q. It follows that there is a number n_0, depending only on x_0 and q, with the property that whenever $n \geq n_0$ then the inequalities

$$p_n \leq n^2 \quad \text{and} \quad \delta_n^{-1} \leq 2\log n \tag{24}$$

hold. These bounds are improved in the following lemma, but they play a role in its proof.

Lemma 3 *Let $x_0, \delta, \mathscr{P}$ and a, q be as in the preceding paragraph. Then there is a number n_0 depending only on x_0 and q such that whenever $n \geq n_0$, one has*

$$\frac{3}{4}\varphi(q)n \log n \leq p_n \leq 2\varphi(q)\delta_n^{-1}n \log n.$$

Within the proof, we may suppose that (5) holds with $\delta = \delta(x)$. But then, for $x \leq p_n$, the bound (5) also holds with $\delta = \delta_n$. Now suppose for contradiction that $p_n > x_0$ and $p_n > 2\varphi(q)\delta_n^{-1}n \log n$ hold simultaneously. We may use (5) with $x = 2\delta_n^{-1}\varphi(q)n \log n$ and then see that

$$n \geq \pi(2\delta_n^{-1}\varphi(q)n \log n; q, a).$$

Using the prime number theorem in arithmetic progressions much as above, this implies via (24) that

$$n \geq \frac{3}{2}n\frac{\log n}{\log \varphi(q)n}.$$

This is certainly false for n large in terms of q. The upper bound for p_n follows.

Next, let ϖ denote the n-th member of the ascending sequence of *all* primes in $\mathscr{P}_{q,a}$. Then $p_n \geq \varpi$, and by the prime number theorem in arithmetic progressions once again, one has $\varpi \geq \frac{3}{4}\varphi(q)n \log n$ for all large n. This completes the proof of Lemma 3.

The lower bound (7) is now immediate. Indeed, by Lemma 3 and (24), we have

$$\delta_n = \delta(p_n) \geq \delta(2\varphi(q)\delta_n^{-1}n \log n) \geq \delta(4\varphi(q)n(\log n)^2),$$

as required.

The next task ahead of us is to establish Theorem 3. For the upper bound, we apply the triangle inequality to (8) and then see from Lemma 3 that

$$\sum_{N<n\leq 2N} \frac{|\Delta_n|}{p_n} \leq 2 \sum_{N<n\leq 2N+1} \frac{p_{n+1}-p_n}{p_{N+1}} \leq \frac{2p_{2N+2}}{p_{N+1}} \leq \frac{11}{\delta_{2N+2}}, \tag{25}$$

provided only that N is large. This already completes the proof of the upper bound, but there is a simple variant of this argument. Suppose the δ-dense set has the additional property that there exists a number A such that

$$\frac{p_{2m}}{p_m} \leq A \tag{26}$$

holds for all large m. Then the inequalities in (25) provide the alternative estimate

$$\sum_{N<n\leq 2N} \frac{|\Delta_n|}{p_n} \leq 2A. \tag{27}$$

This substantiates a remark that we have made in the introductory part of the paper.

The lower bound in Theorem 3 will be deduced from the sieve bounds established in the previous section. It will be useful to introduce the parameters

$$C = 2/\delta_{2N}, \quad B = 10^{-5}C^{-2}. \tag{28}$$

We will use this notation in the remainder of this paper.

Lemma 4 *Fix x_0 and δ as in Theorem 2, and suppose that $\delta(x)^2 \log x$ tends to infinity with x. Then there is a sequence of natural numbers $N_2(q)$ with the property that for all sets of primes $\mathscr{P}$ that are δ-dense relative to x_0 and some $\mathscr{P}_{q,a}$, and for all $N \geq N_2(q)$, the set*

$$\mathscr{B}(N) = \{N < n \leq 2N - 2 : p_{n+2} - p_n \leq 33C\varphi(q)\log N,\ |\Delta_n| \geq B\varphi(q)\log N\}$$

has at least $N/2$ elements.

Note that once this lemma is established, we may apply Lemma 3 to conclude that

$$\sum_{n\in\mathscr{B}(N)} \frac{|\Delta_n|}{p_n} \geq \frac{1}{4}C^{-1}B > 10^{-7}\delta_{2N}^3$$

holds whenever $N \geq N_2(q)$. This includes the lower bound recorded in Theorem 3.

We now give the proof of Lemma 4. Let $\mathscr{S}_1$ be the set of all $n \in (N, 2N - 2]$ where $p_{n+2} - p_n > 33C\varphi(q) \log N$. Then

$$\#\mathscr{S}_1 \leq \sum_{N < n \leq 2N} \frac{p_{n+2} - p_n}{33C\varphi(q) \log N} \leq \frac{p_{2N} + p_{2N-1}}{33C\varphi(q) \log N},$$

and by Lemma 3, for sufficiently large N, we conclude that $\#\mathscr{S}_1 \leq N/8$.

Next, let $\mathscr{S}_2$ be the set of all $n \in (N, 2N - 2]$ where $p_{n+2} - p_n \leq B\varphi(q) \log N$. Then, in view of (8) and Lemma 3, the primes $p = p_n$, $p' = p_{n+1}$ and $p'' = p_{n+2}$ satisfy the conditions (18) with $\alpha = 1$, $x = 2C\varphi(q)N \log 2N$ and $H = \frac{\varphi(q)}{q}B \log N$. Note that (7) implies that $B \log N$ tends to infinity with N. Hence H is large when N is large (in terms of q). Therefore, for large N, Lemma 2 is applicable, and delivers the estimates

$$\#\mathscr{S}_2 \leq U_{1,q,a}(2C\varphi(q)N \log 2N, H)$$

$$\leq (25H^2 + E_\varepsilon H^{1+\varepsilon}) \frac{2C\varphi(q)N \log 2N}{(\varphi(q) \log 2N + O(\log \log N))^3}$$

$$\leq 100B^2CN \leq 10^{-8}N,$$

provided again that N is sufficiently large in terms of q.

For notational convenience, put

$$J_0 = B\varphi(q) \log N, \quad J_1 = 33C\varphi(q) \log N,$$

and let $\mathscr{S}_3$ be the set of all $n \in (N, 2N - 2]$ where

$$J_0 < p_{n+2} - p_n \leq J_1 \quad \text{and} \quad |\Delta_n| < J_0.$$

Note that a number $n \in (N, 2N - 2]$ that is in none of the sets $\mathscr{S}_1$, $\mathscr{S}_2$, $\mathscr{S}_3$ lies in $\mathscr{B}(N)$. Hence, once we have proved that $\#\mathscr{S}_3 \leq N/10$ holds for large N, the proof of Lemma 4 will be complete.

We proceed by a dissection argument. Let $\mathscr{S}_3(H)$ be the subset of $\mathscr{S}_3$ where $\frac{1}{2}qH < p_{n+2} - p_n \leq qH$. The range of H relevant to us is the interval

$$2J_0 \leq qH \leq 2J_1.$$

For such a value of H and $n \in \mathscr{S}_3(H)$ we have $|\Delta_n| \leq J_0 = \alpha(\frac{1}{2}qH) \leq \alpha(p_{n+2}-p_n)$ where we chose

$$\alpha = \frac{2J_0}{qH}.$$

By Lemma 3, we may take $x = 2C\varphi(q)N \log 2N$ in Lemma 2, and then find that

$$\#\mathscr{S}_3(H) \leq U_{\alpha,q,a}(2C\varphi(q)N \log 2N, H)$$

$$\leq (25\alpha H^2 + E_\varepsilon H^{1+\varepsilon}) \frac{2CN \log 2N}{(\log(2C\varphi(q)N \log 2N))^3}.$$

When N is sufficiently large in terms of q, this upper bound simplifies to

$$\#\mathscr{S}_3(H) \leq (50 J_0 H q^{-1} + E_\varepsilon H^{1+\varepsilon}) \frac{3CN}{(\log 2N)^2}.$$

We take $H = 2^j J_0 q^{-1}$ with $2^j \leq J_1/J_0 = 33C/B$ and sum over $j \geq 1$. Since $\mathscr{S}_3$ is covered by the union of these $\mathscr{S}_3(H)$, we conclude that for large N we have

$$\#\mathscr{S}_3 \leq (50 J_0 J_1 q^{-2} + E_\varepsilon (J_1/q)^{1+\varepsilon}) \frac{6CN}{(\log 2N)^2} \leq 10^4 BC^2 N.$$

By hypothesis, we have $10^4 BC^2 = 1/10$ so that indeed $\#\mathscr{S}_3 \leq N/10$, as required.

4   Curvature: Proof of Theorem 2

This section is devoted to the proof of Theorem 2. Recall that $z_n = n + i \log p_n$, and with this choice put

$$k_n = \left| \arg \frac{z_{n+1} - z_n}{z_n - z_{n-1}} \right|.$$

By the elementary properties of the arctan function and its addition theorem, one finds the alternative representations

$$k_n = \left| \arctan \log \frac{p_{n+2}}{p_{n+1}} - \arctan \log \frac{p_{n+1}}{p_n} \right|$$

$$= \arctan \frac{\left| \log \dfrac{p_{n+2} p_n}{p_{n+1}^2} \right|}{1 + \left(\log \dfrac{p_{n+2}}{p_{n+1}} \right)\left(\log \dfrac{p_{n+1}}{p_n} \right)}. \tag{29}$$

It will also be useful to put

$$\Gamma_n = (p_{n+2} - p_{n+1})(p_{n+1} - p_n) p_{n+1}^{-2}. \tag{30}$$

Then, by (8), one has

$$\frac{p_{n+2}p_n}{p_{n+1}^2} = 1 + \frac{\Delta_n}{p_{n+1}} - \Gamma_n. \tag{31}$$

This identity also occurs in Rényi's work (see [8], Eq. (5)), and is crucial to his arguments. Before we proceed further, we note that

$$0 < \Gamma_n \le \frac{p_{n+2} - p_{n+1}}{p_{n+1}}.$$

In particular, Lemma 3 yields $\Gamma_n \le (p_{n+2} - p_{n+1})/(\frac{3}{4}\varphi(q)n\log n)$, and the argument that was used in (25) produces

$$\sum_{N < n \le 2N} \Gamma_n \le \frac{6}{\delta_{2N+2}}, \tag{32}$$

at least when N is large in terms of q, as we assume from now on. For such N we proceed to derive the inequality

$$\sum_{N < n \le 2N} k_n \le \frac{300}{\delta_{2N+2}}. \tag{33}$$

Let $\mathscr{C}$ denote the set of all $n \in (N, 2N]$ where $\Gamma_n > 1/8$, and let $\mathscr{D}$ denote the set of all $n \in (N, 2N]$ where $|\Delta_n| > p_n/8$. Finally, let $\mathscr{E}$ denote the set of the remaining n in $(N, 2N]$. By (32), it follows that $\#\mathscr{C} \le 48/\delta_{2N+2}$, and similarly, one finds from (25) that $\#\mathscr{D} \le 88/\delta_{2N+2}$. The trivial bound $k_n \le \pi/2$ suffices to see that the contribution from all $n \in \mathscr{C} \cup \mathscr{D}$ to the sum in (33) does not exceed $68\pi/\delta_{2N+2}$.

Now suppose that $n \in \mathscr{E}$. Then

$$\left| \frac{\Delta_n}{p_{n+1}} - \Gamma_n \right| \le \frac{1}{4}.$$

Hence, on applying the familiar bound $\arctan t \le t$ that is valid for all $t \ge 0$, we first see from (29) that

$$k_n \le \left| \log \frac{p_{n+2}p_n}{p_{n+1}^2} \right|,$$

and then, by observing that for real numbers t with $|t| \le 1/4$ one has $|\log(1+t)| \le 2|t|$, we conclude via (31) that

$$k_n \le 2\frac{|\Delta_n|}{p_n} + 2\Gamma_n.$$

We sum this over $n \in \mathscr{E}$. Then, by (25) and (32), we see that (33) indeed holds.

The upper bound reported in Theorem 2 is readily deduced from (33). There is a number $N_1(q)$ such that (33) holds for all $N \geq N_1(q)$. But then, if a large M is given, we may take $N = 2^j N_1(q) \leq M$ in (33) and sum over j. Using the trivial bound for $k_n \leq \pi/2$ when $n \leq N_1(q)$ and observing that δ is decreasing, we find in this way that

$$K_M(\mathscr{P}) \leq \frac{\pi}{2} N_1(q) + 499 \frac{\log M}{\delta_M}.$$

When N is sufficiently large in terms of q, this implies the upper bound recorded in Theorem 2. Again, there is a variant of this argument in the case where (26) holds. Then we have (27) available, and for the same reason in (32) the upper bound can be replaced by A. With these estimates in hand, the above argument produces the better bound

$$K_N(\mathscr{P}) \ll_A \log N.$$

Once again, this confirms a claim from the introduction.

The verification of the lower bound in Theorem 2 is somewhat more complex. Throughout the argument below we use the notation as introduced in Lemma 4. Let $N \geq N_0(q)$, and consider a number $n \in \mathscr{B}(N)$. Then by (8) and the triangle inequality,

$$B\varphi(q)\log N \leq |\Delta_n| \leq 33C\varphi(q)\log N.$$

Furthermore, one has $p_{n+2} - p_{n+1} \leq p_{n+2} - p_n \leq 33C\varphi(q)\log N$, and the same inequality holds for $p_{n+1} - p_n$. Hence, by Lemma 3 and (30), we see that $\Gamma_n \leq (44C)^2 N^{-2}$, and that

$$\frac{B}{2CN}\frac{|\Delta_n|}{p_{n+1}} \leq \frac{44C}{N}.$$

Here we have used that N is large. These last inequalities combine with the bound on Γ_n to

$$\frac{B}{3CN} \leq \left|\frac{\Delta_n}{p_{n+1}} - \Gamma_n\right| \leq \frac{45C}{N} \leq \frac{1}{9}.$$

However, when $|t| \leq \frac{1}{2}$ one has $\frac{1}{2}|t| \leq |\log(1+t)| \leq 2|t|$, so that (31) now yields

$$\frac{B}{6CN} \leq \left|\log \frac{p_n p_{n+2}}{p_{n+1}^2}\right| \leq \frac{90C}{N},$$

We apply Lemma 3 again to confirm that for $j = 1$ and 2, one has

$$1 \leq \frac{p_{n+j+1}}{p_{n+j}} = 1 + \frac{p_{n+j+1} - p_{n+j}}{p_{n+j}} \leq 1 + \frac{44C}{N} \leq \frac{10}{9}.$$

Thus

$$0 \leq \log \frac{p_{n+j+1}}{p_{n+j}} \leq \frac{1}{9},$$

and hence,

$$\frac{B}{7CN} \leq \frac{\left| \log \frac{p_{n+2}p_n}{p_{n+1}^2} \right|}{1 + \left(\log \frac{p_{n+2}}{p_{n+1}} \right)\left(\log \frac{p_{n+1}}{p_n} \right)} \leq \frac{90C}{N} \leq \frac{2}{9}.$$

For $0 \leq t \leq \frac{1}{4}$ one has $\arctan t \geq \frac{1}{2}t$. Therefore, by (29), we conclude that $k_n \geq B/(14CN)$ holds for all $n \in \mathcal{B}(N)$. By (28) and Lemma 4 it follows that

$$\sum_{n \in \mathcal{B}(N)} k_n \geq \frac{B}{28C} \geq \frac{1}{10^8 \delta_{2N}^3}.$$

In this estimate, we replace N by $2^{-j}N$ and sum over $1 \leq j \leq \frac{1}{3}\log N$. But then $2^{-j}N \geq \sqrt{N}$ for all j, and we only have to arrange that $\sqrt{N} \geq N_2(q)$, with N_2 as in Lemma 4. We conclude that $K_N(\mathscr{P}) \geq 10^{-8}\delta_N^{-3}\log N$, as required to complete the proof of Theorem 2.

5　A Scattered Sequence

We end with a brief description of a sequence with large curvature. Let $A_1 = 10$ and define A_l by the recursion $A_{l+1} = 2A_l \log 4A_l$ $(l \geq 1)$. Note that the intervals $I_l = [A_l, 4A_l]$ are disjoint. We now construct a set of primes $\mathcal{Q}$ as follows. If the number of primes in I_l is even, then all these primes become elements of $\mathcal{Q}$, and in the contrary case, we put all but the smallest of the primes in I_l in $\mathcal{Q}$. Primes that are not in some I_l are not in $\mathcal{Q}$. Note that we have arranged that the number of elements in $\mathcal{Q} \cap I_l$ is even.

We claim that $\mathcal{Q}$ is δ-dense with $\delta(x) = 1/\log x$. To see this, let $\pi_{\mathcal{Q}}(x)$ denote the number of primes in $\mathcal{Q}$ not exceeding x, and suppose that x is large. Then, there is some l with $\frac{5}{4}A_l \leq x \leq \frac{5}{4}A_{l+1}$. In the case where $\frac{5}{4}A_l \leq x \leq 4A_l$, Chebyshev's estimates give $\pi_{\mathcal{Q}}(x) \gg x/\log x$ which is more　than is required. In the range

$4A_l \le x \le \frac{5}{4}A_{l+1}$ we use the prime number theorem to see that

$$\pi_{\mathscr{Q}}(x) \ge \pi_{\mathscr{Q}}(4A_l) \ge \frac{3A_l}{\log 4A_l}(1 + o(1))$$

$$= \left(\frac{3}{2} + o(1)\right)\frac{A_{l+1}}{(\log A_{l+1})^2} = \left(\frac{6}{5} + o(1)\right)\frac{\frac{5}{4}A_{l+1}}{(\log \frac{5}{4}A_{l+1})^2}.$$

In particular, this confirms (5) with $\delta(x) = 1/\log x$, as desired.

Let (q_j) denote the sequence of the elements of $\mathscr{Q}$ in ascending order. Now let l be large. By construction, $\pi_{\mathscr{Q}}(4A_l)$ is even, say $2N$. Then $q_{2N} < 4A_l$ but $q_{2N+1} > A_{l+1}$. Also, by the prime number theorem, $q_{2N+2} - q_{2N+1} = o(A_{l+1})$ so that we now have $|\Delta_{2N}| \ge A_{l+1}(1 + o(1))$, and hence, again by the prime number theorem,

$$\frac{|\Delta_{2N}|}{q_{2N}} \ge \frac{A_{l+1}}{4A_l}(1 + o(1)) = \left(\frac{1}{2} + o(1)\right)\log A_l.$$

Further, the equation $2N = \pi_{\mathscr{Q}}(4A_l)$ and the straightforward bounds

$$\frac{3A_l}{\log A_l}(1 + o(1)) \le \pi_{\mathscr{Q}}(4A_l) \le \frac{4A_l}{\log A_l}(1 + o(1))$$

imply that $\log A_l = (1 + o(1))\log N$, so that we arrive at

$$\frac{|\Delta_{2N}|}{q_{2N}} \ge \frac{1}{3}\log N.$$

In particular, we see that the sum considered in Theorem 3 contains a single term exceeding $\frac{1}{3}\log N$, which is of the order of δ_{2N+2}^{-1}.

Acknowledgements The authors are grateful to CIRM at Marseille Luminy for creating a stimulating working atmosphere. The second author also likes to thank Forschungsinstitut Mathematik (FIM) at ETH Zürich for a very pleasant stay.

References

1. P. Erdős, A. Rényi, Some problems and results on consecutive primes. Simon Stevin **27**, 115–125 (1950)
2. P. Erdős, P. Turán, On some new questions on the distribution of prime numbers. Bull. Am. Math. Soc. **54**, 371–378 (1948)
3. P.X. Gallagher, On the distribution of primes in short intervals. Mathematika **23**, 4–9 (1976)
4. H. Halberstam, H.-E. Richert, *Sieve Methods*. London Mathematical Society Monographs, vol. 4 (Academic, London/New York, 1974)

5. G.H. Hardy, J.E. Littlewood, Some problems of "Partitio Numerorum": III. On the expression of a number as a sum of primes. Acta Math. **44**, 1–70 (1922)
6. C. Pomerance, The prime number graph. Math. Comp. **33**, 399–408 (1979)
7. K. Prachar, Bemerkung zu einer Arbeit von Erdős und Rényi und Berichtigung. Monatsh. Math. **58**, 117 (1954)
8. A. Rényi, On a theorem of Erdős and Turán. Proc. Am. Math. Soc. **1**, 7–10 (1950)

Sums of the Digits in Bases 2 and 3

Jean-Marc Deshouillers, Laurent Habsieger, Shanta Laishram, and Bernard Landreau

To Robert Tichy, for his 60th birthday

Abstract Let $b \geq 2$ be an integer and let $s_b(n)$ denote the sum of the digits of the representation of an integer n in base b. For sufficiently large N, one has

$$\mathrm{Card}\{n \leq N : |s_3(n) - s_2(n)| \leq 0.1457205 \log n\} > N^{0.970359}.$$

The proof only uses the separate (or marginal) distributions of the values of $s_2(n)$ and $s_3(n)$.

AMS 2010 Classification number: 11K16

J.-M. Deshouillers (✉)
Institut Mathématique de Bordeaux, UMR 5251, Bordeaux INP, Université de Bordeaux, CNRS, 33405 Talence, France
e-mail: jean-marc.deshouillers@math.u-bordeaux.fr

L. Habsieger
Institut Camille Jordan, Université de Lyon, CNRS UMR 5208, Université Claude Bernard Lyon 1, 69622 Villeurbanne Cedex, France
e-mail: habsieger@math.univ-lyon1.fr

S. Laishram
Indian Statistical Institute, 7 SJS Sansanwal Marg, 110016 New Delhi, India
e-mail: shanta@isid.ac.in

B. Landreau
LAREMA Laboratoire Angevin de REcherche en MAthématiques, UMR 6093, FR 2962, Université d'Angers, CNRS, 49045 Angers, France
e-mail: bernard.landreau@univ-angers.fr

© Springer International Publishing AG 2017
C. Elsholtz, P. Grabner (eds.), *Number Theory – Diophantine Problems, Uniform Distribution and Applications*, DOI 10.1007/978-3-319-55357-3_9

1 Introduction

For integers $b \geq 2$ and $n \geq 0$, we denote by "the sum of the digits of n in base b" the quantity

$$s_b(n) = \sum_{j \geq 0} \varepsilon_j, \text{ where } n = \sum_{j \geq 0} \varepsilon_j b^j \text{ with } \forall j : \varepsilon_j \in \{0, 1, \ldots, b-1\}.$$

Our attention on the question of the proximity of $s_2(n)$ and $s_3(n)$ comes from the apparently non-related question of the distribution of the last non-zero digit of $n!$ in base 12 (cf. [3] and [2]).[1]

Computation shows that there are 48 266 671 607 positive integers up to 10^{12} for which $s_2(n) = s_3(n)$, but it seems to be unknown whether there are infinitely many integers n for which $s_2(n) = s_3(n)$ or even for which $|s_2(n) - s_3(n)|$ is significantly small.

We do not know the first appearance of the result we quote as Theorem 1; in any case, it is a straightforward application of the fairly general main result of Bassily and Kátai [1]. We recall that a sequence $\mathcal{A} \subset \mathbb{N}$ of integers is said to have asymptotic natural density 1 if

$$\text{Card}\{n \leq N : n \in \mathcal{A}\} = N + o(N).$$

Theorem 1 *Let ψ be a function tending to infinity with its argument. The sequence of natural numbers n for which*

$$\left(\frac{1}{\log 3} - \frac{1}{\log 4} \right) \log n - \psi(n) \sqrt{\log n} \leq s_3(n) - s_2(n)$$

$$\leq \left(\frac{1}{\log 3} - \frac{1}{\log 4} \right) \log n + \psi(n) \sqrt{\log n}$$

has asymptotic natural density 1.

Our main result is that there exist infinitely many n for which $|s_3(n) - s_2(n)|$ is significantly smaller than $\left(\frac{1}{\log 3} - \frac{1}{\log 4} \right) \log n = 0.18889\ldots \log n$. More precisely we have the following:

Theorem 2 *For sufficiently large N, one has*

$$\text{Card}\{n \leq N : |s_3(n) - s_2(n)| \leq 0.1457205 \log n\} > N^{0.970359}. \tag{1}$$

---

[1] Indeed, if the last non-zero digit of $n!$ in base 12 belongs to $\{1, 2, 5, 7, 10, 11\}$ then $|s_3(n) - s_2(n)| \leq 1$; this seems to occur infinitely many times.

The mere information we use in proving Theorem 2 is the knowledge of the separate (or marginal) distributions of $(s_2(n))_n$ and $(s_3(n))_n$, without using any further information concerning their joint distribution.

In Sect. 2, we provide a heuristic approach to Theorems 1 and 2; the actual distribution of $(s_2(n))_n$ and $(s_3(n))_n$ is studied in Sect. 3. The proof of Theorem 2 is given in Sect. 4.

Let us formulate three remarks as a conclusion to this introductory section.

It seems that our present knowledge of the joint distribution of s_2 and s_3 (cf. for example Stewart [5] for a Diophantine approach or Drmota [4] for a probabilistic one) does not permit us to improve on Theorem 2.

Theorem 2 can be extended to any pair of distinct bases, say q_1 and q_2: more than computation, the authors have deliberately chosen to present an idea to the Dedicatee.

Although we could not prove it, we believe that Theorem 2 represents the limit of our method.

2 A Heuristic Approach

As a warm-up for the actual proofs, we sketch a heuristic approach. A positive integer n may be expressed as

$$n = \sum_{j=0}^{J(n)} \varepsilon_j(n) b^j, \text{ with } J(n) = \left\lfloor \frac{\log n}{\log b} \right\rfloor.$$

If we consider an interval of integers around N, the smaller is j the more equidistributed are the $\varepsilon_j(n)$'s, and the smaller are the elements of a family $\mathcal{J} = \{j_1 < j_2 < \cdots < j_s\}$ the more independent are the $\varepsilon_j(n)$'s for $j \in \mathcal{J}$. Thus a first model for $s_b(n)$ for n around N is to consider a sum of $\left\lfloor \frac{\log N}{\log b} \right\rfloor$ independent random variables uniformly distributed in $\{0, 1, \ldots, b - 1\}$. Thinking of the central limit theorem, we even consider a continuous model, representing $s_b(n)$, for n around N by a Gaussian random variable $S_{b,N}$ with expectation and variance given by

$$\mathbb{E}\,(S_{b,N}) = \frac{(b-1)\log N}{2\log b} \text{ and } \mathbb{V}\,(S_{b,N}) = \frac{(b^2-1)\log N}{12\log b}.$$

In particular

$$\mathbb{E}\,(S_{2,N}) = \frac{\log N}{\log 4} \text{ and } \mathbb{E}\,(S_{3,N}) = \frac{\log N}{\log 3},$$

and their standard deviations have the order of magnitude $\sqrt{\log N}$.

Towards Theorem 1 In [1], it is proved that a central limit theorem actually holds for s_b; more precisely, the following proposition is the special case of the first relation in the main Theorem of [1], with $f(n) = s_b(n)$ and $P(X) = X$.

Proposition 1 *For any positive y, as x tend to infinity, one has*

$$\frac{1}{x} \text{Card} \left\{ n < x : |s_b(n) - \mathbb{E}(S_{b,n})| < y (\mathbb{V}(S_{b,n}))^{1/2} \right\} \to \frac{1}{\sqrt{2\pi}} \int_{-y}^{y} e^{-t^2/2} dt.$$

Theorem 1 easily follows from Proposition 1: the set under our consideration is the intersection of 2 sets of density 1.

Towards Theorem 2 If we wish to deal with a difference $|s_3(n) - s_2(n)| < u \log n$ for some $u < \left(\frac{1}{\log 3} - \frac{1}{\log 4} \right)$ we must, by what we have seen above, consider events of asymptotic probability zero, which means that a heuristic approach must be substantiated by a rigorous proof. Our key remark is that the variance of $S_{3,N}$ is larger than that of $S_{2,N}$; this implies the following: the probability that $S_{3,N}$ is at a distance d from its mean is larger that the probability that $S_{2,N}$ is at a distance d from its mean. So, we have the hope to find some $u < \left(\frac{1}{\log 3} - \frac{1}{\log 4} \right)$ such that the probability that $|S_{2,N} - \mathbb{E}(S_{2,N})| > u \log N$ is smaller than the probability that $S_{3,N}$ is very close to $\mathbb{E}(S_{2,N})$. This will imply that for some ω we have $|S_{3,N}(\omega) - S_{2,N}(\omega)| \le u \log N$.

3 On the Distribution of the Values of $s_2(n)$ and $s_3(n)$

In order to prove Theorem 2 we need

- an upper bound for the tail of the distribution of s_2,
- a lower bound for the tail of the distribution of s_3.

3.1 Upper Bound for the Tail of the Distribution of s_2

Proposition 2 *Let $\lambda \in (0, 1)$. For any*

$$v > 1 - ((1 - \lambda) \log(1 - \lambda) + (1 + \lambda) \log(1 + \lambda)) / \log 4$$

and any sufficiently large integer H, we have

$$\text{Card}\{n < 2^{2H} : |s_2(n) - H| \ge \lambda H\} \le 2^{2Hv}. \tag{2}$$

Proof When $b = 2$, the distribution of the values of $s_2(n)$ is simply binomial; we thus get

$$\text{Card}\,\{0 \le n < 2^{2H} : s_2(n) = m\} = \binom{2H}{m}.$$

Using the fact that the sequence (in m) $\binom{2H}{m}$ is symmetric and unimodal plus Stirling's formula, we obtain that when $m \le (1 - \lambda)H$ or $m \ge (1 + \lambda)H$, one has

$$\binom{2H}{m} \le H^{O(1)} \frac{(2H)^{2H}}{((1 - \lambda)H)^{(1-\lambda)H}((1 + \lambda)H)^{(1+\lambda)H}}$$

$$\le H^{O(1)} \left(\frac{2^2}{(1 - \lambda)^{(1-\lambda)}(1 + \lambda)^{(1+\lambda)}} \right)^H$$

$$\le H^{O(1)} \left(2^{(1-((1-\lambda)\log(1-\lambda)+(1+\lambda)\log(1+\lambda))/2\log 2)} \right)^{2H}.$$

Relation (2) comes from the above inequality and the fact that the left-hand side of (2) is the sum of at most $2H$ such terms. $\square$

3.2 *Lower Bound for the Tail of the Distribution of s_3*

Proposition 3 *Let L be sufficiently large an integer. We have*

$$\text{Card}\{n < 3^L : s_3(n) = \lfloor L \log 3/ \log 4 \rfloor\} \ge 3^{0.970359238L}. \tag{3}$$

Proof The positive integer L being given, we write any integer $n \in [0, 3^L)$ in its non-necessarily proper representation, as a chain of exactly L characters, $\ell_i(n)$ of them being equal to i, for $i \in \{0, 1, 2\}$, the sum $\ell_0(n) + \ell_1(n) + \ell_2(n)$ being equal to L, the total number of digits in this representation.[2] One has

$$\text{Card}\,\{0 \le n < 3^L : s_3(n) = m\} = \sum_{\substack{\ell_0+\ell_1+\ell_2=L \\ \ell_1+2\ell_2=m}} \frac{L!}{\ell_0!\ell_1!\ell_2!}. \tag{4}$$

In order to get a lower bound for the left-hand side of (4), it is enough to select one term in its right-hand side. We choose

$$\mathfrak{l}_2 = \lfloor 0.235001144L \rfloor \,;\, \mathfrak{l}_1 = \lfloor L \log 3/ \log 4 \rfloor - 2\,\mathfrak{l}_2 \,;\, \mathfrak{l}_0 = L - \mathfrak{l}_1 - \mathfrak{l}_2.$$

---

[2]For example, when $L = 5$, the number "sixty" will be represented as 02020. Happy palindromic birthday, Robert!

A straightforward application of Stirling's formula, similar to the one used in the previous subsection, leads to (3). $\qquad\square$

4 Proof of Theorem 2

Let N be sufficiently large an integer. We let $K = \lfloor \log N / \log 3 \rfloor - 2$ and $H = \lfloor (K-1) \log 3 / \log 4 \rfloor + 2$. We notice that we have

$$N/81 \leq 3^{K-1} < 3^K < 2^{2H} \leq N. \tag{5}$$

We use Proposition 2 with $\lambda = 0.14572049 \log 4$, which leads to

$$\mathrm{Card}\{n \leq 2^{2H} : |s_2(n) - H| \geq \lambda H\} \leq 2^{0.970359230 \times 2H} \leq N^{0.970359230}. \tag{6}$$

For any $n \in [2 \cdot 3^{K-1}, 3^K)$ we have $s_3(n) = 2 + s_3(n - 2 \cdot 3^{K-1})$ and so it follows from Proposition 3 that we have

$$
\begin{aligned}
\mathrm{Card}\{n \in [2 \cdot 3^{K-1}, 3^K) &: s_3(n) = H\} \\
&= \mathrm{Card}\{n < 3^{K-1}) : s_3(n) = H - 2\} \\
&= \mathrm{Card}\{n < 3^{K-1}) : s_3(n) = \lfloor (K-1) \log 3 / \log 4 \rfloor\} \\
&\geq 3^{0.970359238(K-1)} \geq N^{0.970359237}.
\end{aligned}
$$

This implies that we have

$$\mathrm{Card}\{n \leq 2^{2H} : s_3(n) = H\} \geq N^{0.970359237}. \tag{7}$$

From (6) and (7), we deduce that for N sufficiently large, we have

$$\mathrm{Card}\{n \leq N : |s_2(n) - s_3(n)| \leq 0.1457205 \log n\} \geq N^{0.970359}.$$

$$\square$$

Acknowledgements The authors are indebted to Bernard Bercu for several discussions on the notion of "spacing" between two random variables, a notion to be developed later. They also thank the Referees for their constructive comments. The first, third, and fourth authors wish to thank the Indo-French centre CEFIPRA for the support permitting them to collaborate on this project (ref. 5401-A). The first named author acknowledges with thank the support of the French-Austrian project MuDeRa (ANR and FWF).

References

1. N.L. Bassily, I. Kátai, Distribution of the values of q-additive functions, on polynomial sequences. Acta Math. Hungar. **68**, 353–361 (1995)
2. J.-M. Deshouillers, A footnote to *The least non zero digit of n! in base 12*. Unif. Distrib. Theory **7**, 71–73 (2012)
3. J.-M. Deshouillers, I. Ruzsa, The least non zero digit of n! in base 12. Publ. Math. Debr. **79**, 395–400 (2011)
4. M. Drmota, The joint distribution of q-additive functions. Acta Arith. **100**, 17–39 (2001)
5. C. Stewart, On the representation of an integer in two different bases. J. Reine Angew. Math. **319**, 63–72 (1980)

On the Discrepancy of Halton–Kronecker Sequences

Michael Drmota, Roswitha Hofer, and Gerhard Larcher

Dedicated to Robert F. Tichy on the occasion of his 60th birthday

Abstract We study the discrepancy D_N of sequences $(\mathbf{z}_n)_{n\geq 1} = ((\mathbf{x}_n, y_n))_{n\geq 0} \in [0,1)^{s+1}$ where $(\mathbf{x}_n)_{n\geq 0}$ is the s-dimensional Halton sequence and $(y_n)_{n\geq 1}$ is the one-dimensional Kronecker-sequence $(\{n\alpha\})_{n\geq 1}$. We show that for α algebraic we have $ND_N = \mathcal{O}(N^\varepsilon)$ for all $\varepsilon > 0$. On the other hand, we show that for α with bounded continued fraction coefficients we have $ND_N = \mathcal{O}\left(N^{\frac{1}{2}}(\log N)^s\right)$ which is (almost) optimal since there exist α with bounded continued fraction coefficients such that $ND_N = \Omega\left(N^{\frac{1}{2}}\right)$.

1 Introduction and Statement of Results

Let $(\mathbf{z}_n)_{n\geq 0}$ be a sequence in the d-dimensional unit-cube $[0,1)^d$, then the discrepancy of the first N points of the sequence is defined by

$$D_N = \sup_{B \subseteq [0,1)^d} \left| \frac{A_N(B)}{N} - \lambda(B) \right|,$$

where

$$A_N(B) := \#\{n : 0 \leq n < N, \mathbf{z}_n \in B\},$$

λ is the d-dimensional volume and the supremum is taken over all axis-parallel boxes $B \subseteq [0,1)^d$.

M. Drmota
Institute of Discrete Mathematics and Geometry, TU Wien, Wien, Austria
e-mail: michael.drmota@tuwien.ac.at

R. Hofer • G. Larcher (⊠)
Institute of Financial Mathematics and Applied Number Theory, University Linz, Linz, Austria
e-mail: roswitha.hofer@jku.at; gerhard.larcher@jku.at

© Springer International Publishing AG 2017
C. Elsholtz, P. Grabner (eds.), *Number Theory – Diophantine Problems,
Uniform Distribution and Applications*, DOI 10.1007/978-3-319-55357-3_10

The sequence $(\mathbf{z}_n)_{n\geq 0}$ is called uniformly distributed if $\lim_{N\to\infty} D_N = 0$.

It is the best known conjecture in the theory of irregularities of distribution that for every sequence $(\mathbf{z}_n)_{n\geq 0}$ in $[0,1)^d$ we have

$$D_N \geq c_d \cdot \frac{(\log N)^d}{N}$$

for a constant $c_d > 0$ and for infinitely many N. Hence sequences whose discrepancy satisfies $D_N = \mathcal{O}\left(\frac{(\log N)^d}{N}\right)$ are called *low-discrepancy sequences*. Note that recent investigations of Bilyk, Lacey et al., see, for example, [1] or [2], have led some people to conjecture that $\frac{(\log N)^{\frac{d+1}{2}}}{N}$ instead of $\frac{(\log N)^d}{N}$ is the best possible order for the discrepancy of sequences in $[0,1)^d$.

Well-known examples of low-discrepancy sequences are the s-dimensional Halton-sequence $(\mathbf{x}_n)_{n\geq 0} \in [0,1)^s$, or the one-dimensional Kronecker sequence $(y_n)_{n\geq 0} = (\{n\alpha\})_{n\geq 0} \in [0,1)$ where α is a given irrational number with bounded continued fraction coefficients.

Therefore, the s-dimensional Halton-sequence satisfies

$$ND_N = \mathcal{O}\left((\log N)^s\right)$$

and the Kronecker sequence with suitable α (namely, if α has bounded continued fraction expansion) satisfies

$$ND_N = \mathcal{O}\left(\log N\right).$$

If α is an algebraic number, then with the help of the Thue–Siegel–Roth Theorem it can be shown that in this case for the discrepancy of the one-dimensional Kronecker-sequence we have $ND_N = \mathcal{O}\left(N^\varepsilon\right)$ for all $\varepsilon > 0$.

For the sake of completeness we recall the definition of the Halton sequence $(\mathbf{x}_n)_{n\geq 0}$:

We choose a basis $b_1, b_2, \ldots, b_s$ of pairwise relatively prime integers larger than 1. To construct the i-th coordinate $x_n^{(i)} \in [0,1)$ of the n-th sequence point $\mathbf{x}_n = \left(x_n^{(1)}, \ldots, x_n^{(s)}\right) \in [0,1)^s$ we represent $n = n_0^{(i)} + n_1^{(i)} b_i + n_2^{(i)} b_i^2 + n_3^{(i)} b_i^3 + \cdots$. in base b_i and set

$$x_n^{(i)} := \frac{n_0^{(i)}}{b_i} + \frac{n_1^{(i)}}{b_i^2} + \frac{n_2^{(i)}}{b_i^3} + \cdots .$$

In the following we will be interested in the discrepancy of the combination

$$\mathbf{z}_n = (\mathbf{x}_n, y_n)_{n\geq 0} \in [0,1)^{s+1} =: [0,1)^d$$

in the $d := s + 1$-dimension unit-cube. For this sequence (we will call it d-dimensional Halton–Kronecker sequence) it was shown (see [8] or [9]) that this sequence is uniformly distributed for all irrational α.

In [7] (see also [5] for an earlier, slightly weaker result) it was shown that for almost all choices of α the Halton–Kronecker sequence is almost a low-discrepancy sequence, i.e., for almost all α we have

$$ND_N = \mathcal{O}\left((\log N)^{s+1+\varepsilon}\right) = \mathcal{O}\left((\log N)^{d+\varepsilon}\right)$$

for all $\varepsilon > 0$.

However until now no explicit choice for α such that this discrepancy bound is attained could be given. When searching for explicit examples of α providing a small discrepancy for the Halton–Kronecker sequence, then two possible ideas are near at hand:

- maybe algebraic α generates a small discrepancy of order $ND_N = \mathcal{O}(N^\varepsilon)$ as in the pure Kronecker case,
- maybe α with bounded continued fraction coefficients generate a low-discrepancy Halton–Kronecker sequence, i.e., $ND_N = \mathcal{O}\left((\log N)^{s+1}\right)$ as in the pure Kronecker case.

We will show in the following that the first assertion is true (see Theorem 1) and that the second assertion in general is not true (see Theorem 2).

So, our results are:

Theorem 1 *Let α be irrational and algebraic, then for the discrepancy D_N of the $(s + 1)$-dimensional Halton–Kronecker sequence $(\mathbf{z}_n)_{n \geq 0} = ((\mathbf{x}_n, y_n))_{n \geq 0}$ we have*

$$ND_N = \mathcal{O}(N^\varepsilon)$$

for all $\varepsilon > 0$.

For the proof of this result we will essentially use Ridout's p-adic version of the Thue–Siegel–Roth-Theorem. Maybe it is possible to prove an analog to Theorem 1 for t-dimensional vectors $\boldsymbol{\alpha}$, i.e., for an $s + t$-dimensional Halton–Kronecker sequence, probably based on multidimensional variants of Ridout's Theorem, as were given, for example, by Schlickewei in [10]. However at the moment we are still not able to give such a proof and leave this as an open problem.

Concerning α with bounded continued fraction coefficients we show:

Theorem 2 *Let α be irrational with bounded continued fraction coefficients. Then the discrepancy D_N of the $(s + 1)$-dimensional Halton–Kronecker sequence $(\mathbf{z}_n)_{n \geq 0} = ((\mathbf{x}_n, y_n))_{n \geq 0}$ satisfies*

$$ND_N = \mathcal{O}\left(N^{\frac{1}{2}}(\log N)^s\right).$$

On the other hand, there exists an irrational number α with bounded continued fraction coefficients and $c > 0$ such that for infinitely many N we have

$$ND_N \geq cN^{\frac{1}{2}}.$$

We want to mention that the logarithmic factor $(\log N)^s$ is certainly not optimal. For example, with slightly more care we can prove $ND_N = \mathcal{O}\left((N \log N)^{\frac{1}{2}}\right)$ in the case $s = 1$. We leave the determination of the precise threshold as an open problem.

2 Proofs of the Results

Proof of Theorem 1 We have $(\mathbf{z}_n) = (\mathbf{x}_n, y_n)_{n\geq 0}$ where $(\mathbf{x}_n)_{n\geq 0}$ is the s-dimensional Halton sequence in bases $b_1, \ldots, b_s$ and y_n is the one-dimensional Kronecker-sequence $(\{n\alpha\})_{n\geq 0}$. Let $I = [0, \beta) \times [0, \gamma) \subseteq [0, 1)^d$, with $d = s + 1, \beta = (\beta_1, \ldots, \beta_s) \in [0, 1)^s$ and $\gamma \in [0, 1)$. We will choose in the following certain disjoint subsets I_{int} and I_{bor} of $[0, 1)^d$ such that $I_{\text{int}} \subseteq I \subseteq I_{\text{int}} \cup I_{\text{bor}}$. Then with

$$A_N(I) := \#\{n : 0 \leq n < N, \mathbf{z}_n \in I\},$$

we obviously have

$$|A_N(I) - N\lambda(I)| \leq |A_N(I_{\text{int}}) - N\lambda(I_{\text{int}})| + A_N(I_{\text{bor}}) + N\lambda(I_{\text{bor}}). \tag{1}$$

The interval I_{int} is constructed as follows:
  Let

$$\beta_i = \frac{\beta_1^{(i)}}{b_i^1} + \frac{\beta_2^{(i)}}{b_i^2} + \cdots$$

with $\beta_j^{(i)} \in \{0, 1, \ldots, b_i - 1\}$. Then let

$$I(j_1, \ldots, j_s, k_1, \ldots, k_s, \gamma) := \prod_{i=1}^{s} \left[\sum_{l=1}^{j_i-1} \frac{\beta_l^{(i)}}{b_i^l} + \frac{k_i}{b_i^{j_i}}, \sum_{l=1}^{j_i-1} \frac{\beta_l^{(i)}}{b_i^l} + \frac{k_i+1}{b_i^{j_i}} \right) \times [0, \gamma)$$

for positive integers $j_1, \ldots, j_s$ and $k_i \in \{0, 1, \ldots, b_i - 1\}$ for $i = 1, \ldots, s$. By the construction of the Halton sequence there is a unique

$$r = r(j_1, \ldots, j_s, k_1, \ldots, k_s) \in \left\{0, 1, \ldots, b_1^{j_1} b_2^{j_2} \ldots b_s^{j_s} - 1\right\}$$

such that $\mathbf{z}_n \in I\,(j_1, \ldots, j_s, k_1, \ldots, k_s, \gamma)$ if and only if

$$n \equiv r \mod \left(b_1^{j_1} b_2^{j_2} \ldots b_s^{j_s}\right) \text{ and } y_n \in [0, \gamma)\,. \tag{2}$$

For $x \in \mathbb{R}$ let $\lfloor x \rfloor$ denote the largest integer less than or equal to x. Then let $L_i := \lfloor \log_{b_i} N \rfloor + 1$ and define I_{int} as union of disjoint intervals by

$$I_{\mathrm{int}} := \bigcup_{j_1=1}^{L_1} \cdots \bigcup_{j_s=1}^{L_s} \bigcup_{k_1=0}^{\beta_{j_1}^{(1)}-1} \cdots \bigcup_{k_s=0}^{\beta_{j_s}^{(s)}-1} I\,(j_1, \ldots, j_s, k_1, \ldots, k_s, \gamma)\,.$$

Further let

$$I_{\mathrm{bor}} := \bigcup_{i=1}^{s} \left(\prod_{j=1}^{i-1} [0,1) \times \left[\sum_{l=1}^{L_i} \frac{\beta_l^{(i)}}{b_i^l}, \sum_{l=1}^{L_i} \frac{\beta_l^{(i)}}{b_i^l} + \frac{1}{b_i^{L_i}} \right) \times \prod_{j=i+1}^{s} [0,1) \right) \times [0, \gamma)\,.$$

Then indeed we have

$$I_{\mathrm{int}} \subseteq I \subseteq I_{\mathrm{int}} \cup I_{\mathrm{bor}}$$

and by (1) and (2) (where we use the notation $\mathbf{j} := (j_1, \ldots, j_s)$, $\mathbf{k} := (k_1, \ldots, k_s)$, $\theta\,(\mathbf{j}, \mathbf{k}) := r\,(\mathbf{j}, \mathbf{k}) \cdot \alpha$, $b(\mathbf{j}) := b_1^{j_1} b_2^{j_2} \ldots b_s^{j_s}$, $N\,(\mathbf{j}) := \lfloor N/b\,(\mathbf{j}) \rfloor$)

$$|A_N(I) - N\lambda(I)| \le \sum_{\mathbf{j}, \mathbf{k}} \# \left\{ m \,\middle|\, 0 \le n = r\,(\mathbf{j}, \mathbf{k}) + mb(\mathbf{j}) < N \text{ and } \{n\alpha\} \in [0, \gamma) \right\}$$

$$\left. - N \frac{1}{b(\mathbf{j})} \lambda\,([0, \gamma)) \right|$$

$$+ \sum_{i=1}^{s} \left(\left[N \frac{1}{b_i^{L_i}} \right] + 1 \right) + N \sum_{i=1}^{s} \frac{1}{b_i^{L_i}}$$

$$\le \left| \sum_{\mathbf{j}} \sum_{\mathbf{k}} \# \{ 0 \le m < N(\mathbf{j}) \mid \{mb(\mathbf{j})\alpha\} \in [\theta\,(\mathbf{j}, \mathbf{k}), \gamma + \theta\,(\mathbf{j}, \mathbf{k})) \} \right.$$

$$\left. - N\,(\mathbf{j})\,\lambda\,([\theta\,(\mathbf{j}, \mathbf{k}), \gamma + \theta\,(\mathbf{j}, \mathbf{k}))) \right|$$

$$+ \sum_{\mathbf{j}} \sum_{\mathbf{k}} 1 + \sum_{i=1}^{s} 3$$

$$\le c\,(\alpha, s, b_1, \ldots, b_s) \cdot \left(\sum_{\mathbf{j}} N\,(\mathbf{j})\, D_{N(\mathbf{j})}\,(b\,(\mathbf{j})\,\alpha) + (\log N)^s \right)\,.$$

and the sums are always interpreted as

$$\sum_{\mathbf{j}} := \sum_{j_1=0}^{L_1} \cdots \sum_{j_s=0}^{L_s}, \qquad \sum_{\mathbf{k}} := \sum_{k_1=1}^{\beta_{j_1}^{(1)}-1} \cdots \sum_{k_s=1}^{\beta_{j_s}^{(s)}-1}.$$

Hence, to prove our Theorem 1 it suffices to show that

$$\sum_{\mathbf{j}} N(\mathbf{j}) D_{N(\mathbf{j})}(b(\mathbf{j})\alpha) = \mathcal{O}(N^{\varepsilon}). \tag{3}$$

To provide this estimate we use the well-known Koksma–Erdös–Turan inequality (see [6] or [4]) together with Ridout's p-adic version of the Thue–Siegel–Roth-Theorem. The discrepancy D_M of a point set $x_0, \ldots, x_{M-1}$ in $[0, 1)$ can be estimated with the Koksma–Erdös–Turan inequality by

$$D_M \le c_1 \cdot \left(\frac{1}{H} + \sum_{h=1}^{H} \frac{1}{h} \left| \frac{1}{M} \sum_{n=0}^{M-1} e^{2\pi i h x_n} \right| \right),$$

for arbitrary $H \ge 1$.

If $x_n = \{n\alpha\}$, then

$$\left| \sum_{n=0}^{M-1} e^{2\pi i h x_n} \right| \le c_2 \cdot \frac{1}{\|h\alpha\|}.$$

Here, and in the following $c_1, c_2, \ldots$ are absolute constants, and $\|x\|$ denotes the distance of x to the nearest integer. Hence we have (choosing $H = N(\mathbf{j})$)

$$N(\mathbf{j}) D_{N(\mathbf{j})}(b(\mathbf{j})\alpha) \le c_3 \cdot \sum_{h=1}^{N(\mathbf{j})} \frac{1}{h} \cdot \frac{1}{\|hb(\mathbf{j})\alpha\|}, \tag{4}$$

and it suffices to show that

$$\sum_{\mathbf{j}} \sum_{h=1}^{N(\mathbf{j})} \frac{1}{h} \cdot \frac{1}{\|hb(\mathbf{j})\alpha\|} = \mathcal{O}(N^{\varepsilon}). \tag{5}$$

Now we use a result which was shown in [3] with the help of Ridout's theorem:

Suppose that ϕ is algebraic and that $q_1, q_2, \ldots, q_s \ge 2$ are pairwise coprime integers. Then for every $\varepsilon > 0$ there exists a constant $C = C(\phi, \varepsilon, q_1, \ldots, q_s)$ such that for all integers $j_1, \ldots, j_s \ge 0$ and $H \ge 1$

$$\sum_{h=1}^{H} \frac{1}{h} \frac{1}{\|q_1^{j_1} \ldots q_s^{j_s} h\phi\|} \le c \cdot \left(q_1^{j_1} \ldots q_s^{j_s} H \right)^{\varepsilon}.$$

Using this result we obtain

$$\sum_{\mathbf{j}} \sum_{h=1}^{N(\mathbf{j})} \frac{1}{h} \cdot \frac{1}{\|hb(\mathbf{j})\alpha\|} \le C \cdot N^{\varepsilon} \cdot \sum_{j_1=1}^{L_1} \cdots \sum_{j_s=1}^{L_s} b_1^{\varepsilon j_1} \cdots b_s^{\varepsilon j_s} \le C \cdot N^{(s+1)\varepsilon},$$

and the result follows. $\qquad\qquad\square$

Proof of Theorem 2 In order to prove the upper bound we proceed similarly to the proof of Theorem 1. However, instead of (4) we use the trivial estimate

$$N(\mathbf{j}) \, D_{N(\mathbf{j})}(b(\mathbf{j})\alpha) \le \sqrt{N}$$

if $N(\mathbf{j}) \le \sqrt{N}$ and

$$N(\mathbf{j}) \, D_{N(\mathbf{j})}(b(\mathbf{j})\alpha) \le c_1 \frac{N(\mathbf{j})}{H(\mathbf{j})} + c_1 c_3 \sum_{h=1}^{H(\mathbf{j})} \frac{1}{h} \cdot \frac{1}{\|hb(\mathbf{j})\alpha\|},$$

if $N(\mathbf{j}) > \sqrt{N}$, where we set $H(\mathbf{j}) = \lfloor N(\mathbf{j})/\sqrt{N}) \rfloor$.

If α has bounded continued fraction coefficients, then we have $\|\alpha h\| \ge C/h$ for all positive integers. Hence it follows (in the case $N(\mathbf{j}) > \sqrt{N}$) that

$$\sum_{h=1}^{H(\mathbf{j})} \frac{1}{h} \cdot \frac{1}{\|hb(\mathbf{j})\alpha\|} \le \frac{1}{C} H(\mathbf{j}) \, b(\mathbf{j}) \le c_4 \sqrt{N}$$

and consequently

$$N(\mathbf{j}) \, D_{N(\mathbf{j})}(b(\mathbf{j})\alpha) \le c_5 \sqrt{N}.$$

Thus we certainly have

$$\sum_{\mathbf{j}} N(\mathbf{j}) \, D_{N(\mathbf{j})}(b(\mathbf{j})\alpha) = \mathcal{O}\left(\sqrt{N} \, (\log N)^s\right)$$

which proves the upper bound.

In order to obtain the lower bound we use the real number $\alpha = \sum_{m=1}^{\infty} \frac{1}{b_1^{2^m}}$, where we first suppose that $b_1 = \max\{b_1, \ldots, b_s\} \ge 3$. By Shallit [11] it is known that α has bounded continued fraction coefficients. Let $k \ge 1$ be fixed and $N = b_1^{2^{k+1}-1}$. We consider the interval $B := \left[0, \frac{1}{b_1^{2^k}}\right) \times [0, 1)^{s-1} \times \left[0, \frac{1}{2}\right)$. By definition it is clear that $x_n^{(1)} \in [0, b_1^{-2^k})$ if and only if $n = \ell b_1^{2^k}$ for some $\ell \le \lfloor N/b_1^{2^k} \rfloor = b_1^{2^k-1}$.

However, for all these n we have

$$n\alpha - \lfloor n\alpha \rfloor = \ell \sum_{m>k} b_1^{2^k - 2^m} \leq b_1^{2^k - 1} \frac{b_1^{-2^k}}{1 - b_1^{-2^{-k}}} \leq \frac{1}{2}$$

provided that k is sufficiently large. Hence, for this interval B we have

$$A_N(B) - N \cdot \lambda(B) \geq b_1^{2^k - 1} - b_1^{2^{k+1} - 1} b_1^{-2^k} \frac{1}{2} = \frac{\sqrt{N}}{2\sqrt{b_1}}.$$

This proves the result in the case $b_1 \geq 3$.

If $b_1 = 2$, then we can proceed in precisely the same way by using $\alpha = \sum_{m=1}^{\infty} \frac{1}{4^{2^m}}$. $\qquad\square$

Acknowledgements Michael Drmota is supported by the Austrian Science Fund (FWF): Project F5502-N26, which is part of the Special Research Program "Quasi-Monte Carlo Methods: Theory and Applications". Roswitha Hofer is supported by the Austrian Science Fund (FWF): Project F5505-N26, which is part of the Special Research Program "Quasi-Monte Carlo Methods: Theory and Applications". Gerhard Larcher is supported by the Austrian Science Fund (FWF): Project F5507-N26, which is part of the Special Research Program "Quasi-Monte Carlo Methods: Theory and Applications".

References

1. D. Bilyk, M.T. Lacey, On the small ball inequality in three dimensions. Duke Math. J. **143**, 81–115 (2008)
2. D. Bilyk, M.T. Lacey, A. Vagharshakyan, On the small ball inequality in all dimensions. J. Funct. Anal. **254**, 2470–2502 (2008)
3. M. Drmota, The generalized Van-der-Corput-Halton sequences. Indag. Math. **26**(5), 748–759 (2015)
4. M. Drmota, R. Tichy, *Sequences, Discrepancies and Applications*. Lecture Notes in Mathematics, vol. 1651 (Springer, Berlin, 1997)
5. R. Hofer, G. Larcher, Metrical results on the discrepancy of Halton–Kronecker sequences. Math. Z. **271**, 1–11 (2012)
6. L. Kuipers, H. Niederreiter, *Uniform Distribution of Sequences* (Wiley, New York, 1974)
7. G. Larcher, Probabilistic Diophantine approximation and the distribution of Halton–Kronecker sequences. J. Complex. **29**, 397–423 (2013)
8. H. Niederreiter, On the discrepancy of some hybrid sequences. Acta Arith. **138**, 373–398 (2009)
9. H. Niederreiter, Further discrepancy bounds and an Erdös-Turan-Koksma inequality for hybrid sequences. Monatsh. Math. **161**, 193–222 (2010)
10. H.-P. Schlickewei, On products of special linear forms with algebraic coefficients. Acta Arith. **31**, 389–398 (1976)
11. J. Shallit, Simple continued fractions for some irrational numbers. J. Number Theory **11**, 209–217 (1979)

More on Diophantine Sextuples

Andrej Dujella and Matija Kazalicki

Abstract A rational Diophantine m-tuple is a set of m nonzero rationals such that the product of any two of them increased by 1 is a perfect square. The first rational Diophantine quadruple was found by Diophantus, while Euler proved that there are infinitely many rational Diophantine quintuples. In 1999, Gibbs found the first example of a rational Diophantine sextuple, and Dujella, Kazalicki, Mikić and Szikszai recently proved that there exist infinitely many rational Diophantine sextuples.

In this paper, generalizing the work of Piezas, we describe a method for generating new parametric formulas for rational Diophantine sextuples.

2010 *Mathematics Subject Classification* 11D09, 11G05, 11Y50

1 Introduction

A Diophantine m-tuple is a set of m positive integers with the property that the product of any two of its distinct elements is one less than a square. If a set of nonzero rationals has the same property, then it is called a rational Diophantine m-tuple. Diophantus of Alexandria found the first example of a rational Diophantine quadruple $\{1/16, 33/16, 17/4, 105/16\}$, while the first Diophantine quadruple in integers was found by Fermat, and it was the set $\{1, 3, 8, 120\}$. It is well known that there exist infinitely many integer Diophantine quadruples (e.g., $\{k, k + 2, 4k + 4, 16k^3 + 48k^2 + 44k + 12\}$ for $k \geq 1$), while it was proved in [3] that an integer Diophantine sextuple does not exist and that there are only finitely many such quintuples. A folklore conjecture is that there does not exist an integer Diophantine quintuple. There is an even stronger conjecture which predicts that all integer Diophantine quadruples $\{a, b, c, d\}$ satisfy the equation $(a + b - c - d)^2 = 4(ab + 1)(cd + 1)$ (such quadru-

A. Dujella (✉) • M. Kazalicki
Department of Mathematics, University of Zagreb, Bijenička cesta 30, 10000 Zagreb, Croatia
e-mail: duje@math.hr; matija.kazalicki@math.hr

C. Elsholtz, P. Grabner (eds.), *Number Theory – Diophantine Problems, Uniform Distribution and Applications*, DOI 10.1007/978-3-319-55357-3_11

ples are called regular). However, in the rational case, there exist larger sets with the same property. Euler found infinitely many rational Diophantine quintuples, e.g. he was able to extend the Fermat quadruple to the rational quintuple $\{1, 3, 8, 120, 777480/8288641\}$. Gibbs [5] found the first rational Diophantine sextuple

$$\{11/192, 35/192, 155/27, 512/27, 1235/48, 180873/16\},$$

while Dujella et al. [4] recently proved that there exist infinitely many rational Diophantine sextuples. No example of a rational Diophantine septuple is known. Moreover, we do not know any rational Diophantine quadruple which can be extended to two different rational Diophantine sextuples. On the other hand, by the construction from [4], we know that there exist infinitely many rational Diophantine triples, each of which can be extended to rational Diophantine sextuples in infinitely many ways. In particular, there are infinitely many rational Diophantine sextuples containing the triples $\{15/14, -16/21, 7/6\}$ and $\{3780/73, 26645/252, 7/13140\}$. The construction from [4] uses elliptic curves induced by Diophantine triples, i.e. curves of the form $y^2 = (x+ab)(x+ac)(x+bc)$ where $\{a, b, c\}$ is a rational Diophantine triple, with torsion group $\mathbb{Z}/2\mathbb{Z} \times \mathbb{Z}/6\mathbb{Z}$ over $\mathbb{Q}$.

Piezas [7] studied Gibbs's examples of rational Diophantine sextuples which do not fit into the construction from [4] and realized that most of them follow a common pattern: they contain two regular subquadruples with two common elements (see Proposition 1). By studying sextuples of that special form, he obtained new simpler parametric formulas for rational Diophantine sextuples, and also obtained infinitely many sextuples $\{a, b, c, d, e, f\}$ with fixed products ab and cd (e.g., $ab = 24$ and $cd = 9/16$).

In this paper, we will reformulate results from [7] in terms of the geometry of a certain algebraic variety parameterizing rational Diophantine quadruples, in fact the fiber product of three Edwards curves over $\mathbb{Q}(t)$, and obtain a method for generating (new) parametric formulas for rational Diophantine sextuples.

2 Construction

2.1 Correspondence

Let $\{a, b, c, d\}$ be a rational Diophantine quadruple with elements in $\mathbb{Q}$ or $\mathbb{Q}(t)$, and let

$$ab + 1 = t_{12}^2 \quad ac + 1 = t_{13}^2 \quad ad + 1 = t_{14}^2$$
$$bc + 1 = t_{23}^2 \quad bd + 1 = t_{24}^2 \quad cd + 1 = t_{34}^2.$$

It follows that $(t_{12}, t_{34}, t_{13}, t_{24}, t_{14}, t_{23}, m' = abcd)$ defines a point on an algebraic variety C defined by the following equations:

$$(t_{12}^2 - 1)(t_{34}^2 - 1) = m'$$

$$(t_{13}^2 - 1)(t_{24}^2 - 1) = m'$$

$$(t_{14}^2 - 1)(t_{23}^2 - 1) = m'.$$

Conversely, the points $(\pm t_{12}, \pm t_{34}, \pm t_{13}, \pm t_{24}, \pm t_{14}, \pm t_{23}, m')$ on C determine two rational Diophantine quadruples $\pm(a, b, c, d)$ (for example, $a^2 = (t_{12}^2 - 1)(t_{13}^2 - 1)/(t_{23}^2 - 1)$) provided that the elements $a, b, c,$ and d are rational, distinct, and non-zero. Note that if one element is rational, then all the elements are rational.

The projection $(t_{12}, t_{34}, t_{13}, t_{24}, t_{14}, t_{23}, m') \mapsto m'$ defines a fibration of C over the projective line, and a generic fiber is the product of three curves $\mathcal{D} : (x^2 - 1)(y^2 - 1) = m'$. Any point on C corresponds to the three points $Q_1 = (t_{12}, t_{34})$, $Q_2 = (t_{13}, t_{24})$ and $Q_3 = (t_{14}, t_{23})$ on $\mathcal{D}$. The elements of the quadruple corresponding to these three points are distinct if and only if no two of these points can be transformed from one to another by changing signs and switching coordinates, e.g. for the points (t_{12}, t_{34}), $(-t_{34}, t_{12})$, and (t_{14}, t_{23}), we have that $a = d$.

2.2 Extending Quadruples to Sextuples

The following proposition gives a criterion for extending quadruples to sextuples.

Proposition 1 (Piezas [7]) *Let $\{a, b, c, d\}$ be a rational Diophantine quadruple, and x_1 and x_2 the roots of*

$$(abcdx + 2abc + a + b + c - d - x)^2 = 4(ab + 1)(ac + 1)(bc + 1)(dx + 1).$$

If $x_1 x_2 \neq 0$ and

$$(abcd - 3)^2 = 4(ab + cd + 3), \tag{1}$$

then $\{a, b, c, d, x_1, x_2\}$ is a Diophantine sextuple. Furthermore,

$$(a + b - x_1 - x_2)^2 = 4(ab + 1)(x_1 x_2 + 1)$$

$$(c + d - x_1 - x_2)^2 = 4(cd + 1)(x_1 x_2 + 1).$$

Note that x_1 and x_2 coincide with the extensions of rational Diophantine quadruples given in [1, Theorem 1], and the condition (1) implies that $x_1 x_2 + 1 = \left(\frac{a+b-c-d}{abcd-1}\right)^2$.

In this section, we will reformulate Proposition 1 in terms of the geometry of the algebraic variety $\mathcal{C}$.

The condition (1) is equivalent to $t_{12}t_{34} = \pm t_{12} \pm t_{34}$, or $t_{34} = \pm t_{12}/(t_{12} \pm 1)$. For the rest of the paper, we set $t_{12} = t$, $t_{34} = t/(t-1)$ and $m' = (t^2-1)(\frac{t^2}{(t-1)^2} - 1) = \frac{2t^2+t-1}{t-1}$, and thus condition (1) is satisfied.

The curve $\mathcal{D}$ over $\mathbb{Q}(t)$

$$\mathcal{D} : (x^2-1)(y^2-1) = \frac{2t^2+t-1}{t-1}$$

is birationally equivalent to the elliptic curve

$$E : S^2 = T^3 - 2 \cdot \frac{2t^2-t+1}{t-1}T^2 + \frac{(2t-1)^2(t+1)^2}{(t-1)^2}T.$$

The map is given by $T = 2(x^2-1)y + 2x^2 - (2 - m')$, and $S = 2Tx$, where $m' = \frac{2t^2+t-1}{t-1}$.

Denote by $P = \left[\dfrac{(2t-1)^2(t+1)}{t-1}, \dfrac{2t(2t-1)^2(t+1)}{t-1} \right] \in E(\mathbb{Q}(t))$ a point of infinite order on E, and by $R = \left[\dfrac{(t+1)(2t-1)}{(t-1)}, \dfrac{2(t+1)(2t-1)}{t-1} \right]$ a point of order 4. The point $(t_{12}, t_{34}) \in \mathcal{D}(\mathbb{Q}(t))$ corresponds to the point $P \in E(\mathbb{Q}(t))$.

Proposition 2 *The Mordell–Weil group of $E(\mathbb{Q}(t))$ is generated by P and R.*

Proof It is enough to prove that the specialization homomorphism at $t_0 = 6$ is injective. Then one can easily check that the specializations of points P and R generate the Mordell–Weil group of $E_{t_0}(\mathbb{Q})$.

We use the injectivity criterion from Theorem 1.3 in [6]. It states that given an elliptic curve $y^2 = x^3 + A(t)x^2 + B(t)x$, where $A, B \in \mathbb{Z}[t]$, with exactly one nontrivial 2-torsion point over $\mathbb{Q}(t)$, the specialization homomorphism at $t_0 \in \mathbb{Q}$ is injective if the following condition is satisfied: for every nonconstant square-free divisor $h(t) \in \mathbb{Z}[t]$ of $B(t)$ or $A(t)^2 - 4B(t)$ the rational number $h(t_0)$ is not a square in $\mathbb{Q}$.

The claim follows (after clearing out the denominators in the defining equation of E). $\qquad\qquad\square$

If $Q \in E$ is the point that corresponds to the point $(x, y) \in \mathcal{D}$, then the points $-Q$ and $Q + R$ correspond to the points $(-x, y)$ and $(y, -x)$. Hence the triple $(Q_1, Q_2, Q_3) \in E(\mathbb{Q}(t))^3$ corresponds to the quadruple whose elements are not distinct if and only if there are two points, say Q_i and Q_j, such that $Q_i = \pm Q_j + kR$, where $k \in \{0, 1, 2, 3\}$.

If instead of m', we fix on $\mathcal{C}$ coordinates t_{12}, t_{13}, t_{23} we will obtain an elliptic curve on $\mathcal{C}$ consisting of points $(t_{34}, t_{24}, t_{14}, m')$ which satisfy

$$(t_{34}^2 - 1) = \frac{m'}{(t_{12}^2 - 1)}$$

$$(t_{24}^2 - 1) = \frac{m'}{(t_{13}^2 - 1)}$$

$$(t_{14}^2 - 1) = \frac{m'}{(t_{23}^2 - 1)}.$$

Thus, to the point $(t_{12}, t_{34}, t_{13}, t_{24}, t_{14}, t_{23}, m')$ on $\mathcal{C}$ that corresponds to the rational quadruple $\{a, b, c, d\}$, we associate the elliptic curve $E_{abc} : y^2 = (x+ab)(x+ac)(x+bc)$ together with the point $W = [abcd, abc \cdot t_{14}t_{24}t_{34}]$. A short calculation shows that if we denote by $V = [1, t_{12}t_{13}t_{23}]$ a point on E_{abc}, then x_1 and x_2 from Proposition 1 are given by

$$x_1 = \frac{x(W + V)}{abc} \quad \text{and} \quad x_2 = \frac{x(W - V)}{abc}.$$

For more details on using the elliptic curve E_{abc} for extending rational Diophantine triples and quadruples, see [1, Theorem 1], [2, Theorem 1] and [4, Proposition 2.1].

2.3 Degenerate Case

In this subsection we fix $Q_1 = P$ and investigate conditions under which the point $(Q_1, Q_2, Q_3) \in E(\mathbb{Q}(t)) \times E(\mathbb{Q}(t)) \times E(\mathbb{Q}(t))$ corresponds to the degenerate Diophantine sextuple (i.e., $x_1 x_2 = 0$). We call such triple degenerate. Following the notation from the previous section, we see that the triple is degenerate if and only if $\pm W \pm V = [0, abc] \in E_{abc}(\mathbb{Q}(t))$ for some choice of the signs.

Proposition 3 Let $Q_2, Q_3 \in E(\mathbb{Q}(t))$. The triple $(Q_1, Q_2, Q_3) \in E(\mathbb{Q}(t)) \times E(\mathbb{Q}(t)) \times E(\mathbb{Q}(t))$ is degenerate if and only if $\pm Q_1 \pm Q_2 \pm Q_3 = R$ for some choice of the signs.

Proof Let $r = x(Q_2)$ and $s = x(Q_3)$. Direct calculation shows that the constant term of the polynomial from Proposition 1 is zero if and only if $g(r, s)h(r, s) = 0$ where

$$g(r, s) = \left((-1 + t)^2 rs - (1 + t)^2(-1 + 2t)(r + s) + (1 + t)^2(-1 + 2t)^2\right)^2$$
$$- 16rst^2(1 + t)^2(-1 + 2t),$$

$$h(r, s) = \left((-1 + t)^2 rs - (1 - t)^2(-1 + 2t)(r + s) + (1 + t)^2(-1 + 2t)^2\right)^2$$
$$- 16rst^2(1 - t)^2(-1 + 2t).$$

One can check that r and s satisfy this equation if $\pm Q_1 \pm Q_2 \pm Q_3 = R$ for some choice of the signs.

Conversely, both $g(r,s) = 0$ and $h(r,s) = 0$ define a curve that is birationally equivalent to E. Hence, we have a degree four map from "the degeneracy locus" in $E \times E$ to E given by $(Q_2, Q_3) \mapsto (x(Q_2), x(Q_3))$. Since we already have 8 irreducible components in "the degeneracy locus" (one for the each choice of the signs), the claim follows. $\qquad\square$

2.4 Rationality

Given a triple $(Q_1, Q_2, Q_3) \in E(\mathbb{Q}(t)) \times E(\mathbb{Q}(t)) \times E(\mathbb{Q}(t))$, where $Q_1 = P$, we want to know if the corresponding Diophantine quadruple is rational. It is enough to prove that one element is rational.

A short calculation shows that for the point $(S,T) \in E(\mathbb{Q}(t))$ we have

$$x^2 - 1 = \left(\frac{S}{2T}\right)^2 - 1 = T\left(\frac{T - \frac{2t^2+t-1}{t-1}}{2T}\right)^2 =: f(T). \tag{2}$$

Since

$$a^2 = \frac{f(Q_1)f(Q_2)f(Q_3)}{m'} \equiv x(Q_1)x(Q_2)x(Q_3)m' \equiv (2t-1)x(Q_2)x(Q_3)$$
$$\equiv x(-P+R)x(Q_2)x(Q_3) \pmod{\mathbb{Q}(t)^{\times 2}}$$

for the rationality of a it is enough to prove that $x(-P+R)x(Q_2)x(Q_3)$ is a square in $\mathbb{Q}(t)$.

Since the point $(0,0) \in E(\mathbb{Q}(t))$ is a point of order 2, the usual 2-descent homomorphism $E(\mathbb{Q}(t)) \to \mathbb{Q}(t)^{\times}/\mathbb{Q}(t)^{\times 2}$, which is for non-torsion points defined by $(T,S) \mapsto T$ (note that $(0,0) \mapsto 1$), implies the following proposition.

Proposition 4 *Let* $Q_2, Q_3 \in E(\mathbb{Q}(t))$.

a) *If* $Q_2 + Q_3 \equiv \mathcal{O} \bmod 2E(\mathbb{Q}(t))$, *then* $a^2 \equiv (2t-1) \bmod \mathbb{Q}(t)^{\times 2}$.
b) *If* $Q_2 + Q_3 \equiv R \bmod 2E(\mathbb{Q}(t))$, *then* $a^2 \equiv (t-1)(t+1) \bmod \mathbb{Q}(t)^{\times 2}$.
c) *If* $Q_2 + Q_3 \equiv P \bmod 2E(\mathbb{Q}(t))$, *then* $a^2 \equiv (t-1)(t+1)(2t-1) \bmod \mathbb{Q}(t)^{\times 2}$.
d) *If* $Q_2 + Q_3 \equiv P + R \bmod 2E(\mathbb{Q}(t))$, *then* $a^2 \equiv 1 \bmod \mathbb{Q}(t)^{\times 2}$.

Remark 1 In the cases a) and b) we can still obtain parametric families of Diophantine sextuples if we specialize to those $t's$ for which $2t-1$ and $(t-1)(t+1)$ are squares (e.g., if we specialize t to $\frac{t^2+1}{2}$ and $\frac{t^2+1}{2t}$). Concerning the case c), the elliptic curve $y^2 = (x-1)(x+1)(2x-1)$ has Mordell–Weil group isomorphic to $\mathbb{Z}/2\mathbb{Z} + \mathbb{Z}/4\mathbb{Z}$.

Remark 2 The proposition covers all the possibilities, since the Mordell–Weil group of $E(\mathbb{Q}(t))$ is generated by P and R (see Proposition 2).

3 Examples

3.1 Family Corresponding to $(P, 2P, 4P)$

For an illustration, we calculate a parametric family $\{a, b, c, d, e, f\}$ of rational Diophantine sextuples that corresponds to the triple $(P, 2P, 4P)$. It follows from Proposition 3 that the triple is not degenerate. The rationality of the sextuple will follow if we replace t by $\frac{t^2+1}{2}$ (see part a) of Proposition 4). Then, the corresponding Diophantine quadruple is equal to

$$a = \frac{(t^8 - 8t^6 - 14t^4 + 32t^2 - 27) \cdot (t^8 + 26t^4 - 40t^2 - 3)}{64 \cdot (t - 1) \cdot t \cdot (t + 1) \cdot (t^4 - 2t^2 + +5) \cdot (t^4 + 6t^2 - 3)},$$

$$b = \frac{16 \cdot t \cdot (t - 1)^2 \cdot (t + 1)^2 \cdot (t^2 + 3) \cdot (t^4 - 2t^2 + 5) \cdot (t^4 + 6t^2 - 3)}{(t^8 - 8t^6 - 14t^4 + 32t^2 - 27) \cdot (t^8 + 26t^4 - 40t^2 - 3)},$$

$$c = \frac{t \cdot (t^8 - 8t^6 - 14t^4 + 32t^2 - 27) \cdot (t^8 + 26t^4 - 40t^2 - 3)}{(t - 1) \cdot (t + 1) \cdot (t^2 - 3)^2 \cdot (t^2 + 1)^2 \cdot (t^4 - 2t^2 + 5) \cdot (t^4 + 6t^2 - 3)},$$

$$d = \frac{4 \cdot t \cdot (t^2 - 3)^2 \cdot (t^2 + 1)^2 \cdot (t^4 - 2t^2 + 5) \cdot (t^4 + 6t^2 - 3)}{(t - 1) \cdot (t + 1) \cdot (t^8 - 8t^6 - 14t^4 + 32t^2 - 27) \cdot (t^8 + 26t^4 - 40t^2 - 3)}.$$

Using Proposition 1 (let $e = x_1$ and $f = x_2$), we find that $e = e_1/e_2$ and $f = f_1/f_2$ are equal to

$$e_1 = (t+1) \cdot (t^2 - 2t+3) \cdot (t^2+2t - 1) \cdot (t^6 - 2t^5+t^4+12t^3y7t^2 - 2t - 9)$$
$$\cdot (t^6+2t^5 - 3t^4y4t^3 - 17t^2+18t+3)$$
$$\cdot (t^{12} - 4t^{11}+6t^{10}+20t^9 - t^8+24t^7 - 12t^6 - 88t^5 - 177t^4+364t^3 - 90t^2 - 60t+81)$$
$$\cdot (t^{12}+4t^{11} - 2t^{10} - 4t^9 - 41t^8+40t^7+100t^6 - 72t^5+63t^4+212t^3 - 66t^2 - 180t+9),$$

$$e_2 = 64 \cdot (t - 1) \cdot t \cdot (t^2 - 3)^2 \cdot (t^2+1)^4 \cdot (t^4 - 2t^2+5) \cdot (t^4+6t^2 - 3)$$
$$\cdot (t^8 - 8t^6 - 14t^4+32t^2 - 27) \cdot (t^8+26t^4 - 40t^2 - 3),$$

$$f_1 = (t - 1) \cdot (t^2 - 2t - 1) \cdot (t^2+2t+3) \cdot (t^6 - 2t^5 - 3t^4 - 4t^3 - 17t^2 - 18t+3)$$
$$\cdot (t^6+2t^5+t^4 - 12t^3+7t^2+2t - 9)$$
$$\cdot (t^{12} - 4t^{11} - 2t^{10}+4t^9 - 41t^8 - 40t^7+100t^6+72t^5+63t^4 - 212t^3 - 66t^2+180t+9)$$
$$\cdot (t^{12}+4t^{11}+6t^{10} - 20t^9 - t^8 - 24t^7 - 12t^6+88t^5 - 177t^4 - 364t^3 - 90t^2+60t+81),$$

$$f_2 = 64 \cdot t \cdot (t+1) \cdot (t^2 - 3)^2 \cdot (t^2+1)^4 \cdot (t^4 - 2t^2+5) \cdot (t^4+6t^2 - 3)$$
$$\cdot (t^8 - 8t^6 - 14t^4+32t^2 - 27)$$
$$\cdot (t^8+26t^4 - 40t^2 - 3).$$

3.2 Rank Two Examples

If we specialize t to t^2+1, the elliptic curve E will have another point of infinite order (independent of P), $S = \left[\frac{(2+t^2)^2}{t^2}, \frac{(2+t^2)^2(1+t^2)}{t^3} \right]$. Now the triple $(P, 2P+S, R+S-P)$ is not degenerate and satisfies the condition of Proposition 4(d). Our construction gives the following family of rational Diophantine sextuples

$$a = \frac{(t^3 + 3t^2 + t + 1) \cdot (t^3 + t^2 + 3t + 1) \cdot (2t - 1)}{2 \cdot (t - 2) \cdot (t - 1) \cdot (t^2 + t + 1) \cdot (t + 1)},$$

$$b = \frac{2 \cdot (t - 1) \cdot (t^2 + t + 1) \cdot (t + 1) \cdot (t^2 + 2) \cdot (t - 2) \cdot t^2}{(t^3 + 3t^2 + t + 1) \cdot (t^3 + t^2 + 3t + 1) \cdot (2t - 1)},$$

$$c = \frac{(t^3 + 3t^2 + t + 1) \cdot (t^3 + t^2 + 3t + 1) \cdot (t - 2)}{2 \cdot (2t - 1) \cdot (t - 1) \cdot (t^2 + t + 1) \cdot (t + 1) \cdot t^2},$$

$$d = \frac{2 \cdot (2t^2 + 1) \cdot (2t - 1) \cdot (t - 1) \cdot (t^2 + t + 1) \cdot (t + 1)}{t^2 \cdot (t^3 + 3t^2 + t + 1) \cdot (t^3 + t^2 + 3t + 1) \cdot (t - 2)},$$

$$e = \frac{8 \cdot t^2 \cdot (t - 1) \cdot (2t + 1) \cdot (t + 2) \cdot (t + 1) \cdot (t^2 + 1)}{(t - 2) \cdot (2t - 1) \cdot (t^2 + t + 1) \cdot (t^3 + t^2 + 3t + 1) \cdot (t^3 + 3t^2 + t + 1)},$$

$$f = \frac{3 \cdot (3t^2 + 2t + 1) \cdot (t^2 + 2t + 3) \cdot (t^4 + 1) \cdot (t^4 + 4t^2 + 1)}{2 \cdot (t - 1) \cdot (t - 2) \cdot (2t - 1) \cdot (t + 1) \cdot (t^2 + t + 1) \cdot (t^3 + t^2 + 3t + 1) \cdot (t^3 + 3t^2 + t + 1)}.$$

If we further require $2(t^2 + 1)$ to be a square, then the resulting parametrization $t \mapsto 1 + \left(\frac{4t^2 - 8t - 4}{4t^2 + 8t - 4} \right)^2$ yields a point K on E

$$K = \left[\frac{(t^2 - 2t + 1) \cdot (t^2 + 2t + 3) \cdot (t^2 + 2t + 1)^2}{(t^2 - 2t - 1)^2 \cdot (t^2 + 2t - 1)^2}, \right.$$
$$\left. \frac{4(t^2 - 2t + 1) \cdot (t^2 + 1) \cdot (t^2 + 2t + 3) \cdot (t^2 + 2t + 1)^2}{(t^2 + 2t - 1)^2 \cdot (t^2 - 2t - 1)^3} \right], \tag{3}$$

with the property that $2K = S$. When we apply our construction to the triple $(P, K, -2K + R)$, we obtain a very simple family of sextuples also found by Piezas [7]

$$a = \frac{(t^2 - 2t - 1) \cdot (t^2 + 2t + 3) \cdot (3t^2 - 2t + 1)}{4t \cdot (t^2 - 1) \cdot (t^2 + 2t - 1)},$$

$$b = \frac{4t \cdot (t^2 - 1) \cdot (t^2 - 2t - 1)}{(t^2 + 2t - 1)^3},$$

$$c = \frac{4t \cdot (t^2 - 1) \cdot (t^2 + 2t - 1)}{(t^2 - 2t - 1)^3},$$

$$d = \frac{(t^2 + 2t - 1) \cdot (t^2 - 2t + 3) \cdot (3t^2 + 2t + 1)}{4t \cdot (t^2 - 1) \cdot (t^2 - 2t - 1)},$$

$$e = \frac{-t \cdot (t^2 + 4t + 1) \cdot (t^2 - 4t + 1)}{(t - 1) \cdot (t + 1) \cdot (t^2 + 2t - 1) \cdot (t^2 - 2t - 1)},$$

$$f = \frac{(t - 1) \cdot (t + 1) \cdot (3t^2 - 1) \cdot (t^2 - 3)}{4t \cdot (t^2 + 2t - 1) \cdot (t^2 - 2t - 1)}.$$

Acknowledgements The authors acknowledge support from the QuantiXLie Center of Excellence. A.D. was supported by the Croatian Science Foundation under the project no. 6422.

References

1. A. Dujella, On Diophantine quintuples. Acta Arith. **81**, 69–79 (1997)
2. A. Dujella, Diophantine m-tuples and elliptic curves. J. Théor. Nombres Bordeaux **13**, 111–124 (2001)
3. A. Dujella, There are only finitely many Diophantine quintuples, J. Reine Angew. Math. **566**, 183–214 (2004)
4. A. Dujella, M. Kazalicki, M. Mikić, M. Szikszai, There are infinitely many rational Diophantine sextuples. Int. Math. Res. Not. **2017**(2), 490–508 (2017)
5. P. Gibbs, Some rational Diophantine sextuples. Glas. Mat. Ser. III **41**, 195–203 (2006)
6. I. Gusić, P. Tadić, Injectivity of the specialization homomorphism of elliptic curves. J. Number Theory **148**, 137–152 (2015)
7. T. Piezas, Extending rational Diophantine triples to sextuples (2016). http://mathoverflow.net/questions/233538/extending-rational-diophantine-triples-to-sextuples

Effective Results for Discriminant Equations over Finitely Generated Integral Domains

Jan-Hendrik Evertse and Kálmán Győry

To Professor Robert Tichy on the occasion of his 60th birthday

Abstract Let A be an integral domain with quotient field K of characteristic 0 that is finitely generated as a $\mathbb{Z}$-algebra. Denote by $D(F)$ the discriminant of a polynomial $F \in A[X]$. Further, given a finite étale K-algebra Ω, denote by $D_{\Omega/K}(\alpha)$ the discriminant of α over K. For non-zero $\delta \in A$, we consider equations

$$D(F) = \delta$$

to be solved in monic polynomials $F \in A[X]$ of given degree $n \geq 2$ having their zeros in a given finite extension field G of K, and

$$D_{\Omega/K}(\alpha) = \delta \ \text{ in } \alpha \in O,$$

where O is an A-order of Ω, i.e., a subring of the integral closure of A in Ω that contains A as well as a K-basis of Ω.

In the series of papers (Győry, Acta Arith 23:419–426, 1973; Győry, Publ Math Debrecen 21:125–144, 1974; Győry, Publ Math Debrecen 23:141–165, 1976; Győry, Publ Math Debrecen 25:155–167, 1978; Győry, Acta Math Acad Sci Hung 32:175–190, 1978; Győry, J Reine Angew Math 324:114–126, 1981), Győry proved that when K is a number field, A the ring of integers or S-integers of K, and Ω a finite field extension of K, then up to natural notions of equivalence the above equations have, without fixing G, finitely many solutions, and that moreover, if K, S, Ω, O, and δ are effectively given, a full system of representatives for the equivalence classes can be effectively determined. Later, Győry (Publ Math Debrecen 29:79–94, 1982) generalized in an ineffective way the above-mentioned finiteness results to the case when A is an integrally closed integral domain with quotient field K of

J.-H. Evertse
Mathematical Institute, Leiden University, P.O.Box 9512, 2300 RA, Leiden, The Netherlands
e-mail: evertse@math.leidenuniv.nl

K. Győry (✉)
Institute of Mathematics, University of Debrecen, H-4032 Debrecen, Egyetem Tér 1, Hungary
e-mail: gyory@science.unideb.hu

© Springer International Publishing AG 2017

C. Elsholtz, P. Grabner (eds.), *Number Theory – Diophantine Problems, Uniform Distribution and Applications*, DOI 10.1007/978-3-319-55357-3_12

characteristic 0 which is finitely generated as a $\mathbb{Z}$-algebra and G is a finite extension of K. Further, in Győry (J Reine Angew Math 346:54–100, 1984) he made these results effective for a special class of integral domains A containing transcendental elements. In Evertse and Győry (Discriminant equations in diophantine number theory, Chap. 10. Cambridge University Press, 2016) we generalized in an effective form the results of Győry (Publ Math Debrecen 29:79–94, 1982) mentioned above to the case where A is an arbitrary integrally closed domain of characteristic 0 which is finitely generated as a $\mathbb{Z}$-algebra, where Ω is a finite étale K-algebra, and where A, δ, and G, respectively Ω, O are effectively given (in a well-defined sense described below).

In the present paper, we extend these effective results further to integral domains A that are not necessarily integrally closed.

2010 Mathematics Subject Classification: 11D99; Secondary 11D41

1 Introduction

We define the discriminant of a monic polynomial $F = X^n + a_1 X^{n-1} + \cdots + a_n = (X - \alpha_1) \cdots (X - \alpha_n)$ of degree $n \geq 2$ by

$$D(F) := \prod_{1 \leq i < j \leq n} (\alpha_i - \alpha_j)^2.$$

Recall that $D(F) \in \mathbb{Z}[a_1, \ldots, a_n]$; in fact, it is a polynomial of total degree $2n - 2$ in $a_1, \ldots, a_n$.

In this paper we consider discriminant equations over integral domains of characteristic 0 that are finitely generated as a $\mathbb{Z}$-algebra. Let A be such a domain, with quotient field K, let n be an integer with $n \geq 2$, let δ be a non-zero element of A, and let G be a finite extension of K. We consider the equation

$$D(F) = \delta \ \text{ in monic polynomials } F \in A[X] \text{ of degree } n$$
$$\text{having all their zeros in } G. \tag{1}$$

Two monic polynomials $F_1, F_2 \in A[X]$ are called *A-equivalent* if there is $a \in A$ such that $F_2(X) = F_1(X + a)$. Clearly, A-equivalent polynomials have the same discriminant, and so the solutions of equation (1) can be divided into A-equivalence classes. We proved [5, Theorem 10.1.1] the following result:

Theorem A *Assume in addition to the above that A is integrally closed. Then the solutions of Eq. (1) lie in finitely many A-equivalence classes. If moreover A, δ and*

G are effectively given (in a sense defined in the next section), then a full system of representatives of these A-equivalence classes can be determined effectively.

The ineffective finiteness statement of this is a consequence of [12, Theorem 4]. The effective part is a culmination of Győry's earlier results from [6–10], mentioned in the abstract.

In the present paper we prove a generalization of Theorem A, where again n is an integer with $n \geq 2$ and where instead of integrally closed domains we consider integral domains A such that

$$\left(\frac{1}{n}A^+ \cap A_K^+\right)/A^+ \text{ is finite,} \tag{2}$$

see Theorem 2.1 below. Here B^+ denotes the additive group of a ring B, and A_K denotes the integral closure of A in K. The class of domains with (2) contains all integrally closed domains, the integral domains that contain n^{-1}, and also all finitely generated subrings of $\overline{\mathbb{Q}}$. We do not know if condition (2) is the weakest possible. We will give an example of an integral domain A and a field extension G for which the finiteness part of Theorem A is false. So some condition will have to be imposed on the domain A.

As suggested by one of the anonymous referees, there are various variations on Eq. (1) that are worth being considered. First of all, instead of (1) one could consider the equation

$$D(F) \in \delta A^* \text{ in monic polynomials } F \in A[X] \text{ of degree } n$$
$$\text{having all their zeros in } G.$$

where A^* denotes the unit group of A and $\delta A^* := \{\delta u : u \in A^*\}$. Secondly, in Eq. (1) one could consider polynomials F that do not have their zeros in a prescribed finite extension G of K. For both variations one can, in certain special cases, prove certain sensible effective finiteness results, see Győry [10, 11]. But it may be hard to obtain such results in full generality. We will return to this at the end of the next section.

We also consider discriminant equations where the unknowns are elements of orders of finite étale K-algebras. Let for the moment K be any field of characteristic 0 and Ω a *finite étale K-algebra*, i.e., $\Omega = K[X]/(P) = K[\theta]$, where $P \in K[X]$ is some separable polynomial and $\theta := X \bmod P$. We write $[\Omega : K] := \dim_K \Omega$; then clearly $[\Omega : K] = \deg P$. Let $\overline{K}$ be an algebraic closure of K. By a *K-homomorphism* of Ω to $\overline{K}$ we mean a non-trivial K-algebra homomorphism. There are precisely $n := [\Omega : K]$ K-homomorphisms of Ω to $\overline{K}$, which map θ to the n distinct zeros of P in $\overline{K}$. We denote these by $x \mapsto x^{(i)}$ ($i = 1, \ldots, n$). The *discriminant* of $\alpha \in \Omega$ over K is given by

$$D_{\Omega/K}(\alpha) := \prod_{1 \leq i < j \leq n} (\alpha^{(i)} - \alpha^{(j)})^2,$$

where $\alpha^{(i)}$ denotes the image of α under $x \mapsto x^{(i)}$. This is an element of K. It is not difficult to show that $D_{\Omega/K}(\alpha + a) = D_{\Omega/K}(\alpha)$ for $\alpha \in \Omega$, $a \in K$. Further, $D_{\Omega/K}(\alpha) \neq 0$ if and only if $\Omega = K[\alpha]$. For more details on finite étale K-algebras, we refer to [5, Chap. 1].

Now let as above A be an integral domain with quotient field K of characteristic 0 that is finitely generated over $\mathbb{Z}$. We denote by A_Ω the integral closure of A in Ω. An *A-order* is a subring of A_Ω that contains A as well as a K-basis of Ω. In particular A_Ω itself is an A-order of Ω. We consider equations of the type

$$D_{\Omega/K}(\alpha) = \delta \ \text{ in } \alpha \in O \tag{3}$$

where δ is a non-zero element of A, and O is an A-order of Ω. We call $\alpha_1, \alpha_2 \in O$ *A-equivalent* if $\alpha_1 - \alpha_2 \in A$. Then the solutions of (3) can be divided into A-equivalence classes. We recall Theorem 10.1.3 of [5]:

Theorem B *Assume in addition to the above that A is integrally closed. Then the solutions of equation* (3) *lie in finitely many A-equivalence classes. If moreover A, δ, and Ω are effectively given as defined in the next section, then a full system of representatives for these A-equivalence classes can be determined effectively.*

The ineffective part of this theorem is a consequence of [12, Theorem 5]. The effective part was proved in [7–13] in the special cases mentioned in the abstract, in more precise, explicit forms.

In this paper we extend Theorem B to integral domains A such that

$$(O \cap K)^+/A^+ \ \text{ is finite,} \tag{4}$$

see Theorem 2.2 below. Notice that $O \cap K = A$ if A is integrally closed. It is shown that the finiteness result becomes false if we weaken condition (4).

In Sect. 2 we give the precise statements of our results. Section 3 contains some tools from effective commutative algebra. Much of them have been taken from [5, §10.7]. In Sect. 4 we recall from [5, Chap. 10] a central proposition, which is the basis of the proofs of the results in the present paper. The main tool in the proof of that proposition is Corollary 1.2 of [4] on unit equations over finitely generated integral domains. In the remaining sections we deduce our theorems.

2   Statements of the Results

We start with the necessary definitions. Let A be an integral domain of characteristic 0 which is finitely generated over $\mathbb{Z}$ (i.e., finitely generated as a $\mathbb{Z}$-algebra). Let K be its quotient field. Suppose $A = \mathbb{Z}[x_1, \ldots, x_r]$ and define the ideal

$$I := \{f \in \mathbb{Z}[X_1, \ldots, X_r] : f(x_1, \ldots, x_r) = 0\}.$$

Thus, A is isomorphic to $\mathbb{Z}[X_1, \ldots, X_r]/I$ and x_i corresponds to the residue class of X_i mod I. Following [5, §10.7], we say that A is *given effectively* if a finite set of generators for the ideal I is given. We call such a set of generators for I an *ideal representation* for A. For A to be an integral domain of characteristic 0 it is necessary and sufficient that I be a prime ideal of $\mathbb{Z}[X_1,\ldots,X_r]$ with $1 \notin \mathbb{Q}I$. This can be checked, for instance, by means of [1, Proposition 4.10] and [20, §4]. By a representation for an element y of A we mean a polynomial $f \in \mathbb{Z}[X_1,\ldots,X_r]$ such that $y = f(x_1,\ldots,x_r)$ and we say that y is *effectively given/computable* if such f is given/can be computed. We can check whether two polynomials $f,g \in \mathbb{Z}[X_1,\ldots,X_r]$ represent the same element of A by checking whether their difference $f - g$ is in I, using an ideal membership algorithm for I (see e.g., [1]). By a representation for $y \in K$ we mean a pair (f,g) with $f,g \in \mathbb{Z}[X_1,\ldots,X_r]$ and $g \notin I$ such that $y = f(x_1,\ldots,x_r)/g(x_1,\ldots,x_r)$. By saying that a polynomial with coefficients in A or K is given/can be determined effectively we mean that its coefficients are given/can be determined effectively. We say that a finite extension G of K is effectively given, if it is given in the form $K[X]/(P)$, where P is an effectively given monic, irreducible polynomial in $K[X]$. We should mention here that for a given polynomial $P \in K[X]$ it can be decided effectively whether it is irreducible, see, for instance, [20, §§33–35].

In what follows, A is an integral domain finitely generated over $\mathbb{Z}$, K its quotient field, and G a finite extension of K. Further, δ is a non-zero element of A and n an integer with $n \geq 2$. Consider the equation

$$D(F) = \delta \quad \text{in monic polynomials } F \in A[X] \text{ of degree } n \tag{1}$$
$$\text{having all their zeros in } G.$$

Our first result is as follows.

Theorem 2.1 *Let n be an integer ≥ 2 and A an integral domain finitely generated over $\mathbb{Z}$ with quotient field K such that*

$$\left(\frac{1}{n}A^+ \cap A_K^+\right)/A^+ \text{ is finite.} \tag{2}$$

Further, let G be a finite extension of K and δ a non-zero element of A.

Then the set of monic polynomials $F \in A[X]$ with (1) is a union of finitely many A-equivalence classes. Moreover, for any effectively given n, A, G, δ as above, a full system of representatives for these equivalence classes can be determined effectively.

By Corollary 3.9 in Sect. 3 below, for any effectively given integral domain A finitely generated over $\mathbb{Z}$ it can be decided effectively whether it satisfies (2). As said in the introduction, we do not know whether condition (2) can be relaxed. Below, we show that Theorem 2.1 is not true for arbitrary finitely generated domains of characteristic 0.

We now turn to elements of orders in finite étale algebras. We start again with some definitions. Let again A be an integral domain finitely generated over $\mathbb{Z}$ and K its quotient field. We say that a finite étale K-algebra Ω is given effectively, if it is given in the form $K[X]/(P)$, where $P \in K[X]$ is an effectively given monic polynomial without multiple zeros. Elements of Ω can be expressed uniquely as $\sum_{i=0}^{n-1} a_i \theta^i$ with $a_0, \ldots, a_{n-1} \in K$, where $\theta := X \bmod P$. We say that an element of Ω is given/can be determined effectively if $a_0, \ldots, a_{n-1}$ are given/can be determined effectively.

Recall that an A-order of Ω is an A-subalgebra of the integral closure of A in Ω, which spans Ω as a K-vector space. By a result of Nagata [18], the integral closure of A in Ω is finitely generated as an A-module. Since the integral domain A is Noetherian, any A-order of Ω is finitely generated as an A-module as well. We say that an A-order O of Ω is given effectively if a finite set of A-module generators $\{\omega_1 = 1, \omega_2, \ldots, \omega_m\}$ of O is given effectively. We say that an element α of O is given/can be determined effectively, if $a_1, \ldots, a_m \in A$ are given/can be determined effectively such that $\alpha = \sum_{i=1}^{m} a_i \omega_i$. Using Corollary 3.10 (i) below one can verify that $\omega_1, \ldots, \omega_m$ do indeed generate an A-order of Ω.

Let Ω be a finite étale K-algebra with $[\Omega : K] =: n \geq 2$, let O be an A-order in Ω, and let δ be a non-zero element of A. We consider the equation

$$D_{\Omega/K}(\alpha) = \delta \ \text{ in } \alpha \in O. \tag{3}$$

We prove the following result.

Theorem 2.2 *Let A be an integral domain finitely generated over $\mathbb{Z}$ with quotient field K, Ω a finite étale K-algebra, O an A-order in Ω, and $\delta \in A$, $\delta \neq 0$. Assume that*

$$(O \cap K)^+/A^+ \text{ is finite.} \tag{4}$$

Then the set of $\alpha \in O$ with (3) is a union of finitely many A-equivalence classes. Further, for any effectively given A, Ω, O, δ as above, a full system of representatives for these classes can be determined effectively.

Using Corollary 3.10 (ii) below, for given A, Ω, O it can be decided effectively whether condition (4) is satisfied. At the end of the present section we show that Theorem 2.2 becomes false if we relax condition (4).

We now show that Theorem 2.1 cannot be true for arbitrary finitely generated domains A. More precisely, we show that for every integer $n \geq 2$, there are an integral domain A finitely generated over $\mathbb{Z}$, a finite extension G of the quotient field of A, and a $\delta \in A \setminus \{0\}$, such that there are infinitely many A-equivalence classes of monic polynomials $F \in A[X]$ of degree n with $D(F) = \delta$ having all their roots in G.

Let n be an integer ≥ 2, let t be transcendental over $\mathbb{Q}$ and define the integral domain

$$A := \mathbb{Z}\left[nt, \binom{n}{2}t^2, \binom{n}{3}t^3, \ldots, t^n\right].$$

Notice that A is a subring of $\mathbb{Z}[t]$ and that A has quotient field $K := \mathbb{Q}(t)$. We can express elements of A as $\sum_{k=0}^{m} s_k t^k \in A$ with $s_k \in \mathbb{Z}$ for all k. We show that s_k is divisible by n if k is coprime with n. Indeed, assume that $\gcd(k, n) = 1$. Notice that s_k is a $\mathbb{Z}$-linear combination of terms

$$\prod_{j=1}^{n} \binom{n}{j}^{l_j} \quad \text{with } l_1, \ldots, l_n \in \mathbb{Z}_{\geq 0}, \quad l_1 + 2l_2 + \cdots + nl_n = k. \tag{5}$$

Let p^r be a prime power occurring in the prime factorization of n. For each term in (5), there is $j \in \{1, \ldots, n-1\}$ such that j is coprime with p and $l_j > 0$, since k is not divisible by p. From well-known divisibility properties of binomial coefficients, it follows that $\binom{n}{j}$ is divisible by p^r. Hence all terms in (5) are divisible by p^r. Consequently, s_k is divisible by each of the prime powers p^r in the factorization of n, hence it is divisible by n.

Now fix a non-zero $c \in A$, let δ be the discriminant of $X^n - c$, and let G be the splitting field of $X^n - c$ over K. Consider the polynomials

$$F_m := (X + t^{mn+1})^n - c = \sum_{j=0}^{n} \binom{n}{j} t^j \cdot (t^n)^{mj} X^{n-j} - c$$

where m runs through the positive integers. Clearly, for every m we have $F_m \in A[X]$, F_m has splitting field G over K, and $D(F_m) = \delta$. We show that the polynomials F_m lie in distinct A-equivalence classes; it then follows that (1) has infinitely many A-equivalence classes of solutions. Let m, m' be two distinct positive integers. Suppose that $F_m, F_{m'}$ are A-equivalent; then there is $a \in A$ such that $F_{m'}(X) = F_m(X + a)$. It follows that there is an n-th root of unity ρ such that $X + t^{m'n+1} = \rho(X + t^{mn+1} + a)$. Consequently, $\rho = 1$ and $t^{m'n+1} - t^{mn+1} \in A$. But this is impossible, since the exponents of both terms are coprime with n, while the coefficients are not divisible by n.

We next show that if, with the notation of Theorem 2.2, the integral domain A and the A-order O of Ω do not satisfy (4), then there is a non-zero $\delta \in A$ such that (3) has infinitely many A-equivalence classes of solutions. Indeed, suppose (4) does not hold. Then there is an infinite sequence $b_1, b_2, \ldots$ of elements of $O \cap K$ such that none of the differences $b_i - b_j$ ($i > j \geq 1$) belongs to A. Pick $\alpha \in O$ such that $\Omega = K[\alpha]$ and put $\delta := D_{\Omega/K}(\alpha)$. Then $D_{\Omega/K}(\alpha + b_i) = \delta$ and $\alpha + b_i \in O$ for $i = 1, 2, \ldots$, and the elements $\alpha + b_i$ ($i = 1, 2, \ldots$) lie in different A-equivalence classes.

We finish this section with some remarks on certain variations on Eq. (1), following a suggestion of one of the referees.

Remark 2.3 Let A, K, n, δ, G be as in Theorem 2.1. Instead of (1) we consider the equation

$$D(F) \in \delta A^* \quad \text{in monic polynomials } F \in A[X] \text{ of degree } n \tag{6}$$
$$\text{having all their zeros in } G,$$

where A^* denotes the unit group of A and $\delta A^* := \{\delta u : u \in A^*\}$. We can partition the solutions of (6) into so-called *weak A-equivalence classes*, where two monic polynomials $F_1, F_2 \in A[X]$ of degree n are called weakly A-equivalent if $F_2(X) = u^{-n}F_1(uX + a)$ for some $u \in A^*, a \in A$.

By a theorem of Roquette [19], any integral domain that is finitely generated over $\mathbb{Z}$ has a finitely generated unit group. Hence for every positive integer m there is a finite subset $\mathcal{W}_m$ of A^* such that every element of A^* can be expressed as wv^m with $w \in \mathcal{W}_m$ and $v \in A^*$. Let $F \in A[X]$ be a solution of (6). Thus, $D(F) = \delta u$ with $v \in A^*$. Write $u = wv^{n(n-1)}$, with $w \in \mathcal{W}_{n(n-1)}$ and $v \in A^*$. Then the polynomial $F_v(X) := v^{-n}F(vX)$ has discriminant $D(F_v) = \delta w$. By Theorem 2.1 and the finiteness of $\mathcal{W}_{n(n-1)}$, the polynomials F_v lie in only finitely many A-equivalence classes, hence the polynomials F with (6) lie in only finitely many weak A-equivalence classes. In case of integrally closed domains A, this was proved in [12].

In certain special cases, for instance when A is the ring of S-integers of a number field K for some finite set of places S of K, we can effectively compute sets $\mathcal{W}_m$ as above, but we do not know of an algorithm that computes such sets $\mathcal{W}_m$ for *arbitrary* effectively given integral domains A that are finitely generated over $\mathbb{Z}$. So at least with the above argument, we cannot in general effectively determine a full system of representatives for the weak A-equivalence classes of polynomials F with (6).

Remark 2.4 Let again A, K, n, δ, G be as in Theorem 2.1. We consider again Eq. (1) but for polynomials not necessarily having their zeros in a prescribed extension G of K, i.e., we consider

$$D(F) = \delta \text{ in monic polynomials } F \in A[X] \text{ of degree } n. \tag{7}$$

Let A be the ring of S-integers of an algebraic number field K, where S is a finite set of places of K. If F is a polynomial satisfying (7), then the discriminant of its splitting field G over K is composed of prime ideals from S and those occurring in the prime ideal factorization of δ. By a consequence of the Hermite–Minkowski Theorem, there are only finitely many possibilities for G, and these can be determined effectively. Together with Theorem 2.1, or with the results of Győry from [10, 11] or [13], this implies that the polynomials F with (7) lie in only finitely many A-equivalence classes, a full system of representatives of which can be determined effectively. Perhaps this effective result can be extended to certain finitely generated integral domains of low transcendence degree. But extending this to arbitrary domains that are finitely generated over $\mathbb{Z}$ seems to be very hard and beyond the scope of this paper.

3   Tools from Effective Commutative Algebra

For the definitions of a domain, étale algebra, order, etc. and elements of those being effectively given/computable, we refer to Sect. 2. We start with some effective results on systems of linear equations in polynomials.

Proposition 3.1 *Let* $\mathbf{k} = \mathbb{Q}$ *or* $\mathbb{F}_p$ *for some prime p. Then for any given positive integer r and any given polynomials* $f_1, \ldots, f_s \in \mathbf{k}[X_1, \ldots, X_r]$ *we can:*

(i) *determine effectively whether a given polynomial g from* $\mathbf{k}[X_1, \ldots, X_r]$ *belongs to the ideal* $I = (f_1, \ldots, f_s)$ *and if so, determine effectively polynomials* $g_1, \ldots, g_s$ *such that* $g = g_1 f_1 + \cdots + g_s f_s$ *(ideal membership problem);*
(ii) *determine effectively whether I is a prime ideal.*

Proof See Seidenberg [20]: §4, p. 277 for (i) and §46, p. 293 for (ii) (in fact Seidenberg gives a method to determine the prime ideals associated to a given ideal *I*, which certainly enables one to decide whether *I* is a prime ideal). The main ideas in the proofs of these results originate from Hermann [14] but her arguments contain gaps. □

For a polynomial f with integer coefficients, we denote by $H(f)$ its height (maximum of the absolute values of its coefficients) and by $\mathrm{Deg}\,f$ its total degree. Further, we define the polynomial ring $R := \mathbb{Z}[X_1, \ldots, X_r]$.

Proposition 3.2 *Let M be an* $m \times n$-*matrix with entries from R, and* $\mathbf{b}$ *a vector from* R^m, *such that the entries of M and* $\mathbf{b}$ *have total degrees at most d and heights at most H.*

(i) *The R-module*

$$\{\mathbf{x} \in R^n : M\mathbf{x} = \mathbf{0}\}$$

is generated by vectors, of which the coordinates are polynomials, whose total degrees are bounded above by an effectively computable number C_1 *depending only on* m, n, d, r *and whose heights are bounded above by an effectively computable number* C_2 *depending only on* $m, n, d, r,$ *and H.*
(ii) *Suppose that the system*

$$M\mathbf{x} = \mathbf{b}$$

is solvable in $\mathbf{x} \in R^n$. *Then this system has a solution* $\mathbf{x}_0 \in R^n$ *whose coordinates have total degrees bounded above by* C_3 *and heights bounded above by* C_4, *where both* C_3, C_4 *are effectively computable numbers depending only on* $m, n, d, r,$ *and H.*

Proof Aschenbrenner [1] proved the above with $C_1 = (2md)^{2^{c_1 r \log 2r}}$, $C_2 = \exp\left((2m(d+1))^{2^{c_2(1+r\log 2r)}}(1 + \log H)\right)$ (cf. his Proposition 5.2), and $C_3 =$

$(2md)^{2^{c_3 r \log 2r}} (1 + \log H)$ (cf. his Theorem 6.5), where c_1, c_2, c_3 are effectively computable absolute constants. In (ii), thanks to our upper bound for the total degrees, the problem to find a solution to $M\mathbf{x} = \mathbf{b}$ reduces to solving a finite system of inhomogeneous linear equations over $\mathbb{Z}$, and then we obtain a value for C_4 by invoking, for instance, a result from [2]. $\qquad\square$

Corollary 3.3 (Ideal Membership over $\mathbb{Z}$) *Let $I = (f_1, \ldots, f_s)$ be an ideal in R and $g \in R$. Suppose that $f_1, \ldots, f_s$ and g have total degrees at most d and heights at most H. If $g \in I$, there exist $g_1, \ldots, g_s \in R$ of total degrees and heights bounded above by effectively computable numbers depending only on r, d, s, and H, such that $g = \sum_{i=1}^{s} g_i f_i$.*

Proof Apply part (ii) of Proposition 3.2 with $m = 1, n = s$. $\qquad\square$

In what follows, A is an integral domain with quotient field K of characteristic 0 that is finitely generated over $\mathbb{Z}$. We assume that A is effectively given, i.e., we have $A = \mathbb{Z}[x_1, \ldots, x_r]$, and we are given a finite set of generators $f_1, \ldots, f_s$ of the ideal

$$I = \{ f \in \mathbb{Z}[X_1, \ldots, X_r] : f(x_1, \ldots, x_r) = 0 \}.$$

Corollary 3.4 *Given an $m \times n$-matrix M with entries in K and $\mathbf{b} \in K^m$ one can:*

(i) effectively determine a finite set of A-module generators $\mathbf{a}_1, \ldots, \mathbf{a}_t$ for the A-module of $\mathbf{x} \in A^n$ with $M\mathbf{x} = \mathbf{0}$;

(ii) decide effectively whether $M\mathbf{x} = \mathbf{b}$ has a solution $\mathbf{x} \in A^n$ and if so, find a solution.

Proof (i) After clearing denominators, one may assume that the entries of M and the coordinates of $\mathbf{b}$ lie in A. Let $m_{ij} \in R$ ($i = 1, \ldots, m, j = 1, \ldots, n$ be representatives for the elements of M. Writing $y_1, \ldots, y_n$ for representatives in R for the coordinates of $\mathbf{x}$ we can rewrite the system $M\mathbf{x} = \mathbf{0}$ as

$$m_{i1} y_1 + \cdots + m_{in} y_n \in I \quad (i = 1, \ldots, m)$$

or as

$$m_{i1} y_1 + \cdots + m_{in} y_n = f_1 y_{i1} + \cdots + f_s y_{is} \quad (i = 1, \ldots, m)$$

in $y_i, y_{ij} \in R$, which is a system of equations as in part (i) of Proposition 3.2. Likewise $M\mathbf{x} = \mathbf{b}$ can be rewritten as a system of equations as in part (ii) of Proposition 3.2. Now one simply has to apply Proposition 3.2 to these systems. $\qquad\square$

We say that a finitely generated A-module $\mathcal{M} \subset K$ is effectively given if a finite set of A-module generators for $\mathcal{M}$ is effectively given. We denote the A-module generated by $a_1, \ldots, a_u$ by $(a_1, \ldots, a_u)$.

Corollary 3.5 *For any two effectively given A-submodules $\mathcal{M}_1$, $\mathcal{M}_2$ of K, one can*

(i) effectively decide whether $\mathcal{M}_1 \subseteq \mathcal{M}_2$;
(ii) effectively compute a finite set of A-module generators for $\mathcal{M}_1 \cap \mathcal{M}_2$.

Proof Let $\mathcal{M}_1 = (a_1, \ldots, a_u)$, $\mathcal{M}_2 = (b_1, \ldots, b_v)$ with the $a_i, b_j \in K$ effectively given. Then (i) comes down to checking whether $a_1, \ldots, a_u \in \mathcal{M}_2$, which is a special case of part (ii) of Corollary 3.4. To determine a finite set of A-module generators for $\mathcal{M}_1 \cap \mathcal{M}_2$, one first determines a finite set of A-module generators for the solution set of $(x_1, \ldots, x_u, y_1, \ldots, y_v) \in A^{u+v}$ of $\sum_{i=1}^{u} x_i a_i = \sum_{j=1}^{v} y_j b_j$ and then for each generator one takes the coordinates $x_1, \ldots, x_u$. $\square$

Probably the following results are well-known but we could not find a proof for it.

Proposition 3.6 *Assume that A is effectively given and let $\mathcal{M}_1$, $\mathcal{M}_2$ be two effectively given finitely generated A-submodules of K with $\mathcal{M}_1 \subset \mathcal{M}_2$. Then it can be decided effectively whether $\mathcal{M}_2/\mathcal{M}_1$ is finite. If this is the case, a full system of representatives for $\mathcal{M}_2/\mathcal{M}_1$ can be determined effectively.*

We use the following simple lemma.

Lemma 3.7 *Suppose we are given a sequence $\mathcal{N}_1 \subseteq \cdots \subseteq \mathcal{N}_r$ of finitely generated A-modules contained in K. Then $\mathcal{N}_r/\mathcal{N}_1$ is finite if and only if for $i = 1, \ldots, r-1$, the quotient $\mathcal{N}_{i+1}/\mathcal{N}_i$ is finite. Further, if this is the case, we obtain a full system of representatives for $\mathcal{N}_r/\mathcal{N}_1$ by taking all sums $a_1 + \cdots + a_{r-1}$ where a_i runs through a full system of representatives for $\mathcal{N}_{i+1}/\mathcal{N}_i$ for $i = 1, \ldots, r-1$.*

Proof Obvious. $\square$

Proof of Proposition 3.6 We may assume that A is given in the form

$$\mathbb{Z}[X_1, \ldots, X_r]/(f_1, \ldots, f_s),$$

with given polynomials $f_1, \ldots, f_s \in \mathbb{Z}[X_1, \ldots, X_r]$, and that x_i is the residue class of X_i modulo $(f_1, \ldots, f_s)$, for $i = 1, \ldots, r$. Then the elements of K may be represented as quotients $g(x_1, \ldots, x_r)/h(x_1, \ldots, x_r)$, where $g, h \in \mathbb{Z}[X_1, \ldots, X_r]$ and $h \notin (f_1, \ldots, f_s)$. After multiplying the given generators of $\mathcal{M}_1$ and $\mathcal{M}_2$ with the product of their denominators, we may assume that $\mathcal{M}_1, \mathcal{M}_2 \subseteq A$. There is clearly no loss of generality to assume that $\mathcal{M}_1, \mathcal{M}_2$ are given as $\mathcal{M}_1 = (a_1, \ldots, a_u)$, $\mathcal{M}_2 = (a_1, \ldots, a_v)$ with $v > u$. In fact, it suffices to prove our Theorem in the special case $v = u + 1$. Then the general case with arbitrary v can be deduced from Lemma 3.7.

So we assume henceforth that $v = u + 1$. Let

$$J := \{x \in A : x \cdot a_{u+1} \in \mathcal{M}_1\} = A \cap a_{u+1}^{-1}\mathcal{M}_1;$$

then $\mathcal{M}_2/\mathcal{M}_1$ is isomorphic to the additive group of A/J. By Corollary 3.5 we can compute a finite set of generators for J, which we may represent as residue classes modulo $(f_1, \ldots, f_s)$ of polynomials $f_{s+1}, \ldots, f_t$ from $\mathbb{Z}[X_1, \ldots, X_r]$. Then

$$\mathfrak{M}_2/\mathfrak{M}_1 \cong \mathbb{Z}[X_1, \ldots, X_r]/I,$$

where $I = (f_1, \ldots, f_s, \ldots, f_t)$. So it suffices to prove that it can be decided effectively whether $\mathbb{Z}[X_1, \ldots, X_r]/I$ is finite and that in this case a full system of representatives can be computed effectively.

A necessary condition for $\mathbb{Z}[X_1, \ldots, X_r]/I$ to be finite is that $I \cap \mathbb{Z} \neq (0)$. This in turn is equivalent to the existence of $g_1, \ldots, g_t \in \mathbb{Q}[X_1, \ldots, X_r]$ such that $g_1 f_1 + \cdots + g_t f_t = 1$. By Proposition 3.1 it can be decided effectively whether such $g_1, \ldots, g_t$ exist and if so, they can be computed. Supposing such $g_1, \ldots, g_t$ exist, by clearing the denominators of their coefficients we find non-zero $b \in \mathbb{Z}$ in $I \cap \mathbb{Z}$. Using Corollary 3.3 we can check, for every divisor $a \in \mathbb{Z}$ of b whether $a \in I$. In this manner we eventually find a with $I \cap \mathbb{Z} = (a)$.

If $a = 1$, then $I = \mathbb{Z}[X_1, \ldots, X_r]$ and we are done. Suppose that $a \neq 1$. We make a reduction to the case that $a = p$ is a prime number. Suppose that $a = p_1 \cdots p_k$ where $p_1, \ldots, p_k$ are not necessarily distinct prime numbers. We may write $I = (p_1 \cdots p_k, f_1, \ldots, f_t)$. For $i = 1, \ldots, k$, put $I_i := (p_1 \cdots p_i, f_1, \ldots, f_t)$ and for $i \in \{1, \ldots, k-1\}$ define

$$J_i := \{f \in \mathbb{Z}[X_1, \ldots, X_r] : p_1 \cdots p_i f \in I_{i+1}\}.$$

Then I_i/I_{i+1} is isomorphic to the additive group of $\mathbb{Z}[X_1, \ldots, X_r]/J_i$. Now if we are able to decide, for $i = 1, \ldots, k-1$, whether $\mathbb{Z}[X_1, \ldots, X_r]/J_i$ is finite and find a full system of representatives for this quotient, we can do the same for I_i/I_{i+1} and then, thanks to Lemma 3.7, for $\mathbb{Z}[X_1, \ldots, X_r]/I$.

Using Proposition 3.2 we find a set of generators for J_i. By what has been explained above, from this we can compute $b_i \in \mathbb{Z}$ with $J_i \cap \mathbb{Z} = (b_i)$. Clearly, $p_{i+1} \in J_i$; hence $J_i \cap \mathbb{Z} = (1)$ or (p_{i+1}). The case $J_i = (1)$ being obvious, it remains to check whether $\mathbb{Z}[X_1, \ldots, X_r]/J_i$ is finite if $J_i \cap \mathbb{Z} = (p_{i+1})$.

Changing notation, we see that it suffices to show, for any given ideal I of $\mathbb{Z}[X_1, \ldots, X_r]$ with $I \cap \mathbb{Z} = (p)$ for some prime p, whether $\mathbb{Z}[X_1, \ldots, X_r]/I$ is finite and if so, to compute a full system of representatives for $\mathbb{Z}[X_1, \ldots, X_r]$ modulo I. We may assume that I is given in the form $I = (p, f_1, \ldots, f_t)$, with $f_1, \ldots, f_t \in \mathbb{Z}[X_1, \ldots, X_r]$. Given $f \in \mathbb{Z}[X_1, \ldots, X_r]$, denote by $\bar{f}$ its reduction modulo p, and put $\bar{I} = (\bar{f}_1, \ldots, \bar{f}_t)$. Then $\mathbb{Z}[X_1, \ldots, X_r]/I \cong \mathbb{F}_p[X_1, \ldots, X_r]/\bar{I}$. So we have to decide whether this latter residue class ring is finite and if so, to compute a full system of representatives for the residue classes.

For any positive integer m, denote by V_m the set of residue classes modulo $\bar{I}$ of all polynomials of degree $\leq m$ in $\mathbb{F}_p[X_1, \ldots, X_r]$. This is a finite dimensional $\mathbb{F}_p$-vector space. Recall that the *Hilbert function* $H_{\bar{I}}$ of $\bar{I}$ is defined by $H_{\bar{I}}(m) := \dim_{\mathbb{F}_p} V_m$. It is known that there are an integer $m_{\bar{I}}$, and a polynomial $p_{\bar{I}} \in \mathbb{Q}[X]$, called the *Hilbert polynomial* of $\bar{I}$, such that $H_{\bar{I}}(m) = p_{\bar{I}}(m)$ for $m \geq m_{\bar{I}}$. Now $\mathbb{F}_p[X_1, \ldots, X_r]/\bar{I}$ is finite if and only if $p_{\bar{I}}$ is constant, and this being the case, every residue class of $\mathbb{F}_p[X_1, \ldots, X_r]$ modulo $\bar{I}$ is represented by a polynomial of degree at most $m_{\bar{I}}$. There is a general procedure, based on Gröbner basis theory, to compute $m_{\bar{I}}$ and $p_{\bar{I}}$, given a set of generators for $\bar{I}$, see [3, §§15.1.1, 15.10.2]. With this procedure one can decide whether $\mathbb{F}_p[X_1,\ldots,X_r]/\bar{I}$ is finite. Subsequently, using Proposition 3.1, one can select a full system of representatives modulo $\bar{I}$ from the polynomials of degree $\leq m_{\bar{I}}$.

This completes the proof of Proposition 3.6. $\square$

For a finite extension G of K, we denote by A_G the integral closure of A in G. In particular, A_K is the integral closure of A in its quotient field K.

Proposition 3.8 *Assume that A and a finite extension G of K are effectively given. Then one can effectively compute a finite set of A-module generators for A_G. Moreover, one can compute an ideal representation for A_G.*

Proof The computability of a finite set of A-module generators for A_G follows from results of Nagata [18], de Jong [15], Matsumura [17], and Matsumoto [16]. For more details, see [5, Corollary 10.7.18]. Then an ideal representation for A_G can be computed using [5, Theorems 10.7.13, 10.7.16]. $\square$

Corollary 3.9 *Assume that A is effectively given. Then one can effectively decide whether $(\frac{1}{n}A^+ \cap A_K^+)/A^+$ is finite and if so, compute a full system of representatives for $(\frac{1}{n}A^+ \cap A_K^+)/A^+$.*

Proof Immediate consequence of Proposition 3.8, Corollary 3.5, (ii) and Proposition 3.6. $\square$

Corollary 3.10 *Assume that A and a finite étale K-algebra Ω are effectively given. Further, let $\omega_2,\ldots,\omega_u \in \Omega$ be effectively given and let O be the A-module generated by $1,\omega_2,\ldots,\omega_u$.*

(i) It can be effectively decided whether O is an A-order of Ω.
(ii) If O is an A-order of Ω, one can effectively decide whether $(O \cap K)^+/A^+$ is finite, and if so, compute a full system of representatives for $(O \cap K)^+/A^+$.

Proof We assume that $\Omega = K[X]/(P)$ where $P \in K[X]$ is an effectively given, separable monic polynomial. Let $n := [\Omega : K] = \deg P$ and $\theta := X \bmod P$. Then $\{1,\theta,\ldots,\theta^{n-1}\}$ is a K-basis of Ω. Further, we assume that $\omega_2,\ldots,\omega_u$ are effectively given as K-linear combinations of $1,\theta,\ldots,\theta^{n-1}$. Then we may express elements of O as $\sum_{k=0}^{n-1} l_k(\mathbf{x})\theta^k$ with $\mathbf{x} \in A^u$, where $l_0,\ldots,l_{n-1}$ are linear forms from $K[X_1,\ldots,X_u]$.

(i) We first verify that the linear forms $l_0,\ldots,l_{n-1}$ have rank n over K, to make sure that O contains a K-basis of Ω. The next thing to verify is whether $\omega_i\omega_j$ is an A-linear combination of $1,\omega_2,\ldots,\omega_u$ for $i,j = 2,\ldots,u$. Compute $b_{ij} \in K$ such that $\omega_i\omega_j = \sum_{k=0}^{n-1} b_{ijk}\theta^k$. Then we have to verify whether the system $l_k(\mathbf{x}) = b_{ijk}$ ($k = 0,\ldots,n-1$) is solvable in $\mathbf{x} \in A^u$, for $i,j = 2,\ldots,u$, and this can be done by means of Corollary 3.4 (ii). Lastly, it is a standard fact from algebra, that if A is a subring of a commutative ring B that is finitely generated as an A-module, then B

is in fact integral over A. So in particular, if we have verified that O is closed under multiplication, then it is automatically contained in A_Ω.

(ii) Using Corollary 3.4 (i) we can compute a finite set of A-module generators, say $\mathbf{x}_1, \ldots, \mathbf{x}_v$ for the A-module of $\mathbf{x} \in A^u$ with $l_i(\mathbf{x}) = 0$ for $i = 1, \ldots, n-1$. Then $(O \cap K)^+$ is generated as an A-module by $l_0(\mathbf{x}_1), \ldots, l_0(\mathbf{x}_v)$. With these generators for $(O \cap K)^+$ and Proposition 3.6, we can check whether $(O \cap K)^+/A^+$ is finite, and if so, compute a full system of representatives. $\square$

4 The Main Proposition

We recall from [5] a central proposition from which Theorems 2.1 and 2.2 are deduced. We keep the notation from Sect. 2.

Proposition 4.1 *For any integral domain A of characteristic 0 that is finitely generated over $\mathbb{Z}$, any finite extension G of the quotient field of A, any non-zero $\delta \in A$, and any integer $n \geq 2$, all effectively given, one can determine effectively a finite subset $\mathscr{F} = \mathscr{F}_{A,G,n,\delta}$ of G with the following property: if F is any monic polynomial from $A[X]$ of degree n and discriminant δ having all its zeros, say $\alpha_1, \ldots, \alpha_n$, in G, then*

$$\alpha_i - \alpha_j \in \mathscr{F} \ \text{for } i,j \in \{1, \ldots, n\}, i \neq j. \tag{8}$$

Proof This is Proposition 10.2.1 of [5]. Its proof is based on Corollary 1.2 of [4] on unit equations over finitely generated integral domains. $\square$

5 Proof of Theorem 2.1

We start with a preliminary lemma.

Lemma 5.1 *For every integral domain A finitely generated over $\mathbb{Z}$ and every two monic polynomials $F_1, F_2 \in A[X]$, all effectively given, we can determine effectively whether F_1, F_2 are A-equivalent.*

Proof It suffices to consider the case when F_1, F_2 have equal degrees. Write $F_1(X) = X^n + a_1 X^{n-1} + \cdots$, $F_2(X) = X^n + b_1 X^{n-1} + \cdots$. We have to check whether there exists $a \in A$ with $F_2(X) = F_1(X + a)$. Comparing the coefficients of X^{n-1} we see that for such a we must have $na = b_1 - a_1$. Using Corollary 3.4 (ii) we can check whether $a \in A$ and then whether indeed $F_2(X) = F_1(X + a)$. $\square$

Henceforth, the integral domain A is given effectively in the form

$$\mathbb{Z}[X_1, \ldots, X_r]/(f_1, \ldots, f_s) = \mathbb{Z}[x_1, \ldots, x_r]$$

where x_i is the residue class of X_i mod $(f_1, \ldots, f_s)$ for $i = 1, \ldots, r$. Further the finite extension G of the quotient field K of A is given in the form $K[X]/(P)$ or $K(w)$, where w is the residue class of X mod P. The polynomial P may be represented as $b_0^{-1} \sum_{i=0}^{d} b_i X^{d-i}$ with $b_0, \ldots, b_d$ given as polynomials in $x_1, \ldots, x_r$ with integer coefficients. Define

$$u := b_0 w.$$

Then u has minimal polynomial

$$Q = X^d + \sum_{i=1}^{d} b_i b_0^{d-1-i} X^{d-i} =: X^d + \sum_{i=1}^{d} c_i X^{d-i} \in A[X] \tag{9}$$

over K. Now clearly, $G = K(u)$, u is integral over A, and every element of G can be expressed in the form $\sum_{i=0}^{d-1} (a_i/b) u^i$ with $a_0, \ldots, a_{d-1}, b \in A$, given as polynomials with integer coefficients in $x_1, \ldots, x_r$.

Proof of Theorem 2.1 Let A, G, n, δ be effectively given and satisfy the conditions of Theorem 2.1. Further, let $\mathscr{F}$ be the finite effectively determinable set from Proposition 4.1.

Take a monic polynomial F from $A[X]$ with (1). Then F has all its zeros in G, say $F(X) = (X - \alpha_1) \cdots (X - \alpha_n)$, with $\alpha_1, \ldots, \alpha_n \in G$. By Proposition 4.1 we have

$$\alpha_i - \alpha_j \in \mathscr{F} \text{ for } i, j \in \{1, \ldots, n\} \text{ with } i \neq j.$$

Recall that $\mathscr{F}$ is finite, and effectively determinable in terms of A, G, n, δ. For each tuple $\left(\gamma_{ij} : i, j \in \{1, \ldots, n\}, i \neq j\right)$ with elements from $\mathscr{F}$ we consider the polynomials F with (1) and with $\alpha_i - \alpha_j = \gamma_{ij}$ for $i, j \in \{1, \ldots, n\}, i \neq j$. That is, we consider polynomials F such that

$$\begin{cases} F \in A[X], F \text{ monic}, \deg F = n, D(F) = \delta, \\ F = (X - \alpha_1) \cdots (X - \alpha_n) \text{ for some } \alpha_1, \ldots, \alpha_n \in G \\ \text{such that } \alpha_i - \alpha_j = \gamma_{ij} \text{ for } i, j \in \{1, \ldots, n\}, i \neq j. \end{cases} \tag{10}$$

Our proof will be completed as follows. We show that for each tuple $\{\gamma_{ij}\}$ it can be decided effectively whether a polynomial F with (10) exists. If so, we show that the polynomials with (10) lie in finitely many A-equivalence classes, and determine effectively a full system of representatives for them. Then from the union of these systems, we extract a full system of representatives for the A-equivalence classes of solutions of (1).

Fix elements γ_{ij} from $\mathscr{F}$ $(1 \leq i, j \leq n, i \neq j)$. Suppose there is a polynomial F with (10). For this polynomial we have

$$n\alpha_i = y + \gamma_i \text{ for } i = 1, \ldots, n, \tag{11}$$

with $y = \alpha_1 + \cdots + \alpha_n$, $\gamma_i = \sum_{j=1}^n \gamma_{ij}$ for $i = 1, \ldots, n$, where we have put $\gamma_{ii} := 0$ for $i = 1, \ldots, n$. Here $\gamma_1, \ldots, \gamma_n$ are fixed and $y, \alpha_1, \ldots, \alpha_n$ are unknowns. The number y is a coefficient of F, so $y \in A$. Further, if there is a polynomial F with (10), then

$$(X - \gamma_1) \cdots (X - \gamma_n) = n^n F\left(\frac{X + y}{n}\right) \in A[X]. \tag{12}$$

The coefficients of $(X - \gamma_1) \cdots (X - \gamma_n)$ belong to G. It can be checked whether they belong to K, and then by means of Corollary 3.4 (ii), it can be checked whether they belong to A. If not so, there is no polynomial with (10). So we assume henceforth that $(X - \gamma_1) \cdots (X - \gamma_n) \in A[X]$. Then $\gamma_1, \ldots, \gamma_n \in A_G$.

Using Proposition 3.8, we compute a finite set of A-module generators for A_G. From this, we deduce a system $\{\mathbf{a}_1, \ldots, \mathbf{a}_t\}$ of A-module generators for A_G^n. The numbers $\alpha_1, \ldots, \alpha_n$ from (10) are in A_G. So there are $x_1, \ldots, x_t \in A$ such that

$$\begin{pmatrix} \alpha_1 \\ \vdots \\ \alpha_n \end{pmatrix} = x_1 \mathbf{a}_1 + \cdots + x_t \mathbf{a}_t, \tag{13}$$

and we can rewrite (11) as

$$x_1(n\mathbf{a}_1) + \cdots + x_t(n\mathbf{a}_t) = y \begin{pmatrix} 1 \\ \vdots \\ 1 \end{pmatrix} + \begin{pmatrix} \gamma_1 \\ \vdots \\ \gamma_n \end{pmatrix}. \tag{14}$$

By linear algebra, we can determine a maximal K-linearly independent subset of $\{n\mathbf{a}_1, \ldots, n\mathbf{a}_t, (1, \ldots, 1)^T, (\gamma_1, \ldots, \gamma_n)^T\}$, say $\{\mathbf{b}_1, \ldots, \mathbf{b}_m\}$. Further, we can compute expressions for $n\mathbf{a}_1, \ldots, n\mathbf{a}_t, (1, \ldots, 1)^T, (\gamma_1, \ldots, \gamma_n)^T$ as K-linear combinations of $\mathbf{b}_1, \ldots, \mathbf{b}_m$. By substituting these into (14) and equating the coordinates of (14), we obtain a system of inhomogeneous linear equations:

$$M\mathbf{x} = \mathbf{b} \text{ in } \mathbf{x} = (x_1, \ldots, x_t, y)^T \in A^{t+1} \tag{15}$$

where the matrix M and vector $\mathbf{b}$ have their entries in K. Then using Corollary 3.4 we can decide whether (15) is solvable and if so, compute a solution. Translating this back to (14), we can decide whether (14) is solvable and if so, compute a solution.

If (14) is unsolvable, then there is no polynomial F satisfying (10). Assume (14) is solvable and compute a solution, say $(x_{10}, \ldots, x_{t0}, y_0) \in A^{t+1}$. Thus, $\sum_{i=1}^t x_{i0}(n\mathbf{a}_i) - y_0(1, \ldots, 1)^T = (\gamma_1, \ldots, \gamma_n)^T$. Put

$$\begin{pmatrix} \alpha_{10} \\ \vdots \\ \alpha_{n0} \end{pmatrix} := x_{10}\mathbf{a}_1 + \cdots + x_{n0}\mathbf{a}_t. \tag{16}$$

Then

$$n\alpha_{i0} = y_0 + \gamma_i \text{ for } i = 1, \ldots, n \text{ with } y_0 \in A. \tag{17}$$

Now let again F be an arbitrary polynomial with (10) and let y be as in (11). From (11), (17) we infer that

$$\alpha_i - \alpha_{i0} = \frac{y - y_0}{n} =: a \text{ for } i = 1, \ldots, n. \tag{18}$$

Clearly, $a \in \frac{1}{n}A$. Identity (16) implies that $\alpha_{10}, \ldots, \alpha_{n0} \in A_G$. Hence a is integral over A. So in fact, $a \in \frac{1}{n}A \cap A_K$.

By Corollary 3.9, we can compute a full system of representatives, say $\{\theta_1, \ldots, \theta_h\}$ for $(\frac{1}{n}A^+ \cap \overline{A}^+)/A^+$. For $j = 1, \ldots, h$, put

$$F_j(X) := (X - \alpha_{10} - \theta_j) \cdots (X - \alpha_{n0} - \theta_j).$$

For some $j \in \{1, \ldots, h\}$ we have $a = \theta_j + c$ for some $c \in A$. Then (18) implies that $\alpha_i = \alpha_{i0} + \theta_j + c$ for $i = 1, \ldots, n$, and so $F(X) = F_j(X - c)$. Hence F is A-equivalent to F_j.

The polynomials $F_1, \ldots, F_h$ can be determined effectively. Their coefficients belong to K and using Corollary 3.4 we can select those polynomials that have their coefficients in A. Thus, for each tuple $\{\gamma_{ij}\}$ with $\gamma_{ij} \in \mathscr{F}$ we can compute a finite system of polynomials from $A[X]$ such that every polynomial with (10) is A-equivalent to one of them. By taking the union of these systems for all tuples $\{\gamma_{ij}\}$, we effectively determine a finite list of polynomials from $A[X]$ such that every polynomial with (1) is A-equivalent to at least one of them. For each polynomial from the list we can effectively decide whether it satisfies (1) and if not so, remove it. Finally, assuming the list is ordered, by means of Lemma 5.1 we can effectively decide whether a polynomial from the list is A-equivalent to an earlier polynomial in the list and if so, remove it. This leaves us with a full system of representatives for the A-equivalence classes of polynomials with (1). This completes the proof of Theorem 2.1. $\qquad\square$

6 Proof of Theorem 2.2

Let A be an integral domain finitely generated over $\mathbb{Z}$, effectively given as usual in the form $\mathbb{Z}[X_1, \ldots, X_r]/(f_1, \ldots, f_s) = \mathbb{Z}[x_1, \ldots, x_r]$, where $f_1, \ldots, f_s \in \mathbb{Z}[X_1, \ldots, X_r]$ and where x_i is the residue class of X_i mod $(f_1, \ldots, f_s)$ for $i = 1, \ldots, r$. Denote by K the quotient field of A. Let Ω be a finite étale K-algebra, effectively given in the form $K[X]/(P) = K[\theta]$, where $P \in K[X]$ is a monic polynomial without multiple zeros, and $\theta = \operatorname{mod} P$.

We need some results from [5, §10.7]. Using [5, Corollary 10.7.7] we can construct the splitting field of P over K; call this G. By means of [5, Corollary 10.7.8] we can compute w such that $G = K(w)$, together with the minimal polynomial of w over K. As was explained in Sect. 5, we can compute from this another representation for G of the form $K(u)$, where u is integral over A, together with the monic minimal polynomial Q of u over K. Elements of G are always given in the form $\sum_{i=0}^{d-1} (a_i/b)u^i$ where $d = [G : K]$ and $a_0, \ldots, a_{d-1}, b$ are elements of A.

The polynomial P factorizes as $(X - \theta^{(1)}) \cdots (X - \theta^{(n)})$ in G, and by Evertse and Győry [5, Corollary 10.7.8] we can compute expressions of $\theta^{(1)}, \ldots, \theta^{(n)}$ as K-linear combinations of $1, u, \ldots, u^{d-1}$. With these expressions we can compute, for any element $\alpha = \sum_{i=0}^{n-1} c_i \theta^i \in \Omega$ with $c_0, \ldots, c_{n-1} \in K$, its images $\alpha^{(j)} = \sum_{i=0}^{n-1} c_i (\theta^{(j)})^i$ $(j = 1, \ldots, n)$ under the K-homomorphisms of Ω to G.

We start with a lemma.

Lemma 6.1 *For any two effectively given $\alpha_1, \alpha_2 \in O$ with $K[\alpha_1] = K[\alpha_2] = \Omega$, we can decide effectively whether α_1, α_2 are A-equivalent.*

Proof Compute expressions $\alpha_1 = \sum_{i=0}^{n-1} c_i \theta^i$, $\alpha_2 = \sum_{i=0}^{n-1} d_i \theta^i$ with $c_i, d_i \in K$, and check if $c_0 - d_0 \in A$, $c_i = d_i$ for $i = 1, \ldots, n-1$. $\qquad\square$

Proof of Theorem 2.2 Let A, Ω, O be the effectively given integral domain, finite étale K-algebra, and A-order in Ω. Assume that $(O \cap K)^+/A^+$ is finite. Let $\{\omega_1 = 1, \ldots, \omega_m\}$ be the effectively given system of A-module generators for O. Further, let $n = [\Omega : K]$, $n \geq 2$ and let δ be the given element of A. Lastly, let G be the field defined above, given in the form $K(u)$ with u integral over A.

Recall that by Proposition 3.6, we can compute an ideal representation for the integral closure A_K of A, i.e., A_K is effectively given as an integral domain in the usual sense. So we can apply Proposition 4.1 with A_K instead of A. Let $\mathscr{F}'$ be the finite set $\mathscr{F}$ from Proposition 4.1 but taken with A_K instead of A. This set can be computed effectively in terms of A_K, G, δ, hence in terms of A, Ω, δ. Now if α is an element of O with (3), i.e., $D_{\Omega/K}(\alpha) = \delta$, then $\alpha \in A_\Omega$, hence $F_\alpha(X) := (X - \alpha^{(1)}) \cdots (X - \alpha^{(n)})$ has its coefficients in A_K, we have $D(F_\alpha) = \delta$, and F_α has its zeros in G. Hence

$$\alpha^{(i)} - \alpha^{(j)} \in \mathscr{F}' \ \text{for} \ i, j \in \{1, \ldots, n\}, i \neq j.$$

We now pick elements γ_{ij} from $\mathscr{F}'$ and consider the elements α with

$$\begin{cases} \alpha \in O, \ D_{\Omega/K}(\alpha) = \delta, \\ \alpha^{(i)} - \alpha^{(j)} \in \gamma_{ij} \ \text{for} \ i, j \in \{1, \ldots, n\}, i \neq j. \end{cases} \tag{19}$$

We show that it can be decided effectively whether (19) is solvable and if so, compute a solution of (19). Notice that (19) is certainly unsolvable if $\prod_{1 \leq i < j \leq n} \gamma_{ij}^2 \neq \delta$. Assume that $\prod_{1 \leq i < j \leq n} \gamma_{ij}^2 = \delta$. Then the condition $D_{\Omega/K}(\alpha) = \delta$ can be dropped.

Writing α as $\sum_{k=1}^{m} x_k \omega_k$ with $x_1, \ldots, x_m \in A$, we can rewrite (19) as

$$\sum_{k=1}^{m} x_k \left(\omega_k^{(i)} - \omega_k^{(j)} \right) = \gamma_{ij} \text{ for } i, j \in \{1, \ldots, n\}, i \neq j. \tag{20}$$

Clearly, $(x_1, \ldots, x_m)$ is a solution of (20) in A^m if and only if $\alpha := \sum_{k=1}^{m} x_k \omega_k$ is a solution of (19).

By expressing $\omega_k^{(i)} - \omega_k^{(j)}$ and the numbers γ_{ij} as K-linear combinations of $1, u, \ldots, u^{d-1}$ where $d = [G : K]$ and u is the generating element of G over K, we can rewrite (20) as a system of inhomogeneous linear equations in A^m like in Corollary 3.4 (ii). Thus, it can be decided effectively whether (20) is solvable, and if so, a solution can be computed. Equivalently, it can be decided effectively whether (19) is solvable and if so, a solution can be computed.

For each choice of $\gamma_{ij} \in \mathscr{F}'$ ($1 \leq i, \leq n$, $i \neq j$), we check if (19) is solvable and if so, we compute a solution. Let $\mathcal{T} = \{\alpha_1, \ldots, \alpha_g\}$ be the finite set obtained in this manner. By Corollary 3.10 (ii) we can compute a full system of representatives $\{\theta_1, \ldots, \theta_h\}$ for $(O \cap K)^+ / A^+$. Using Lemma 6.1, we can compute a maximal subset $\mathcal{U}$ of $\{\alpha_i + \theta_j : i = 1, \ldots, g, j = 1, \ldots, h\}$ such that any two distinct elements of $\mathcal{U}$ are not A-equivalent. We show that $\mathcal{U}$ is a full system of representatives for the A-equivalence classes of solutions of (3).

Let α be a solution of (3). Then α satisfies (19) for certain $\gamma_{ij} \in \mathscr{F}'$. Let α_0 be an element from $\mathcal{T}$ satisfying (19) for these γ_{ij}. Then $\alpha^{(i)} - \alpha^{(j)} = \alpha_0^{(i)} - \alpha_0^{(j)}$ for $i, j \in \{1, \ldots, n\}$, hence

$$\alpha^{(1)} - \alpha_0^{(1)} = \cdots = \alpha^{(n)} - \alpha_0^{(n)}.$$

It follows that $\alpha - \alpha_0 =: a \in O \cap K$. Hence $a = \theta_j + c$ for some $j \in \{1, \ldots, h\}$ and $c \in A$, and so, $\alpha = \alpha_0 + \theta_j + c$. Now clearly, α is A-equivalent to an element of $\mathcal{U}$. This completes our proof of Theorem 2.2. $\qquad\square$

Acknowledgements We would like to thank the two anonymous referees for their careful scrutiny of our paper and their valuable comments and corrections.

References

1. M. Aschenbrenner, Ideal membership in polynomial rings over the integers. J. Am. Math. Soc. **17**, 407–442 (2004)
2. I. Borosh, M. Flahive, D. Rubin, B. Treybig, A sharp bound for solutions of linear diophantine equations. Proc. Am. Math. Soc. **105**, 844–846 (1989)
3. D. Eisenbud, *Commutative Algebra with a View Toward Algebraic Geometry* (Springer, Berlin, 1994)
4. J.-H. Evertse, K. Győry, Effective results for unit equations over finitely generated domains. Math. Proc. Camb. Philos. Soc. **154**, 351–380 (2013)

5. J.-H. Evertse, K. Győry, *Discriminant Equations in Diophantine Number Theory* (Cambridge University Press, 2016)
6. K. Győry, Sur les polynômes à coefficients entiers et de discriminant donné. Acta Arith. **23**, 419–426 (1973)
7. K. Győry, Sur les polynômes à coefficients entiers et de discriminant donné II. Publ. Math. Debrecen **21**, 125–144 (1974)
8. K. Győry, Sur les polynômes à coefficients entiers et de discriminant donné III. Publ. Math. Debrecen **23**, 141–165 (1976)
9. K. Győry, On polynomials with integer coefficients and given discriminant IV. Publ. Math. Debrecen **25**, 155–167 (1978)
10. K. Győry, On polynomials with integer coefficients and given discriminant V, p-adic generalizations. Acta Math. Acad. Sci. Hung. **32**, 175–190 (1978)
11. K. Győry, On discriminants and indices of integers of an algebraic number field. J. Reine Angew. Math. **324**, 114–126 (1981)
12. K. Győry, On certain graphs associated with an integral domain and their applications to diophantine problems. Publ. Math. Debrecen **29**, 79–94 (1982)
13. K. Győry, Effective finiteness theorems for polynomials with given discriminant and integral elements with given discriminant over finitely generated domains. J. Reine Angew. Math. **346**, 54–100 (1984)
14. G. Hermann, Die Frage der endlich vielen Schritte in der Theorie der Polynomideale. Math. Ann. **95**, 736–788 (1926)
15. T. de Jong, An algorithm for computing the integral closure. J. Symb. Comput. **26**, 273–277 (1998)
16. R. Matsumoto, On computing the integral closure. Commun. Algebra **28**, 401–405 (2000)
17. H. Matsumura, *Commutative Ring Theory* (Cambridge University Press, Cambridge, 1986)
18. M. Nagata, A general theory of algebraic geometry over Dedekind domains I. Am. J. Math. **78**, 78–116 (1956)
19. P. Roquette, Einheiten und Divisorenklassen in endlich erzeugbaren Körpern. Jahresber. Deutsch. Math. Verein **60**, 1–21 (1957)
20. A. Seidenberg, Constructions in algebra. Trans. Am. Math. Soc. **197**, 273–313 (1974)

Quasi-Equivalence of Heights and Runge's Theorem

Philipp Habegger

Abstract Let P be a polynomial that depends on two variables X and Y and has algebraic coefficients. If x and y are algebraic numbers with $P(x, y) = 0$, then by work of Néron $h(x)/q$ is asymptotically equal to $h(y)/p$ where p and q are the partial degrees of P in X and Y, respectively. In this paper we compute a completely explicit bound for $|h(x)/q - h(y)/p|$ in terms of P which grows asymptotically as $\max\{h(x), h(y)\}^{1/2}$. We apply this bound to obtain a simple version of Runge's Theorem on the integral solutions of certain polynomial equations.

Mathematics Subject Classification. Primary: 11G50; Secondary: 11D41, 11G30, 14H25, 14H50

1   Introduction

Suppose P is an irreducible polynomial in two variables X and Y and whose coefficients are in $\overline{\mathbb{Q}}$, an algebraic closure of $\mathbb{Q}$. If x and y are algebraic numbers with $P(x, y) = 0$, we investigate the relation between the absolute logarithmic Weil heights $h(x)$ and $h(y)$; this height is defined in Sect. 2.

Say $p = \deg_X P \geq 1$ and $q = \deg_Y P \geq 1$. By work of Néron [11] there exists a constant $c(P)$ such that if x and y are algebraic numbers with $P(x, y) = 0$, then

$$\left| \frac{h(x)}{q} - \frac{h(y)}{p} \right| \leq c(P) \max \left\{ \frac{h(x)}{q}, \frac{h(y)}{p} \right\}^{1/2}. \tag{1}$$

See Corollary 9.3.10 in Bombieri and Gubler's book [5] or Theorem B.5.9 in Hindry and Silverman's book [9] for a highbrow approach to this bound. One sometimes says that $h(x)/q$ and $h(y)/p$ are quasi-equivalent.

P. Habegger (✉)
Department of Mathematics and Computer Science, University of Basel, Spiegelgasse 1, 4051 Basel, Switzerland
e-mail: philipp.habegger@unibas.ch

© Springer International Publishing AG 2017

C. Elsholtz, P. Grabner (eds.), *Number Theory – Diophantine Problems, Uniform Distribution and Applications*, DOI 10.1007/978-3-319-55357-3_13

Our aim is to determine an admissible constant $c(P)$ which is completely explicit in terms of P. We will strive for a good dependency in the projective height $h_p(P)$ of P, which we also define in Sect. 2, and the partial degrees p and q.

Theorem 1 *Let $P \in \overline{\mathbb{Q}}[X, Y]$ be irreducible with $p = \deg_X P \geq 1$ and $q = \deg_Y P \geq 1$. If $P(x, y) = 0$ with $x, y \in \overline{\mathbb{Q}}$, then (1) holds with*

$$c(P) = 5 \left(\log \left(2^{\min\{p,q\}} (p + 1)(q + 1) \right) + h_p(P) \right)^{1/2}.$$

Abouzaid [1] proved a related height estimate. In his bound, the dependence on the partial degrees and the numerical constants are slightly worse. Quasi-equivalence of heights also follows from Bartolome's Theorem 1.3 [2], but again with larger numerical constants and worse dependency on p and q.

It is essential that P is irreducible. For example, both partial degrees of $(X^2 - Y)(X - Y^2)$ equal 3. But (1) cannot hold for this polynomial as $h(x^2) = 2h(x)$ for all algebraic x. It is possible to formulate a version of Theorem 1 when $K \subset \overline{\mathbb{Q}}$ is a number field and if $P \in K[X, Y]$ is irreducible. In this case P is up to a scalar factor the product of polynomials which are irreducible in $\overline{\mathbb{Q}}[X, Y]$ and conjugated over K. Thus all factors have equal partial degrees.

Sometimes it is useful to bound $h(y)$ uniformly in terms of $h(x)$ if the height of x is large. We do this in the following corollary.

Corollary 1 *Let $P, p,$ and q be as in Theorem 1. If $P(x, y) = 0$ with $x, y \in \overline{\mathbb{Q}}$ and*

$$\max \left\{ \frac{h(x)}{q}, \frac{h(y)}{p} \right\} \geq 100 \left(\log \left(2^{\min\{p,q\}} (p + 1)(q + 1) \right) + h_p(P) \right), \qquad (2)$$

then $h(y) \leq 2\frac{p}{q}h(x)$.

The proof of Theorem 1 depends on the theory of functions fields and on the Absolute Siegel Lemma by Zhang [17]. Roy and Thunder's [12, 13] absolute Siegel Lemma would also suffice but would lead to different numerical constants. Using Bombieri and Vaaler's classical version of Siegel's Lemma instead would come at the cost of introducing a dependency in $c(P)$ on a number field containing the coefficients of P.

We give a short sketch of the proof of Theorem 1. Let m and n be large integers such that n/m is approximately equal to p/q. In Sect. 2 we use the Absolute Siegel Lemma to construct polynomials A and B in X and Y with algebraic coefficients of bounded height, not both zero, such that P divides $AY^m - B$ as a polynomial. By choosing the parameters appropriately, we can arrange that $P \nmid A$. Say $P(x, y) = 0$, then $A(x, y)y^m = B(x, y)$. If we assume for the moment $A(x, y) \neq 0$, then we may bound the height of y in terms of the height of x by using the product formula. In Sect. 3 we show that a suitable vanishing order cannot be too large. We then apply an appropriate differential operator and replace A, B by new polynomials A', B' with controlled projective height and degree such that $A'(x, y)y^m = B'(x, y)$ and $A'(x, y) \neq 0$. Thus again we get a bound for $h(y)$ in terms of $h(x)$. By swapping x

and y we get an estimate in the other direction and this completes the proof if (x, y) is not singular point on the vanishing locus of P. Singular points can be handled directly. A novel aspect of our approach is that it does not depend on Eisenstein's Theorem which bounds the coefficients of a power series of an algebraic function. An explicit version of Eisenstein's Theorem was used in the work of Abouzaid and Bartolome.

Runge [14] proved that $P(x, y) = 0$ admits only finitely many solutions $(x, y) \in \mathbb{Z}^2$ if $P \in \mathbb{Q}[X, Y]$ is irreducible with $\deg_X P = \deg_Y P = \deg P$ and if the homogeneous part of P of maximal degree is not a rational multiple of the power of an irreducible polynomial in $\mathbb{Q}[X, Y]$. Runge's method is effective and explicit upper bounds for $\max\{|x|, |y|\}$ were obtained, for example, by Hilliker-Straus [8] and Walsh [15, 16]. They rely on Eisenstein's Theorem.

We will prove a simple and explicit version of Runge's Theorem using Theorem 1.

Theorem 2 *Let $P \in \mathbb{Z}[X, Y]$ be irreducible in $\overline{\mathbb{Q}}[X, Y]$ and assume $d = \deg_X P = \deg_Y P = \deg P$. Furthermore, assume that the homogeneous part of P of degree d is not a rational multiple of the power of an irreducible polynomial in $\mathbb{Q}[X, Y]$. If $P(x, y) = 0$ with $x, y \in \mathbb{Z}$, then*

$$\log \max\{1, |x|, |y|\} \le 115 d^4 (\log(2d) + h_p(P)).$$

Walsh's [15, 16] result holds for a larger class of polynomials but our dependency on the degree is d^4 instead of his d^6.

It would be interesting to see if a more sophisticated version of Runge's Theorem, such as Bombieri's on page 304 [4], can be proved using our Theorem 1.

A variation of this paper appeared in the appendix of the author's 2007 Ph.D. thesis. He thanks his supervisor David Masser for support throughout those years. He is also grateful to Umberto Zannier for pointing out Zhang's version of the Absolute Siegel Lemma and to the referee for helpful comments and corrections.

2   Construction Using the Absolute Siegel Lemma

We begin by setting up notation, our reference for heights is Chap. 1.5 in Bombieri and Gubler's book [5].

Let K be a number field. A place v of K is an absolute value $|\cdot|_v : K \to [0, \infty)$ such that either $|x|_v = \max\{x, -x\}$ for all $x \in \mathbb{Q}$ or $|\cdot|_v$ coincides with the p-adic absolute value on $\mathbb{Q}$ for a prime p and $|p|_v = 1/p$. In the former case we call v infinite and in the latter we call it finite. If v is infinite, then $|x|_v = |\sigma(x)|$ for a ring homomorphism $\sigma : K \to \mathbb{C}$ that is uniquely determined up-to complex conjugation. We set $d_v = 1$ if $\sigma(K) \subset \mathbb{R}$ and $d_v = 2$ else wise. A finite place v is induced by a maximal ideal in the ring of integers of K. We set d_v to be the product of the ramification index and the residue degree attached to this prime ideal. We let M_K denote the set of all places of K.

The absolute logarithmic Weil height of $x \in K$ is

$$h(x) = \frac{1}{[K:\mathbb{Q}]} \sum_{v \in M_K} d_v \log \max\{1, |x|_v\}. \tag{3}$$

It is well-defined and attains the same value at x when evaluated using any number field $F \supset K$.

For $a \in K^N$ we define $|a|_v$ to be the maximum of the absolute values of the coordinates of a with respect to v. If P is a polynomial in any number of variables with coefficients in K, then $|P|_v$ denotes the maximum of the absolute values of the coefficients of P with respect to v. The projective height of $a \neq 0$ is

$$h_p(a) = \frac{1}{[K:\mathbb{Q}]} \sum_{v \in M_K} d_v \log |a|_v. \tag{4}$$

By the product formula, cf. Proposition 1.4.4 [5], $h_p(a)$ is invariant under replacing a by a non-zero scalar multiple of itself, so $h_p(a) \geq 0$. If P is a non-zero polynomial with algebraic coefficients, we set $h_p(P)$ to be the projective height of the vector whose coordinates are the non-zero coefficients of P. If Q is a further polynomial with algebraic coefficients, it will be useful to set $h_p(P, Q) = h_p(P + TQ)$ where T is an unknown that does not appear in P or Q.

For any place v of K and integer $n \geq 1$ it is convenient to define $\delta_v(n) = \max\{1, |n|_v\}$.

We start by proving a lemma concerning simple properties of places and heights. It corresponds to Lemma A.1 in the author's thesis, whose part (i) is incorrect.

Lemma 1 *Let K be a field and $A, B \in K[X, Y]$.*

(i) If K is a number field and if $v \in M_K$, then

$$|A + B|_v \leq \delta_v(2) \max\{|A|_v, |B|_v\},$$

$$|AB|_v \leq \delta_v((\min\{\deg_X A, \deg_X B\} + 1)(\min\{\deg_Y A, \deg_Y B\} + 1))|A|_v |B|_v.$$

(ii) If $K = \overline{\mathbb{Q}}$ and $x, y \in \overline{\mathbb{Q}}$ with $A(x, y) \neq 0$, then

$$h(B(x, y)/A(x, y)) \leq h_p(A, B) + \max\{\deg_X A, \deg_X B\}h(x)$$

$$+ \max\{\deg_Y A, \deg_Y B\}h(y)$$

$$+ \log \max\{(\deg_X A + 1)(\deg_Y A + 1), (\deg_X B + 1)(\deg_Y B + 1)\}$$

(iii) Say $K = \overline{\mathbb{Q}}$ and $x, y \in \overline{\mathbb{Q}}$ with $A(x, y) = 0$. If A is not divisible in $\overline{\mathbb{Q}}[X, Y]$ by any $X - \alpha$ with $\alpha \in \overline{\mathbb{Q}}$, then

$$h(y) \leq h_p(A) + (\deg_X A)h(x) + \log((\deg_X A + 1)\deg_Y A). \tag{5}$$

Proof The first inequality in (i) follows from the triangle inequality. To prove the second inequality we write $A = \sum_{i,j} a_{ij} X^i Y^j$ and $B = \sum_{i,j} b_{ij} X^i Y^j$. Then $AB = \sum_{i,j} c_{ij} X^i Y^j$ with

$$c_{ij} = \sum_{\substack{i'+i''=i \\ j'+j''=j}} a_{i'j'} b_{i''j''}.$$

The sum above involves at most $(1+\min\{\deg_X A, \deg_X B\})(1+\min\{\deg_Y A, \deg_Y B\})$ non-zero terms. Hence the desired inequality follows from the triangle inequality.

Now we prove part (ii). The product formula implies

$$h(B(x, y)/A(x, y)) = \frac{1}{[F : \mathbb{Q}]} \sum_{v \in M_F} d_v \log \max\{|A(x, y)|_v, |B(x, y)|_v\} \tag{6}$$

where F is a number field containing x, y, and the coefficients of A and B. Note that the polynomial A involves at most $(\deg_X A + 1)(\deg_Y A + 1)$ non-zero coefficients, hence the triangle inequality gives

$$|A(x, y)|_v$$

$$\leq \delta_v((\deg_X A + 1)(\deg_Y A + 1))|A|_v \max\{1, |x|_v\}^{\deg_X A} \max\{1, |y|_v\}^{\deg_Y A}$$

for each $v \in M_F$. Of course a similar inequality holds for $|B(x, y)|_v$. These inequalities inserted into (6) conclude this part of the lemma.

Our proof for Part (iii) follows the lines of Proposition 5 [3]. Say $A = a_q Y^q + \cdots + a_0$ with $a_i \in \overline{\mathbb{Q}}[X]$ and $a_q \neq 0$, so $q = \deg_Y A$. By hypothesis there exists a maximal $q' \geq 1$ such that $a_{q'}(x) \neq 0$. Let F be a number field that contains x, y, and the coefficients of A. If $v \in M_F$, then

$$|a_{q'}(x)y^{q'}|_v \leq \delta_v(q') \max_{0 \leq k \leq q'-1}\{|a_k(x)|_v\} \max\{1, |y|_v\}^{q'-1}$$

and so

$$\max\{1, |y|_v\} \leq \delta_v(q') \max_{0 \leq k \leq q'}\{|a_k(x)|_v/|a_{q'}(x)|_v\}.$$

We use the last inequality, $q' \leq q$, the product formula, and (3) to deduce

$$h(y) \leq \log q + \frac{1}{[F : \mathbb{Q}]} \sum_{v \in M_F} d_v \log \max_{0 \leq k \leq q} \{|a_k(x)|_v\}. \tag{7}$$

The triangle inequality implies

$$|a_k(x)|_v \leq \delta_v (\deg_X A + 1) \max\{1, |x|_v\}^{\deg_X A} |A|_v$$

for any $v \in M_F$. We apply this inequality to (7) to complete the proof.

We will sometimes apply property (iii) of the previous lemma to a non-zero $A \in \overline{\mathbb{Q}}[Y]$ and $x = 0$. Inequality (5) then reduces to $h(y) \leq h_p(A) + \log(\deg_Y A)$.

We introduce a basic notion of a sparsity. If $\mathcal{A} = (a_{ij})$ is an $M \times N$ matrix, then we set

$$S(\mathcal{A}) = \max_{1 \leq i \leq M} \#\{j : a_{ij} \neq 0\}.$$

If $\mathcal{A}$ has coefficients in a number field K and is non-zero, we define $h_p(\mathcal{A})$ to be the projective height of $\mathcal{A}$ taken as an element of $K^{MN} \smallsetminus \{0\}$, as in (4).

Next we adapt Zhang's Absolute Siegel Lemma, as it was used by David and Philippon [7], to our notation. See also Bartolome's Sect. 2.4 [2].

Lemma 2 *Let $\mathcal{A} \in \mathrm{Mat}_{MN}(\overline{\mathbb{Q}})$ have rank $M < N$. Then there exists $v \in \overline{\mathbb{Q}}^N \smallsetminus \{0\}$ such that $\mathcal{A}v = 0$ and*

$$h(v) \leq \frac{M}{N - M} \left(\frac{1}{2} \log S(\mathcal{A}) + h_p(\mathcal{A}) \right) + \frac{\log(N - M)}{2}.$$

Proof Let $\epsilon > 0$. We apply David and Philippon's Lemme 4.7 [7] by which there exists $v \in \overline{\mathbb{Q}}^N \smallsetminus \{0\}$ with $\mathcal{A}v = 0$ and

$$h(v) \leq \frac{1}{N - M} h(V) + \frac{1}{N - M} \sum_{i=1}^{N-M-1} \sum_{j=1}^{i} \frac{1}{2j} + \epsilon$$

where $V \subset \overline{\mathbb{Q}}^N$ is the kernel of $\mathcal{A}$ and $h(V)$ is the logarithmic height of the vector space V as defined just before Lemme 4.7 [7]. We have $(N - M)^{-1} \sum_{i=1}^{N-M-1} \sum_{j=1}^{i} (2j)^{-1} < 2^{-1} \log(N - M)$ if $N - M > 1$, so

$$h(v) \leq \frac{1}{N - M} h(V) + \frac{\log(N - M)}{2} \tag{8}$$

if $\epsilon > 0$ is small enough. This bound also holds if $N - M = 1$ as then ϵ can be omitted in David and Philippon's Lemma 4.7; however, we will not encounter this case. We note that the height used by David and Philippon uses the Euclidean norm at the infinite places. It is at least as large as $h(v)$ which uses the supremum norm at all places. By Corollary 2.8.12 [5] the height $h(V)$ equals $h_{\mathrm{Ar}}(\mathcal{A}^t)$ as in Remark 2.8.7 *loc.cit.* where $\mathcal{A}^t$ is the transpose of $\mathcal{A}$. In other words, $h(V)$ is the height of the vector in $\overline{\mathbb{Q}}^{\binom{N}{M}}$ whose entries are the determinants of all $M \times M$ minors of $\mathcal{A}$ with the Euclidean norm taken at the infinite places and maximum norm at the finite places.

By Fischer's Inequality, cf. Remark 2.8.9 and 2.9.8 [5], we find $h(V) \leq (M/2) \log S(\mathcal{A}) + M h_p(\mathcal{A})$; here we used that each row of $\mathcal{A}$ contains at most $S(\mathcal{A})$ non-zero entries.

Lemma 3 *Let $P \in \overline{\mathbb{Q}}[X, Y]$ with $p = \deg_X P \geq 1$ and $q = \deg_Y P \geq 1$. Furthermore, let m and n be integers with $m \geq 2q + 1$ and $n \geq p$. If $t = q(n + 1) - mp \geq 1$, there exist $A, B \in \overline{\mathbb{Q}}[X, Y]$ with $P \nmid A$,*

$$AY^m - B \in P \cdot \overline{\mathbb{Q}}[X, Y] \smallsetminus \{0\}, \quad \deg_X A, \deg_X B \leq n, \quad \deg_Y A, \deg_Y B \leq q - 1, \tag{9}$$

and

$$h_p(A, B) \leq \frac{m(n - p + 1)}{t} \left(\log\left((p + 1)(q + 1)\right) + h_p(P) \right) + \frac{\log(2nq)}{2}. \tag{10}$$

Proof Let $\mathfrak{Q} = \sum_{k,l} q_{kl} X^k Y^l \in \mathbb{Z}[X, Y, q_{kl}]$ with $\deg_X \mathfrak{Q} = n - p$, $\deg_Y \mathfrak{Q} = m - 1$, and where the q_{kl} are treated as unknowns. We define linear forms $f_{ij} \in \overline{\mathbb{Q}}[q_{kj}]$ for $0 \leq i \leq n, 0 \leq j \leq m + q - 1$ by

$$P\mathfrak{Q} = \sum_{i,j} f_{ij} X^i Y^j.$$

Each non-zero coefficient of f_{ij} is a coefficient of P. So

$$f_{ij} = 0 \quad (0 \leq i \leq n, \quad q \leq j \leq m - 1) \tag{11}$$

is a system of linear equations of a certain rank M in the $N = (n - p + 1)m$ unknowns q_{kl}. We have

$$M \leq (n + 1)(m - q) = N - t. \tag{12}$$

Because $N - M \geq t \geq 1$ there is a non-trivial solution. Any such solution gives rise to a non-zero polynomial $Q \in \overline{\mathbb{Q}}[X, Y]$ such that the coefficients of PQ satisfy (11) and hence $PQ = AY^m - B$ for unique polynomials $A, B \in \overline{\mathbb{Q}}[X, Y]$ with $\deg_X A, \deg_X B \leq n$ and $\deg_Y A, \deg_Y B \leq q - 1$. The terms in AY^m and B do not overlap, hence $h_p(A, B) = h_p(PQ)$.

The final term in the upper bound (8) works against us if $N - M$ is large. We now work out a lower bound for M. A non-trivial $\overline{\mathbb{Q}}$-linear combination of $X^k Y^l P$

where $0 \leq k \leq n - p$ and $q \leq l \leq m - q - 1$ is not of the form $AY^m - B$ with A and B satisfying the degree bounds in (9). Recall that $m \geq 2q + 1$, so $M \geq (n - p + 1)(m - 2q) \geq n - p + 1 \geq 1$ and hence

$$N - M \leq (n - p + 1)(m - (m - 2q)) \leq 2nq. \tag{13}$$

We will apply Siegel's Lemma to find a solution Q with small projective height. We choose a subset of the linear forms f_{ij} ($0 \leq i \leq n, q \leq j \leq m-1$) with rank M and use the coefficients of each such linear form to define a row in an $M \times N$ matrix $\mathcal{A}$. The non-zero entries of $\mathcal{A}$ are coefficients of P, hence $h_p(\mathcal{A}) \leq h_p(P)$. Furthermore, by definition each f_{ij} involves at most $(p + 1)(q + 1)$ non-zero coefficients and hence $S(\mathcal{A}) \leq (p + 1)(q + 1)$. By Lemma 2 and our discussion above there exists a non-zero solution $Q \in \overline{\mathbb{Q}}[X, Y]$ of (11) that satisfies

$$h_p(Q) \leq \frac{M}{N - M} \left(\frac{1}{2} \log((p + 1)(q + 1)) + h_p(P) \right) + \frac{\log(N - M)}{2}. \tag{14}$$

Lemma 1(i) implies $h_p(PQ) \leq \log((p+1)(q+1))+h_p(P)+h_p(Q)$. Furthermore, we use the inequalities (14), (12), and (13) to conclude that $h_p(PQ)$ is at most

$$\log((p + 1)(q + 1)) + h_p(P) + \frac{M}{t} \left(\log((p + 1)(q + 1)) + h_p(P) \right) + \frac{\log(2nq)}{2}$$

$$= \frac{M + t}{t} \left(\log((p + 1)(q + 1)) + h_p(P) \right) + \frac{\log(2nq)}{2}$$

$$\leq \frac{N}{t} \left(\log ((p + 1)(q + 1)) + h_p(P) \right) + \frac{\log(2nq)}{2}.$$

This inequality completes the proof of (10) because $N = m(n - p + 1)$.

Finally, we must verify $P \nmid A$. Indeed assuming the contrary, then P also divides B. Because $\deg_Y A, \deg_Y B \leq q - 1$ we have $A = B = 0$, a contradiction to $AY^m - B \neq 0$.

3 Multiplicity Estimates

We need some facts about function fields which we recall here for the reader's convenience. We refer to Chevalley's book [6] for proofs.

For a field F we write $F^\times = F \smallsetminus \{0\}$. Suppose F contains an algebraically closed subfield L and that there exists an element $t \in F$ that is transcendental over L such that F is a finite field extension of $L(t)$. Then F is a function field over L. We define M_F to be the set of the maximal ideals of all the proper valuation rings of F containing L. This set is the function field analogue of M_K for a number field K. Observe that its elements, the places of F, have degree 1 since L is algebraically closed. We will identify an element of M_F with the valuation function it induces.

Hence an element of M_F is a surjective map $v : F \to \mathbb{Z} \cup \{\infty\}$ such that for all $a, b \in F$ we have $v(ab) = v(a) + v(b)$ and $v(a + b) \geq \min\{v(a), v(b)\}$, $v(a) = \infty$ if and only if $a = 0$, and $v(a) = 0$ if $a \in L^\times$; we use the convention $\infty + x = x + \infty = \infty$ and $\min\{\infty, x\} = \min\{x, \infty\} = x$ for all $x \in \mathbb{Z} \cup \{\infty\}$.

If $a \in F^\times$, then $v(a) = 0$ for all but finitely many $v \in M_F$ and

$$\sum_{v \in M_F} v(a) = 0.$$

Furthermore, if $a \in F \smallsetminus L$, then

$$\sum_{v \in M_F} \max\{0, v(a)\} = [F : L(a)].$$

Suppose $P \in \overline{\mathbb{Q}}[X, Y]$ is irreducible and let F denote the field of fractions of the domain $\overline{\mathbb{Q}}[X, Y]/(P)$. Then F is a function field over $L = \overline{\mathbb{Q}}$. By abuse of notation we shall consider polynomials in $\overline{\mathbb{Q}}[X, Y]$ as elements of F via the quotient map. Note that any polynomial in $\overline{\mathbb{Q}}[X, Y]$ that is not divisible by P maps to $F^\times$.

Let $\pi = (x, y) \in \overline{\mathbb{Q}}^2$ with $P(\pi) = 0$ such that $\frac{\partial P}{\partial X}$, $\frac{\partial P}{\partial Y}$ do not both vanish at π, then we call π a *regular zero* of P. Let us assume for the moment that $\frac{\partial P}{\partial Y}(\pi) \neq 0$, then there exists a unique $v_\pi \in M_F$ with $v_\pi(X - x) = 1$ and $v_\pi(Y - y) \geq 1$. Moreover, there exists E in $\overline{\mathbb{Q}}[[T]]$, the ring of formal power series with coefficients in $\overline{\mathbb{Q}}$, such that $E(0) = 0$ and $P(x + T, y + E) = 0$. For any $A \in \overline{\mathbb{Q}}[X, Y]$ not divisible by P we have

$$\operatorname{ord} A(x + T, y + E) = v_\pi(A)$$

where ord is the standard valuation on $\overline{\mathbb{Q}}[[T]]$. Therefore $v_\pi(A) \geq 1$ if and only if $A(x, y) = 0$. If $\frac{\partial P}{\partial X}(\pi) \neq 0$, then these properties hold with the roles of X and Y reversed.

Let $A \in \overline{\mathbb{Q}}[X, Y]$, we define

$$D(A) = \frac{\partial P}{\partial Y} \frac{\partial A}{\partial X} - \frac{\partial P}{\partial X} \frac{\partial A}{\partial Y} \in \overline{\mathbb{Q}}[X, Y].$$

We also set $D^0(A) = A$ and inductively $D^s(A) = D(D^{s-1}(A))$ for all positive integers s. A formal verification yields $D(P) = 0$ and $D(AB) = D(A)B + AD(B)$ for all $B \in \overline{\mathbb{Q}}[X, Y]$. Thus we have Leibniz's rule

$$D^s(AB) = \sum_{k=0}^{s} \binom{s}{k} D^k(A) D^{s-k}(B) \quad \text{and} \quad D^s(PA) = PD^s(A) \text{ if } s \geq 0. \tag{15}$$

Lemma 4 *Let K be a number field and $P \in K[X, Y]$ with $p = \deg_X P \geq 1$ and $q = \deg_Y P \geq 1$. Furthermore, assume $A \in K[X, Y]$ such that $\deg_X A \leq n$, $\deg_Y A \leq$*

$q - 1$. *Then for any non-negative $s \in \mathbb{Z}$ we have*

$$\deg_X D^s(A) \le n + (p-1)s \quad and \quad \deg_Y D^s(A) \le (q-1)(s+1). \tag{16}$$

Moreover, if $r = \max\{p, q\}$ and $v \in M_K$, then

$$|D^s(A)|_v \le \delta_v(2(p+1)(q+1)r(n+rs))^s |P|_v^s |A|_v. \tag{17}$$

Proof We note $\deg_X D(A) \le \deg_X(A) + p - 1$ and so the first inequality in (16) follows by induction on s. The second inequality is proved similarly.

We now show (17) by induction on s. The case $s = 0$ being trivial we may assume $s \ge 1$ and also $D^s(A) \ne 0$. For brevity set $|\cdot| = |\cdot|_v$. We apply Lemma 1(i) to deduce

$$|D^s(A)| \le \delta_v(2(p+1)(q+1)) \max \left\{ \left|\frac{\partial P}{\partial Y}\right| \left|\frac{\partial D^{s-1}(A)}{\partial X}\right|, \left|\frac{\partial P}{\partial X}\right| \left|\frac{\partial D^{s-1}(A)}{\partial Y}\right| \right\}.$$

By bounding the partial derivatives of the polynomials in the usual manner we get

$$|D^s(A)| \le \delta_v(2(p+1)(q+1)r \max\{\deg_X D^{s-1}(A), \deg_Y D^{s-1}(A)\})|P||D^{s-1}(A)|.$$

The inequalities in (16) imply

$$|D^s(A)| \le \delta_v(2(p+1)(q+1)r \max\{n + (p-1)(s-1), (q-1)s\})|P||D^{s-1}(A)|.$$

The expressions inside the maximum are bounded from above by $n + rs$. Applying the induction hypothesis completes the proof.

Lemma 5 *Suppose $P \in \overline{\mathbb{Q}}[X, Y]$ is irreducible, let $\pi \in \overline{\mathbb{Q}}^2$ be a regular zero of P and let $v = v_\pi \in M_F$ be the valuation described above. If $A \in \overline{\mathbb{Q}}[X, Y]$ is not divisible by P and $A(\pi) = 0$, then $D(A)$ is not divisible by P and*

$$v(D(A)) = v(A) - 1.$$

Proof We assume $\frac{\partial P}{\partial Y}(\pi) \ne 0$, the case $\frac{\partial P}{\partial X}(\pi) \ne 0$ is similar. Say $\pi = (x, y)$. There exists $E \in T\overline{\mathbb{Q}}[[T]]$ such that $P(x+T, y+E) = 0$ and $v(A) = \operatorname{ord} A(x+T, y+E) \ge 1$. By the chain rule we have

$$0 = \frac{d}{dT}P(x+T, y+E) = \frac{\partial P}{\partial X}(x+T, y+E) + \frac{dE}{dT}\frac{\partial P}{\partial Y}(x+T, y+E).$$

We use this and the definition of D to obtain

$$\operatorname{ord} D(A)(x+T, y+E)$$

$$= \operatorname{ord}\left(\left(\frac{\partial P}{\partial Y}\frac{\partial A}{\partial X} - \frac{\partial P}{\partial X}\frac{\partial A}{\partial Y}\right)(x+T, y+E)\right)$$

$$= \operatorname{ord}\frac{\partial P}{\partial Y}(x+T, y+E) + \operatorname{ord}\left(\frac{\partial A}{\partial X}(x+T, y+E) + \frac{dE}{dT}\frac{\partial A}{\partial Y}(x+T, y+E)\right).$$

By our assumption we have $\operatorname{ord}\frac{\partial P}{\partial Y}(x+T, y+E) = 0$ which we insert into the equality above and use the chain rule again as well as $A(x, y) = 0$ to get

$$\operatorname{ord} D(A)(x+T, y+E) = \operatorname{ord}\frac{d}{dT}A(x+T, y+E) = \operatorname{ord} A(x+T, y+E) - 1.$$

Hence $v(D(A)) = v(A) - 1$. In particular, P does not divide $D(A)$.

We now prove a multiplicity estimate which will be useful later on.

Lemma 6 *Let A, B, P, m, p, q, and t be as in Lemma 3 with $t \geq 1$. Furthermore, assume P is irreducible and $\deg P = p + q$. If $\pi \in \overline{\mathbb{Q}}^2$ is a regular zero of P, there exists an integer s with $0 \leq s \leq t + pq - p - q$ such that $D^s(A)(\pi) \neq 0$ and $D^k(A)(\pi) = 0$ for all $0 \leq k < s$.*

Proof Let F be as above Lemma 4. For brevity set $v = v_\pi$. Clearly $X, Y \in F \smallsetminus \overline{\mathbb{Q}}$ since p and q are both positive; also $v(X), v(Y) \geq 0$. Furthermore, $A \neq 0$ in F by Lemma 3.

We first claim that for any $v' \in M_F$ at least one of the two $v'(X), v'(Y)$ is non-negative. Indeed we argue by contradiction so let us assume $v'(X) < 0$ and $v'(Y) < 0$. Then for any integers i, j with $0 \leq i \leq p, 0 \leq j \leq q$ and $i+j < p+q$ we have

$$iv'(X) + jv'(Y) > pv'(X) + qv'(Y). \tag{18}$$

Now by hypothesis $P = \alpha X^p Y^q + \tilde{P}$ with $\alpha \neq 0$ and $\deg\tilde{P} < p+q$. We apply the ultrametric inequality and (18) to get

$$pv'(X) + qv'(Y) \geq \min_{\substack{0 \leq i \leq p, 0 \leq j \leq q \\ i+j<p+q}} \{iv'(X) + jv'(Y)\} > pv'(X) + qv'(Y),$$

a contradiction.

Now assume $v' \in M_F$ such that $v'(Y) < 0$. Then $v'(X) \geq 0$ by the discussion above and

$$v'(A) = v'(BY^{-m}) = v'(B) - mv'(Y). \tag{19}$$

Now $\deg_Y B \leq q - 1$ so the ultrametric inequality implies $v'(B) \geq (q-1)v'(Y)$. We insert this last inequality into (19) to find

$$v'(A) \geq (q - 1 - m)v'(Y) \geq -v'(Y) > 0 \tag{20}$$

because $q \leq m$. Hence

$$\sum_{v' \in M_F} \max\{0, v'(A)\} \geq \max\{0, v(A)\} + \sum_{\substack{v' \in M_F \\ v'(Y) < 0}} \max\{0, v'(A)\}$$

$$\geq v(A) + (m + 1 - q) \sum_{\substack{v' \in M_F \\ v'(Y) < 0}} \max\{0, -v'(Y)\} \tag{21}$$

where the last inequality follows from (20). Next we insert the equality

$$\sum_{\substack{v' \in M_F \\ v'(Y) < 0}} \max\{0, v'(Y^{-1})\} = [F : \overline{\mathbb{Q}}(Y^{-1})] = [F : \overline{\mathbb{Q}}(Y)] = p$$

into (21) to find

$$\sum_{v' \in M_F} \max\{0, v'(A)\} \geq v(A) + (m + 1 - q)p > 0. \tag{22}$$

In particular $A \notin \overline{\mathbb{Q}}$.

We continue by bounding the left-hand side of (22) from above. If $v' \in M_F$ with $v'(X) \geq 0$ and $v'(Y) \geq 0$, then $v'(A) \geq 0$ because A is a polynomial in X and Y. Hence

$$\sum_{v' \in M_F} \max\{0, -v'(A)\} \leq \sum_{v' \in M_F, v'(X) < 0} \max\{0, -v'(A)\}$$
$$+ \sum_{v' \in M_F, v'(Y) < 0} \max\{0, -v'(A)\}. \tag{23}$$

Actually equality holds above because at most one $v'(X)$, $v'(Y)$ can be negative, but this is not important here. Around (20) we showed that if $v'(Y) < 0$ then $v'(A) > 0$, hence the second term on the right-hand side of (23) is zero. Now recall that $\deg_X A \leq n$; if $v'(X) < 0$, then $v'(Y) \geq 0$ and the ultrametric inequality leads us to $v'(A) \geq nv'(X)$. If we insert this inequality into (23), we get

$$\sum_{v' \in M_F} \max\{0, -v'(A)\} \leq n \sum_{v' \in M_F, v'(X) < 0} \max\{0, -v'(X)\} \tag{24}$$

$$= n[F : \overline{\mathbb{Q}}(X^{-1})] = n[F : \overline{\mathbb{Q}}(X)] = nq.$$

The left-hand sides of (22) and (24) are both equal to $[F : \overline{\mathbb{Q}}(A)]$, so we get

$$v(A) \leq nq + (q - m - 1)p = t + pq - p - q.$$

If we set $s = v(A)$, then Lemma 5 and induction give $v(D^k(A)) = v(A) - k$ for $0 \leq k \leq s$. Hence $D^s(A)(\pi) \neq 0$ and $D^k(A)(\pi) = 0$ for $0 \leq k < s$.

4   Completion of Proof

Lemma 7 *Say $\kappa > 0$ and $\lambda > 0$ are real numbers with $\kappa\lambda \geq 2$. Let $P \in \overline{\mathbb{Q}}[X, Y]$ be irreducible with $p = \deg_X P \geq 1$, $q = \deg_Y P \geq 1$, and $\deg P = p + q$. For a regular zero $(x, y) \in \overline{\mathbb{Q}}^2$ of P we define*

$$k = \max \left\{ \frac{h(x)}{q}, \frac{h(y)}{p} \right\} \quad and \quad h = \log\left((p + 1)(q + 1)\right) + h_p(P). \tag{25}$$

If $k \geq \lambda^2 h$, then

$$h(y) \leq \frac{p}{q} h(x) + p \left(\kappa + \frac{1}{\lambda} + \frac{4}{\kappa} \right) (hk)^{1/2}$$
$$+ \frac{1}{\kappa\lambda} \left(\frac{\log(16\kappa\lambda)}{8} + \frac{\log 3}{24} + 9\log(p + 1) + \log\left(1 + \frac{\kappa\lambda}{2}\right) \right).$$

Proof We first handle the case $q = 1$ using Lemma 1(iii). Indeed, as P is irreducible, it cannot be divisible by a polynomial that is linear in X. As $h \geq h_p(P)$ and $h \geq \log(p + 1)$, we find

$$h(y) \leq ph(x) + \log(p + 1) + h_p(P) \leq ph(x) + h \leq ph(x) + \lambda^{-1}(hk)^{1/2} \tag{26}$$

where we also used $h \leq \lambda^{-2}k$. Observe that $\lambda^{-1} \leq p\kappa$ because $\kappa\lambda \geq 1$. The bound (26) is better than our claim. Hence we now assume $q \geq 2$.

As $h > 0$ we may define

$$m = q \left\lceil \kappa pq \left(\frac{k}{h}\right)^{1/2} \right\rceil \quad and \quad n = m\frac{p}{q} + p - 1 \tag{27}$$

where $\lceil z \rceil$ is the least integer greater or equal to $z \in \mathbb{R}$. So m and n are integers and

$$m \geq \kappa pq^2 \left(\frac{k}{h}\right)^{1/2} \geq \kappa\lambda pq^2 \tag{28}$$

and

$$m \le q\left(\kappa pq\left(\frac{k}{h}\right)^{1/2} + 1\right) = \kappa pq^2\left(\frac{k}{h}\right)^{1/2} + q. \tag{29}$$

The lower bound for m implies $m > 2q$ as $\kappa\lambda > 1$ by hypothesis and since $q \ge 2$. Therefore, $m \ge 2q + 1$ and $n \ge p$.

Let $t = q(n + 1) - mp$ be as in Lemma 3, then $t = pq \ge 1$. So the said lemma provides $A, B \in \overline{\mathbb{Q}}[X, Y]$ as therein. Because of Lemma 6 there exists an integer s with

$$0 \le s \le t + pq - p - q \le 2pq - 1$$

such that $D^s(A)(x, y) \ne 0$ and $D^k(A)(x, y) = 0$ for all $0 \le k < s$. We apply Leibniz's rule (15) to $D^s(AY^m - B)$ and use the fact that $AY^m - B$ is divisible by P to conclude

$$y^m = \frac{D^s(B)(x, y)}{D^s(A)(x, y)}.$$

Lemma 1(ii) and $h(y^m) = mh(y)$ gives

$$mh(y) \le h_p(D^s(A), D^s(B)) + \max\{\deg_X D^s(A), \deg_X D^s(B)\}h(x) \tag{30}$$
$$+ \max\{\deg_Y D^s(A), \deg_Y D^s(B)\}h(y)$$
$$+ \log\max\{(\deg_X D^s(A) + 1)(\deg_Y D^s(A) + 1),$$
$$(\deg_X D^s(B) + 1)(\deg_Y D^s(B) + 1)\}.$$

We use Lemma 4 applied to A and B to deduce

$$\max\{\deg_X D^s(A), \deg_X D^s(B)\} \le n + (p - 1)s,$$
$$\max\{\deg_Y D^s(A), \deg_Y D^s(B)\} \le (q - 1)(s + 1)$$

and

$$h_p(D^s(A), D^s(B)) \le h_p(A, B) + sh_p(P) + s\log\big(2(p + 1)(q + 1)r(n + rs)\big)$$

with $r = \max\{p, q\}$; the last line follows from summing up the local bounds in (17). We insert these bounds in (30) to see

$$mh(y) \le h_p(A, B) + sh_p(P) + (n + (p - 1)s)h(x) + (q - 1)(s + 1)h(y)$$
$$+ s\log\big(2(p + 1)(q + 1)r(n + rs)\big) + \log\big((1 + n + (p - 1)s)q(s + 1)\big).$$

Next we use the bound given for $h_p(A, B)$ in Lemma 3 and recall $t = pq$ to get

$$mh(y) \leq \frac{m(n-p+1)}{pq}\left(\log\left((p+1)(q+1)\right) + h_p(P)\right) + \frac{\log(2nq)}{2} \tag{31}$$

$$+ sh_p(P) + (n + (p-1)s)h(x) + (q-1)(s+1)h(y)$$

$$+ s\log\left(2(p+1)(q+1)r(n+rs)\right) + \log\left((1+n+(p-1)s)q(s+1)\right).$$

We use $n \leq mp/q + p$, which follows from (27), and find

$$(n + (p-1)s)h(x) + (q-1)(s+1)h(y)$$

$$\leq m\frac{p}{q}h(x) + (p + (p-1)s)qk + (q-1)(s+1)pk$$

$$\leq m\frac{p}{q}h(x) + \left(pq + (p-1)sq + (q-1)(s+1)p\right)k$$

$$\leq m\frac{p}{q}h(x) + 4p^2q^2k$$

using the definition of k and since $s \leq 2pq-1$. We insert this bound into (31), divide by m, and use the definition of h to get

$$h(y) \leq \frac{p}{q}h(x) + \left(\frac{n-p+1}{pq} + \frac{s}{m}\right)h + \frac{4p^2q^2}{m}k + \frac{\log(2nq)}{2m}$$

$$+ \frac{2pq-1}{m}\log\left(2(p+1)(q+1)r(n+rs)\right) + \frac{1}{m}\log\left((1+n+(p-1)s)q(s+1)\right)$$

Suppose for the moment $s \geq 1$, then $1 + n + (p-1)s \leq n + rs$. We also have $q(s+1) \leq 2pq^2 \leq 2(p+1)(q+1)r$. Therefore, $(1+n+(p-1)s)q(s+1) \leq 2(p+1)(q+1)r(n+rs)$ and this inequality also holds for $s = 0$ as $n \geq 1$. We find

$$h(y) \leq \frac{p}{q}h(x) + \left(\frac{n-p+1}{pq} + \frac{s}{m}\right)h + \frac{4p^2q^2}{m}k + \frac{\log(2nq)}{2m}$$

$$+ \frac{2pq}{m}\log\left(2(p+1)(q+1)r(n+rs)\right) \tag{32}$$

We now continue by bounding each term on the right-hand side of inequality (32).

The first term $\frac{p}{q}h(x)$ is the main contribution to $h(y)$.

To bound the second term we recall our choice (27) which yields $(n - p + 1)/(pq) = m/q^2$. Observe that

$$\frac{n-p+1}{pq} + \frac{s}{m} \leq \frac{m}{q^2} + \frac{2}{\kappa\lambda q} \leq \frac{m}{q^2} + \frac{1}{q}$$

by (28), $s \le 2pq$, and $\kappa\lambda \ge 2$. So the second term on the right of (32) satisfies

$$\left(\frac{n-p+1}{pq} + \frac{s}{m}\right) h \le \left(\frac{m}{q^2} + \frac{1}{q}\right) h \le \left(\kappa p \left(\frac{k}{h}\right)^{1/2} + \frac{2}{q}\right) h \le \kappa p (hk)^{1/2} + h$$

because of (29) and $q \ge 2$. Using $h \le \lambda^{-2} k$ we find

$$\left(\frac{n-p+1}{pq} + \frac{s}{m}\right) h \le p \left(\kappa + \frac{1}{p\lambda}\right) (hk)^{1/2} \le p \left(\kappa + \frac{1}{\lambda}\right) (hk)^{1/2}. \tag{33}$$

The third term in (32) can be bounded from above using the first inequality in (28) as follows:

$$\frac{4p^2 q^2}{m} k \le \frac{4p}{\kappa} (hk)^{1/2}. \tag{34}$$

We move on to the fourth term and recall $n \le mp/q + p$. As $m \ge \kappa\lambda pq^2 \ge 4\kappa\lambda p \ge 4\kappa\lambda, m \ge q$, since $z \mapsto z^{-1} \log(4z)$ is decreasing on $[1, \infty)$, and because $p^{-1} \log p \le (\log 3)/3$ we find

$$\begin{aligned}
\frac{\log(2nq)}{2m} &\le \frac{\log(2mp + 2pq)}{2m} \le \frac{\log(4mp)}{2m} \le \frac{\log(16\kappa\lambda)}{8\kappa\lambda} + \frac{\log p}{8\kappa\lambda p} \\
&\le \frac{\log(16\kappa\lambda)}{8\kappa\lambda} + \frac{\log 3}{24\kappa\lambda}.
\end{aligned} \tag{35}$$

For the fifth term we use $n \le mp/q + p, s \le 2pq - 1$, and $q \ge 2$ to bound

$$n + rs \le m\frac{p}{q} + p + r(2pq - 1) \le m\frac{p}{2} + 2pqr \le 2pqr \left(1 + \frac{m}{4pq}\right)$$

and thus

$$\begin{aligned}
\frac{2pq}{m} \log(n + rs) &\le \frac{2\log(2pqr)}{\kappa\lambda q} + \frac{2pq}{m} \log\left(1 + \frac{m}{4pq}\right) \\
&\le \frac{2\log(2pqr)}{\kappa\lambda q} + \frac{1}{\kappa\lambda} \log\left(1 + \frac{\kappa\lambda}{2}\right)
\end{aligned}$$

as $m/(pq) \ge \kappa\lambda q \ge 2\kappa\lambda$ and since $z \mapsto z^{-1} \log(1 + z/2)$ is decreasing on $(0, \infty)$. We deduce

$$\begin{aligned}
\frac{2pq}{m} \log(2(p+1)(q+1)r(n+rs)) \le {} &\frac{2}{\kappa\lambda}\left(\frac{\log(2(p+1)r)}{q} + \frac{\log(q+1)}{q}\right. \\
&\left. + \frac{\log(2pqr)}{q} + \frac{1}{2} \log\left(1 + \frac{\kappa\lambda}{2}\right)\right).
\end{aligned} \tag{36}$$

Below we will use the estimates $q^{-1}\log(q+1) \le (\log 3)/2$ and $q^{-1}\log(2q) \le \log 2$, which hold since $q \ge 2$.

If $q > p$, then $r = q$ and so $q^{-1}\log(2(p+1)r) \le 0.5\log(p+1) + \log 2$. We have $q^{-1}\log(2q^2) \le \log\sqrt{8}$. Hence the right-hand side of (36) is at most

$$\frac{2}{\kappa\lambda}\left(\log(p+1) + \log 2 + \frac{1}{2}\log 3 + \log\sqrt{8} + \frac{1}{2}\log\left(1 + \frac{\kappa\lambda}{2}\right)\right)$$

$$\le \frac{1}{\kappa\lambda}\left(9\log(p+1) + \log\left(1 + \frac{\kappa\lambda}{2}\right)\right). \tag{37}$$

Now suppose $p \ge q$, then $r = p \ge 2$. We bound $0.5\log(2(p+1)p) \le 0.5\log(2) + \log(p+1) \le 1.5\log(p+1)$, $q^{-1}\log(q+1) \le 0.5\log(p+1)$, and $q^{-1}\log(2pqr) \le 0.5\log(p^2) + \log 2 \le 2\log(p+1)$. In this case, the right-hand side of (36) is at most

$$\frac{2}{\kappa\lambda}\left(4\log(p+1) + \frac{1}{2}\log\left(1 + \frac{\kappa\lambda}{2}\right)\right) = \frac{1}{\kappa\lambda}\left(8\log(p+1) + \log\left(1 + \frac{\kappa\lambda}{2}\right)\right).$$

In both cases we find that the fifth term is bounded by (37). We insert this upper bound as well as (33)–(35) into (32) to find

$$h(y) \le \frac{p}{q}h(x) + p\left(\kappa + \frac{1}{\lambda} + \frac{4}{\kappa}\right)(hk)^{1/2} + \frac{1}{8\kappa\lambda}\left(\log(16\kappa\lambda) + \frac{\log 3}{3}\right)$$

$$+ \frac{1}{\kappa\lambda}\left(9\log(p+1) + \log\left(1 + \frac{\kappa\lambda}{2}\right)\right),$$

as required.

Lemma 8 *Let $P \in \overline{\mathbb{Q}}[X,Y]$ be irreducible with $p = \deg_X P \ge 1$ and $q = \deg_Y P \ge 1$. If $(x,y) \in \overline{\mathbb{Q}}^2$ with $P(x,y) = 0$ is not a regular zero of P, then*

$$\max\left\{\frac{h(x)}{q}, \frac{h(y)}{p}\right\} \le 2h_p(P) + 4\log((p+1)(q+1)). \tag{38}$$

Proof By symmetry we may assume $h(y)/p \ge h(x)/q$. Let $D \in \overline{\mathbb{Q}}[Y]$ be the resultant of the two polynomials $P, \partial P/\partial X \in \overline{\mathbb{Q}}(Y)[X]$ (cf. Chap. IV, §8 [10]). Then $D \ne 0$ because P is irreducible in $\overline{\mathbb{Q}}(Y)[X]$. The resultant D is the determinant of a $(2p-1) \times (2p-1)$ matrix whose entries, denoted here by m_{ij}, are polynomials in Y with degrees bounded by q. We find $\deg D \le (2p-1)q \le 2pq$ and, using the structure of the resultant matrix, that D is a sum of at most $(p+1)^{p-1}p^p \le (p+1)^{2p-1}$ products of the m_{ij}. If K is a number field containing the coefficients of the m_{ij} and $v \in M_K$, then by Lemma 1(i) we find

$$|D|_v \le \delta_v((p+1)^{2p-1}) \max_\sigma\{|m_{1,\sigma(1)} \cdots m_{2p-1,\sigma(2p-1)}|_v\} \tag{39}$$

where σ runs over all permutations of the first $2p - 1$ positive integers. The second bound of Lemma 1(i) applied to the univariate m_{ij} yields

$$|m_{1,\sigma(1)} \cdots m_{2p-1,\sigma(2p-1)}|_v \le \delta_v((q + 1)^{2p-2})|m_{1,\sigma}|_v \cdots |m_{2p-1,\sigma(2p-1)}|_v.$$

This inequality and $|m_{ij}|_v \le \delta_v(p)|P|_v$ inserted into (39) gives

$$|D|_v \le \delta_v((p + 1)^{2p-1}(q + 1)^{2p-2}p^{2p-1})|P|_v^{2p-1} \le \delta_v((p + 1)^{4p}(q + 1)^{2p})|P|_v^{2p-1}.$$

We take the sum over all places of K to find $h_p(D) \le 2ph_p(P) + 4p \log(p + 1) + 2p \log(q + 1)$.

Now if $(x, y) \in \overline{\mathbb{Q}}^2$ with $P(x, y) = 0$ is not a regular zero of P, then $D(y) = 0$. By Lemma 1(iii) we can bound $h(y) \le h_p(D) + \log \deg D \le h_p(D) + \log(2pq)$. The bound (38) follows from this inequality together with the bound for $h_p(D)$ and since $p^{-1} \log(2pq) \le \log(2q) \le 2 \log(q + 1)$.

Lemma 9 *Say $\kappa > 0$ and $\lambda > 0$ are real numbers with $\kappa\lambda \ge 4$. Let $P \in \overline{\mathbb{Q}}[X, Y]$ be irreducible with $p = \deg_X P \ge 1$, $q = \deg_Y P \ge 1$, and $\deg P = p + q$. For $(x, y) \in \overline{\mathbb{Q}}^2$ with $P(x, y) = 0$ we define k and h as in (25). If $k \ge \lambda^2 h$, then*

$$\left| \frac{h(x)}{q} - \frac{h(y)}{p} \right| \le \left(\kappa + \frac{1}{\lambda} + \frac{4}{\kappa} \right) (hk)^{1/2} + \frac{9}{8\kappa\lambda} \log(363\kappa\lambda). \tag{40}$$

Proof By symmetry we may suppose $h(y)/p \ge h(x)/q$. Let us assume that (x, y) is a regular zero of P. We divide the bound in Lemma 7 by p, use $p^{-1} \log(p+1) \le \log 2$, $1 + \kappa\lambda/2 \le \kappa\lambda$, and

$$\frac{\log(16\kappa\lambda)}{8p} + \frac{\log 3}{24p} + 9 \log 2 + \frac{\log(\kappa\lambda)}{p} \le \frac{9}{8} \log \left(2^{76/9}3^{1/27}\kappa\lambda \right) \le \frac{9}{8} \log(363\kappa\lambda)$$

to conclude (40).

The upper bound given by Lemma 8 is at most $4h \le 4\lambda^{-1}(hk)^{1/2}$. If (x, y) is not a regular zero, this is also an upper bound for $|h(x)/q - h(y)/p|$ so the current lemma follows as $4\lambda^{-1} \le \kappa$.

Lemma 10 *Let $P \in \overline{\mathbb{Q}}[X, Y]$ be irreducible with $p = \deg_X P \ge 1$ and $q = \deg_Y P$. Then there is a root of unity ξ such that the polynomial $\tilde{P} = X^p P(X^{-1} + \xi, Y)$ has total degree $p + q$, is irreducible in $\overline{\mathbb{Q}}[X, Y]$, and satisfies*

$$\deg_X \tilde{P} = p, \quad \deg_Y \tilde{P} = q, \quad h_p(\tilde{P}) \le p \log 2 + h_p(P).$$

Proof We may write $P = \sum_j a_j Y^j$ with $a_j = \sum_i a_{ij} X^i \in \overline{\mathbb{Q}}[X]$. By hypothesis we have $a_q \ne 0$ and a_q has degree at most p as a polynomial in X. We choose a root of unity ξ such that $a_q(\xi) \ne 0$. A direct computation using the irreducibility of P

shows that $\tilde{P}$ is irreducible. By construction $\deg_X \tilde{P} \le p$, $\deg_Y \tilde{P} \le q$ and therefore $\deg \tilde{P} \le p + q$.

Say $0 \le i \le p$ and $0 \le j \le q$, then the coefficient of $X^i Y^j$ in $\tilde{P}$ equals

$$\sum_{k=p-i}^{p} a_{kj} \binom{k}{p-i} \xi^{i-p+k}. \tag{41}$$

So if $i = p$ and $j = q$ we see that the coefficient of $X^p Y^q$ is $a_q(\xi) \ne 0$ and conclude that $\deg_X \tilde{P} = p$, $\deg_Y \tilde{P} = q$, and $\deg \tilde{P} = p + q$.

Now say K is a number field that contains ξ and all coefficients of P. If $v \in M_K$, then by (41) and standard facts on binomial coefficients we get

$$|\tilde{P}|_v \le \max_{0 \le i \le p} \delta_v \left(\sum_{k=p-i}^{p} \binom{k}{p-i} \right) |P|_v = \max_{0 \le i \le p} \delta_v \left(\binom{p+1}{p-i+1} \right) |P|_v$$

$$\le \delta_v(2^p) |P|_v.$$

We sum over the locals bounds to complete the proof.

Proof (of Theorem 1) By symmetry, we may suppose $p = \deg_X P \le \deg_Y P = q$.

First suppose $q = 1$. Then $p = 1$. Clearly, $|h(x) - h(y)| \le \max\{h(x), h(y)\}$ and Lemma 1(iii) implies $|h(x) - h(y)| \le \log(2) + h_p(P)$. So $|h(x) - h(y)|$ is bounded from above by the geometric mean

$$(\log(2) + h_p(P))^{1/2} \max\{h(x), h(y)\}^{1/2}$$

which is less than the bound in the assertion.

So we may assume $q \ge 2$, in particular $(p+1)(q+1) \ge 6$. We make the choice $\kappa = 2.25$ and $\lambda = 4.98$. Let $k = \max\{h(x)/q, h(y)/p\}$.

We let ξ and $\tilde{P}$ be as in Lemma 10.

Say (x, y) is as in the hypothesis. If $x = \xi$, then $h(x) = 0$ and $h(y) \le \log((p+1)q) + h_p(P)$ by Lemma 1(iii). Thus

$$\left| \frac{h(x)}{q} - \frac{h(y)}{p} \right| = \frac{h(y)}{p} \le \left(\log((p+1)q) + h_p(P) \right)^{1/2} \left(\frac{h(y)}{p} \right)^{1/2}.$$

This is better than our claim.

We now assume $x \ne \xi$ and define $\tilde{x} = (x - \xi)^{-1}$. Hence $\tilde{P}(\tilde{x}, y) = 0$. By basic height properties, cf. Lemma 1.5.18 and Proposition 1.5.15 [5], we find $h(\tilde{x}) = h(x - \xi)$ and $|h(\tilde{x}) - h(x)| \le \log 2$. Thus

$$\left| \frac{h(x)}{q} - \frac{h(y)}{p} \right| \le \left| \frac{h(\tilde{x})}{q} - \frac{h(y)}{p} \right| + \frac{\log 2}{q} \le \left| \frac{h(\tilde{x})}{q} - \frac{h(y)}{p} \right| + \frac{\log 2}{2}. \tag{42}$$

First, we suppose $k - (\log 2)/q \geq \lambda^2 \tilde{h}$ with $\tilde{h} = \log((p+1)(q+1)) + h_p(\tilde{P})$. Then $\max\{h(\tilde{x})/q, h(y)/p\} \geq \lambda^2 \tilde{h}$. Lemma 9 applied to $\tilde{P}$ and (42) imply

$$\left| \frac{h(x)}{q} - \frac{h(y)}{p} \right| \leq \left(\kappa + \frac{1}{\lambda} + \frac{4}{\kappa} \right) (\tilde{h}\tilde{k})^{1/2} + \frac{9}{8\kappa\lambda} \log(363\kappa\lambda) + \frac{\log 2}{2}$$

$$\leq 4.229(\tilde{h}\tilde{k})^{1/2} + 1.181$$

with our choice of κ and λ and with $\tilde{k} = \max\{h(\tilde{x})/q, h(y)/p\}$. Now

$$\tilde{h} \leq \log((p+1)(q+1)) + p\log 2 + h_p(P) = \log(2^p(p+1)(q+1)) + h_p(P). \tag{43}$$

We have $k \geq \lambda^2 \tilde{h} \geq \lambda^2 \log 6 > 1$ and therefore

$$\tilde{k} \leq k + \frac{\log 2}{q} \leq k + \frac{\log 2}{2} \leq k + \frac{\log 2}{2\lambda^2 \log 6} k \leq 1.008k.$$

We conclude $4.229(\tilde{h}\tilde{k})^{1/2} + 1.181 \leq 4.246(\log(2^p(p+1)(q+1)) + h_p(P))^{1/2}k^{1/2} + 1.181$. Observe that $\log(2^p(p+1)(q+1)) \geq \log 12$. The theorem follows in this case as $k \geq 1$ and $4.246 + 1.181(\log 12)^{-1/2} < 5$.

Second, we must treat the case $k - (\log 2)/q < \lambda^2 \tilde{h}$. The bound (43) continues to hold and we find

$$\left| \frac{h(x)}{q} - \frac{h(y)}{p} \right| \leq \max\left\{ \frac{h(x)}{q}, \frac{h(y)}{p} \right\} = k \leq \left(\frac{\log 2}{2} + \lambda^2 \tilde{h} \right)^{1/2} k^{1/2}$$

$$\leq \left(\frac{\log 2}{2} + \lambda^2 \log(2^p(p+1)(q+1)) + \lambda^2 h_p(P) \right)^{1/2} k^{1/2}$$

$$\leq \left(\frac{\log 2}{2\log 12} + \lambda^2 \right)^{1/2} \left(\log(2^p(p+1)(q+1)) + h_p(P) \right)^{1/2} k^{1/2}$$

where we used $\log(2^p(p+1)(q+1)) \geq \log 12$ again. This is better than our claim as $(\log(2)/(2\log 12) + \lambda^2)^{1/2} < 5$.

Corollary 1 is easy to prove using Theorem 1. We set $k = \max\{h(x)/q, h(y)/p\}$. Say $P(x, y) = 0$ with $x, y \in \overline{\mathbb{Q}}$ and $h(y) > 2\frac{p}{q}h(x)$. Then

$$\frac{k}{2} \leq \max\left\{ \frac{h(x)}{q}, \frac{h(y)}{2p} \right\} = \frac{h(y)}{2p} < \frac{h(y)}{p} - \frac{h(x)}{q}.$$

By Theorem 1 we have

$$\frac{k}{2} < 5(\log(2^{\min\{p,q\}}(p+1)(q+1)) + h_p(P))^{1/2}k^{1/2}$$

and so

$$k^{1/2} < 10 \left(\log(2^{\min\{p,q\}}(p+1)(q+1)) + h_p(P) \right)^{1/2}.$$

This contradicts the bound in (2).

5 On a Theorem of Runge

In this section we will prove Theorem 2.

Let $P = \sum_{i,j} p_{ij} X^i Y^j \in \mathbb{Z}[X,Y]$ be irreducible in $\overline{\mathbb{Q}}[X,Y]$ with $d = \deg P = \deg_X P = \deg_Y P$. Furthermore, since Theorem 2 applies only to polynomials of degree at least 2, we assume $d \geq 2$. We use $P_d = \sum_{i+j=d} p_{ij} X^i Y^j$ to denote the homogeneous part of degree d. Finally, we can factor P_d as $p_{d0} \prod_{s=1}^{d}(X - t_s Y)$ with $t_s \in \overline{\mathbb{Q}}^{\times}$ for each $s \in \{1,\ldots,d\}$.

Lemma 11 *Set $t = t_s$ for some $s \in \{1,\ldots,d\}$. We have $h(t) \leq h_p(P) + \log d$. Furthermore, if $R_t(X,Z) = P(X, t^{-1}(X-Z)) \in \overline{\mathbb{Q}}[X,Z]$, then $\deg_Z R = d$, $\deg_X R \leq d - 1$, and $h_p(R_t) \leq d(\log(2d) + 2h_p(P))$.*

Proof The bound for $h(t)$ follows from Lemma 1(iii); indeed t is a zero of $P_d(X,1) \in \overline{\mathbb{Q}}[X]$ whose coefficients are coefficients of P.

Let $i,j \geq 0$ be integers, the coefficient of $X^i Z^j$ in R_t is

$$(-1)^j \sum_{k=0}^{d-j} \binom{j+k}{k} p_{i-k,j+k} t^{-j-k}.$$

Clearly, $\deg_Z R_t \leq d$. The coefficient of Z^d is nonzero and that of $X^i Z^j$ is zero provided $i \geq d$. The lemma now follows from local inequalities as in the proof of Lemma 10 together with basic height properties.

If $P(x,y) = 0$ with x, y algebraic numbers, then $R_t(x, x - ty) = 0$.

Lemma 12 *Let $P(x,y) = 0$ with $x, y \in \mathbb{Q}$ and*

$$h(x) \geq 100d \left(\log \left((4d)^d d(d+1) \right) + 2dh_p(P) \right). \tag{44}$$

Set $t = t_s$ for some $s \in \{1,\ldots,d\}$, there exists an embedding $\sigma : \mathbb{Q}(t) \to \mathbb{C}$ such that

$$\log \max\{1, |x - \sigma(t)y|\} \leq \frac{d-1}{d} h(x) + 10.4d(\log(2d) + h_p(P))^{1/2} h(x)^{1/2}.$$

Proof Let $R = R_t$ be as in Lemma 11, then R is irreducible. Now $\deg_X R \geq 1$, indeed if $\deg_X R = 0$, then by construction P has degree 1, which contradicts our assumption $d \geq 2$. Let $z = x - ty$, so $R(x, z) = 0$.

We recall that $\deg_X R \leq d - 1$ and $\deg_z R = d$. So

$$\log(2^{\deg_X R}(\deg_X R + 1)(\deg_z R + 1)) + h_p(R) \leq \log\left((4d)^d d(d + 1)\right) + 2dh_p(P) \tag{45}$$

by Lemma 11. Our hypothesis (44) together with Corollary 1 yields $h(z) \leq 2(\deg_X R)h(x)/d$ and hence

$$\max\left\{\frac{h(x)}{d}, \frac{h(z)}{\deg_X R}\right\} \leq 2\frac{h(x)}{d}. \tag{46}$$

Next we apply Theorem 1 to R and use (46) as well as (45) to find

$$\frac{h(z)}{\deg_X R} \leq \frac{h(x)}{d} + \frac{5\sqrt{2}}{d^{1/2}}\left(\log\left((4d)^d d(d + 1)\right) + 2dh_p(P)\right)^{1/2} h(x)^{1/2}.$$

We multiply with $\deg_X R \leq d - 1$ to obtain

$$h(z) \leq \frac{d - 1}{d}h(x) + 10d\left(\frac{1}{2d}\log\left((4d)^d d(d + 1)\right) + h_p(P)\right)^{1/2} h(x)^{1/2}.$$

As $d \geq 2$ we find $(2d)^{-1}\log\left((4d)^d d(d + 1)\right) \leq 1.08\log(2d)$ and this yields

$$h(z) \leq \frac{d - 1}{d}h(x) + 10.4d\left(\log(2d) + h_p(P)\right)^{1/2} h(x)^{1/2}.$$

By the definition (3) of the height there exists an embedding $\sigma : \mathbb{Q}(t) \to \mathbb{C}$ with $\log\max\{1, |x - \sigma(t)y|\} \leq h(x - \sigma(t)y) = h(z)$. This concludes the proof.

We now prove Theorem 2:

Let P be as in the hypothesis, so $d = \deg_X P \geq 2$ and there are non-conjugated zeros $t', t'' \in \overline{\mathbb{Q}}^\times$ of $P_d(X, 1)$. Suppose $x, y \in \mathbb{Z}$ with $P(x, y) = 0$. By symmetry we may assume $|y| \leq |x|$ and also $x \neq 0$.

To prove this theorem, we may assume $h(x) = \log|x| \geq 100d^4 h$ with $h = \log(2d) + h_p(P)$. Using $d \geq 2$ we find

$$h \geq \frac{\log(4d)}{d^2} + \frac{\log(d(d + 1))}{d^3} + \frac{2}{d^2}h_p(P)$$

and this implies

$$h(x) \geq 100d^4 h \geq 100d\left(\log\left((4d)^d d(d + 1)\right) + 2dh_p(P)\right),$$

the hypothesis of Lemma 12.

Let σ', σ'' be the embeddings given by Lemma 12 applied to t', t'' respectively, then $\sigma'(t') \neq \sigma''(t'')$. For brevity we define $\xi = x - \sigma'(t')y$ and $\eta = x - \sigma''(t'')y$; we eliminate y to get

$$x = \frac{\xi \sigma''(t'') - \eta \sigma'(t')}{\sigma''(t'') - \sigma'(t')}.$$

So

$$|x| \leq 2 \max\{|\xi|, |\eta|\} \max\{|\sigma'(t')|, |\sigma''(t'')|\} |\sigma'(t') - \sigma''(t'')|^{-1} \tag{47}$$

for the complex absolute value $|\cdot|$.

We will bound $|\xi|$ and $|\eta|$ using Lemma 12, but first we bound the remaining absolute values in (47) using height inequalities. If $\alpha \in \overline{\mathbb{Q}}$, then $\log \max\{1, |\alpha|_v\} \leq [\mathbb{Q}(\alpha) : \mathbb{Q}]h(\alpha)$ for any $v \in M_{\mathbb{Q}(\alpha)}$. This inequality follows immediately from the definition of the height. Since, for example, $[\mathbb{Q}(t') : \mathbb{Q}] \leq d$ and $[\mathbb{Q}(t', t'') : \mathbb{Q}] \leq d^2$ we deduce from (47) that

$$\log|x| \leq \log 2 + d \max\{h(t'), h(t'')\} + d^2 h(\sigma'(t') - \sigma''(t'')) + \log \max\{1, |\xi|, |\eta|\}. \tag{48}$$

Lemma 11 implies $\max\{h(t'), h(t'')\} \leq h_p(P) + \log d$. Next we use the bounds for $|\xi|$ and $|\eta|$ from Lemma 12. Together with (48) and standard height inequalities we have

$$\log|x| \leq \log 2 + d(h_p(P) + \log d) + d^2(2h_p(P) + 2\log d + \log 2)$$

$$+ \frac{d-1}{d}h(x) + 10.4dh^{1/2}h(x)^{1/2}$$

$$\leq \frac{d-1}{d}h(x) + 3d^2h + 10.4dh^{1/2}h(x)^{1/2}.$$

As $h(x) = \log|x|$ we find

$$\log|x| \leq 3d^3h + 10.4d^2h^{1/2}(\log|x|)^{1/2}.$$

If $B, C, T \geq 0$ with $T \leq C + B\sqrt{T}$, then

$$T \leq \frac{1}{4}\left((B^2 + 4C)^{1/2} + B\right)^2.$$

We apply this inequality with $B = 10.4d^2h^{1/2}$, $C = 3d^3h$, and $T = \log|x|$ to conclude that

$$\log|x| \leq \frac{1}{4}\left((10.4^2 + 12)^{1/2}d^2h^{1/2} + 10.4d^2h^{1/2}\right)^2 \leq 115d^4h.$$

The upper bound for $\log|y|$ in (2) follows since $|y| \leq |x|$.

References

1. M. Abouzaid, Heights and logarithmic GCD on algebraic curves. Int. J. Number Theory **4**(2), 177–197 (2008)
2. B. Bartolome, The Skolem-Abouzaïd theorem in the singular case. Atti Accad. Naz. Lincei Rend. Lincei Mat. Appl. **26**(3), 263–289 (2015)
3. Y.F. Bilu, D. Masser, A quick proof of Sprindzhuk's decomposition theorem, in *More Sets, Graphs and Numbers*. Bolyai Society Mathematical Studies, vol. 15 (Springer/Berlin, 2006), pp. 25–32
4. E. Bombieri, On Weil's "théorème de décomposition". Am. J. Math. **105**(2), 295–308 (1983)
5. E. Bombieri, W. Gubler, *Heights in Diophantine Geometry* (Cambridge University Press, Cambridge, 2006)
6. C. Chevalley, *Introduction to the Theory of Algebraic Functions of One Variable*. Mathematical Surveys, vol. VI (American Mathematical Society, New York, 1951)
7. S. David, P. Philippon, Minorations des hauteurs normalisées des sous-variétés des tores. Ann. Scuola Norm. Sup. Pisa Cl. Sci. (4) **28**(3), 489–543 (1999)
8. D.L. Hilliker, E.G. Straus, Determination of bounds for the solutions to those binary Diophantine equations that satisfy the hypotheses of Runge's theorem. Trans. Am. Math. Soc. **280**(2), 637–657 (1983)
9. M. Hindry, J.H. Silverman, *Diophantine Geometry An Introduction* (Springer, Berlin, 2000)
10. S. Lang, *Algebra*. Graduate Texts in Mathematics, vol. 211, 3d edn. (Springer, New York, 2002)
11. A. Néron, *Quasi-fonctions et hauteurs sur les variétés abéliennes*. Ann. Math. **82**(2), 249–331 (1965)
12. D. Roy, J.L. Thunder, *An absolute Siegel's lemma*. J. Reine Angew. Math. **476**, 1–26 (1996)
13. D. Roy, J.L. Thunder, Addendum and erratum to: "An absolute Siegel's lemma". J. Reine Angew. Math. **508**, 47–51 (1999)
14. C. Runge, Ueber ganzzahlige Lösungen von Gleichungen zwischen zwei Veränderlichen. J. Reine Angew. Math. **100**, 425–435 (1887)
15. P.G. Walsh, A quantitative version of Runge's theorem on Diophantine equations. Acta Arith. **62**(2), 157–172 (1992)
16. P.G. Walsh, Corrections to: "A quantitative version of Runge's theorem on Diophantine equations". Acta Arith. **73**(4), 397–398 (1995)
17. S. Zhang, Positive line bundles on arithmetic varieties. J. Am. Math. Soc. **8**(1), 187–221 (1995)

On the Monoid Generated by a Lucas Sequence

Clemens Heuberger and Stephan Wagner

Dedicated to Prof. Robert Tichy on the occasion of his 60th birthday

Abstract A Lucas sequence is a sequence of the general form $v_n = (\phi^n - \overline{\phi}^n)/(\phi - \overline{\phi})$, where ϕ and $\overline{\phi}$ are real algebraic integers such that $\phi + \overline{\phi}$ and $\phi\overline{\phi}$ are both rational. Famous examples include the Fibonacci numbers, the Pell numbers, and the Mersenne numbers. We study the monoid that is generated by such a sequence; as it turns out, it is almost freely generated. We provide an asymptotic formula for the number of positive integers $\leq x$ in this monoid, and also prove Erdős–Kac type theorems for the distribution of the number of factors, with and without multiplicity. While the limiting distribution is Gaussian if only distinct factors are counted, this is no longer the case when multiplicities are taken into account.

2010 *Mathematics Subject Classification* 11N37; 11B39

1 Introduction

Let ϕ and $\overline{\phi}$ be real algebraic integers such that $\phi + \overline{\phi}$ and $\phi\overline{\phi}$ are fixed non-zero coprime rational integers with $\phi > |\overline{\phi}|$. The *Lucas numbers* associated with $(\phi, \overline{\phi})$ are

$$v_n = v_n(\phi, \overline{\phi}) = \frac{\phi^n - \overline{\phi}^n}{\phi - \overline{\phi}}, \qquad n = 1, 2, 3, \ldots.$$

C. Heuberger (✉)
Institut für Mathematik, Alpen-Adria-Universität Klagenfurt, Universitätsstrasse 65–67,
9020 Klagenfurt, Austria
e-mail: clemens.heuberger@aau.at

S. Wagner
Department of Mathematical Sciences, Stellenbosch University, 7602 Stellenbosch, South Africa
e-mail: swagner@sun.ac.za

© Springer International Publishing AG 2017

C. Elsholtz, P. Grabner (eds.), *Number Theory – Diophantine Problems, Uniform Distribution and Applications*, DOI 10.1007/978-3-319-55357-3_14

All these numbers are positive integers, and the sequence is strictly increasing for $n \geq 2$. Famous examples include the Fibonacci numbers ($\phi = \frac{1+\sqrt{5}}{2}, \overline{\phi} = \frac{1-\sqrt{5}}{2}$), the Pell numbers ($\phi = 1 + \sqrt{2}, \overline{\phi} = 1 - \sqrt{2}$), and the Mersenne numbers ($\phi = 2$, $\overline{\phi} = 1$). The multiplicative group generated by such a sequence was studied in a recent paper by Luca et al. [15]; by definition, it consists of all quotients of products of elements of the sequence. In [15], a near-asymptotic formula for the number of integers in this group was given (in the specific case of the Fibonacci sequence, but the method is more general). This continued earlier work of Luca and Porubský [14], who considered the multiplicative group generated by a Lehmer sequence and showed that the number of integers below x in the resulting group is $O(x/(\log x)^\delta)$ for any positive number δ.

As it turns out, the group that is generated by a Lucas sequence is almost a free group; this is due to the existence of *primitive divisors*. A prime number p is a primitive divisor of $v_n(\phi, \overline{\phi})$ if p divides v_n but does not divide $v_1 \ldots v_{n-1}$. It is a classical result, due to Carmichael, that almost all elements of a Lucas sequence have primitive divisors:

Theorem (Carmichael [3, Theorem XXIII]) *If $n \notin \{1, 2, 6, 12\}$, then $v_n(\phi, \overline{\phi})$ has a primitive divisor.*

See also Bilu et al. [2]. Note the slightly different definition in [2] which does not allow a primitive divisor to divide $(\phi - \overline{\phi})^2$.

Let now

$$\mathcal{F} = \{v_n(\phi, \overline{\phi}) \mid v_n(\phi, \overline{\phi}) \text{ has a primitive divisor}\}.$$

By Carmichael's theorem, $\mathcal{F}$ includes all but finitely many v_n. We let $\mathcal{F}_0$ be the set of all $v_n(\phi, \overline{\phi})$ with $n \leq 12$ that have a primitive divisor, so that

$$\mathcal{F} = \mathcal{F}_0 \cup \{v_n(\phi, \overline{\phi}) \mid n \geq 13\}.$$

For example, in the case of the golden ratio $\phi = (1 + \sqrt{5})/2, \overline{\phi} = (1 - \sqrt{5})/2$, we have

$$\mathcal{F}_0 = \{2, 3, 5, 13, 21, 34, 55, 89\}$$

and

$$\mathcal{F} = \{2, 3, 5, 13, 21, 34, 55, 89, 233, 377, \ldots\},$$

which are all the Fibonacci numbers except 1, 8, and 144.

Instead of the full group that was studied in [14] and [15], we consider the free monoid

$$\mathcal{M}(\mathcal{F}) = \{m_1 \ldots m_k \mid k \geq 0, m_j \in \mathcal{F}\}$$

generated by $\mathcal{F}$. It is an easy consequence of the existence of primitive divisors that every element of $\mathcal{M}(\mathcal{F})$ has a *unique* factorisation into elements of $\mathcal{F}$: If an element has two factorisations, consider a primitive divisor of the largest element of $\mathcal{F}$ occurring in either of those factorisations. This prime number has to occur in all factorisations, so the largest element of $\mathcal{F}$ occurring in any factorisation of the element is fixed. The result follows by induction.

To give a concrete example, the elements of our monoid are

$$\mathcal{M}(\mathcal{F}) = \{1, 2, 3, 4, 5, 6, 8, 9, 10, 12, 13, 15, 16, 18, 20, 21, 24, 25, 26, 27, 30, 32, 34, \ldots\}$$

in the case of the Fibonacci numbers (cf. [18, A065108]).

Our first result describes the asymptotic number of elements in $\mathcal{M}(\mathcal{F})$ up to a given bound, paralleling the aforementioned result of [15], but even being more precise. All constants, implicit constants in O-terms, and the Vinogradov notation will depend on ϕ and $\overline{\phi}$.

Theorem 1 *We have*

$$|\mathcal{M}(\mathcal{F}) \cap [1, x]| = k_0 (\log x)^{k_1} \exp\left(\pi \sqrt{\frac{2 \log x}{3 \log \phi}} \right) \left(1 + O\left(\frac{1}{(\log x)^{1/10}} \right) \right)$$

for $x \to \infty$ and suitable constants k_0 and k_1. Specifically,

$$k_1 = \frac{|\mathcal{F}_0| - 13}{2} + \frac{\log(\phi - \overline{\phi})}{2 \log \phi}.$$

Remark An explicit expression for k_0 can be given as well, but it is rather unwieldy.

Since all elements of $\mathcal{M}(\mathcal{F})$ have a unique factorisation into elements of $\mathcal{F}$, it makes sense to consider the number of factors in this factorisation and to study its distribution. The celebrated Erdős–Kac theorem [6] states that the number of distinct prime factors of a randomly chosen integer in $[1, x]$ is asymptotically normally distributed. The same is true if primes are counted with multiplicity. We refer to Chap. 12 of [5] for a detailed discussion of the Erdős–Kac theorem and its generalisations.

Our aim is to prove similar statements for the monoid $\mathcal{M}(\mathcal{F})$. Let n be an element of this monoid. By $\omega_{\mathcal{F}}(n)$ and $\Omega_{\mathcal{F}}(n)$, we denote the number of factors in the factorisation of n into elements of $\mathcal{F}$ without and with multiplicities, respectively.

We first prove asymptotic normality for $\omega_{\mathcal{F}}$, in complete analogy with the Erdős–Kac theorem:

Theorem 2 *Let N be a uniformly random positive integer in $\mathcal{M}(\mathcal{F}) \cap [1, x]$ and let*

$$a_1 = \frac{1}{\pi} \sqrt{\frac{6}{\log \phi}}, \qquad a_2 = \frac{\pi^2 - 6}{2\pi^3} \sqrt{\frac{6}{\log \phi}}.$$

The random variable $\omega_{\mathcal{F}}(N)$ is asymptotically normal: we have

$$\lim_{x \to \infty} \mathbb{P}\left(\frac{\omega_{\mathcal{F}}(N) - a_1 \log^{1/2} x}{\sqrt{a_2} \log^{1/4} x} \le z \right) = \frac{1}{\sqrt{2\pi}} \int_{-\infty}^{z} e^{-y^2/2}\, dy.$$

However, the situation changes when multiplicities are taken into account: unlike the arithmetic function Ω, which counts all prime factors with multiplicity, $\Omega_{\mathcal{F}}$ is not normally distributed. Its limiting distribution can rather be described as a sum of shifted exponential random variables, as the following theorem shows:

Theorem 3 *Let N be a uniformly random positive integer in $\mathcal{M}(\mathcal{F}) \cap [1, x]$, let a_1 be the same constant as in the previous theorem and, with γ denoting the Euler–Mascheroni constant,*

$$b_1 = \frac{\sqrt{6 \log \phi}}{\pi} \left(\frac{2\gamma - \log(\pi^2 \log \phi / 6)}{2 \log \phi} + \sum_{m \in \mathcal{F}_0} \frac{1}{\log m} + \frac{1}{\log v_{13}(\phi, \overline{\phi})} \right.$$

$$\left. + \sum_{k \ge 1} \left(\frac{1}{\log v_{k+13}(\phi, \overline{\phi})} - \frac{1}{k \log \phi} \right) \right),$$

$$b_2 = \frac{\sqrt{6 \log \phi}}{\pi}.$$

The random variable $\Omega_{\mathcal{F}}(N)$, suitably normalised, converges weakly to a sum of shifted exponentially distributed random variables:

$$\frac{\Omega_{\mathcal{F}}(N) - \frac{a_1}{2} \log^{1/2} x \log \log x - b_1 \log^{1/2} x}{b_2 \log^{1/2} x} \xrightarrow{(d)} \sum_{m \in \mathcal{F}} \left(X_m - \frac{1}{\log m} \right),$$

where $X_m \sim \mathrm{Exp}(\log m)$.

This is somewhat reminiscent of the situation for the number of parts in a *partition*: Goh and Schmutz [11] proved that the number of distinct parts in a random partition of a large integer n is asymptotically normally distributed, while the total number of parts with multiplicity was shown by Erdős and Lehner [7] to follow a Gumbel distribution (which can also be represented as a sum of shifted exponential random variables). Indeed, our results are multiplicative analogues in a certain sense, since products turn into sums upon applying the logarithm, and $\log v_n(\phi, \overline{\phi}) \sim (\log \phi)n$.

We remark that the monoid $\mathcal{M}(\mathcal{F})$ fits the definition of an *arithmetical semigroup* as studied in abstract analytic number theory (see [12] for a general reference on the subject). However, as Theorem 1 shows, it is very sparse, so it does not satisfy the growth conditions that are typically imposed in this context. For arithmetical semigroups that satisfy such growth conditions, Erdős–Kac type theorems are known as well, see [13, Theorem 7.6.5] and [19, Theorem 3.1].

2 Proof of Theorems 1 and 2

For real u in a neighbourhood of 1, we consider the Dirichlet generating function $d(z, u)$ that is defined as follows:

$$d(z, u) := \sum_{n \in \mathcal{M}(\mathcal{F})} \frac{u^{\omega_{\mathcal{F}}(n)}}{n^z},$$

for all complex z for which the series converges (Lemma 2 will provide detailed information on convergence). Within the region of convergence, we have the product representation

$$d(z, u) = \prod_{m \in \mathcal{F}} (1 + um^{-z} + um^{-2z} + \cdots) = \prod_{m \in \mathcal{F}} \left(1 + \frac{um^{-z}}{1 - m^{-z}}\right). \tag{1}$$

Set $h(n) = u^{\omega_{\mathcal{F}}(n)}$ if $n \in \mathcal{M}(\mathcal{F})$ and $h(n) = 0$ otherwise. We use the Mellin–Perron summation formula in the version

$$\sum_{1 \le n \le x} h(n)\left(1 - \frac{n}{x}\right) = \frac{1}{2\pi i} \int_{r-i\infty}^{r+i\infty} \left(\sum_{n=1}^{\infty} \frac{h(n)}{n^z}\right) x^z \frac{dz}{z(z+1)}, \tag{2}$$

for $x > 0$ where r is in the half-plane of absolute convergence of the Dirichlet series $\sum_{n=1}^{\infty} \frac{h(n)}{n^s}$ (see, e.g., [1, Chap. 13] or [9, Theorem 2.1]). Thus

$$I_{\omega_{\mathcal{F}}}(x, u) := \sum_{\substack{n \in \mathcal{M}(\mathcal{F}) \\ n \le x}} u^{\omega_{\mathcal{F}}(n)}\left(1 - \frac{n}{x}\right) = \frac{1}{2\pi i} \int_{r-i\infty}^{r+i\infty} \frac{d(z, u)}{z(z+1)} x^z \, dz. \tag{3}$$

We will use a saddle-point approach to evaluate this integral. As a first step, we establish an estimate for $d(z, u)$ for $z = r + it$, $r \to 0^+$ and small t. We start with an auxiliary lemma for another Dirichlet generating series.

Lemma 1 *Let v be a complex number such that $|v| < v_0$, where*

$$v_0 := \min \left\{ \log m \;\Big|\; m \in \mathcal{F}_0 \; or \; m = \frac{\phi^{13}}{\phi - \bar{\phi}} \right\} \tag{4}$$

and let $\Lambda(s, v)$ be given by the Dirichlet series

$$\Lambda(s, v) := \sum_{m \in \mathcal{F}} \frac{1}{(\log m - v)^s},$$

which converges for $\Re s > 1$. The function $\Lambda(s, v)$ can be analytically continued to a meromorphic function in s with a single simple pole at $s = 1$ with residue $1/\log\phi$.

We have

$$\Lambda(0, v) = |\mathcal{F}_0| - \frac{25}{2} + \frac{\log(\phi - \bar{\phi}) + v}{\log\phi}, \tag{5}$$

$$\left.\frac{\partial\Lambda(s, v)}{\partial s}\right|_{s=0} = -\sum_{m\in\mathcal{F}}\left(\log\left(1 - \frac{v}{\log m}\right) + \frac{v}{\log m}\right) + \kappa_1 v + \kappa_2,$$

where κ_1 and κ_2 are constants, and κ_1 is given by

$$\kappa_1 := \frac{\gamma - \log\log\phi}{\log\phi} + \sum_{m\in\mathcal{F}_0}\frac{1}{\log m} + \frac{1}{\log v_{13}(\phi, \bar{\phi})} + \sum_{k\geq 1}\left(\frac{1}{\log v_{k+13}(\phi, \bar{\phi})} - \frac{1}{k\log\phi}\right). \tag{6}$$

Finally,

$$\Lambda(s, v) = O(s^2)$$

for $-3/2 \leq \Re s \leq 2$ with $|s - 1| \geq 1$ and $|s| \geq 1$.

Proof Let

$$\Lambda_0(s, v) := \sum_{m\in\mathcal{F}_0}\frac{1}{(\log m - v)^s},$$

$$\Lambda_1(s, v) := \sum_{j\geq 13}\left(\left(\log\frac{\phi^j - \bar{\phi}^j}{\phi - \bar{\phi}} - v\right)^{-s} - \left(\log\frac{\phi^j}{\phi - \bar{\phi}} - v\right)^{-s}\right),$$

$$\alpha(v) := 13 - \frac{\log(\phi - \bar{\phi}) + v}{\log\phi}.$$

Note that our choice of v_0 guarantees that each summand of Λ, Λ_0, and Λ_1 is well-defined and that $\Re\alpha(v) > 0$.

Thus

$$\Lambda(s, v) = \Lambda_0(s, v) + \sum_{j\geq 13}\frac{1}{(j\log\phi - \log(\phi - \bar{\phi}) - v)^s} + \Lambda_1(s, v)$$

$$= \frac{1}{\log^s\phi}\zeta(s, \alpha(v)) + \Lambda_0(s, v) + \Lambda_1(s, v),$$

where $\zeta(s, \beta) = \sum_{k\geq 0}(k + \beta)^s$ denotes the Hurwitz zeta function (the series converges for $\Re s > 1$ if $\beta \notin \{0, -1, -2, \ldots\}$, and it can be analytically continued), cf. [16, 25.11.1].

The function $\Lambda_0(s, v)$ is obviously an entire function, and it is bounded for $\Re s \geq -3/2$.

Estimating the difference occurring in $\Lambda_1(s, v)$, we see that

$$\Lambda_1(s, v) = O\left(\sum_{j \geq 13} \frac{s(|\overline{\phi}|/\phi)^j}{(j \log \phi - \log(\phi - \overline{\phi}) - v)^{\Re s + 1}} \right).$$

Thus $\Lambda_1(s, v)$ is an entire function, and we have the estimate $\Lambda_1(s, v) = O(s)$ for $\Re s \geq -3/2$.

Moreover, $\zeta(s, \alpha)$ is a meromorphic function with a single simple pole at $s = 1$ with residue 1, and by Whittaker and Watson [20, §13.51], we have

$$\zeta(s, \alpha(v)) = O(s^2)$$

in the given area, which proves the desired asymptotic estimate. It remains to determine the values of the function and its derivative at $s = 0$.

Observe first that $\Lambda_0(0, v) = |\mathcal{F}_0|$ and $\Lambda_1(0, v) = 0$. Moreover, we have

$$\zeta(0, \alpha(v)) = \frac{1}{2} - \alpha(v),$$

cf. [16, 25.11.13]. Now we consider the first derivative. Λ_0 and Λ_1 are simply differentiated term by term:

$$\left.\frac{\partial \Lambda_0(s, v)}{\partial s}\right|_{s=0} = - \sum_{m \in \mathcal{F}_0} \log(\log m - v) = - \sum_{m \in \mathcal{F}_0} \left(\log \log m + \log \left(1 - \frac{v}{\log m}\right) \right)$$

and

$$\left.\frac{\partial \Lambda_1(s, v)}{\partial s}\right|_{s=0} = - \sum_{j \geq 13} \left(\log \left(\log \frac{\phi^j - \overline{\phi}^j}{\phi - \overline{\phi}} - v \right) - \log \left(\log \frac{\phi^j}{\phi - \overline{\phi}} - v \right) \right)$$

$$= - \sum_{j \geq 13} \left(\log \left(\log v_j(\phi, \overline{\phi}) - v \right) - \log \log \phi - \log(j - 13 + \alpha(v)) \right).$$

Finally, it is well known (see [16, 25.11.18]) that

$$\left.\frac{\partial \zeta(s, \alpha)}{\partial s}\right|_{s=0} = \log \Gamma(\alpha) - \frac{1}{2} \log(2\pi),$$

hence

$$\left.\frac{\partial}{\partial s} \frac{1}{\log^s \phi} \zeta(s, \alpha(v))\right|_{s=0} = - \log \log \phi \left(\frac{1}{2} - \alpha(v) \right) + \log \Gamma(\alpha(v)) - \frac{1}{2} \log(2\pi).$$

Now we make use of the product representation of the Gamma function, which yields

$$\frac{\partial}{\partial s}\frac{1}{\log^s \phi}\zeta(s,\alpha(v))\Big|_{s=0} = -\log\log\phi\left(\frac{1}{2}-\alpha(v)\right)-\frac{1}{2}\log(2\pi)-\gamma\alpha(v)-\log\alpha(v)$$
$$-\sum_{k\geq 1}\left(\log(k+\alpha(v))-\log k-\frac{\alpha(v)}{k}\right).$$

The infinite sum can be combined with the sum in the derivative of Λ_1 to give us

$$\frac{\partial\Lambda(s,v)}{\partial s}\Big|_{s=0} = -\sum_{m\in\mathcal{F}_0}\left(\log\log m + \log\left(1-\frac{v}{\log m}\right)\right)-\log\log\phi\left(\frac{1}{2}-\alpha(v)\right)$$
$$-\frac{1}{2}\log(2\pi)-\gamma\alpha(v)-\log\left(1-\frac{v}{\log v_{13}(\phi,\overline{\phi})}\right)$$
$$-\log\log v_{13}(\phi,\overline{\phi})+\log\log\phi$$
$$-\sum_{k\geq 1}\left(\log(\log v_{k+13}(\phi,\overline{\phi})-v)-\log\log\phi-\log k-\frac{\alpha(v)}{k}\right),$$

which can finally be rewritten as

$$\frac{\partial\Lambda(s,v)}{\partial s}\Big|_{s=0} = -\sum_{m\in\mathcal{F}}\left(\log\left(1-\frac{v}{\log m}\right)+\frac{v}{\log m}\right)$$
$$-\sum_{m\in\mathcal{F}_0}\log\log m + \sum_{m\in\mathcal{F}_0}\frac{v}{\log m}+\log\log\phi\left(\frac{27}{2}-\frac{\log(\phi-\overline{\phi})+v}{\log\phi}\right)$$
$$-\frac{1}{2}\log(2\pi)-\gamma\left(13-\frac{\log(\phi-\overline{\phi})+v}{\log\phi}\right)-\log\log v_{13}(\phi,\overline{\phi})$$
$$+\frac{v}{\log v_{13}(\phi,\overline{\phi})}+v\sum_{k\geq 1}\left(\frac{1}{\log v_{k+13}(\phi,\overline{\phi})}-\frac{1}{k\log\phi}\right)$$
$$-\sum_{k\geq 1}\left(\log\log v_{k+13}(\phi,\overline{\phi})-\log\log\phi-\log k-\frac{13}{k}+\frac{\log(\phi-\overline{\phi})}{k\log\phi}\right).$$

Apart from the first sum, all terms are indeed either constant or linear in v. Collecting the linear terms gives the stated formula for κ_1, completing our proof.

$\square$

Lemma 2 *Let $r > 0$, $z = r + it$ with $|t| \leq r^{7/5}$, and $|1 - u| < 1$.*

The Dirichlet series $d(z, u)$ converges absolutely for $\Re z > 0$, and we have the asymptotic estimates

$$d(z, u) = d(r, u) \exp\left(-\frac{ia(u)t}{r^2} - \frac{a(u)t^2}{r^3} + O(r^{1/5}) \right),$$

$$d(r, u) = \exp\left(\frac{a(u)}{r} + b \log r + c(u) + O(r) \right) \tag{7}$$

for $r \to 0^+$ and

$$a(u) = \frac{\pi^2/6 - \mathrm{Li}_2(1 - u)}{\log \phi},$$

where Li denotes the polylogarithm,

$$b = -|\mathcal{F}_0| + \frac{25}{2} - \frac{\log(\phi - \overline{\phi})}{\log \phi}$$

and $c(u)$ is a function which is analytic in a complex neighbourhood of 1. The estimates are uniform in u on compact sets.

Proof Let

$$g(z, u) := \sum_{m \in \mathcal{F}} \log\left(1 + \frac{um^{-z}}{1 - m^{-z}} \right),$$

which implies $d(z, u) = \exp(g(z, u))$ by (1). For $\Re z > 0$ and fixed u, we have

$$\log\left(1 + \frac{um^{-z}}{1 - m^{-z}} \right) \sim um^{-z}$$

for $m \to \infty$. Since the elements of $\mathcal{F}$ follow the asymptotic formula $v_\ell = v_\ell(\phi, \overline{\phi}) \sim \frac{\phi^\ell}{\phi - \overline{\phi}}$, the series $g(z, u)$ converges absolutely.

We rewrite $g(z, u)$ as

$$g(z, u) = \sum_{m \in \mathcal{F}} \log\left(1 + \frac{ue^{-z \log m}}{1 - e^{-z \log m}} \right) = \sum_{m \in \mathcal{F}} f(z \log m, u)$$

for $f(z, u) = \log\left(1 + \frac{ue^{-z}}{1 - e^{-z}} \right)$.

For $\Re s > 1$, we consider the Mellin transform $g^\star(s, u)$ of the harmonic sum $g(z, u)$ and obtain

$$g^\star(s, u) = f^\star(s, u) \sum_{m \in \mathcal{F}} \frac{1}{(\log m)^s} = f^\star(s, u)\Lambda(s, 0) \tag{8}$$

with Λ as defined in Lemma 1.

We compute $f^\star(s, u)$. We have

$$f(z, u) = \log\left(1 + \frac{ue^{-z}}{1 - e^{-z}}\right) = \log(1 - (1 - u)e^{-z}) - \log(1 - e^{-z})$$

$$= \sum_{j=1}^{\infty} \frac{1 - (1 - u)^j}{j} e^{-zj}.$$

Thus the Mellin transform is

$$f^\star(s, u) = \left(\sum_{j \geq 1} \frac{1 - (1 - u)^j}{j^{1+s}}\right)\Gamma(s) = (\zeta(s + 1) - \mathrm{Li}_{s+1}(1 - u))\Gamma(s).$$

Note that the polylogarithm $\mathrm{Li}_{s+1}(1 - u)$ is an entire function in s for $|1 - u| < 1$.

We conclude that

$$g^\star(s, u) = (\zeta(s + 1) - \mathrm{Li}_{s+1}(1 - u))\Gamma(s)\Lambda(s, 0).$$

This is a meromorphic function in s with a simple pole at $s = 1$, a double pole at $s = 0$ and at most simple poles when s is a negative integer. When s runs along vertical lines, $g^\star(s, u)z^{-s}$ decreases exponentially for $|\arg(z)| < \pi/4$ due to the factor $\Gamma(s)$. We have

$$\mathrm{Res}_{s=1}\, g^\star(s, u) = \frac{\pi^2/6 - \mathrm{Li}_2(1 - u)}{\log \phi} =: a(u).$$

In addition, we have the singular expansion

$$g^\star(s, u) = -\frac{b}{s^2} + \frac{c(u)}{s} + O(1)$$

around 0, where $b = -\Lambda(0, 0)$ and

$$c(u) = \Lambda(0, 0) \log u + \left.\frac{\partial \Lambda(s, 0)}{\partial s}\right|_{s=0}$$

is an analytic function of u for $|1 - u| < 1$. Note that b is independent of u because the only component of $g^\star(s, u)$ depending on u is $\mathrm{Li}_{s+1}(1 - u)$, which does not contribute to the term $1/s^2$.

By Flajolet et al. [10, Theorem 4] (which remains valid for complex z with $|\arg(z)| < \pi/4$ because of the exponential decay observed above; cf. [8]), we get

$$g(z, u) = \frac{a(u)}{z} + b \log z + c(u) + O(z)$$

for $z \to 0$, $|\arg(z)| < \pi/4$.

As the Mellin transform of $z \frac{\partial g(z,u)}{\partial z}$ is $(-s)g^\star(s, u)$ by general properties of the Mellin transform, we immediately deduce that

$$z\frac{\partial g(z, u)}{\partial z} = -\frac{a(u)}{z} + O(1) \tag{9}$$

for $z \to 0$, $|\arg(z)| < \pi/4$, because the Mellin transform now has a simple pole at $s = 0$. Repeating the argument, we get

$$z^2\frac{\partial^2 g(z, u)}{\partial^2 z} + z\frac{\partial g(z, u)}{\partial z} = \frac{a(u)}{z} + O(z), \tag{10}$$

$$z^3\frac{\partial^3 g(z, u)}{\partial^3 z} + 3z^2\frac{\partial^2 g(z, u)}{\partial^2 z} + z\frac{\partial g(z, u)}{\partial z} = -\frac{a(u)}{z} + O(z) \tag{11}$$

for $z \to 0$, $|\arg(z)| < \pi/4$. Solving the linear system consisting of (9)–(11) yields

$$\frac{\partial g(z, u)}{\partial z} = -\frac{a(u)}{z^2} + O\left(\frac{1}{z}\right),$$

$$\frac{\partial^2 g(z, u)}{\partial^2 z} = \frac{2a(u)}{z^3} + O\left(\frac{1}{z^2}\right),$$

$$\frac{\partial^3 g(z, u)}{\partial^3 z} = -\frac{6a(u)}{z^4} + O\left(\frac{1}{z^3}\right)$$

for $z \to 0$, $|\arg(z)| < \pi/4$.

Thus we can approximate $g(z, u)$ by Taylor expansion around $z = r$ as

$$g(z, u) = g(r, u) - \frac{ia(u)t}{r^2} + O\left(\frac{t}{r}\right) - \frac{a(u)t^2}{r^3} + O\left(\frac{t^2}{r^2}\right) + O\left(\frac{t^3}{r^4}\right).$$

With our choice of the upper bound for t, we get

$$g(z, u) = g(r, u) - \frac{ia(u)t}{r^2} - \frac{a(u)t^2}{r^3} + O(r^{1/5}).$$

This concludes the proof of the lemma. $\qquad\qquad\square$

As a next step, we show that $|d(r + it, u)|$ is exponentially smaller than $d(r, u)$ for all t which are not too close to integer multiples of $2\pi/\log\phi$.

Lemma 3 *Let $r > 0$ and $z = r + it$ with $|t| \geq r^{7/5}$. Then*

$$\log d(r, u) - \Re \log d(z, u) \gg \frac{1}{r^{1/5}}$$

for $u \in (1/2, 3/2)$ and $r \to 0^+$ unless there is a non-zero integer k such that $|t - 2k\pi/\log\phi| < r^{3/4}$.

Proof Using the function $g(z, u)$ from the beginning of the proof of Lemma 2, we have

$$g(z, u) = \sum_{m \in \mathcal{F}} \log\left(\frac{1 - (1 - u)m^{-z}}{1 - m^{-z}}\right) = \sum_{m \in \mathcal{F}} \sum_{k \geq 1} \frac{1 - (1 - u)^k}{k} m^{-kz}$$

$$= \sum_{m \in \mathcal{F}} \sum_{k \geq 1} \frac{1 - (1 - u)^k}{k} m^{-kr}(\cos(kt \log m) - i \sin(kt \log m)).$$

This implies that

$$\log d(r, u) - \Re \log d(z, u) = \sum_{m \in \mathcal{F}} \sum_{k \geq 1} \frac{1 - (1 - u)^k}{k} m^{-kr}(1 - \cos(kt \log m)).$$

Obviously, all summands are non-negative, so we take the first summand as a lower bound and obtain

$$\log d(r, u) - \Re \log d(z, u) \geq u \sum_{m \in \mathcal{F}} m^{-r}(1 - \cos(t \log m)).$$

For $\ell \geq 13$, we use the estimate $v_\ell = \frac{\phi^\ell - \bar{\phi}^\ell}{\phi - \bar{\phi}} \leq K_0 \phi^\ell$ for a suitable $K_0 > 1$. In the following, we assume that $r < 1$. If $13 \leq \ell \leq 1/r$, then

$$v_\ell^r \leq K_0^r \phi^{\ell r} \leq K_0 \phi.$$

Thus restricting the sum to those ℓ and the corresponding v_ℓ yields

$$\log d(r, u) - \Re \log d(z, u) \gg \sum_{13 \leq \ell \leq 1/r} (1 - \cos(t \log v_\ell)). \tag{12}$$

We first consider the case that $|t| \leq r/\log\phi$. In this case, we have

$$|t| \log v_\ell \leq \frac{r}{\log\phi}(\ell \log\phi + \log K_0) \leq 1 + \frac{r \log K_0}{\log\phi} < \frac{\pi}{2}$$

for sufficiently small r. Thus we may use the inequality $1 - \cos\theta = 2\sin^2(\theta/2) \geq (4/\pi^2)\theta^2$ (which is a consequence of concavity of sin) to obtain

$$\log d(r, u) - \Re \log d(z, u) \gg \sum_{13 \leq \ell \leq 1/r} t^2 \log^2 v_\ell \gg t^2 \sum_{13 \leq \ell \leq 1/r} \ell^2 \gg \frac{t^2}{r^3} \geq \frac{1}{r^{1/5}}.$$

Next, we consider the case that $r/\log\phi \leq |t| \leq r^{1/5}$. Here we omit all summands with $\ell > 1/(|t|\log\phi)$ in (12) and obtain

$$\log d(r, u) - \Re \log d(z, u) \gg \sum_{13 \leq \ell \leq 1/(|t|\log\phi)} (1 - \cos(t \log v_\ell))$$

$$\gg t^2 \sum_{13 \leq \ell \leq 1/(|t|\log\phi)} \log^2 v_\ell \gg \frac{t^2}{|t|^3} = \frac{1}{|t|} \geq \frac{1}{r^{1/5}}$$

by the same arguments as above.

Finally, we turn to the case that $|t| > r^{1/5}$. We estimate the sum of the cosines by a geometric sum:

$$\sum_{13 \leq \ell \leq 1/r} \cos(t \log v_\ell) = \Re \sum_{13 \leq \ell \leq 1/r} \exp(it \log v_\ell)$$

$$\leq \left| \sum_{13 \leq \ell \leq 1/r} \exp\left(it\left(\ell \log\phi - \log(\phi - \bar\phi) + O\left(\left(\frac{|\bar\phi|}{\phi}\right)^\ell\right)\right)\right)\right|$$

$$= \left| \sum_{13 \leq \ell \leq 1/r} \left(\exp(it\ell \log\phi) + O\left(\left(\frac{|\bar\phi|}{\phi}\right)^\ell\right)\right)\right|$$

$$\leq \frac{2}{|\exp(it \log\phi) - 1|} + O(1).$$

Choose an integer k such that $|t \log\phi - 2k\pi|$ is minimal. Then $|t \log\phi - 2k\pi| \geq r^{3/4} \log\phi$ in view of the assumption made on t in the statement of the lemma (if $k \neq 0$) and because $|t| > r^{1/5}$ (if $k = 0$). We therefore have $|\exp(it \log\phi) - 1| \gg r^{3/4}$. Combining this with (12), we conclude that

$$\log d(r, u) - \Re \log d(z, u) \gg \frac{1}{r} - \frac{1}{r^{3/4}} \gg \frac{1}{r},$$

as required. $\qquad\square$

We are now able to estimate the integral in (3), choosing r appropriately.

Lemma 4 *Let $u \in (1/2, 3/2)$ and $r = \sqrt{a(u)/\log x}$. Then*

$$\frac{1}{2\pi i} \int_{r-i\infty}^{r+i\infty} \frac{d(z,u)}{z(z+1)} x^z \, dz = \frac{d(r,u)x^r}{2\pi} \frac{r^{1/2}\sqrt{\pi}}{\sqrt{a(u)}} (1 + O(r^{1/5})) \tag{13}$$

for $x \to \infty$.

Proof We first compute the integral over

$$M_0 = \{r + it \mid |t| \le r^{7/5}\}.$$

For $z \in M_0$, we have

$$\frac{1}{z} = \frac{1}{r}\left(1 + O\left(\frac{t}{r}\right)\right) = \frac{1}{r}(1 + O(r^{2/5})), \qquad \frac{1}{z+1} = 1 + O(r).$$

By Lemma 2, we have

$$\frac{d(z,u)x^z}{z(z+1)} = \frac{d(r,u)x^r}{r} \exp\left(-\frac{ia(u)t}{r^2} - \frac{a(u)t^2}{r^3} + it\log x + O(r^{1/5})\right).$$

The value $r = \sqrt{a(u)/\log x}$ has been chosen in such a way that the linear terms vanish. Thus we get

$$\frac{d(z,u)x^z}{z(z+1)} = \frac{d(r,u)x^r(1 + O(r^{1/5}))}{r} \exp\left(-\frac{a(u)t^2}{r^3}\right).$$

We have

$$\frac{1}{2\pi i} \int_{z \in M_0} \frac{d(z,u)}{z(z+1)} x^z \, dz = \frac{d(r,u)x^r(1 + O(r^{1/5}))}{2\pi r} \int_{-r^{7/5}}^{r^{7/5}} \exp\left(-\frac{a(u)t^2}{r^3}\right) dt$$

$$= \frac{d(r,u)x^r(1 + O(r^{1/5}))}{2\pi r} \int_{-\infty}^{\infty} \exp\left(-\frac{a(u)t^2}{r^3}\right) dt$$

because adding the tails induces an exponentially small error (note that $a(u) > 0$). Computing the integral yields

$$\frac{1}{2\pi i} \int_{z \in M_0} \frac{d(z,u)}{z(z+1)} x^z \, dz = \frac{d(r,u)x^r}{2\pi} \frac{r^{1/2}\sqrt{\pi}}{\sqrt{a(u)}} (1 + O(r^{1/5})). \tag{14}$$

Next, we compute the integral over

$$M_1 = \{r + it \in \mathbb{C} \mid \exists k \in \mathbb{Z} \setminus \{0\}: |t - 2k\pi/\log\phi| < r^{3/4}\}.$$

We use the trivial bound $|d(z, u)| \leq d(r, u)$, which follows from the definition of $d(z, u)$ as a Dirichlet series. Using the estimates $|z| \geq |\Im z|$ and $|z + 1| \geq |\Im z|$ yields

$$\left| \frac{1}{2\pi i} \int_{z \in M_1} \frac{d(z, u)}{z(z + 1)} x^z \, dz \right| \leq \frac{d(r, u) x^r}{2\pi} \sum_{k \in \mathbb{Z} \setminus \{0\}} \frac{2 r^{3/4}}{\left(\left| \frac{2k\pi}{\log \phi} \right| - r^{3/4} \right)^2}$$

$$\ll \frac{d(r, u) x^r}{2\pi} r^{3/4}.$$

Thus this integral can be absorbed by the error term of (14).

Finally, we compute the integral over

$$M_2 := \{r + it \in \mathbb{C}\} \setminus (M_0 \cup M_1).$$

For $z \in M_2$, we have

$$|d(z, u)| \leq d(r, u) \exp\left(- \frac{K_1}{r^{1/5}} \right)$$

for a suitable positive constant K_1 by Lemma 3. Thus

$$\left| \frac{1}{2\pi i} \int_{z \in M_2} \frac{d(z, u)}{z(z + 1)} x^z \, dz \right| \ll d(r, u) x^r \exp\left(- \frac{K_1}{r^{1/5}} \right) \int_{z \in M_2} \frac{1}{|z(z + 1)|} \, d|z|$$

$$\ll d(r, u) x^r \exp\left(- \frac{K_1}{r^{1/5}} \right) \log \frac{1}{r}.$$

As this integral is also absorbed by the error term of (14), we get (13). $\qquad\square$

In view of (3), Lemma 4 immediately gives us

$$I_{\omega_{\mathcal{F}}}(x, u) = \sum_{\substack{n \in \mathcal{M}(\mathcal{F}) \\ n \leq x}} u^{\omega_{\mathcal{F}}(n)} \left(1 - \frac{n}{x} \right) \sim \frac{d(r, u) x^r}{2\pi} \frac{r^{1/2} \sqrt{\pi}}{\sqrt{a(u)}}.$$

However, we are actually interested in an expression for the sum without the additional factor $(1 - \frac{n}{x})$. This is achieved in the following lemma.

Lemma 5 *We have*

$$\sum_{\substack{n \in \mathcal{M}(\mathcal{F}) \\ n \leq x}} u^{\omega_{\mathcal{F}}(n)}$$

$$= \frac{1}{2\sqrt{\pi}} \exp\left(2\sqrt{a(u)} \sqrt{\log x} - \frac{2b + 1}{4} \log \log x + \frac{2b - 1}{4} \log a(u) + c(u) \right)$$

$$\times \left(1 + O\left(\frac{1}{(\log x)^{1/10}} \right) \right)$$

for $x \to \infty$ and $1/2 < u < 3/2$.

Proof Trivially, the inequality

$$I_{\omega_{\mathcal{F}}}(x, u) \le \sum_{\substack{n \in \mathcal{M}(\mathcal{F}) \\ n \le x}} u^{\omega_{\mathcal{F}}(n)} \tag{15}$$

holds for positive u. On the other hand, we also have

$$\frac{I_{\omega_{\mathcal{F}}}(x \log x, u)}{1 - \frac{1}{\log x}} = \frac{1}{1 - \frac{1}{\log x}} \sum_{\substack{n \in \mathcal{M}(\mathcal{F}) \\ n \le x \log x}} u^{\omega_{\mathcal{F}}(n)} \left(1 - \frac{n}{x \log x}\right)$$

$$\ge \frac{1}{1 - \frac{1}{\log x}} \sum_{\substack{n \in \mathcal{M}(\mathcal{F}) \\ n \le x}} u^{\omega_{\mathcal{F}}(n)} \left(1 - \frac{n}{x \log x}\right) \tag{16}$$

$$\ge \sum_{\substack{n \in \mathcal{M}(\mathcal{F}) \\ n \le x}} u^{\omega_{\mathcal{F}}(n)}$$

for positive u and $x > e$.

Now we choose $r = \sqrt{a(u)/\log x}$ as in Lemma 4 and use (3) as well as Lemma 4 to obtain

$$I_{\omega_{\mathcal{F}}}(x, u) = \frac{d(r, u) x^r r^{1/2}}{2\sqrt{\pi a(u)}}(1 + O(r^{1/5})).$$

Finally, the asymptotic formula (7) for $d(r, u)$ gives us

$$I_{\omega_{\mathcal{F}}}(x, u) = \frac{1}{2\sqrt{\pi}} \exp\left(\frac{a(u)}{r} + b \log r + c(u) + r \log x + \frac{1}{2}(\log r - \log a(u))\right)$$

$$\times (1 + O(r^{1/5}))$$

$$= \frac{1}{2\sqrt{\pi}} \exp\left(2\sqrt{a(u)}\sqrt{\log x} - \frac{2b+1}{4} \log \log x + \frac{2b-1}{4} \log a(u) + c(u)\right)$$

$$\times \left(1 + O\left(\frac{1}{(\log x)^{1/10}}\right)\right).$$

If we replace x by $x \log x$ and divide by $(1 - 1/\log x)$, we obtain exactly the same asymptotic expansion, the difference being absorbed by the error term. Combining this with (15) and (16) yields the result. $\qquad\square$

We are now able to prove Theorems 1 and 2.

Proof of Theorem 1 Setting $u = 1$ in Lemma 5 yields the result. $\qquad\square$

Proof of Theorem 2 We consider the moment generating function

$$\mathbb{E}(e^{\omega_{\mathcal{F}}(N)t}) = \frac{\sum_{\substack{n \in \mathcal{M}(\mathcal{F}) \\ n \leq x}} e^{\omega_{\mathcal{F}}(n)t}}{\sum_{\substack{n \in \mathcal{M}(\mathcal{F}) \\ n \leq x}} 1}.$$

Lemma 5 yields

$$\mathbb{E}(e^{\omega_{\mathcal{F}}(N)t}) = \exp\left(2(\sqrt{a(e^t)} - \sqrt{a(1)})\sqrt{\log x} + O(t)\right)\left(1 + O\left(\frac{1}{(\log x)^{1/10}}\right)\right)$$

for $\log \frac{1}{2} < t < \log \frac{3}{2}$. We compute the Taylor expansion of $2(\sqrt{a(e^t)} - \sqrt{a(1)})$ around 0 as

$$2(\sqrt{a(e^t)} - \sqrt{a(1)}) = a_1 t + \frac{a_2}{2}t^2 + O(t^3)$$

for the constants a_1, a_2 given in the theorem. Thus the moment generating function of the renormalised random variable $Z = (\omega_{\mathcal{F}}(N) - a_1 \log^{1/2} x)/(\sqrt{a_2} \log^{1/4} x)$ is

$$\mathbb{E}(e^{Zt}) = \exp\left(\frac{t^2}{2} + O\left(\frac{t^3 + t}{\log^{1/4} x}\right)\right)\left(1 + O\left(\frac{1}{(\log x)^{1/10}}\right)\right).$$

For all real t, this moment generating function converges pointwise to the moment generating function $e^{t^2/2}$ of the standard normal distribution. By Curtiss' theorem [4], the random variable Z converges weakly to the standard normal distribution for $x \to \infty$. $\qquad\square$

3 Proof of Theorem 3: Counting with Multiplicities

Let $r > 0$ and consider the interval $U = U(r) = (\exp(-v_0 r/2), \exp(v_0 r/2))$, where v_0 has been defined in (4).

For $u \in U$ and $\Re z > r/2$, we study the Dirichlet generating function

$$D(z, u) = \sum_{n \in \mathcal{M}(\mathcal{F})} \frac{u^{\Omega_{\mathcal{F}}(n)}}{n^z},$$

which has a product representation

$$D(z, u) = \prod_{m \in \mathcal{F}} (1 + um^{-z} + u^2 m^{-2z} + \cdots) = \prod_{m \in \mathcal{F}} \frac{1}{1 - um^{-z}}$$

for all such z and u.

Lemma 6 *Let $r > 0$, $|t| \le r^{7/5}$ and $u \in U$. Then*

$$D(r + it, u) = D(r, u) \exp\left(- \frac{iAt}{r^2} - \frac{At^2}{r^3} + O(r^{1/5}) \right),$$

$$D(r, u) = \exp\left(\frac{A}{r} + B\left(\frac{\log u}{r} \right) \log r + C\left(\frac{\log u}{r} \right) + O(r) \right) \tag{17}$$

for $A = \pi^2/(6 \log \phi)$, $B(v) = -\Lambda(0, v)$, and $C(v) = \left. \frac{\partial \Lambda(s,v)}{\partial s} \right|_{s=0}$ in a neighbourhood of the origin.

Proof Let $\Re z > r/2$ and $v = (\log u)/z$. The assumption $u \in U$ implies that $|v| < rv_0/(2|z|) < v_0$.

We consider the sum

$$G(z, v) = - \sum_{m \in \mathcal{F}} \log(1 - e^{vz}m^{-z}).$$

Expanding the logarithm yields

$$G(z, v) = \sum_{m \in \mathcal{F}} \sum_{k \ge 1} \frac{e^{kvz}m^{-kz}}{k} = \sum_{m \in \mathcal{F}} \sum_{k \ge 1} \frac{\exp(-(\log m - v)kz)}{k}.$$

Its Mellin transform is

$$G^\star(s, v) = \sum_{m \in \mathcal{F}} \sum_{k \ge 1} \frac{1}{k^{1+s}} \frac{1}{(\log m - v)^s} \Gamma(s) = \Gamma(s)\zeta(1 + s)\Lambda(s, v),$$

where Λ has been defined in Lemma 1.

Again, $G^\star(s, v)$ has a simple pole at $s = 1$ and a double pole at $s = 0$. At $s = 1$, the local expansion is

$$G^\star(s, v) = \frac{A}{s - 1} + O(1).$$

The local expansion around $s = 0$ is

$$G^\star(s, v) = -\frac{B(v)}{s^2} + \frac{C(v)}{s} + O(1),$$

with $B(v)$ and $C(v)$ as in the statement of the lemma.

The rest of the proof follows along the lines of the proof of Lemma 2. $\qquad\square$

Lemma 7 *Let $r > 0$ and $z = r + it$ with $|t| \ge r^{7/5}$. Then*

$$\log D(r, u) - \Re \log D(z, u) \gg \frac{1}{r^{1/5}}$$

for $u \in (\exp(-v_0 r/2), \exp(v_0 r/2))$ and $r \to 0^+$ unless there is a non-zero integer k such that $|t - 2k\pi/\log\phi| < r^{3/4}$.

Proof We have

$$\Re \log D(z, u) = -\Re \sum_{m \in \mathcal{F}} \log(1 - um^{-z}) = \Re \sum_{m \in \mathcal{F}} \sum_{k \geq 1} \frac{u^k}{k} m^{-kz}$$

$$= \sum_{m \in \mathcal{F}} \sum_{k \geq 1} \frac{u^k}{k} m^{-kr} \cos(kt \log m).$$

This implies that

$$\log D(r, u) - \Re \log D(z, u) = \sum_{m \in \mathcal{F}} \sum_{k \geq 1} \frac{u^k}{k} m^{-kr}(1 - \cos(kt \log m))$$

$$\geq u \sum_{m \in \mathcal{F}} m^{-r}(1 - \cos(t \log m)).$$

The remainder of the proof is exactly the same as that of Lemma 3. $\qquad\square$

Proof of Theorem 3 We now consider asymptotic expansions for $x \to \infty$; we set $r = \sqrt{A/\log x}$. The statements and proofs of Lemmata 4 and 5 carry over (only the range of u has to be adapted). So we have

$$\sum_{\substack{n \in \mathcal{M}(\mathcal{F}) \\ n \leq x}} u^{\Omega_{\mathcal{F}}(n)} \tag{18}$$

$$= \frac{1}{2\sqrt{\pi}} \exp\left(2\sqrt{A}\sqrt{\log x} - \frac{2B(v) + 1}{4} \log\log x + \frac{2B(v) - 1}{4} \log A + C(v)\right)$$

$$\times\left(1 + O\left(\frac{1}{(\log x)^{1/10}}\right)\right)$$

for $x \to \infty$, $u \in U(\sqrt{A/\log x})$ and $v = \log u \sqrt{\log x/A}$.

We now consider the moment generating function

$$\mathbb{E}(e^{\Omega_{\mathcal{F}}(N)t}) = \frac{\sum_{\substack{n \in \mathcal{M}(\mathcal{F}) \\ n \leq x}} e^{\Omega_{\mathcal{F}}(n)t}}{\sum_{\substack{n \in \mathcal{M}(\mathcal{F}) \\ n \leq x}} 1}.$$

Equation (18) yields

$$\mathbb{E}(e^{\Omega_{\mathcal{F}}(N)t}) = \exp\left(\frac{1}{2}(\log\log x - \log A)(B(0) - B(v)) + C(v) - C(0)\right)$$

$$\times\left(1 + O\left(\frac{1}{(\log x)^{1/10}}\right)\right),$$

where $v = t\sqrt{\log x}/\sqrt{A}$, for all t such that $|t| \le v_0\sqrt{A}/(2\sqrt{\log x})$. Since $B(v) = -\Lambda(0, v)$, Eq. (5) gives us $B(0) - B(v) = v/\log\phi$. Likewise, Lemmata 6 and 1 yield

$$C(v) - C(0) = -\sum_{m \in \mathcal{F}} \left(\log\left(1 - \frac{v}{\log m}\right) + \frac{v}{\log m} \right) + \kappa_1 v,$$

so finally

$$\mathbb{E}(e^{\Omega_{\mathcal{F}}(N)t})$$

$$= \exp\left(t\left(\frac{a_1}{2}\sqrt{\log x}\log\log x + b_1\sqrt{\log x}\right)\right) \prod_{m \in \mathcal{F}} e^{-v/(\log m)}\left(1 - \frac{v}{\log m}\right)^{-1}$$

$$\times\left(1 + O\left(\frac{1}{(\log x)^{1/10}}\right)\right),$$

where $b_1 = A^{-1/2}(\kappa_1 - (\log A)/(2\log\phi))$. Since $(1 - v/\lambda)^{-1}$ is exactly the moment generating function of an $\mathrm{Exp}(\lambda)$-distributed random variable, Theorem 3 follows immediately from Curtiss's theorem in the same way as Theorem 2. $\qquad\square$

Acknowledgements C. Heuberger is supported by the Austrian Science Fund (FWF): P 24644-N26. Parts of this paper have been written while he was a visitor at Stellenbosch University.

S. Wagner is supported by the National Research Foundation of South Africa, grant number 96236.

References

1. T.M. Apostol, *Introduction to Analytic Number Theory*. Undergraduate Texts in Mathematics (Springer, New York/Heidelberg, 1976)
2. Y. Bilu, G. Hanrot, P.M. Voutier, Existence of primitive divisors of Lucas and Lehmer numbers. J. Reine Angew. Math. **539**, 75–122 (2001). With an appendix by M. Mignotte (2002j:11027)
3. R.D. Carmichael, On the numerical factors of the arithmetic forms $\alpha^n \pm \beta^n$. Ann. Math. **15**, 30–70 (1913)
4. J.H. Curtiss, A note on the theory of moment generating functions. Ann. Math. Stat. **13**, 430–433 (1942)
5. P.D.T.A. Elliott, *Probabilistic Number Theory. II. Central Limit Theorems*. Grundlehren der Mathematischen Wissenschaften, vol. 240 (Springer, Berlin/New York, 1980)
6. P. Erdős, M. Kac, The Gaussian law of errors in the theory of additive number theoretic functions. Am. J. Math. **62**, 738–742 (1940)
7. P. Erdős, J. Lehner, The distribution of the number of summands in the partitions of a positive integer. Duke Math. J. **8**, 335–345 (1941)
8. P. Flajolet, H. Prodinger, Register allocation for unary-binary trees. SIAM J. Comput. **15**(3), 629–640 (1986) (87j:68052)
9. P. Flajolet, P. Grabner, P. Kirschenhofer, H. Prodinger, R.F. Tichy, Mellin transforms and asymptotics: digital sums. Theor. Comput. Sci. **123**, 291–314 (1994)

10. P. Flajolet, X. Gourdon, P. Dumas, Mellin transforms and asymptotics: harmonic sums. Theor. Comput. Sci. **144**, 3–58 (1995)
11. W.M.Y. Goh, E. Schmutz, The number of distinct part sizes in a random integer partition. J. Comb. Theory Ser. A **69**(1), 149–158 (1995)
12. J. Knopfmacher, *Abstract Analytic Number Theory* (North-Holland/American Elsevier, Amsterdam/Oxford/New York, 1975)
13. J. Knopfmacher, W.-B. Zhang, *Number Theory Arising from Finite Fields*. Monographs and Textbooks in Pure and Applied Mathematics, vol. 241 (Marcel Dekker, New York, 2001)
14. F. Luca, Š. Porubský, The multiplicative group generated by the Lehmer numbers. Fibonacci Quart. **41**(2), 122–132 (2003)
15. F. Luca, C. Pomerance, S. Wagner, Fibonacci integers. J. Number Theory **131**(3), 440–457 (2011)
16. NIST Digital library of mathematical functions, Release 1.0.10 of 2015-08-07 (2015). Online companion to [17]. http://dlmf.nist.gov/
17. F.W.J. Olver, D.W. Lozier, R.F. Boisvert, C.W. Clark (eds.), *NIST Handbook of Mathematical Functions* (Cambridge University Press, New York, 2010)
18. The On-Line Encyclopedia of Integer Sequences (2016) http://oeis.org
19. S. Wehmeier, The Erdős-Kac theorem for additive arithmetical semigroups. Lith. Math. J. **47**(3), 352–360 (2007)
20. E.T. Whittaker, G.N. Watson, *A Course of Modern Analysis* (Cambridge University Press, Cambridge, 1996). Reprint of the fourth edition (1927)

Measures of Pseudorandomness: Arithmetic Autocorrelation and Correlation Measure

Richard Hofer, László Mérai, and Arne Winterhof

Dedicated to Robert F. Tichy on the occasion of his 60th birthday

Abstract We prove a relation between two measures of pseudorandomness, the arithmetic autocorrelation, and the correlation measure of order k. Roughly speaking, we show that any binary sequence with small correlation measure of order k up to a sufficiently large k cannot have a large arithmetic correlation. We apply our result to several classes of sequences including Legendre sequences defined with polynomials.

1 Introduction

Pseudorandom numbers are generated by deterministic algorithms and are not random at all. However, in contrast to truly random numbers they guarantee certain randomness properties. Their desirable features depend on the application area. For example, unpredictable sequences are needed for cryptography and uncorrelated sequences for wireless communication or radar. Some corresponding quality measures are linear complexity and expansion complexity for unpredictability and autocorrelation or more general correlation measure of order k.

Finding relations between different measures of pseudorandomness is an important goal. For example, the linear complexity provides essentially the same quality measure as certain lattice tests coming from the area of Monte Carlo methods, see [4, 20]. The correlation measure of order k is a rather general measure of pseudorandomness introduced by Mauduit and Sárközy [16]. A relation between linear complexity and the correlation measure of order k is given in [2]. Hence, we may roughly say that correlation measure is a stronger measure than linear complexity. Expansion complexity introduced in [3] is another measure which is essentially the same as linear complexity in the periodic case but finer in the

R. Hofer • L. Mérai • A. Winterhof (⊠)

Johann Radon Institute for Computational and Applied Mathematics, Austrian Academy of Sciences, Altenberger Str. 69, 4040 Linz, Austria

e-mail: richard.hofer@oeaw.ac.at; laszlo.merai@oeaw.ac.at; arne.winterhof@oeaw.ac.at

© Springer International Publishing AG 2017

C. Elsholtz, P. Grabner (eds.), *Number Theory – Diophantine Problems, Uniform Distribution and Applications*, DOI 10.1007/978-3-319-55357-3_15

aperiodic case [19] (see also [14, 18]). There are many other related measures of pseudorandomness for sequences, see [11, 21, 22], and analyzing their hierarchy is very important.

In this paper we analyze the relation between another figure of merit coming from coding theory, the *arithmetic autocorrelation*, and the *correlation measures of higher orders*. Roughly speaking, we show that any binary sequence with small correlation measure of order k up to a sufficiently large k cannot have a large arithmetic autocorrelation. Correlation measure of order k and arithmetic autocorrelation are defined in the following paragraphs.

For a (purely) T-periodic binary sequence (a_n),

$$C_k(a_n) = \max_{0 < d_1 < \ldots < d_{k-1} < T} \left| \sum_{n=0}^{T-1} (-1)^{a_n + a_{n+d_1} + \cdots + a_{n+d_{k-1}}} \right|.$$

denotes the (periodic) *correlation measure of order* $k \geq 1$ of (a_n). $C_2(a_n)$ is also called autocorrelation of (a_n).

In this article we study a different notion of autocorrelation, the *arithmetic autocorrelation* introduced by Mandelbaum [15]. Sequences with small arithmetic autocorrelation can be used to define good error-correcting codes over the integers (instead of finite fields). Also, see the recent monograph by Goresky and Klapper [9] for more background and results on arithmetic correlations.

For an eventually T-periodic binary sequence (s_n) with preperiod T_0, that is $s_{n+T} = s_n$ for all $n \geq T_0$, the *imbalance* $Z(s_n)$ is defined by

$$Z(s_n) = N_0 - N_1,$$

where

$$N_i = |\{T_0 \leq n \leq T_0 + T - 1 : s_n = i\}|, \qquad i = 0, 1.$$

(Note that for a purely periodic sequence (a_n) we have $C_1(a_n) = |Z(a_n)|$.) The *arithmetic autocorrelation function* $A(t)$ of a (purely) T-periodic binary sequence (a_n) is defined as follows. For $t \in \{1, 2, \ldots, T-1\}$ let (a_{n+t}) be the shift of (a_n) by lag t. Put

$$x_t = \sum_{n=0}^{T-1} a_{n+t} 2^n \quad \text{and} \quad \alpha_t = \sum_{n=0}^{\infty} a_{n+t} 2^n, \qquad 0 \leq t < T.$$

Since α_t converges in the 2-norm of $\mathbb{Q}$ it holds (see [13])

$$\alpha_t = -\frac{x_t}{2^T - 1}, \qquad 0 \leq t < T.$$

We write

$$\alpha_0 - \alpha_t = \sum_{n=0}^{\infty} s_{n,t} 2^n \tag{1}$$

with unique $s_{n,t} \in \{0, 1\}$. Note that $(s_{n,t})$ is (purely) periodic with period T if $x_0 \geq x_t$ and eventually periodic with period T from T on if $x_0 < x_t$ (see [13] for more details). In both cases we define

$$A(t) = Z(s_{n,t}), \qquad 1 \leq t \leq T - 1.$$

In Sect. 2 we estimate the arithmetic autocorrelation of a binary sequence of period T in terms of correlation measures. In Sect. 3 we apply this result to several classes of sequences including Legendre sequences defined with polynomials.

2 A Bound on the Arithmetic Autocorrelation

Theorem 1 *Put*

$$\Gamma_s = \max_{1 \leq l \leq s} C_l(a_n).$$

Then the arithmetic autocorrelation function of a T-periodic binary sequence (a_n) satisfies

$$A(t) \ll \min \left\{ T^{1/2} \Gamma_{\lfloor \log T \rfloor}^{1/2}, 2^r \Gamma_{\lfloor \log T \rfloor} \log T \right\}$$

where $r = \min\{t, T - t\}$ for $1 \leq t \leq T - 1$.

Proof By the symmetry $A(t) = -A(T - t)$ of the arithmetic autocorrelation (see [13, Proposition 2.1]) we may assume $1 \leq t \leq \lfloor T/2 \rfloor$. In the following we derive a lower bound on the number N_1 of ones in a period of the T-periodic sequence $(s_{n,t})$ defined by (1).

For a binary vector $(i_0, i_1, \ldots, i_{k-1}) \in \{0, 1\}^k$ and lags $0 < d_1 < \cdots < d_{k-1} < T$, put

$$N = |\{0 \leq n \leq T - 1 : a_n = i_0, a_{n+d_1} = i_1, \ldots, a_{n+d_{k-1}} = i_{k-1}\}|.$$

Similarly as in [17, Theorem 2] it can be shown that

$$\left| N - \frac{T}{2^k} \right| \leq \frac{1}{2^k} \sum_{l=1}^{k} \binom{k}{l} C_l(a_n). \tag{2}$$

Now take $c \in \{0, 1\}$. For some $k = 1, \ldots, m - 1$ and $n = T, \ldots, 2T - 1$ assume

$$(a_{n-k}, a_{n-k+t}) = (c, 1 - c),$$

$$a_{n-k+j} = a_{n-k+j+t}, \quad j = 1, \ldots, k - 1, \tag{3}$$

$$(a_n, a_{n+t}) \in \{0, 1\}^2.$$

First we assume $m + 1 \le t \le \lfloor T/2 \rfloor$. From (2) we know that (for fixed c) the number of patterns

$$\begin{pmatrix} a_{n-k} & a_{n-k+1} & \cdots & a_{n-1} & a_n \\ a_{n-k+t} & a_{n-k+t+1} & \cdots & a_{n-1+t} & a_{n+t} \end{pmatrix} \tag{4}$$

satisfying the assumptions (3) in

$$\begin{matrix} a_{T-k} & a_{T-k+1} & \cdots a_{T-1} & a_T & \cdots a_{2T-2} & a_{2T-1} \\ a_{t+T-k} & a_{t+T-k+1} & \cdots a_{t+T-1} & a_{t+T} & \cdots a_{t+2T-2} & a_{t+2T-1} \end{matrix} \tag{5}$$

is at least $T/2^{2k+2} - \Gamma_{2k+2}$. We have to distinguish between two cases.

If $c = 1$, then $(a_{n-k}, a_{n-k+t}) = (1, 0)$. The subtraction of 0 from 1 gives no carry, no matter if there was a carry in the previous step. Hence

$$s_{n,t} = \begin{cases} 1 & \text{if } a_n \ne a_{n+t}, \\ 0 & \text{if } a_n = a_{n+t}. \end{cases}$$

Since there are 2^k possible choices for the pattern (4) we count at least $T/2^{k+2} - 2^k \Gamma_{2k+2}$ different $T \le n < 2T$ with $s_{n,t} = 1$.

If $c = 0$, then $(a_{n-k}, a_{n-k+t}) = (0, 1)$. The subtraction of 1 from 0 gives a carry, no matter if there was a carry in the previous step. Hence

$$s_{n,t} = \begin{cases} 1 & \text{if } a_n = a_{n+t}, \\ 0 & \text{if } a_n \ne a_{n+t}. \end{cases}$$

Just as before there are 2^k possible choices for the pattern (4) and so we get at least $T/2^{k+2} - 2^k \Gamma_{2k+2}$ additional n with $s_{n,t} = 1$.

Thus in total we have at least $T/2^{k+1} - 2^{k+1} \Gamma_{2k+2}$ different $T \le n < 2T$ with $a_{n-k} \ne a_{n-k+t}$, $(a_{n-k+j}, a_{n-k+j+t}) \in \{(0, 0), (1, 1)\}$ for $j = 1, \ldots, k - 1$ and $s_{n,t} = 1$.

Summing up all the contributions we get

$$N_1 \ge \frac{1}{2} \left(\sum_{k=1}^{m-1} 2^{-k} \right) T - 2 \left(\sum_{k=1}^{m-1} 2^k \Gamma_{2k+2} \right) \ge \frac{1}{2} \left(\sum_{k=1}^{m-1} 2^{-k} \right) T - 2 \Gamma_{2m} \left(\sum_{k=1}^{m-1} 2^k \right)$$

$$\ge \frac{T}{2} - 2^{-m} T - 2^{m+1} \Gamma_{2m}$$

where we used $\Gamma_s = \Gamma_{2k+2} \leq \Gamma_{2m}$ since $k \leq m-1$. Analogously N_0 can be bounded below by

$$N_0 \geq \frac{T}{2} - 2^{-m}T - 2^{m+1}\Gamma_{2m}$$

and therefore

$$|A(t)| = |N_0 - N_1| \leq 2^{-m+1}T + 2^{m+2}\Gamma_{2m}$$

since

$$N_0 - N_1 = T - 2N_1 \leq 2^{-m+1}T + 2^{m+2}\Gamma_{2m},$$
$$N_0 - N_1 = 2N_0 - T \geq -(2^{-m+1}T + 2^{m+2}\Gamma_{2m}).$$

Now we assume $1 \leq t \leq m$, that means some indices in (4) coincide and so we have to deal with shorter patterns. From (2) we know that (for fixed c) the number of patterns (4) satisfying the assumptions (3) in (5) is at least

$$T/2^{2k+2} - \Gamma_{2k+2}, \qquad k \leq t-1,$$
$$T/2^{k+t+1} - \Gamma_{k+t+1}, \qquad k \geq t.$$

Similarly as before if $c = 1$, then

$$s_{n,t} = \begin{cases} 1 & \text{if } a_n \neq a_{n+t}, \\ 0 & \text{if } a_n = a_{n+t}, \end{cases}$$

and if $c = 0$, then

$$s_{n,t} = \begin{cases} 1 & \text{if } a_n = a_{n+t}, \\ 0 & \text{if } a_n \neq a_{n+t}. \end{cases}$$

For each case we have 2^k possible choices for the pattern (4) if $k \leq t-1$ and 2^{t-1} possible choices if $k \geq t$ and thus in total we count at least

$$T/2^{k+1} - 2^{k+1}\Gamma_{2k+2}, \qquad k \leq t-1,$$
$$T/2^{k+1} - 2^t\Gamma_{k+t+1}, \qquad k \geq t,$$

different $T \leq n < 2T$ with $a_{n-k} \neq a_{n-k+t}$, $(a_{n-k+j}, a_{n-k+j+t}) \in \{(0,0),(1,1)\}$ for $j = 1, \ldots, k-1$ and $s_{n,t} = 1$.

Put $m' = 2m - t$. Summing up all the contributions we get

$$N_1 \geq \frac{1}{2} \left(\sum_{k=1}^{m'-1} 2^{-k} \right) T - 2 \left(\sum_{k=1}^{t-1} 2^k \Gamma_{2k+2} + 2^{t-1} \sum_{k=t}^{m'-1} \Gamma_{k+t+1} \right)$$

$$\geq \frac{1}{2} \left(\sum_{k=1}^{m'-1} 2^{-k} \right) T - 2\Gamma_{2m}(2^t - 2) - 2^t \Gamma_{m+t}(m' - t)$$

$$\geq \frac{T}{2} - 2^{-m'} T - 2^{t+1} \Gamma_{2m} - 2^t \Gamma_{2m}(m' - t)$$

$$\geq \frac{T}{2} - 2^{-2m+t} T - 2^{t+1}(m - t + 1)\Gamma_{2m}$$

where we used $\Gamma_{2k+2} \leq \Gamma_{2m}$ or $\Gamma_{k+t+1} \leq \Gamma_{m+t} \leq \Gamma_{2m}$, respectively. Analogously N_0 can be bounded below by

$$N_0 \geq \frac{T}{2} - 2^{-2m+t} T - 2^{t+1}(m - t + 1)\Gamma_{2m}$$

and therefore

$$|A(t)| = |N_0 - N_1| \leq 2^{-2m+t+1} T + 2^{t+2}(m - t + 1)\Gamma_{2m}.$$

Choosing

$$m = \left\lfloor \frac{1}{2} \log \frac{T}{\Gamma_{\lfloor \log T \rfloor}} \right\rfloor$$

we obtain the result. (Note that we may assume $\Gamma_{\lfloor \log T \rfloor} = o(T)$ and thus $m \geq 2$ since otherwise the result is trivial.) $\qquad\square$

3 Applications

Now we apply our result to several classes of sequences.

For a prime $p > 2$ and a squarefree polynomial $f \in \mathbb{F}_p[x]$ with positive degree d let the p-periodic sequences (ℓ_n) be defined by

$$\ell_n = \begin{cases} 1 & \text{if } \left(\frac{f(n)}{p} \right) = 1, \\ 0 & \text{otherwise}, \end{cases} \quad n \geq 0, \tag{6}$$

where $\left(\frac{\cdot}{p}\right)$ is the Legendre symbol modulo p. For $f(n) = n$ these sequences are the Legendre sequences.

Corollary 1 *If $d < 0.5 \log p / \log \log p$ or 2 is a primitive root modulo p, then the arithmetic autocorrelation function of the p-periodic sequences (ℓ_n) defined by (6) satisfies*

$$A(t) \ll \min\left\{d^{1/2} p^{3/4} \log^{1/2} p, 2^r d p^{1/2} \log^2 p\right\}$$

where $r = \min\{t, p - t\}$ for $1 \le t \le p - 1$.

This result immediately follows from Theorem 1 and from the following proposition.

Proposition 1 *If f has no multiple zeros in the algebraic closure of $\mathbb{F}_p$ and*

(i) $k < p$ and 2 is a primitive root modulo p, or
(ii) $(4k)^d < p$,

then the (periodic) correlation measure of order k satisfies $C_k(\ell_n) \ll k d p^{1/2}$.

Proof Similar to the proof of Theorem 1 in [10] (since we consider the periodic correlation measure of order k instead of the aperiodic one we lose the $\log p$ term).
$\square$

Let q be the power of an odd prime, α a primitive element of $\mathbb{F}_q$, and let η denote the quadratic character of $\mathbb{F}_q$. We denote by (a_n) the $(q-1)$-periodic *Sidelnikov–Lempel–Cohn–Eastman sequence* defined by

$$a_n = \begin{cases} 1 & \text{if } \eta(\alpha^n + 1) = 1, \\ 0 & \text{otherwise,} \end{cases} \quad n \ge 0.$$

Similar to the proof in [2, Lemma 1] it follows that the (periodic) correlation measure of order k satisfies $C_k \ll k q^{1/2}$. From Theorem 1 we get

$$A(t) \ll \min\left\{q^{3/4} \log^{1/2} q, 2^r q^{1/2} \log^2 q\right\},$$

where $r = \min\{t, q - 1 - t\}$ for $1 \le t \le q - 2$.

Let p be an odd prime, $\lambda \in \mathbb{F}_p^*$ of multiplicative order T, and $f \in \mathbb{F}_p[x]$ a polynomial of positive degree d not of the form $c x^\alpha (g(x))^2$ with $c \in \mathbb{F}_p$, $\alpha \in \mathbb{N}$ and $g(x) \in \mathbb{F}_p[x]$. Define the T-periodic sequence (b_n) by

$$b_n = \begin{cases} 1 & \text{if } \left(\frac{f(\lambda^n)}{p}\right) = 1, \\ 0 & \text{otherwise,} \end{cases} \quad n \ge 0.$$

Similar to the proof in [12, Theorem 2] it follows that if T is a prime and either $(4k)^d \le T$ or 2 is a primitive root modulo T, then the (periodic) correlation measure of order k satisfies $C_k \ll kdp^{1/2}$. From Theorem 1 we get that if $d \le \log p / \log \log p$ or 2 is a primitive root modulo T, then

$$A(t) \ll \min \left\{ d^{1/2} p^{1/4} T^{1/2} \log^{1/2} T, 2^r dp^{1/2} \log^2 T \right\}$$

where $r = \min\{t, T - t\}$ for $1 \le t \le T - 1$.

Let $q = 2^k$ such that $q - 1$ is prime, $f \in \mathbb{F}_q[x]$ be a polynomial of positive odd degree $d \ge \log q$ such that the coefficients of its terms are zero if and only if the term has an even exponent. Let $\mathrm{Tr} : \mathbb{F}_{2^k} \to \mathbb{F}_2$ denote the absolute trace function of $\mathbb{F}_{2^k}$ and α be a primitive element of $\mathbb{F}_{2^k}$. Define the $q - 1$-periodic sequence (d_n) by

$$d_n = \mathrm{Tr}(f(\alpha^n)), \quad n \ge 0.$$

Similar to the proof in [5, Theorem 4] it follows that if $k \le d + 1$ the (periodic) correlation measure of order k satisfies $C_k \ll dq^{1/2}$. In the same way as Corollary 1 we get

$$A(t) \ll \min \left\{ d^{1/2} q^{3/4} \log^{1/2} q, 2^r dq^{1/2} \log^2 q \right\}$$

where $r = \min\{t, q - 1 - t\}$ for $1 \le t \le q - 2$.

4 Final Remarks

For fixed $1 \le t < T$, Goresky and Klapper [7, 8] proved that the expected arithmetic autocorrelation, averaged over all binary sequences of period T, is

$$\frac{T}{2^{T - \gcd(t,T)}}.$$

Sequences with *ideal arithmetic autocorrelation* equal to zero for all nontrivial shifts t are known, see [6]. However, the maximum absolute value of the (classical) autocorrelation of these so-called *ℓ-sequences* equals the period since the second half of a period is the bit-wise complement of the first half [6, Proposition 1]. Hence, these sequences are far away from looking random. In contrast to these sequences, the sequences studied in Sect. 3 still guarantee a rather small arithmetic autocorrelation with respect to their periods if the period is sufficiently large.

Actually, the correlation measure of order k was defined for finite sequences. Analogs of our results for finite sequences can be easily obtained with the obvious definition of arithmetic autocorrelation. Moreover, for a truly random sequence of length T, Alon et al. [1] showed that the correlation measure of order k is of order of magnitude $k^{1/2}T^{1/2}\log^{1/2}T$ and thus its arithmetic autocorrelation is at most of order of magnitude $T^{3/4}\log^{1/2}T$.

The correlation measure of order k was also introduced and analyzed for non-binary sequences, see [17], as well as the arithmetic autocorrelation for non-binary sequences, see [7]. A similar relation as Theorem 1 can be easily obtained in a similar way, that is, reduce first the problem of estimating the arithmetic autocorrelation to the pattern distribution of the sequence which can be estimated by the correlation measure of order k.

Acknowledgements The authors are partially supported by the Austrian Science Fund FWF Project 5511-N26 which is part of the Special Research Program "Quasi-Monte Carlo Methods: Theory and Applications."

References

1. N. Alon, Y. Kohayakawa, C. Mauduit, C.G. Moreira, V. Rödl, Measures of pseudorandomness for finite sequences: typical values. Proc. Lond. Math. Soc. (3) **95**, 778–812 (2007)
2. N. Brandstätter, A. Winterhof, Linear complexity profile of binary sequences with small correlation measure. Period. Math. Hungar. **52**, 1–8 (2006)
3. C. Diem, On the use of expansion series for stream ciphers. LMS, J. Comput. Math. **15**, 326–340 (2012)
4. G. Dorfer, A. Winterhof, Lattice structure and linear complexity profile of nonlinear pseudo-random number generators. Appl. Algebra Eng. Comm. Comput. **13**, 499–508 (2003)
5. J. Folláth, Construction of pseudorandom binary sequences using additive characters over $GF(2^k)$. Period. Math. Hungar. **57**, 73–81 (2008)
6. M. Goresky, A. Klapper, Arithmetic crosscorrelations of feedback with carry shift register sequences. IEEE Trans. Inf. Theory **43**, 1342–1345 (1997)
7. M. Goresky, A. Klapper, Some results on the arithmetic correlation of sequences (extended abstract). *Sequences and Their Applications–SETA 2008*. Lecture Notes in Computer Science, vol. 5203 (Springer, Berlin, 2008), pp. 71–80
8. M. Goresky, A. Klapper, Statistical properties of the arithmetic correlation of sequences. Int. J. Found. Comput. Sci. **22**, 1297–1315 (2011)
9. M. Goresky, A. Klapper, *Algebraic Shift Register Sequences* (Cambridge University Press, Cambridge, 2012)
10. L. Goubin, C. Mauduit, A. Sárközy, Construction of large families of pseudorandom binary sequences. J. Number Theory **106**, 56–69 (2004)
11. K. Gyarmati, Measures of pseudorandomness, in *Finite Fields and Their Applications*, ed. by P. Charpin, A. Pott, A. Winterhof. Radon Series on Computational and Applied Mathematics, vol. 11 (de Gruyter, Berlin, 2013), pp. 43–64
12. K. Gyarmati, A. Pethő, A. Sárközy, On linear recursion and pseudorandomness. Acta Arith. **118**, 359–374 (2005)
13. R. Hofer, A. Winterhof, On the arithmetic autocorrelation of the Legendre sequence. Adv. Math. Commun. **11**, 237–244 (2017)
14. R. Hofer, A. Winterhof, Linear complexity and expansion complexity of some number theoretic sequences, in *Arithmetic of Finite Fields (WAIFI 2016)*. Lecture Notes in Computer Science, vol. 10064 (Springer, Cham, 2016), pp. 67–74
15. D. Mandelbaum, Arithmetic codes with large distance. IEEE Trans. Inf. Theory **13**, 237–242 (1967)

16. C. Mauduit, A. Sárközy, On finite pseudorandom binary sequences. I. Measure of pseudorandomness, the Legendre symbol. Acta Arith. **82**, 365–377 (1997)
17. C. Mauduit, A. Sárközy, On finite pseudorandom sequences of k symbols. Indag. Math. **13**, 89–101 (2002)
18. L. Mérai, A. Winterhof, On the Nth linear complexity of p-automatic sequences over $\mathbb{F}_p$ (preprint 2016)
19. L. Mérai, H. Niederreiter, A. Winterhof, Expansion complexity and linear complexity of sequences over finite fields. Cryptogr. Commun. (2016). doi:10.1007/s12095-016-0189-2
20. H. Niederreiter, A. Winterhof, Lattice structure and linear complexity of nonlinear pseudorandom numbers. Appl. Algebra Eng. Commun. Comput. **13**, 319–326 (2002)
21. J. Rivat, A. Sárközy, On pseudorandom sequences and their application, in *General Theory of Information Transfer and Combinatorics*. Lecture Notes in Computer Science, vol. 4123 (Springer, Berlin, 2006), pp. 343–361
22. A. Topuzoğlu, A. Winterhof, Pseudorandom sequences, in *Topics in Geometry, Coding Theory and Cryptography*. Algebra and Applications, vol. 6 (Springer, Dordrecht, 2007), pp. 135–166

On Multiplicative Independent Bases for Canonical Number Systems in Cyclotomic Number Fields

Manfred G. Madritsch, Paul Surer, and Volker Ziegler

Dedicated to Professor Robert F. Tichy on the occasion of his 60th birthday

Abstract In the present paper we are interested in number systems in the ring of integers of cyclotomic number fields in order to obtain a result equivalent to Cobham's theorem. For this reason we first search for potential bases. This is done in a very general way in terms of canonical number systems. In a second step we analyse pairs of bases in view of their multiplicative independence. In the last part we state an appropriate variant of Cobham's theorem.

2010 *Mathematics Subject Classification* 11R18, 11Y40, 11A63

1 Introduction

Let $q \geq 2$ be a positive integer. Then every $n \in \mathbb{N}$ admits a unique q-adic representation of the form

$$n = \sum_{j=0}^{\ell} a_j q^j \qquad \text{with } a_j \in \mathcal{N} := \{0, \ldots, q-1\} \text{ for } 0 \leq j \leq \ell.$$

M.G. Madritsch (✉)
Institut Elie Cartan de Lorraine, Université de Lorraine, UMR 7502, Vandoeuvre-lès-Nancy F-54506, France

Institut Elie Cartan de Lorraine, CNRS, UMR 7502, Vandoeuvre-lès-Nancy F-54506, France
e-mail: manfred.madritsch@univ-lorraine.fr

P. Surer
Institut für Mathematik, Universtiät für Bodenkultur (BOKU), Gregor-Mendel-Straße 33, 1180 Wien, Austria http://www.palovsky.com
e-mail: me@palovsky.com

V. Ziegler
Institute of Mathematics, University of Salzburg, Hellbrunnerstrasse 34/I, 5020 Salzburg, Austria
e-mail: volker.ziegler@sbg.ac.at

© Springer International Publishing AG 2017
C. Elsholtz, P. Grabner (eds.), *Number Theory – Diophantine Problems, Uniform Distribution and Applications*, DOI 10.1007/978-3-319-55357-3_16

Fig. 1 The automaton
recognising the congruence
classes modulo 3 in base 2

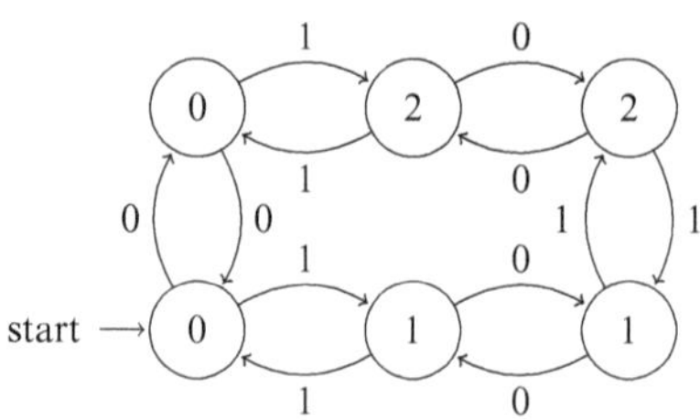

We call the pair $(q, \mathcal{N})$ a number system with base q and set of digits $\mathcal{N}$. A set
of positive integers $E \subset \mathbb{N}$ is called q-recognisable, if the language (that is the
set of the occurring digit strings) of q-adic representations of the elements of E
is recognisable by an automaton (i.e. the language is regular). As an example we
consider the automaton in Fig. 1, which determines the congruence class modulo 3
of integers in binary representation. In particular, starting with the least significant
bit of the base-2 expansion of n we follow the path whose labels correspond to the
digit string. Then the number in the final state yields the congruence class modulo 3
of n. It is easy to construct a similar automaton that recognises any given arithmetic
progression with respect to any given base. Therefore ultimately periodic sets, i.e.
unions of a finite set and a finite union of arithmetic progressions, are the easiest
examples of sets that are q-recognisable for every $q \geq 2$.

The converse direction, however, is more involved. Suppose that a set $E \subset
\mathbb{N}$ is q-recognisable as well as q'-recognisable with $q \neq q'$. If q and q' are
multiplicatively dependent, i.e. the equation $q^p = q'^{p'}$ has non-trivial solutions
over the non-negative integers (solutions of the form $(p, p') \neq (0, 0)$), then the
q-adic representation is a recoding of the q'-adic one and vice versa. Thus, any
set that is q-recognisable is trivially q'-recognisable, too. Therefore, we suppose
that q and q' are multiplicatively independent. In this case Cobham's theorem [18]
states that arithmetic progressions and unions thereof are the only sets which are
q-recognisable as well as q'-recognisable.

Theorem 1.1 ([18, Cobham's Theorem]) *Let $q, q' \geq 2$ be two multiplicatively
independent integers. A set $E \subset \mathbb{N}$ is both, q-recognisable and q'-recognisable, if
and only if E is a finite union of arithmetic progressions.*

Cobham's theorem has been generalised in several directions as U-numerations
[8], fractals [1], automata [9], and substitutions [19]. A more detailed overview can
be found in the survey of Durand and Rigo [20].

The aim of the present article is to show a variant of Cobham's theorem in
the spirit of Hansel and Safer [24] for number systems in the ring of integers
of cyclotomic fields. This will be realised in three parts: firstly we have to find
bases that induce a number system, secondly we will study their multiplicative
independence, and finally we have to make minor adaptations in order to transfer the
proof of Cobham's theorem to our generalised notion of number systems. Especially

the first two parts are done in a very general way and yield interesting results on their own that we want to outline quickly.

In the first part we are concerned with appropriate generalisations of number systems. Knuth [30] suggested to use the base $2i$ and the set of digits $\{0, 1, 2, 3\}$. Since $(2i)^2 = -4$ this corresponds to an expansion of the real and imaginary part in base -4 (see page 205f in volume 2 of Knuth's famous monograph "The art of computer programming" [31]). This idea was independently extended to $-1 \pm i$ by Khmelnik [28] and Penney [37]. In particular, they show that every Gaussian integer z can be represented uniquely as

$$z = \sum_{j=0}^{\ell} a_j(-1 + i)^j \qquad \text{with } a_j \in \{0, 1\} \text{ for } 0 \leq j \leq \ell.$$

Analogously to above, we call $(-1 + i, \{0, 1\})$ a number system in the Gaussian integers $\mathbb{Z}[i]$. This concept can be generalised in a straightforward way. Indeed, some years later Kátai and Szabó [27] proved that $(q, \mathcal{N})$ is a number system for $\mathbb{Z}[i]$ if and only if $q = -m \pm i$ with $m \geq 1$, and $\mathcal{N} = \{0, \ldots, m^2\}$. In subsequent researches, Gilbert [23] and Kátai and Kovács [25, 26] independently characterised all number systems for the ring of integers in arbitrary quadratic fields.

As already mentioned we are interested in number systems in $\mathbb{Z}[\zeta]$ where ζ is a primitive kth root of unity. Since the Gaussian integers and the Eisenstein integers are the ring of integers in the fourth and sixth cyclotomic field, respectively, this can be seen as a generalisation of the above-mentioned results for quadratic number fields.

Every base of a number system in the ring of integers of a number field is clearly also a power base of the ring of integers. Thus, the most obvious bases have the shape $a \pm \zeta$ with $a \in \mathbb{Z}$. Less obvious are bases of the shape $a \pm \eta$ and $a \pm \theta$ where $\eta = (1 + \zeta)^{-1}$ and $\theta = (1 - \zeta)^{-1}$. This is motivated by a conjecture of Bremner [10] and Robertson [43] stating that these are the only possible power integral bases (see also [22, 40–42, 44] for details and partial results). Hence, we concentrate on these three types of potential bases and have to find out for which integers a we obtain number systems.

We consider the problem from the point of view of *canonical number system*. This generalisation of the concept of number systems has been presented in 1991 by Pethő [38]: For a given monic polynomial $P \in \mathbb{Z}[x]$ let $\mathcal{N} := \{0, \ldots, |P(0)| - 1\}$. We call $(P, \mathcal{N})$ a canonical number system (CNS for short) if for each $Q \in \mathbb{Z}[x]$ there exists a polynomial $A \in \mathcal{N}[x]$ such that $Q \equiv A \pmod{P}$. For irreducible polynomials P we clearly have that $(P, \mathcal{N})$ is a CNS if and only if $(q, \mathcal{N})$ is a number system in $\mathbb{Z}[q]$ for each $q \in \mathbb{C}$ with $P(q) = 0$.

Due to their complex structure the characterisation of CNS polynomials (i.e. polynomials that induce a CNS) is a challenging task. While the quadratic case is completely solved by the above cited results, there exist only partial results and algorithms for polynomials of higher degree (e.g., [2–4, 7, 11–17, 32, 33, 47]). Furthermore, canonical number systems are closely related with the so-called *shift radix systems* (SRS for short, see [5]). From this interaction one obtains further characterisation results for CNS. For an overview concerning shift radix systems we refer the interested reader to the survey of Kirschenhofer and Thuswaldner [29].

In view of our initial problem we have to check for which $a \in \mathbb{Z}$ the minimal polynomials of $a \pm \zeta$, $a \pm \eta$ and $a \pm \theta$ are CNS-polynomials. The problem is completely solved for the quadratic cases $k = 3, 4, 6$ in [23, 27]. Concerning cyclotomic fields of higher degree the first and the last author showed in [35] that $\Phi_k(a + x)$ induces a CNS for $a \leq -\varphi(k) - 1$, where Φ_k is the kth cyclotomic polynomial and φ denotes Euler's totient function. We show analogous results for the minimal polynomials of η and θ. In fact, this will follow from our main result of the first part which is a much stronger statement. For a given polynomial $R \in \mathbb{Z}[x]$ we will give a bound for a such that $R(a + x)$ induces a CNS. As a further generalisation we will not only consider the case $a \in \mathbb{Z}$ but also the case $a \in \mathbb{Q}$. This is justified by a recently published research [48] that shows that the notion of canonical number system can be easily generalised to non-monic polynomials.

In the second part we analyse under which conditions pairs of bases are multiplicatively independent. The concept of multiplicative independence appears in other contexts, too. Senge and Straus [49] showed that the set of integers whose sum of digits is bounded with respect to different bases is finite if and only if these bases are multiplicatively independent. Steward [50] obtained an effective version of this result and Schlickewei [46] extended this to $t > 2$ multiplicative independent bases. Finally, Pethő and Tichy [39] solved the CNS analog.

The present article is organised in the following way. In Sect. 2 we give all necessary definitions and state the main results. Then, in Sect. 3, we show the results on canonical number systems. The statements concerning the multiplicative independence of the bases are proved in Sect. 4. Finally, Sect. 5 contains the proof of our variant of Cobham's theorem for number systems in the ring of integers of cyclotomic fields.

2 Definitions and Statement of Results

In the introduction we already defined the notion of canonical number system and noted that it can be generalised to non-monic polynomials. Therefore we want to give a unique definition which we will use in the sequel.

Definition 2.1 (Canonical Number System, cf. [38, 48]) Let $P = p_d x^d + p_{d-1} x^{d-1} + \cdots + p_0 \in \mathbb{Z}[x]$ be a (not necessarily monic) polynomial and $\mathcal{Z} := \mathbb{Z}[x]/(P)$ the factor ring. If each element $\gamma \in \mathcal{Z}$ can be represented as

$$\gamma = \sum_{j=0}^{\ell} a_j X^j \text{ with } a_j \in \mathcal{N} := \{0, 1, \ldots, |p_0| - 1\} \tag{1}$$

(where $X \in \mathcal{Z}$ denotes the image of $x \in \mathbb{Z}[x]$ under the canonical epimorphism) then the pair $(P, \mathcal{N})$ is called a *canonical number system* (CNS for short) and P is a *CNS polynomial*.

Kovacs [32] remarked that for each (monic, irreducible) polynomial $R \in \mathbb{Z}[x]$ there exists a (sufficiently large) integer a such that the polynomial $R(x-a)$ induces a CNS. In our first main theorem we restate this result in a more general way and give an explicit bound for a.

Theorem 2.2 *Let $R \in \mathbb{Z}[x]$ and $p, q \in \mathbb{N}$ with $(p, q) = 1$. If*

$$p/q \geq \deg(R) + \max\{\mathrm{Re}(\lambda) : R(\lambda) = 0\},$$

then $P := q^{\deg(R)} R\left(-p/q + x\right)$ is a CNS-Polynomial.

For a better understanding of the background we want to state several details concerning CNS. Consider a polynomial P and an element $\gamma \in \mathcal{Z}$. We obtain the representation (1) by using the backward division algorithm, i.e. by successive application of the maps

$$\delta_{\mathcal{N}} : \mathcal{Z} \longrightarrow \mathcal{N}, \gamma \longmapsto a \text{ the unique element of } \mathcal{N} \text{ with } \gamma \equiv a \, (\mathrm{mod}\, X),$$

$$T_P : \mathcal{Z} \longrightarrow \mathcal{Z}, \gamma \longmapsto \frac{\gamma - \delta_{\mathcal{N}}(\gamma)}{X}.$$

In particular, if $T_P^\ell(\gamma) = 0$ for some $\ell \in \mathbb{N}$, then we have

$$\gamma = \sum_{j=0}^{\ell} \delta_{\mathcal{N}}(T_P^j(\gamma)) X^j.$$

Thus, the pair $(P, \mathcal{N})$ is a CNS if and only if for each $\gamma \in \mathcal{Z}$ there exists an $\ell \in \mathbb{N}$ such that $T_P^\ell(\gamma) = 0$.

It is well known that for P an expanding polynomial (that is, all roots are outside the complex unit circle) the backward division algorithm is eventually periodic for each $\gamma \in \mathcal{Z}$. On the other hand, if there is at least one root inside the unit circle, then we can find elements of $\mathcal{Z}$ with infinite T-orbit. Thus, CNS polynomials are necessarily expanding polynomials. For polynomials whose roots have absolute value greater or equal to 1 and equality holds at least once it is up to now not clear whether all T_P-orbits are finite or not. An overview of existing results in this context can be found in the survey [29] (from the point of view of SRS).

The ring $\mathcal{Z}$ is a finitely generated free $\mathbb{Z}$-module if and only if P is monic. This fact seems to cause difficulties in the non-monic case. However, the next lemma states that in the context of CNS polynomials the monic and the non-monic case can be treated quite analogously.

Lemma 2.3 (cf. [48, Theorem 4.9]) *Let $P = p_d x^d + p_{d-1} x^{d-1} + \cdots + p_0 \in \mathbb{Z}[x]$ and $\mathcal{N} := \{0, \ldots, |p_0| - 1\}$. For each $m \in \{0, \ldots, d-1\}$ let $\omega_m := \sum_{j=0}^{m} p_{d-j} X^j$. Then the following statements are equivalent:*

(1) The pair $(P, \mathcal{N})$ is a canonical number system.

(2) For each $\gamma \in \mathcal{Z}_1$ there exists an ℓ such that $T_P^\ell(\gamma) = 0$ where $\mathcal{Z}_1$ is the free $\mathbb{Z}$-module generated by $\{1, X, \ldots, X^{d-1}\}$.

(3) For each $\gamma \in \mathcal{Z}_2$ there exists an ℓ such that $T_P^\ell(\gamma) = 0$ where $\mathcal{Z}_2$ is the free $\mathbb{Z}$-module generated by $\{\omega_0, \ldots, \omega_{d-1}\}$ (this basis is sometimes referred to as Brunotte basis).

Observe that $\mathcal{Z}_2 \subseteq \mathcal{Z}_1 \subseteq \mathcal{Z}$ where equality holds if and only if P is monic. Furthermore, $\mathcal{Z}_1$ as well as $\mathcal{Z}_2$ are closed with respect to the application of T_P.

Our most important tool for proving Theorem 2.2 will be the following monotonicity condition which appears in several articles on CNS in slightly different formulations and contexts (cf. [15]). The present form is, in fact, a non-monic version of [15, Corollary 7] which follows immediately from [48, Theorem 5.3] and [6, Theorem 3.5].

Proposition 2.4 *Let $P = p_d x^d + p_{d-1} x^{d-1} + \cdots + p_0 \in \mathbb{Z}[x]$. If $p_0 \geq 2$ and $0 < p_d \leq p_{d-1} \leq \cdots \leq p_1 < p_0$, then P is a CNS polynomial.*

Note that the bound provided in Theorem 2.2 ensures that P satisfies the condition of this proposition. But observe that Theorem 2.2 is only a sufficiency result, that is, the monotonicity condition is fulfilled even for smaller rationals p/q. However, without additional assumptions (for examples on the roots of R) it seems to be difficult to improve this bound (see also [35, Remark 1]). Furthermore we want to note that Proposition 2.4 itself provides sufficient conditions on polynomials in order to induce a CNS, that is a polynomial P can be a CNS-polynomial even when P does not satisfy the monotonicity condition. We want to summarise these considerations in the following result (which is actually a corollary to Theorem 2.2).

Proposition 2.5 *Let $P \in \mathbb{Z}[x]$ such that for each root λ we have $\mathrm{Re}(\lambda) \leq -\deg(P)$ $(\mathrm{Re}(\lambda) < -\deg(P)$ if P does not have at least 2 distinct roots). Then P satisfies the conditions of Proposition 2.4, especially, P induces a CNS.*

Finally we should mention that for a given polynomial $R \in \mathbb{Z}[x]$ it is absolutely not clear that there exists a bound $K \in \mathbb{R}$ such that $q^{\deg(R)} R(-p/q + x)$ induces a CNS for $p/q > K$ while for $p/q < K$ it does not.

Let $k \in \mathbb{N}$, denote by ζ a primitive kth root of unity, and by $\Phi_k \in \mathbb{Z}[x]$ its minimal polynomial (the kth cyclotomic polynomial). Then Theorem 2.2 immediately yields that $\Phi_k(-a + x)$ is a CNS-polynomial for an integer $a \geq \varphi(k) + 1$ (cf. [35, Theorem 1.1]).

In view of Bremner's conjecture we are also interested in the two other power integral bases induced by

$$\eta = \eta(\zeta) := (1 + \zeta)^{-1} \quad \text{and} \quad \theta = \theta(\zeta) := (1 - \zeta)^{-1}.$$

Since $\text{Re}(\eta) = \text{Re}(\theta) = 1/2$ (cf. Lemma 4.1) we obtain as corollary to Theorem 2.2.

Corollary 2.6 *Let $k \in \mathbb{N}$ (with $k \geq 3$) and $a \in \mathbb{Q}$. If $a \geq \varphi(k) + 1/2$, then the minimal polynomials of $-a + \eta$ and $-a + \theta$ are CNS polynomials. If $a \geq \varphi(k) - 1/2$, then the minimal polynomials of $-a - \eta$ and $-a - \theta$ are CNS polynomials.*

Remark 1 Observe that for odd k the kth cyclotomic field equals the $(2k)$th one. More precisely, we have $\Phi_{2k}(x) = \Phi_k(-x)$. On the other hand, if $k \equiv 0 \pmod 4$, then for each primitive kth root of unity ζ, $-\zeta$ is also a primitive kth root of unity. Finally observe that $\theta(\zeta) = \eta(-\zeta)$ for each primitive kth root of unity ζ. Thus, the results up to now already give us a quite good overview of bases for number systems in the ring of integers of the kth cyclotomic field, namely, $-a \pm \zeta$, $-a + \eta(\zeta)$ and $-a + \theta(\zeta)$ for integers $a \geq \varphi(k) + 1$ as well as $-a - \eta(\zeta)$ and $-a - \theta(\zeta)$ for integers $a \geq \varphi(k)$.

Now let us turn to conditions for the multiplicative independence of two bases.

Definition 2.7 *Let α and β be two algebraic numbers. Then we call α and β multiplicatively independent if $(p, p') = (0, 0)$ is the only pair of integers that solves the equation $\alpha^p = \beta^{p'}$.*

Note that if α and β are algebraic integers or $|\alpha|, |\beta| > 1$ or $|\alpha|, |\beta| < 1$, then it is sufficient to restrict to pairs of non-negative integers (p, p') that solve $\alpha^p = \beta^{p'}$.

In [35, 36] the authors ask whether $a + \zeta$ and $a' + \zeta$ are multiplicatively independent numbers for negative integers $a \neq a'$ and ζ a primitive kth root of unity. In fact, the two papers already contain partial results concerning this problem. The following considerations complete this question. Actually, we are able to show much more with our idea.

Theorem 2.8 *Let ζ be a root of the kth cyclotomic polynomial $\Phi_k(x)$ with $k \notin \{1, 2, 3, 4, 6\}$ and $a, a' \in \mathbb{Q}$ rational numbers with $a' < a$ such that*

$$|a + \zeta|^p = |a' + \zeta|^{p'} \tag{2}$$

holds for a pair of non-negative integers $(p, p') \neq (0, 0)$. Then one of the following conditions is necessarily satisfied.

(i) $a' < a < -\delta_2$ and $aa' < 1$;
(ii) $-\delta_2 < a' < a < 0$ and $a + a' > -\delta_2$;
(iii) $0 < a' < a < -\delta_1$ and $a + a' < -\delta_1$;
(iv) $-\delta_1 < a' < a$ and $aa' < 1$,

where

$$\delta_1 := \begin{cases} 2\cos((k-1)\pi/k) & \textit{if } k \equiv 1 \pmod 2 \\ 2\cos((k-2)\pi/k) & \textit{if } k \equiv 0 \pmod 4 \\ 2\cos((k-4)\pi/k) & \textit{if } k \equiv 2 \pmod 4 \end{cases} < 0,$$

$$\delta_2 := 2\cos(2\pi/k) > 0.$$

This theorem allows several conclusions. In particular we want to note that:

- The multiplicative independence of $|a + \zeta|$ and $|a' + \zeta|$ implies that of $a + \zeta$ and $a' + \zeta$.
- Multiplicative independence follows as soon as we have that $|a + \zeta|$ and $|a' + \zeta|$ are both larger than 1.
- For a, a' integers with $a, a' \neq 0$ the numbers $a + \zeta$ and $a' + \zeta$ are always multiplicatively independent (which answers the initial question—the excluded cases $2, 3, 4, 6$ are treated in [35, 36]).
- For $k \neq 5$ the numbers $|a + \zeta|$ and $|a' + \zeta|$ are multiplicatively independent as soon as a and a' are smaller than $-\delta_2$ since the first case can only occur for $k = 5$.
- For $k \neq 10$ the numbers $|a + \zeta|$ and $|a' + \zeta|$ are multiplicatively independent as soon as a and a' are larger than $-\delta_1$ since the last case can only occur for $k = 10$.

Finally, we want to remark that it would be interesting whether there exist a pair of rationals $(a, a') \neq (0, 0)$ (that necessarily satisfy one of the conditions (i)–(iv) such that $a + \zeta$ and $a' + \zeta$ are multiplicatively dependent.

Now we are interested in a similar result for $\eta(\zeta)$ and $\theta(\zeta)$. The situation is easier due to the fact that all Galois conjugates have real part $a + 1/2$ and $a' + 1/2$, respectively (see Lemma 4.1 in Sect. 4).

Theorem 2.9 *Let k be a positive integer with $k \notin \{1, 2, 3, 4, 6\}$, ζ a root of Φ_k, $\theta = (1 - \zeta)^{-1}$, and $a, a' \in \mathbb{Q}$ rational numbers. Then $|a + \theta|$ and $|a' + \theta|$ are multiplicatively independent provided that*

(i) $a + a' + 1 \neq 0$ and

(ii) if $k = 5$, then we assume that $(2a + 1)^2 > 3 + \frac{2\sqrt{5}}{5}$ and $(2a' + 1)^2 > 3 + \frac{2\sqrt{5}}{5}$.

Due to the close relation of $\eta(\zeta)$ and $\theta(\zeta)$ (see Remark 1) an analogue result for $\eta(\zeta)$ follows immediately.

Corollary 2.10 *Let k be a positive integer with $k \notin \{1, 2, 3, 4, 6\}$, ζ a root of Φ_k, and $a, a' \in \mathbb{Q}$ rational numbers. Then $|a + \eta|$ and $|a' + \eta|$ are multiplicatively independent provided that*

(i) $a + a' + 1 \neq 0$ and

(ii) if $k = 10$, then we assume that $(2a + 1)^2 > 3 + \frac{2\sqrt{5}}{5}$ and $(2a' + 1)^2 > 3 + \frac{2\sqrt{5}}{5}$.

Observe that in both cases the assumption that a, a' are non-zero integers ensures multiplicative independence. As before, the multiplicative independence of $|a + \theta|$ and $|a' + \theta|$ ($|a + \eta|$ and $|a' + \eta|$, respectively) immediately implies the multiplicative independence of $a + \theta$ and $a' + \theta$ ($a + \eta$ and $a' + \eta$, respectively).

Remark 2 Let ζ be a primitive kth root of unity and remind that our initial purpose was to find multiplicative independent bases of number systems in $\mathbb{Z}[\zeta]$ (see Remark 1). From Theorem 2.8 we see that numbers of the shape $a + \zeta$ and $a' + \zeta$

with integers a, a' such that $a, a' \neq 0$ are multiplicative independent. Now observe that $a + \zeta$ and $-a - \zeta$ are clearly multiplicatively dependent and multiplicative dependence is a transitive property. Therefore, all bases of the form $a \pm \zeta$ are pairwise multiplicatively independent since $a = 0$ does not yield a base of a number system.

With the same argumentation we see that all bases of the shape $a \pm \theta(\zeta)$ are pairwise multiplicative independent (where $a \neq 0$ is an integer). The same holds for all bases of the shape $a \pm \eta(\zeta)$.

For completeness it would be interesting whether numbers $a + \zeta$, $a' + \eta(\zeta)$ and $a'' + \theta(\zeta)$ are multiplicatively independent. We also do not know anything about the multiplicative independence of $a + \zeta$ and $a' + \zeta'$ for two different primitive kth roots of unity ζ and ζ' (and, in an analogous way, for the corresponding values θ and η).

With these infinite families of multiplicative independent bases we want to turn to the last part and show a result analogue to Cobham's theorem for number systems in the ring of integers $\mathbb{Z}[\zeta]$ where ζ is a primitive kth root of unity.

Consider two multiplicatively independent bases α and β. A main ingredient in the proof of Cobham's theorem is that the set of all numbers of the form $\alpha^m \beta^{-n}$ with $m, n \in \mathbb{N}$ lie dense in $\mathbb{C}$. For real α and β it is easy to show that the corresponding result holds true. However, in the complex case we do not have such a result and we are not even close to one. We circumvent this issue by using the four exponentials conjecture to obtain the desired density result.

Conjecture 2.11 (Four Exponentials Conjecture)   Let (x_1, x_2) and (y_1, y_2) be two pairs of complex numbers such that each pair is linearly independent over $\mathbb{Q}$. Then at least one of the four numbers

$$e^{x_1 y_1}, \quad e^{x_1 y_2}, \quad e^{x_2 y_1}, \quad e^{x_2 y_2}$$

is transcendental.

For a detailed account to the four exponentials conjecture we refer the reader to the book of Waldschmidt [51, Chapter 1.3 resp. Chapter 11].

Theorem 2.12 *Let $\alpha, \beta \in \mathbb{C} \setminus \{0\}$ be algebraic numbers such that $|\alpha|$ and $|\beta|$ are multiplicatively independent. If the four exponentials conjecture is true, then the set*

$$P_{\alpha,\beta} := \left\{ \frac{\alpha^m}{\beta^n} : m, n \in \mathbb{N} \right\}$$

is dense in $\mathbb{C}$.

Theorem 2.12 was proved by Hansel and Safer [24] in the special case that α and β are of the form $\alpha = -a + i$ and $\beta = -a' + i$, with a an a' positive rational integers.

Since transferring our density conjecture one obtains something very similar to the four exponentials conjecture it seems that they are very close or even equivalent. It is tempting to use the six exponentials theorem to obtain a variant of Cobham's theorem for more than two bases, but currently we have no idea how to do that.

As α is a basis of a number system in $\mathbb{Z}[\zeta]$ (with digit set $\mathcal{N}$), each element $z \in \mathbb{Z}[\zeta]$ has a unique representation of the form

$$z = \sum_{j=0}^{\ell} a_j \alpha^j \quad \text{with } a_j \in \mathcal{N} \text{ for } 0 \leq j \leq \ell \text{ and } a_\ell \neq 0.$$

We denote by $\rho_\alpha(z)$ the corresponding digit string over the alphabet $\mathcal{N}$, that is

$$\rho_\alpha(z) := a_0 a_1 \ldots a_\ell \in \mathcal{N}^*.$$

For a set $S \subset \mathbb{Z}[\zeta]$ we define the language $\rho_\alpha(S)$ by

$$\rho_\alpha(S) := \{\rho_\alpha(x) : x \in S\}$$

We call S α-recognizable if the language $\rho_\alpha(S)$ is recognizable by a finite automaton. Moreover let α and β be two multiplicatively independent bases of number systems in $\mathbb{Z}[\zeta]$. Then we call a set (α, β)-recognizable if it is α-recognizable and β-recognizable.

Our final result is a weaker form of Cobham's theorem. In particular, we do not show that the set is ultimately periodic but syndetic. For better understanding of this terminology let S be a subset of the positive integers $\mathbb{N}$. Then we call S syndetic (or with bounded gaps) if there exists $r \in \mathbb{N}$ such that $S \cap [n, n+r] \neq \emptyset$ for each $n \in \mathbb{N}$. The analog for a lattice Λ in $\mathbb{C}$ is that for a given set $S \subset \Lambda$ there exists $r \in \mathbb{R}$ such that $S \cap B(n, r) \neq \emptyset$ for each $n \in \Lambda$, where $B(n, r)$ is the closed disc with center n and radius r.

Theorem 2.13 *Let ζ be a primitive kth root of unity and consider two multiplicatively independent bases α and β for number systems in $\mathbb{Z}[\zeta]$ such that $P_{\alpha,\beta}$ is dense in $\mathbb{C}$. If S is an infinite (α, β)-recognizable subset of $\mathbb{Z}[\zeta]$, then S is syndetic.*

3 Proof of Theorem 2.2

The following lemma estimates the coefficients of a polynomial if we shift the center from 0 to a.

Lemma 3.1 *Let $R(x) \in \mathbb{Z}[x]$ and $a = \frac{p}{q} \in \mathbb{Q}$. Then*

$$R(x - a) = \sum_{n=0}^{\deg(R)} \frac{R^{(n)}(-a)x^n}{n!} \in \mathbb{Q}[x]$$

(where $R^{(n)}$ denotes the nth derivation of $R = R^{(0)}$). If R is the minimal polynomial of an algebraic $z \in \mathbb{C}$, then $P(x) := q^{\deg(R)}R(x - a)$ is the minimal polynomial of $z + a$.

Proof The first assertion is clear by Taylor's theorem. The second part follows from the observation that P is an integer polynomial, $P(\hat{z} + a) = q^{\deg(R)}R(\hat{z}) = 0$ for each Galois conjugate $\hat{z}$ of z, and the degree of P coincides with the degree of the minimal polynomial of $z - a$. $\qquad\qquad\square$

Proof of Theorem 2.2 Let $a := -p/q$. By assumption we have

$$a \leq -\deg(R) - \max\{\mathrm{Re}(\lambda) : R(\lambda) = 0\}.$$

We will show that

$$\frac{R^{(n+1)}(-a)n!}{R^{(n)}(-a)(n + 1)!} = \frac{R^{(n+1)}(-a)}{R^{(n)}(-a)(n + 1)} < 1$$

holds for all $n \in \{0, \dots, \deg(k) - 1\}$. Then, by Lemma 3.1 and Proposition 2.4, $P(x) = q^{\deg(k)}R(a + x)$ induces a CNS.

We first show the case $n = 0$. Denote by $\xi_1, \dots, \xi_t$ the t (not necessarily distinct) real roots and by $u_1 \pm iv_1, \dots, u_s \pm iv_s$ the s (not necessarily distinct) pairs of complex conjugate roots of R (that is, $2s + t = \deg(R)$). Thus, $R(x) = \prod_{j=1}^{s}(x^2 - 2u_jx + u_j^2 + v_j^2) \cdot \prod_{j=1}^{t}(x - \xi_j)$. From this we easily obtain the logarithmic derivative

$$\frac{R'(-a)}{R(-a)} = \sum_{j=1}^{s} \frac{-2a - 2u_j}{a^2 + 2u_ja + u_j^2 + v_j^2} + \sum_{j=1}^{t} \frac{1}{-a - \xi_j}$$

$$\leq \sum_{j=1}^{s} \frac{2(-a - u_j)}{a^2 + 2u_ja + u_j^2} + \sum_{j=1}^{t} \frac{1}{-a - \xi_j} = \sum_{j=1}^{s} \frac{2}{-a - u_j} + \sum_{j=1}^{t} \frac{1}{-a - \xi_j}.$$

$$(3)$$

Now observe that by the choice of a we have $-a - \xi_j \geq \deg(R)$ for all $j \in \{1, \dots, t\}$ and $-a - u_j \geq \deg(R)$ for all $j \in \{1, \dots, s\}$. Therefore we obtain

$$\sum_{j=1}^{s} \frac{2}{-a - u_j} + \sum_{j=1}^{t} \frac{1}{-a - \xi_j} \leq \sum_{j=1}^{s} \frac{2}{\deg(R)} + \sum_{j=1}^{t} \frac{1}{\deg(R)} = \frac{2s + t}{\deg(R)} = 1. \qquad (4)$$

Note that the inequality in (3) is sharp when $s >= 1$ (hence, R has non-real roots) while the inequality in (4) is sharp when at least two real parts are different. Thus, $\frac{R'(-a)}{R(-a)} < 1$ provided that R has at least 2 different roots.

For $n \geq 1$ the situation is less critical. Again we denote by $\xi_1, \dots, \xi_t$ the real roots of $R^{(n)}$ and by $u_1 \pm iv_1, \dots, u_s \pm iv_s$ the pairs of complex conjugate roots (again, $2s + t = \deg(R) - n$). As before we consider the logarithmic derivative and obtain analogously

$$\frac{R^{(n+1)}(-a)}{R^{(n)}(-a)} \le \sum_{j=1}^{s} \frac{2}{-a - u_j} + \sum_{j=1}^{t} \frac{1}{-a - \xi_j}.$$

By the Gauss–Lucas theorem, the roots of $R^{(n)}(x)$ are contained in the convex hull of the roots of $R^{(n-1)}(x)$. Hence, $\max\{\mathrm{Re}(\lambda) : R(\lambda) = 0\} \ge \max\{\mathrm{Re}(\lambda) : R^{(n)}(\lambda) = 0\}$. Similarly as above this immediately yields

$$\frac{R^{(n+1)}(-a)}{R^{(n)}(-a)(n+1)} < \frac{R^{(n+1)}(-a)}{R^{(n)}(-a)} \le \frac{2s + t}{\deg(R)} = \frac{\deg(R) - n}{\deg(R)} < 1.$$

$\square$

Remark 3 Observe that if R does not have at least 2 different roots (that is, $R(x) = (x - \xi)^n$ for some $n \in \mathbb{N}$), then Theorem 2.2 holds if we require

$$p/q > \deg(R) + \max\{\mathrm{Re}(\lambda) : R(\lambda) = 0\}(= \deg(R) + \xi).$$

4 Multiplicative Independent Bases

In this section we want to collect the proofs of the results concerning multiplicative independence.

Proof of Theorem 2.8 It is obvious that $|a + \zeta|$ and $|a' + \zeta|$ are multiplicatively independent if and only if $|a + \zeta|^2$ and $|a' + \zeta|^2$ are multiplicatively independent. Therefore we may concentrate on $|a + \zeta|^2$ and $|a' + \zeta|^2$, respectively. In particular, we prove the theorem by showing that if a pair a, a' of rational numbers does not satisfy one of the conditions (i)–(iv), then

$$|a + \zeta|^{2p} = |a' + \zeta|^{2p'} \tag{5}$$

cannot hold for a pair $(p, p') \ne (0, 0)$ of non-negative integers.

At first we claim that $aa' = 0$ contradicts (5). Indeed, suppose that one of the both rationals, say a, were 0. Then $|a + \zeta|^2 = 1$ and $|a' + \zeta|^2 = a'^2 + 1 + 2a'\mathrm{Re}(\zeta)$. Since $\mathrm{Re}(\zeta) \notin \mathbb{Q}$ for the considered values of k, (5) cannot hold.

Before we continue we need some further considerations. Denote by ζ_1 and ζ_2, respectively, two roots of $\Phi_k(x)$ such that

$$\mathrm{Re}(\zeta_1) \le \mathrm{Re}\left(\hat{\zeta}\right) \le \mathrm{Re}(\zeta_2)$$

holds for each root $\hat{\zeta}$ of $\Phi_k(x)$. Since the degree of $\Phi_k(x)$ is at least 4 we clearly have $\delta_1/2 = \mathrm{Re}(\zeta_1) < 0 < \mathrm{Re}(\zeta_2) = \delta_2/2$. Define for each $i \in \{1, 2\}$

$$\begin{aligned}
\alpha_i &:= |a + \zeta_i|^2 = a^2 + 2a\mathrm{Re}(\zeta_i) + 1, \\
\alpha_i' &:= |a' + \zeta_i|^2 = a'^2 + 2a'\mathrm{Re}(\zeta_i) + 1.
\end{aligned} \tag{6}$$

Observe that $|a + \zeta|^2$, α_1 and α_2 as well as $|a' + \zeta|^2$, α_1' and α_2' are algebraically conjugate, hence, from (5) we conclude that for $i \in \{1, 2\}$

$$\alpha_i^p = \alpha_i'^{p'}. \tag{7}$$

holds. Furthermore, this clearly implies

$$(\alpha_1/\alpha_2)^p = (\alpha_1'/\alpha_2')^{p'}. \tag{8}$$

With this we show that our initial assumption (5) cannot hold if a and a' have different signs. Indeed, suppose that $a' < 0 < a$. From (6) we easily see that $\alpha_1 < \alpha_2$, hence, $\alpha_1/\alpha_2 < 1$. On the other hand we have $\alpha_1' > \alpha_2'$. Thus, $\alpha_1'/\alpha_2' > 1$. This is a contradiction to (8) and, hence, (5).

Up to now we have shown that (5) implies either $0 < a' < a$ or $a' < a < 0$. We first concentrate on the latter case. We obviously have $0 < \alpha_2 < \alpha_1$ as well as $0 < \alpha_2' < \alpha_1'$. Furthermore, we clearly have $1 < \alpha_1 < \alpha_1'$. Thus, (7) implies

$$p > p'. \tag{9}$$

Now we distinguish three cases.

Case 1: $-\delta_2 < a' < a < 0$ In this case we have $\alpha_2, \alpha_2' < 1$. Suppose that $a + a' < -2\mathrm{Re}(\zeta_2)$. Since $(a - a') > 0$ this implies $(a - a')(a + a') + 2(a - a')\mathrm{Re}(\zeta_2) < 0$ and, hence

$$\alpha_2 = a^2 + 1 + 2a\mathrm{Re}(\zeta_2) < a'^2 + 1 + 2a'\mathrm{Re}(\zeta_2) = \alpha_2' < 1$$

which contradicts (7) since we have $p > p'$. Thus, we necessarily have $a + a' > -2\mathrm{Re}(\zeta_2) = -\delta_2$ (since a and a' are rational numbers, we always have $a + a' \neq -2\mathrm{Re}(\zeta_2)$ for the possible values of n).

Case 2: $a' < -\delta_2 < a < 0$ We see that this yields $\alpha_2 < 1$ and $\alpha_2' > 1$ which contradicts (7).

Case 3: $a' < a < -\delta_2$ We obviously have $\alpha_2 < \alpha'_2$. We calculate

$$
\frac{\alpha_1}{\alpha_2} - \frac{\alpha'_1}{\alpha'_2}
$$

$$
= \frac{\alpha_1 \alpha'_2 - \alpha'_1 \alpha_2}{\alpha_2 \alpha'_2}
$$

$$
= \frac{(a^2 + 1)2a'\mathrm{Re}(\zeta_2) + (a'^2 + 1)2a\mathrm{Re}(\zeta_1) - (a'^2 + 1)2a\mathrm{Re}(\zeta_2) - (a^2 + 1)2a'\mathrm{Re}(\zeta_1)}{\alpha_2 \alpha'_2}
$$

$$
= \frac{(a^2 a' + a' - a'^2 a - a)2\mathrm{Re}(\zeta_2) + (a'^2 a + a - a^2 a' - a')2\mathrm{Re}(\zeta_1)}{\alpha_2 \alpha'_2}
$$

$$
= \frac{(aa' - 1)(a - a')2\mathrm{Re}(\zeta_2) + (aa' - 1)(a' - a)2\mathrm{Re}(\zeta_1)}{\alpha_2 \alpha'_2}
$$

$$
= \frac{(aa' - 1)(a - a')(2\mathrm{Re}(\zeta_2) - 2\mathrm{Re}(\zeta_1))}{\alpha_2 \alpha'_2}
$$

$$
= \frac{1}{\alpha_2 \alpha'_2}(aa' - 1)(a - a')(\delta_2 - \delta_1).
$$

For $aa' > 1$ the latter expression is strictly positive, hence, $1 < \frac{\alpha'_1}{\alpha'_2} < \frac{\alpha_1}{\alpha_2}$. This violates (8) since $p > p'$.

By exploiting symmetries the case $0 < a' < a$ runs analogously. $\qquad\square$

Now we consider two bases of the form $a + \theta$ and $a' + \theta$. We start with the following basic lemma which will be useful in the sequel.

Lemma 4.1 *Let $z \in \mathbb{C} \setminus \{1\}$ with $|z| = 1$ and $a \in \mathbb{R}$. Then*

$$
\mathrm{Re}\left(a + (1 - z)^{-1}\right) = a + \frac{1}{2}, \tag{10}
$$

$$
\mathrm{Im}\left(a + (1 - z)^{-1}\right) = \frac{\mathrm{Im}(\zeta)}{2(1 - \mathrm{Re}(z))}, \tag{11}
$$

$$
\left|a + (1 - z)^{-1}\right|^2 = a^2 + a + \frac{1}{2(1 - \mathrm{Re}(z))}. \tag{12}
$$

Proof Easy exercise. $\qquad\square$

Proof of Theorem 2.9 The idea of the proof is essentially the same as that of Theorem 2.8. We concentrate on $|a + \theta|^2$ and $|a' + \theta|^2$ and we show the assertion indirectly, thus, we suppose that $|a + \theta|^2$ and $|a' + \theta|^2$ were not multiplicatively independent. Then there exists a pair of non-negative integers $(p, p') \neq (0, 0)$ such that

$$
|a + \theta|^{2p} = |a' + \theta|^{2p'}. \tag{13}
$$

At first we claim that under our conditions we always have $p \neq p'$. Indeed, $p = p'$ implies that $|a + \theta|^2 = |a' + \theta|^2$. By observing Lemma 4.1 this is possible if and only if we either have the trivial case $a = a'$ or the excluded one $a + a' + 1 = 0$.

Thus, we can suppose that $p \neq p'$ and, without loss of generality, we may assume that $0 < p < p'$. By the assumption on k there exist roots of $\Phi_k(x)$ with different real parts. We let ζ_1 be a root with the minimal real part and ζ_2 a root with the maximal real part, hence $\mathrm{Re}(\zeta_1) = \delta_1/2 < 0 < \delta_2/2 = \mathrm{Re}(\zeta_2)$ where δ_1 and δ_2 are defined in the statement of Theorem 2.8. We define for each $j \in \{1, 2\}$

$$\alpha_j := \left| a + (1 - \zeta_j)^{-1} \right|^2 = a^2 + a + \frac{1}{2(1 - \mathrm{Re}(\zeta_j))} > 0,$$

$$\alpha'_j := \left| a' + (1 - \zeta_j)^{-1} \right|^2 = a'^2 + a' + \frac{1}{2(1 - \mathrm{Re}(\zeta_j))} > 0,$$

(where we used the results of Lemma 4.1) and note that, $|a + \theta|^2$, α_1 and α_2 (and $|a' + \theta|^2$, α'_1 and α'_2, respectively) are algebraic conjugates. Therefore, from (13) follows that $\alpha_j^p = \alpha_j'^{p'}$ holds for each $j \in \{1, 2\}$. From the assumption $p < p'$ we deduce that either $\alpha_j < \alpha'_j < 1$ or $\alpha_j > \alpha'_j > 1$ holds for both, $j = 1$ and $j = 2$.

We claim that the first case cannot occur within the terms of the theorem. At first we observe that $\alpha'_2 < 1$ only if

$$1 > \mathrm{Im}(a' + (1 - \zeta_2)^{-1})^2 = \frac{\mathrm{Im}(\zeta_2)^2}{4(1 - \mathrm{Re}(\zeta_2))^2} = \frac{1 + \mathrm{Re}(\zeta_2)}{4(1 - \mathrm{Re}(\zeta_2))}.$$

This immediately yields the condition $\mathrm{Re}(\zeta_2) = \frac{\delta_2}{2} < 3/5$ which is satisfied in the case $k = 5$ only. Now, if $k = 5$, then $\alpha_2 < 1$ as well as $\alpha'_2 < 1$ must hold. By using (12) and the well-known identity $\mathrm{Re}(\zeta_2) = \cos(2\pi/5) = (\sqrt{5}-1)/4$ one readily verifies that this implies that $(2a + 1)^2 < 3 + \frac{2\sqrt{5}}{5}$ and $(2a' + 1)^2 < 3 + \frac{2\sqrt{5}}{5}$.

Thus we may concentrate on the case that $\alpha_j > \alpha'_j > 1$ holds for $j = 1$ as well as $j = 2$, which immediately implies

$$0 > \alpha'_j - \alpha_j = a'^2 - a^2 + a' - a \tag{14}$$

for $j \in \{1, 2\}$. Now observe that from (13) we also obtain that

$$(\alpha_2 \alpha_1^{-1})^p = (\alpha'_2 \alpha_1'^{-1})^{p'}. \tag{15}$$

must hold. We compute

$$\alpha_2 - \alpha_1 = \frac{1}{2(1 - \mathrm{Re}(\zeta_2))} - \frac{1}{2(1 - \mathrm{Re}(\zeta_1))} > 0$$

which shows that $\alpha_2\alpha_1^{-1} > 1$, and we estimate

$$\alpha_2\alpha_1' - \alpha_1\alpha_2' = \left(a'^2 + a' - a^2 - a\right) \frac{1}{2(1 - \mathrm{Re}(\zeta_2))} + \left(a^2 + a - a'^2 - a'\right) \frac{1}{2(1 - \mathrm{Re}(\zeta_1))}$$

$$= \frac{1}{2}\left(a'^2 + a' - a^2 - a\right) \left(\frac{1}{1 - \mathrm{Re}(\zeta_2)} - \frac{1}{1 - \mathrm{Re}(\zeta_1)}\right) < 0,$$

where we used (14) for the last inequality. Thus, $1 < \alpha_2\alpha_1^{-1} < \alpha_2'\alpha_1'^{-1}$. This contradicts (15) since we assumed $p < p'$. $\qquad\square$

Remark 4 We have seen that when $a + a' + 1 = 0$, then we possibly have $|a + \theta|^p = |a' + \theta|^p$ for a positive integer p. The question remains whether this is possible for distinct $a, a' \in \mathbb{Q}$? The methods from above do not seem to work here.

5   Complex Bases and Density Properties

This short section is devoted to the proof of Theorem 2.12. Throughout the section we suppose that $\alpha = ae^{i\psi}$ and $\beta = be^{i\omega}$ are bases of number systems in $\mathbb{Z}[\zeta]$ (with ζ a primitive kth root of unity) such that $a = |\alpha|$ and $b = |\beta|$ are multiplicatively independent. With these notations at hand the proof of Theorem 2.12 is an immediate consequence of the following two lemmas.

Lemma 5.1 ([24, Lemme 1]) *The set*

$$P_{\alpha,\beta} = \left\{\frac{\alpha^m}{\beta^n} : m, n \in \mathbb{N}\right\}$$

is dense in $\mathbb{C}$*, if*

$$\frac{\log b}{\log a}, \quad \frac{\psi}{2\pi}\frac{\log b}{\log a} - \frac{\omega}{2\pi}, \quad 1$$

are linearly independent over $\mathbb{Q}$.

Lemma 5.2 ([24, Lemma 2]) *If the four exponentials conjecture, Conjecture 2.11, holds, then*

$$\frac{\log b}{\log a}, \quad \frac{\psi}{2\pi}\frac{\log b}{\log a} - \frac{\omega}{2\pi}, \quad 1$$

are linearly independent over $\mathbb{Q}$.

Recall that a set $S \subset \mathbb{Z}[\zeta]$ is α-recognizable if the set of representations $\rho_\alpha(S) = \{\rho_\alpha(x) : x \in S\}$ is recognizable in $\mathcal{N}^*$. Using the Nerode equivalence (cf. Sakarovitch [45]) this means that a set S is α-recognizable if and only if the equivalence relation

$\mathbb{Z}[\zeta]_S^\alpha$ on $\mathcal{N}^*$ defined by

$$u \mathbb{Z}[\zeta]_S^\alpha v :\Leftrightarrow \left(\forall w \in \mathcal{N}^* : uw \in \rho_\alpha(S) \Leftrightarrow vw \in \rho_\alpha(S) \right)$$

is of finite index (cf. Proposition 9.3.3 of [21]).

Since ρ_α is a bijection from $\mathbb{Z}[\zeta] \setminus \{0\}$ into $(\mathcal{N} \setminus \{0\}) \mathcal{N}^*$, we can pull this definition back to $\mathbb{Z}[\zeta]$. In particular, a set $S \subset \mathbb{Z}[\zeta]$ is α-recognizable if the equivalence relation $\mathbb{Z}[\zeta]_S^\alpha$ on $\mathbb{Z}[\zeta]$ defined by

$$x \mathbb{Z}[\zeta]_S^\alpha y \Leftrightarrow \left(\forall w \in \mathcal{N}^* : \rho_\alpha(x) w \in \rho_\alpha(S) \Leftrightarrow \rho_\alpha(y) w \in \rho_\alpha(S) \right)$$

is of finite index.

For an $s \in \mathbb{N}$ denote by $\mathbb{Z}[\zeta]_s$ the subset of elements of $\mathbb{Z}[\zeta]$ whose expansion with respect to the base α has at most length s, i.e.

$$\mathbb{Z}[\zeta]_s := \left\{ \sum_{j=0}^{s-1} a_j \alpha^j : a_j \in \mathcal{N} \text{ for } 0 \leq j < s \right\}.$$

Proposition 5.3 *Let S' be an infinite equivalence class of the relation $\mathbb{Z}[\zeta]_S^\alpha$ and $s \in \mathbb{N}^*$. There is a finite set of positive integers F_s such that for each $x \in \mathbb{Z}[\zeta]$, there exists $\lambda \in F_s$ and $z \in \mathbb{Z}[\zeta]_{\lambda s}$ such that*

$$x \alpha^{\lambda s} + z \in S'.$$

Proof This is Proposition 8 of Hansel and Safer [24] replacing the estimate for the length of expansion by the corresponding one for canonical number systems due to Kovács and Pethő [34]. $\square$

Proposition 5.4 ([24, Proposition 9]) *Let S' be an infinite equivalence class of the relation $\mathbb{Z}[\zeta]_S^\alpha$. There exist $s \in \mathbb{N}$ and $z \in \mathbb{Z}[\zeta]_s$ such that $S' \alpha^s + z \subset S'$.*

Now we have all the tools needed for the

Proof of Theorem 2.13 Let S' be one of the infinite equivalence classes of $\mathbb{Z}[\zeta]_S^\alpha$ contained in S. It suffices to show the theorem for S'.

An application of Proposition 5.4 yields the existence of $s \in \mathbb{N}$ and $z \in \mathbb{Z}[\zeta]_s$ such that

$$S' \alpha^s + z \subset S'. \tag{16}$$

Let F_s be the finite set whose existence is guaranteed by an application of Proposition 5.3 and set $\overline{\lambda} = \max F_s$. Furthermore let $x \in \mathbb{Z}[\zeta]$. Then there exists $\lambda \in F_s$ and $z_0 \in \mathbb{Z}[\zeta]_{\lambda s}$ such that

$$x \alpha^{\lambda s} + z_0 \in S'. \tag{17}$$

Now we recursively show for each $n \in \mathbb{N}$ the existence of $z_n \in \mathbb{Z}[\zeta]_{(\lambda+n)s}$ such that

$$x\alpha^{(\lambda+n)s} + z_n \in S'. \tag{18}$$

For $n = 0$ the existence follows from (17). Now suppose that we already found $z_0, z_1, \ldots, z_{n-1}$ satisfying (18). Then it follows from (16) that

$$\left(x\alpha^{(\lambda+(n-1)s)} + z_{n-1}\right)\alpha^s + z \in S'.$$

Since $z_{n-1} \in \mathbb{Z}[\zeta]_{(\lambda+(n-1))s}$ and $z \in \mathbb{Z}[\zeta]_s$, we have that $z_{n-1}\alpha^s + z \in \mathbb{Z}[\zeta]_{(\lambda+n)s}$. Therefore (18) is satisfied for $z_n = z_{n-1}\alpha^s + z$.

We set $n = \bar{\lambda} - \lambda$. Then we get that for each $x \in \mathbb{Z}[\zeta]$ there exists a $z_x \in \mathbb{Z}[\zeta]_{\bar{\lambda}s}$ such that

$$x\alpha^{\bar{\lambda}s} + z_x \in S'.$$

Let $y \in \mathbb{Z}[\zeta]$ and let q and r be such that

$$y = q\alpha^{\bar{\lambda}s} + r \quad \text{with} \quad N(r) < N\left(\alpha^{\bar{\lambda}s}\right).$$

Therefore $q\alpha^{\bar{\lambda}s} + z_q \in S'$ and we have that

$$\left|y - \left(q\alpha^{\bar{\lambda}s} + z_q\right)\right| = |r - z_q| < 4|\alpha|^{\bar{\lambda}s+1}.$$

Thus for each $y \in \mathbb{Z}[\zeta]$ we have that $B\left(y, 4|\alpha|^{\bar{\lambda}s+1}\right) \cap S' \neq \emptyset$ and therefore S' is syndetic. $\qquad\square$

Acknowledgements We want to explicitly give thanks to the anonymous referees. Due to their hints and suggestions we were able to improve a lot the quality and readability of this article. The second author's research was supported by the Austrian Research Foundation (FWF), Project P23990.

References

1. B. Adamczewski, J. Bell, An analogue of Cobham's theorem for fractals. Trans. Am. Math. Soc. **363**(8), 4421–4442 (2011). MR2792994
2. S. Akiyama, A. Pethő, On canonical number systems. Theor. Comput. Sci. **270**(1–2), 921–933 (2002)
3. S. Akiyama, H. Rao, New criteria for canonical number systems. Acta Arith. **111**(1), 5–25 (2004). MR2038059 (2005d:11007)
4. S. Akiyama, H. Brunotte, A. Pethő, Cubic CNS polynomials, notes on a conjecture of W.J. Gilbert. J. Math. Anal. Appl. **281**(1), 402–415 (2003). MR1980100 (2004j:11009)

5. S. Akiyama, T. Borbély, H. Brunotte, A. Pethő, J.M. Thuswaldner, Generalized radix representations and dynamical systems. I. Acta Math. Hungar. **108**(3), 207–238 (2005). MR2162561 (2006i:37023)
6. S. Akiyama, H. Brunotte, A. Pethő, J.M. Thuswaldner, Generalized radix representations and dynamical systems. II. Acta Arith. **121**(1), 21–61 (2006). MR2216302
7. S. Akiyama, H. Brunotte, A. Pethő, Reducible cubic CNS polynomials. Period. Math. Hungar. **55**(2), 177–183 (2007). MR2375040 (2008m:11058)
8. A. Bertrand-Mathis, Comment écrire les nombres entiers dans une base qui n'est pas entière. Acta Math. Hungar. **54**(3–4), 237–241 (1989). MR1029085
9. B. Boigelot, J. Brusten, A generalization of Cobham's theorem to automata over real numbers. Theor. Comput. Sci. **410**(18), 1694–1703 (2009). MR2508527
10. A. Bremner, On power bases in cyclotomic number fields. J. Number Theory **28**(3), 288–298 (1988). MR932377
11. H. Brunotte, On trinomial bases of radix representations of algebraic integers. Acta Sci. Math. (Szeged) **67**(3–4), 521–527 (2001)
12. H. Brunotte, Characterization of CNS trinomials. Acta Sci. Math. (Szeged) **68**(3–4), 673–679 (2002). MR1954540 (2003k:11157)
13. H. Brunotte, On cubic CNS polynomials with three real roots. Acta Sci. Math. (Szeged) **70**(3–4), 495–504 (2004). MR2107523 (2005h:11055)
14. H. Brunotte, Symmetric CNS trinomials. Integers **9**(A19), 201–214 (2009). MR2534909 (2010g:11039)
15. H. Brunotte, A unified proof of two classical theorems on CNS polynomials. Integers **12**(4), 709–721 (2012). MR2988542
16. H. Brunotte, Unusual CNS polynomials. Math. Pannon. **24**(1), 125–137 (2013). MR3234910
17. H. Brunotte, A. Huszti, A. Pethő, Bases of canonical number systems in quartic algebraic number fields. J. Théor. Nombres Bordeaux **18**(3), 537–557 (2006). MR2330426 (2008g:11179)
18. A. Cobham, On the base-dependence of sets of numbers recognizable by finite automata. Math. Syst. Theory **3**, 186–192 (1969). MR0250789
19. F. Durand, Cobham's theorem for substitutions. J. Eur. Math. Soc. (JEMS) **13**(6), 1799–1814 (2011). MR2835330
20. F. Durand, M. Rigo, On Cobham's theorem, in *Handbook of Automata: From Mathematics to Applications* (European Mathematical Society Publishing House, Zurich, 2017)
21. S. Eilenberg, *Automata, Languages, and Machines. Vol. A*, Pure and Applied Mathematics, vol. 58 (Academic Press [A subsidiary of Harcourt Brace Jovanovich, Publishers], New York, 1974). MR0530382
22. I. Gaál, L. Robertson, Power integral bases in prime-power cyclotomic fields. J. Number Theory **120**(2), 372–384 (2006). MR2257552
23. W.J. Gilbert, Radix representations of quadratic fields. J. Math. Anal. Appl. **83**(1), 264–274 (1981). MR632342 (83m:12005)
24. G. Hansel, T. Safer, Vers un théorème de Cobham pour les entiers de Gauss. Bull. Belg. Math. Soc. Simon Stevin **10**(Suppl.), 723–735 (2003). MR2073023 (2005c:68236)
25. I. Kátai, B. Kovács, Kanonische Zahlensysteme in der Theorie der quadratischen algebraischen Zahlen. Acta Sci. Math. (Szeged) **42**(1–2), 99–107 (1980). MR576942 (81i:12002)
26. I. Kátai, B. Kovács, Canonical number systems in imaginary quadratic fields. Acta Math. Acad. Sci. Hungar. **37**(1–3), 159–164 (1981). MR616887 (83a:12005)
27. I. Kátai, J. Szabó, Canonical number systems for complex integers. Acta Sci. Math. (Szeged) **37**(3–4), 255–260 (1975). MR0389759 (52 #10590)
28. S.I. Khmelnik, Specialized digital computer for operations with complex numbers. Quest. Radio Electronics **XII**(2) (1964), 60–82; in Russian.
29. P. Kirschenhofer, J.M. Thuswaldner, Shift radix systems—a survey. RIMS Kôkyûroku Bessatsu **B46**, 1–59 (2014). MR3330559
30. D.E. Knuth, A imaginary number system. CACM **3**(4), 245–247 (1960)
31. D.E. Knuth, *The Art of Computer Programming*. Vol. 2: Seminumerical Algorithms (Addison-Wesley, Reading, 1969). MR0286318 (44 #3531)

32. B. Kovács, Canonical number systems in algebraic number fields. Acta Math. Acad. Sci. Hungar. **37**(4), 405–407 (1981). MR619892 (82j:12014)
33. B. Kovács, A. Pethő, Number systems in integral domains, especially in orders of algebraic number fields. Acta Sci. Math. (Szeged) **55**(3–4), 287–299 (1991). MR1152592 (92m:11116)
34. B. Kovács, A. Pethő, On a representation of algebraic integers. Studia Sci. Math. Hungar. **27**(1–2), 169–172 (1992). MR1207568
35. M.G. Madritsch, V. Ziegler, An infinite family of multiplicatively independent bases of number systems in cyclotomic number fields. Acta Sci. Math. (Szeged) **81**(1–2), 33–44 (2015). MR3381872
36. M.G. Madritsch, V. Ziegler, On multiplicatively independent bases in cyclotomic number fields. Acta Math. Hungar. **146**(1), 224–239 (2015). MR3348190
37. W. Penney, A "binary" system for complex numbers. J. ACM **12**(2), 247–248 (April 1965)
38. A. Pethő, On a polynomial transformation and its application to the construction of a public key cryptosystem. *Computational Number Theory (Debrecen, 1989)* (Walter de Gruyter, Berlin, 1991), pp. 31–43. MR1151853 (93e:94011)
39. A. Pethő, R.F. Tichy, S-unit equations, linear recurrences and digit expansions. Publ. Math. Debr. **42**(1–2), 145–154 (1993). MR1208858
40. G. Ranieri, Générateurs de l'anneau des entiers d'une extension cyclotomique. J. Number Theory **128**(6), 1576–1586 (2008). MR2419179
41. L. Robertson, Power bases for cyclotomic integer rings. J. Number Theory **69**(1), 98–118 (1998). MR1611089
42. L. Robertson, Power bases for 2-power cyclotomic fields. J. Number Theory **88**(1), 196–209 (2001). MR1825999
43. L. Robertson, Monogeneity in cyclotomic fields. Int. J. Number Theory **6**(7), 1589–1607 (2010). MR2740723
44. L. Robertson, R. Russell, A hybrid Gröbner bases approach to computing power integral bases. Acta Math. Hungar. **147**(2), 427–437 (2015). MR3420587
45. J. Sakarovitch, Elements of Automata Theory (Cambridge University Press, Cambridge, 2009); Translated from the 2003 French original by Reuben Thomas. MR2567276
46. H.P. Schlickewei, Linear equations in integers with bounded sum of digits. J. Number Theory **35**(3), 335–344 (1990). MR1062338
47. K. Scheicher, J.M. Thuswaldner, On the characterization of canonical number systems. Osaka J. Math. **41**(2), 327–351 (2004). MR2069090 (2005c:11013)
48. K. Scheicher, P. Surer, J.M. Thuswaldner, C.E. van de Woestijne, Digit systems over commutative rings. Int. J. Number Theory **10**(6), 1459–1483 (2014)
49. H.G. Senge, E.G. Straus, PV-numbers and sets of multiplicity. Period. Math. Hungar. **3**, 93–100 (1973); Collection of articles dedicated to the memory of Alfréd Rényi, II. MR0340185
50. C.L. Stewart, On the representation of an integer in two different bases. J. Reine Angew. Math. **319**, 63–72 (1980). MR586115
51. M. Waldschmidt, *Diophantine Approximation on Linear Algebraic Groups*. Grundlehren der Mathematischen Wissenschaften [Fundamental Principles of Mathematical Sciences], vol. 326 (Springer, Berlin, 2000); Transcendence properties of the exponential function in several variables. MR1756786

Refined Estimates for Exponential Sums and a Problem Concerning the Product of Three *L*-Series

Werner Georg Nowak

Abstract This article deals with estimates for *single* exponential sums, combining tools from the classic Van der Corput's theory with an ingredient from M. Huxley's work. Further, a very precise way of balancing terms is applied with gain. The result obtained is used to derive asymptotic estimates for the coefficients of products of three Dirichlet *L*-series, as was initiated by Friedlander and Iwaniec (Can. J. Math. 57(3):494–505, 2005).

2010 *Mathematics Subject Classification* 11M06, 11M41, 11N37

1 Introduction: The Classic Van der Corput Method

This article can be considered as an addendum to, resp., a continuation of the author's previous paper [16]. It is concerned with estimations of *single* exponential sums

$$E_{F,I} := \sum_{n \in I} e^{2\pi i F(n)}$$

where F is a real function with derivatives of sufficiently high order, defined on an interval I of length $M > 1$. Throughout, $T > 0$ will be another real parameter, satisfying $|\log T| \asymp \log M$. Furthermore, for any integer $j \geq 2$, C_j will denote the condition

$$\left| F^{(j)} \right| \asymp \frac{T}{M^j} \tag{C_j}$$

W.G. Nowak (✉)
Institut für Mathematik, Universität für Bodenkultur (BOKU), Wien, Austria
e-mail: nowak@boku.ac.at

© Springer International Publishing AG 2017
C. Elsholtz, P. Grabner (eds.), *Number Theory – Diophantine Problems, Uniform Distribution and Applications*, DOI 10.1007/978-3-319-55357-3_17

throughout I. Van der Corput's classic method basically consists of applying Poisson's formula to $E_{F,I}$ and to estimate, resp., evaluate asymptotically the integrals obtained. Its simplest result tells us that C_2 implies

$$E_{F,I} \ll T^{1/2} + MT^{-1/2} . \tag{1}$$

For fairly recent monograph references on this, see, e.g., Krätzel [13], and Graham and Kolesnik [5], and also the older literature cited there.

The second important tool of the Van der Corput's method is known as the *differencing lemma*, which was discovered independently also by Weyl [21]: For any Q with $1 \le Q \le M$,

$$E_{F,I} \ll \frac{M}{\sqrt{Q}} + \left(\frac{M}{Q} \sum_{1 \le q \le Q} \left| \sum_{n:\, n,n+q \in I} e^{2\pi i (F(n+q)-F(n))} \right| \right)^{1/2} . \tag{2}$$

See, e.g., Krätzel [13, Theorem 2.5], or also Graham and Kolesnik [5, pp. 10–11].

2   The Fine Art of Balancing Terms

An obvious way to combine (1) and (2) is to estimate the inner sum in (2) by (1). Assuming C_3 for F, by the mean-value theorem

$$F''(u + q) - F''(u) \asymp q F'''(\xi) \asymp qT/M^3 ,$$

hence $F(u + q) - F(u)$ satisfies C_2 for T replaced by qT/M. The inner sum in (2) is $\ll (qT/M)^{1/2} + M (qT/M)^{-1/2}$ and, altogether,

$$E_{F,I} \ll MQ^{-1/2} + (MQT)^{1/4} + M^{5/4}(QT)^{-1/4} . \tag{3}$$

In order to minimize this expression by an appropriate choice of Q, we observe that one term on the right-hand side increases with Q while the other two decrease. In the literature frequently (see, e.g., [19, Theorem 5.11] or [13, Theorem 2.6]) the a priori assumption was made that the term $M^{5/4}(QT)^{-1/4}$ is negligible compared to $MQ^{-1/2}$, hence only the latter was balanced against $(MQT)^{1/4}$, which gives $Q \asymp MT^{-1/3}$, and overall

$$E_{F,I} \ll M^{1/2}T^{1/6} + MT^{-1/6} , \tag{4}$$

as a consequence of C_3.

However, it turned out that it is more favorable to take all terms into account for balancing. The underlying principle may be stated in satisfactory generality as follows.

Lemma 1 *For integers $J, K \geq 1$, consider an expression*

$$A(X) = \sum_{j=1}^{J} a_j X^{\alpha_j} + \sum_{k=1}^{K} b_k X^{-\beta_k} \,,$$

where X is a positive real parameter, the α's and β's are positive numerical constants, and the a's and b's are positive variables. Furthermore, let reals $X_2 > X_1 \geq 0$ be given. Then there exists a value $X^* \in [X_1, X_2]$ such that

$$A(X^*) \ll \sum_{j=1}^{J} \sum_{k=1}^{K} \left(b_k^{\alpha_j} a_j^{\beta_k} \right)^{1/(\alpha_j + \beta_k)} + \sum_{j=1}^{J} a_j X_1^{\alpha_j} + \sum_{k=1}^{K} b_k X_2^{-\beta_k} \,,$$

where the $\ll$-constant depends on J, K and the α's, β's only.

Proof A result of this type was first established by van der Corput [20] who concentrated on the case $X_1 = 0$. Surprisingly, he seems to have never used this tool in his later works; it also has been ignored by the monographs [19] and [13] and most of the literature cited there. The general case was first dealt with by Srinivasan [18]. See also [5, Lemma 2.4].   □

By distinguishing cases, it is easy to verify that this estimate is at least not worse than what is obtained by balancing only $a_1 X^{\alpha_1}$ against $b_1 X^{-\beta_1}$ (say).

By the way, if Lemma 1 is used in (3), one gets

$$E_{F,I} \ll M^{1/2} T^{1/6} + M^{3/4} + M T^{-1/4} \,, \tag{5}$$

which in all cases is at least not worse than (4). See [5, Theorem 2.6].

3   Huxley's Contribution and a First Hybrid Result

Based on earlier deep work by Bombieri and Iwaniec [2] and Iwaniec and Mozzochi [11], around the turn of the millennium Huxley [6–8] elaborated a new approach which he called "Discrete Hardy-Littlewood method."

For the single exponential sum, his sharpest result reads (see [8, Prop. 1, formulae (1.10)–(1.13), and Theorem 1]): *If*[1]

$$T^{141/328+\epsilon} \ll M \ll T^{181/328} \,, \tag{6}$$

---

[1] For simplicity, we avoid logarithmic factors and replace them by an ϵ in the exponent where applicable. Throughout, ϵ denotes an arbitrarily small positive constant, not the same at each occurrence.

and if C_j *is fulfilled for* $j = 2, 3, 4$, *then it follows that*[2]

$$E_{F,I} \ll M^{1/2} T^{32/205+\epsilon} . \tag{7}$$

In the previous paper [16], the author combined Van der Corput's and Huxley's bounds, obtaining as the simplest result, for any fixed $\epsilon > 0$,

$$M^{-\epsilon} E_{F,I} \ll M^{\frac{1}{2}} T^{\frac{32}{205}} + T^{\frac{751}{1968}} + M^{\frac{871}{1086}} + M T^{-1/2} , \tag{8}$$

under the condition that C_2, C_3, and C_4 are satisfied. This is the case $r = 4$ of [16, Theorem 1]. The proof of this result is quite simple and natural[3]: If (6) is satisfied, then (7) is used. In all other cases, either (1) or (4) is applied.

Applying the differencing lemma repeatedly to (8), Theorem 1 of [16] has been established. In particular, for $r = 5$ it tells us that C_3, C_4, and C_5 imply that

$$M^{-\epsilon} E_{F,I} \ll M^{\frac{679}{948}} T^{\frac{16}{237}} + M^{\frac{18,997}{45,504}} T^{\frac{3755}{22,752}} + M^{\frac{1957}{2172}} + M^{\frac{2101}{1896}} T^{-\frac{205}{948}} . \tag{9}$$

However, in this argument, for balancing only $\frac{M}{\sqrt{Q}}$ and the term coming ultimately from Huxley's bound (7) have been taken into account.

In the present paper we provide a sharper estimate, using Lemma 1 in order to take care of all terms for balancing.

4   Refined Balancing: Applying Lemma 1

Theorem 1 *For length*$(I) \asymp M > 1$, $T > 0$, *with* $|\log T| \asymp \log M$, *and F a real function with six continuous derivatives on the interval I, for which* C_3, C_4, *and* C_5 *are satisfied, it is true that*

$$M^{-\epsilon} E_{F,I} \ll M^{\frac{679}{948}} T^{\frac{16}{237}} + M^{\frac{1}{2}} T^{\frac{751}{5438}} + M^{\frac{1957}{2172}} + M T^{-1/4} ,$$

for every $\epsilon > 0$.[4]

Proof We start from (2) and estimate the inner exponential sum by (8), noting that $F(u + q) - F(u)$ as a function of u satisfies C_2, C_3, and C_4, with T replaced by $\frac{qT}{M}$.

---

[2]This may be compared with (4) and (5), noting that $\frac{32}{205} = 0.156\cdots < \frac{1}{6}$. Further, since $\frac{141}{328} = 0.429878\ldots$, $\frac{181}{328} = 0.5518\ldots$, the condition (6) roughly restricts the validity of (7) to M not so far away from $\sqrt{T}$.

[3]In fact, the idea basically can be traced back to a much earlier paper by Müller and the author [15]. We note furthermore that (8) cannot be improved by replacing (4) by (5).

[4]By distinguishing cases, it is easy to verify that this result is at least not weaker than (9). See also the remark after the statement of Lemma 1.

Hence,[5] for every $Q \in [1, M]$,

$$E_{F,I} \ll_{(\epsilon)} \frac{M}{\sqrt{Q}} + \left(\frac{M}{Q} \sum_{1 \leq q \leq Q} \left(M^{\frac{1}{2}} \left(\frac{qT}{M} \right)^{\frac{32}{205}} + \left(\frac{qT}{M} \right)^{\frac{751}{1968}} + M^{\frac{871}{1086}} + M \left(\frac{qT}{M} \right)^{-\frac{1}{2}} \right) \right)^{\frac{1}{2}}$$

$$\ll \frac{M}{\sqrt{Q}} + M^{\frac{551}{820}} (QT)^{\frac{16}{205}} + M^{\frac{1217}{3936}} (QT)^{\frac{751}{3936}} + M^{\frac{1957}{2172}} + M^{\frac{5}{4}} (QT)^{-\frac{1}{4}}$$

$$\ll M^{\frac{1957}{2172}} + \sum_{j=1}^{2} a_j Q^{\alpha_j} + \sum_{k=1}^{2} b_k Q^{-\beta_k} \, ,$$

where

$$\alpha_1 = \frac{16}{205}, \ \alpha_2 = \frac{751}{3936}, \ \beta_1 = \frac{1}{2}, \ \beta_2 = \frac{1}{4} \, ,$$

$$a_1 = M^{\frac{551}{820}} T^{\frac{16}{205}} \, , \ a_2 = M^{\frac{1217}{3936}} T^{\frac{751}{3936}} \, , \ b_1 = M \, , \ b_2 = M^{\frac{5}{4}} T^{-\frac{1}{4}} \, .$$

Applying Lemma 1, we see that there is a choice of $Q \in [1, M]$ such that

$$E_{F,I} \ll_{(\epsilon)} M^{\frac{1957}{2172}} + \sum_{j=1}^{2} \sum_{k=1}^{2} \left(b_k^{\alpha_j} a_j^{\beta_k} \right)^{1/(\alpha_j + \beta_k)} + \sum_{j=1}^{2} a_j + \sum_{k=1}^{2} b_k M^{-\beta_k}$$

$$= M^{\frac{1957}{2172}} + M^{\frac{871}{1076}} + M^{\frac{1243}{1735}} + M^{\frac{679}{948}} T^{\frac{16}{237}} \tag{10}$$

$$+ M^{\frac{551}{820}} T^{\frac{16}{205}} + M^{\frac{1217}{3936}} T^{\frac{751}{3936}} + \sqrt{M} \left(T^{\frac{751}{5438}} + 1 \right) + \frac{M}{T^{\frac{1}{4}}} \, .$$

Computing the exponents numerically, it is clear that of the terms which contain a power of M only, $M^{\frac{1957}{2172}}$ is the largest one, hence the others are negligible. Furthermore, $M^{\frac{551}{820}} T^{\frac{16}{205}}$ is larger than $M^{\frac{679}{948}} T^{\frac{16}{237}}$ if and only if $T > M^{269/64}$. In this case, both terms are larger than M which is the trivial bound for the exponential sum. Hence $M^{\frac{551}{820}} T^{\frac{16}{205}}$ can be omitted. Similarly, $M^{\frac{1217}{3936}} T^{\frac{751}{3936}}$ exceeds $\sqrt{M} T^{\frac{751}{5438}}$ if and only if $T > M^{2719/751}$. In this case, again both terms are larger than the trivial bound M, hence $M^{\frac{1217}{3936}} T^{\frac{751}{3936}}$ is negligible in (10). Altogether, (10) thus simplifies to what is stated in Theorem 1 whose proof is thereby complete. $\qquad \square$

Corollary 1.1 *The assertion of Theorem 1 remains true for any interval* $I' \subset I$.[6]

Proof If length$(I') \gg M$, there is nothing to prove. Otherwise, $I \setminus I'$ consists of (at most) two connected components of which at least one—call it I''—is of length

---

[5] In what follows, $A_1 \ll_{(\epsilon)} A_2$ stands for $M^{-\epsilon} A_1 \ll A_2$, for any expressions A_1, A_2.
[6] An analogous corollary is true for Theorem 2 below.

$\gg M$. Applying Theorem 1 to both $I' \cup I''$ and I'' and subtracting, we readily deduce the corollary. $\qquad\square$

5   Applying the Van der Corput Transformation

For certain applications it is useful to submit the exponential sum $E_{F,I}$ first to a process called the Van der Corput transformation, which basically consists of using Poisson's formula and evaluating the arising integrals by the method of stationary phase. This yields a new exponential sum which is then bounded by Theorem 1.

Theorem 2 *For a certain interval $[A, B] \subseteq I$, let $F \in C^6(I)$ and suppose that, for positive parameters $X \gg 1$ and Y, we have $B - A \ll X$ and*

$$F^{(j)} \ll X^{2-j}Y^{-1} \quad for \quad j = 2, 3, 4 \; F'' \geq c_0 Y^{-1} \,,$$

throughout the interval I, with some constant $c_0 > 0$. Let J denote the image of I under F', and φ the inverse function of F'. Suppose also that $F^{(3)}$ has only $O(1)$ sign changes on $[A, B]$, and that there exists a positive parameter T satisfying $|\log T| \asymp |\log(X/Y)|$ such that, for all $v \in J$,

$$\frac{d^j}{dv^j} \left(F(\varphi(v)) - v\varphi(v) \right) = -\varphi^{(j-1)}(v) \asymp \frac{T}{(X/Y)^j} \; for \; j = 3, 4, 5 \,.$$

Then it follows that

$$(1 + T)^{-\epsilon} \sum_{A < n \leq B} e^{2\pi i F(n)} \ll \log(2 + X/Y)$$

$$+ \sqrt{Y} \left(1 + \left(\frac{X}{Y} \right)^{\frac{679}{948}} T^{\frac{16}{237}} \right.$$

$$\left. + \left(\frac{X}{Y} \right)^{\frac{1}{2}} T^{\frac{751}{5438}} + \left(\frac{X}{Y} \right)^{\frac{1957}{2172}} + \frac{X}{Y} T^{-1/4} \right) .$$

Proof In all essentials, the proof is the same as that of [16, Theorem 2]. $\qquad\square$

6   Applications

First of all, it must be admitted that, for many classic problems, like the Gaussian circle problem, the Dirichlet divisor problem, and the order of $\zeta(\frac{1}{2}+it)$, the approach via *single* exponential sums fails to yield the best results, but is superseded by techniques dealing with *averages* over exponential sums or multiple exponential sums methods.

A large class of applications concerns estimates for the *discrepancy of sequences*. Given a finite sequence

$$\omega = (F(1), F(2), \ldots, F(N)),$$

with F as before, then the inequality of Erdős–Turán (see, e.g., the monograph by Drmota and Tichy [3]) tells us that its discrepancy can be bounded by

$$D(\omega) := \sup_{0 \le a < b \le 1} \left| \frac{1}{N} \sum_{n=1}^{N} \mathbf{c}_{[a,b]}(\{F(n)\}) - (b-a) \right|$$

$$\ll \frac{1}{H} + \frac{1}{N} \sum_{h=1}^{H} \frac{1}{h} \left| \sum_{n=1}^{N} e^{2\pi i h F(n)} \right| ,$$

with an integer H at our disposal. Using an appropriate choice of our bounds provided so far to estimate the inner exponential sum, one readily gets discrepancy bounds for a whole class of sequences.

6.1 A Problem Considered by Friedlander and Iwaniec

Taking up anew a topic treated in a previous article [17], we discuss in detail a problem dealt with by Friedlander and Iwaniec [4] which is most appropriate for the application of single exponential sums. Let[7] χ_1, χ_2, χ_3 be (ordinary) primitive Dirichlet characters with moduli D_1, D_2, D_3, and put $D = D_1 D_2 D_3$. Consider, for x large,

$$\Delta(x) = \Delta(\chi_1, \chi_2, \chi_3; x)$$

$$:= \sum_{n_1 n_2 n_3 \le x} \chi_1(n_1) \chi_2(n_2) \chi_3(n_3) - \operatorname*{Res}_{s=1} \left(L(s, \chi_1) L(s, \chi_2) L(s, \chi_3) \frac{x^s}{s} \right) .$$

$$(11)$$

---

[7]Friedlander and Iwaniec concentrated on the case of *three* Dirichlet characters because the case of two factors had been discussed at length by Huxley and Watt [9, 10], while for four or more factors the approach via exponential sums turns out to be no longer useful. Contour integration techniques using bounds for the L-functions involved are the appropriate choice in these cases. By the way, for $D = 1$ this is the three-factor Piltz divisor problem: see Atkinson [1].

By Van der Corput's method, Friedlander and Iwaniec [4] proved that

$$\Delta(\chi_1, \chi_2, \chi_3; x) \ll x^\epsilon D \left(\frac{x}{D}\right)^{37/75} . \tag{12}$$

In [17, formula (2.6)] we refined this to[8]

$$\Delta(\chi_1, \chi_2, \chi_3; x) \ll x^\epsilon D \left(\frac{x}{D}\right)^{2498/5073} , \tag{13}$$

on the basis of (9). Furthermore, we established some refinements under special conditions on the relative size of D_1, D_2, D_3, and x.

In the present paper, we will use Theorem 1 to get sharper bounds in some of these cases.

Theorem 3 *Let χ_1, χ_2, χ_3 be primitive Dirichlet characters with moduli D_1, D_2, D_3. Put $D := D_1 D_2 D_3$, $D_{\max} := \max(D_1, D_2, D_3)$, and define $\Delta(\chi_1, \chi_2, \chi_3; x)$ as in (11). Let further N be a real parameter, $1 \leq N \leq x$, and $\epsilon > 0$ fixed. Then, for $x \geq D$ large, and uniformly in D_1, D_2, D_3,*

$$x^{-\epsilon} \Delta(\chi_1, \chi_2, \chi_3; x) \ll \left(\frac{Dx^2}{N}\right)^{1/3} + \sqrt{D_{\max}} \left(D^{205/1422} N^{743/2844} x^{253/711}\right.$$

$$+ D^{328/2719} N^{1735/8157} x^{2063/5438} \tag{14}$$

$$\left. + D^{1/6} N^{1957/6516} x^{1/3} + (DNx)^{1/4}\right) .$$

Furthermore, with an appropriate choice of $N \in [1, x]$, it follows that

$$\begin{aligned}
\Delta(\chi_1, \chi_2, \chi_3; x) \ll x^\epsilon &\left(D^{1153/5073} D_{\max}^{474/1691} x^{2498/5073}\right. \\
&+ D^{3043/12{,}387} D_{\max}^{1086/4129} x^{6086/12{,}387} \\
&+ D^{2719/13{,}362} D_{\max}^{2719/8908} x^{13{,}129/26{,}724} + \sqrt{D_{\max}} (Dx)^{1/4} \\
&+ (D^2 D_{\max}^2 x^3)^{1/7} + x^{1/3}(D^{1/3} + \sqrt{D_{\max}} D^{1/6}) \\
&\left. + \sqrt{D_{\max}} \left(D^{328/2719} x^{2063/5438} + D^{205/1422} x^{253/711}\right)\right) .
\end{aligned} \tag{15}$$

Proof Starting from [17, formulae (3.1)–(3.2)], we see that

$$\Delta(\chi_1, \chi_2, \chi_3; x) \ll D^{1/6} x^{1/3} \left|\sum\right| + x^\epsilon \left(\frac{Dx^2}{N}\right)^{1/3} , \tag{16}$$

---

[8]Note that $\frac{2498}{5073} = 0.4924\ldots$, while $\frac{37}{75} = 0.4933\ldots$.

where $\sum$ can be decomposed into $O\left((\log x)^3\right)$ subsums $S(N_1, N_2, N_3)$ which satisfy

$$S(N_1, N_2, N_3) \ll P^{-2/3} \sum_{\substack{n_j \sim N_j \\ j=1,2}} \left| \sum_{n_3 \sim N_3} \bar{\chi}_3(n_3) \, e\left(\pm 3(n_1 n_2 n_3 x/D)^{1/3}\right) \right| . \tag{17}$$

Here $P := N_1 N_2 N_3 \ll N$, and $n_j \sim N_j$ means throughout that n_j ranges over some interval contained in $[N_j, 2N_j]$. W.l.o.g., $N_1 \leq N_2 \leq N_3$. Following the deduction in [17] further, we see that the inner sum here is

$$\ll \sqrt{D_{\max}} \max_{m=1,\dots,D_3} \left| \sum_{n_3 \sim N_3} e\left(3(n_1 n_2 n_3 x/D)^{1/3} \mp \frac{m n_3}{D_3}\right) \right| .$$

The exponential sum here is now bounded by means of Theorem 1 whose conditions are satisfied with $M = N_3$, $T = (Px/D)^{1/3}$. In this way we obtain, after a straightforward calculation,

$$S(N_1, N_2, N_3)$$

$$\ll x^\epsilon P^{1/3} \sqrt{D_{\max}} \left(\frac{(Px/D)^{16/711}}{N_3^{269/948}} + (Px/D)^{751/16,314} N_3^{-1/2} \right. \tag{18}$$

$$\left. + N_3^{-215/2172} + (Px/D)^{-1/12} \right) .$$

To get rid of N_3, we use that $N_3 \geq P^{1/3}$. Further, since $P \ll N$,

$$S(N_1, N_2, N_3)$$

$$\ll x^\epsilon \sqrt{D_{\max}} \left((x/D)^{16/711} N^{743/2844} + (x/D)^{751/16,314} N^{1735/8157} \right. \tag{19}$$

$$\left. + N^{1957/6516} + (x/D)^{-1/12} N^{1/4} \right) .$$

Going back to (16), we immediately establish formula (14) of Theorem 3.

To derive (15) from (14), we again use Lemma 1, with $A(N)$ denoting the right-hand side of (14). I.e., in the notation of Lemma 1,[9]

$$\alpha_1 = \frac{743}{2844}, \quad \alpha_2 = \frac{1735}{8157}, \quad \alpha_3 = \frac{1957}{6516}, \quad \alpha_4 = \frac{1}{4},$$

$$\beta = \frac{1}{3},$$

$$a_1 = \sqrt{D_{\max}}\, D^{205/1422} x^{253/711}, \quad a_2 = \sqrt{D_{\max}}\, D^{328/2719} x^{2063/5438},$$

---

[9]Obviously, there is just one term with a negative exponent of N.

$$a_3 = \sqrt{D_{\max}}\, D^{1/6} x^{1/3}\,, \quad a_4 = \sqrt{D_{\max}}\, (Dx)^{1/4}\,,$$

$$b = (Dx^2)^{1/3}\,.$$

By Lemma 1, there exists a value N^* of N so that $1 \le N^* \le x$ and

$$A(N^*) \ll \sum_{j=1}^{4} \left(b^{\alpha_j} a_j^{\beta} \right)^{1/(\alpha_j + \beta)} + \sum_{j=1}^{4} a_j + b x^{-\beta}\,.$$

Apart from the factor x^ϵ, this is just the right-hand side of (15), as an involved but straightforward computation shows. $\qquad\square$

We notice that the result of Theorem 3 neither improves upon (13) from [17, formula (2.6)] nor upon[10] [17, Corollary 1.1], since the first terms on the right-hand sides of (15) and of [17, formula (2.5)] are the same.

However, for other special assumptions on the relative size of D_1, D_2, D_3, more precise estimates can be deduced.

Corollary 3.1 *Let $a_n(\mathbb{F})$ denote the coefficients of the Dirichlet series of the Dedekind zeta-function*

$$\zeta_{\mathbb{F}}(s) = L(s, \chi_1) L(s, \bar{\chi}_1) \zeta(s)$$

of a cubic number field $\mathbb{F}$ which is a normal extension of the rationals, χ_1 a primitive character modulo $D_1 > 1$. Then, for $x > D$ large, $\epsilon > 0$ fixed, $D = D_1^2$, and $\lambda := \frac{\log D}{\log x}$,

$$\sum_{n \le x} a_n(\mathbb{F}) = |L(1, \chi_1)|^2\, x$$

$$+ \begin{cases} O\left(D^{\frac{1864}{5073}} x^{\frac{2498}{5073}+\epsilon} \right) & \textit{if} \quad 0 < \lambda \le \frac{2852}{25{,}487} = 0.1119\ldots, \\[2ex] O\left(D^{\frac{4672}{12{,}387}} x^{\frac{6086}{12{,}387}+\epsilon} \right) & \textit{if} \quad \frac{2852}{25{,}487} < \lambda \le 1\,. \end{cases}$$

Proof By assumption, $D_{\max} = D_1 = D_2 = \sqrt{D}$, $D < x$. Hence it follows from (15) that

$$x^{-\epsilon} \Delta(\chi_1, \bar{\chi}_1, \mathbf{1}; x)$$

$$\ll D^{1864/5073} x^{2498/5073} + D^{4672/12{,}387} x^{6086/12{,}387}$$

$$+ D^{19{,}033/53{,}448} x^{13{,}129/26{,}724} \tag{20}$$

$$+ D^{3/7} x^{3/7} + D^{4031/10{,}876} x^{2063/5438}$$

$$+ D^{1121/2844} x^{253/711} + x^{1/3} D^{5/12}\,,$$

---

[10]This corollary is concerned with the case that only $D_1 > 1$, while $D_2 = D_3 = 1$.

since $(Dx)^{1/3}$ and $\sqrt{D}\,x^{1/4}$ are negligible compared to $x^{1/3}D^{5/12}$. Computing the exponents numerically, it is obvious that the third term in (20) is again negligible. We now put $D = x^\lambda, 0 \le \lambda \le 1$. So the above estimates become

$$x^{-\epsilon}\Delta(\chi_1, \bar{\chi}_1, \mathbf{1}; x)$$

$$\ll x^{2(932\lambda+1249)/5073} + x^{2(2336\lambda+3043)/12,387} + x^{3(\lambda+1)/7} \tag{21}$$

$$+ x^{(4031\lambda+4126)/10,876} + x^{(1121\lambda+1012)/2844} + x^{(5\lambda+4)/12}.$$

So all the exponents are linear polynomials in λ—call them, in the same order as in (21),—$e_1(\lambda), \ldots, e_6(\lambda)$. A straightforward analysis shows that $e_1(\lambda)$ is greater than $e_3(\lambda), \ldots, e_6(\lambda)$ throughout $[0, 1]$, while $e_1(\lambda) = e_2(\lambda)$ at $\lambda = \lambda_1 := \frac{2852}{25,487} = 0.1119\ldots$. On $[0, \lambda_1[$, $e_1(\lambda) > e_2(\lambda)$, while on $]\lambda_1, 1]$, $e_1(\lambda) < e_2(\lambda)$. $\qquad\square$

Remarks

1. By virtue of the proof, the estimation of the error term remains true for the more general case that $D_1 \asymp D_2$ (both exceeding 1), and $D_3 \ll 1$. However, the leading term of order x arises only if $D_3 = 1$.
2. Müller [14] dealt with that problem for an arbitrary but *fixed* cubic number field $\mathbb{F}$, without paying attention to the dependence of the error estimate on the discriminant D. He obtained the bound $O_{\mathbb{F}}(x^{43/96+\epsilon})$, using a much more sophisticated method for the estimation of exponential sums due to Kolesnik [12].

Corollary 3.2 *Let* χ_1, χ_2, χ_3 *be nontrivial primitive Dirichlet characters with moduli* $D_1 \asymp D_2 \asymp D_3 \asymp \sqrt[3]{D}$, *all* $D_j > 1$. *Then, for* $x > D$ *large and* $\epsilon > 0$ *fixed,* $\lambda := \frac{\log D}{\log x}$,

$$\sum_{n_1 n_2 n_3 \le x} \chi_1(n_1)\chi_2(n_2)\chi_3(n_3)$$

$$\ll \begin{cases} D^{\frac{1627}{5073}} x^{\frac{2498}{5073}+\epsilon} & \text{if} \quad 0 < \lambda \le \frac{713}{8258} = 0.08634\ldots, \\ D^{\frac{1}{3}} x^{\frac{6086}{12,387}+\epsilon} & \text{if} \quad \frac{713}{8258} < \lambda \le 1.. \end{cases}$$

Proof Since $D_{\max} \asymp \sqrt[3]{D}$, it readily follows from (15) that

$$x^{-\epsilon} \sum_{n_1 n_2 n_3 \le x} \chi_1(n_1)\chi_2(n_2)\chi_3(n_3)$$

$$\ll D^{1627/5073} x^{2498/5073} + D^{1/3} x^{6086/12,387} + D^{2719/8908} x^{13,129/26,724} \tag{22}$$

$$+ D^{8/21} x^{3/7} + D^{4687/16,314} x^{2063/5438} + D^{221/711} x^{253/711} + D^{1/3} x^{1/3},$$

because $D^{5/12}x^{1/4}$ is negligible compared to $D^{1/3}x^{1/3}$. Evaluating all exponents numerically, it is evident that also the third, fifth, sixth, and seventh terms of (22) are negligible, hence (22) simplifies to

$$x^{-\epsilon} \sum_{n_1 n_2 n_3 \leq x} \chi_1(n_1)\chi_2(n_2)\chi_3(n_3) \tag{23}$$

$$\ll D^{1627/5073} x^{2498/5073} + D^{1/3} x^{6086/12{,}387} + D^{8/21} x^{3/7}.$$

Writing again $D = x^{\lambda}$, we thus get

$$x^{-\epsilon} \sum_{n_1 n_2 n_3 \leq x} \chi_1(n_1)\chi_2(n_2)\chi_3(n_3) \tag{24}$$

$$\ll x^{(1627\lambda+2498)/5073} + x^{(4129\lambda+6086)/12{,}387} + x^{(8\lambda+9)/21}.$$

Call these exponents again[11] $e_1(\lambda), e_2(\lambda), e_3(\lambda)$. Again by straightforward calculations, we see that that $e_1(\lambda) > e_3(\lambda)$ throughout $[0, 1]$, while $e_1(\lambda) = e_2(\lambda)$ at $\lambda = \lambda_2 := \frac{713}{8258} = 0.08634\ldots$. On $[0, \lambda_2[$, $e_1(\lambda) > e_2(\lambda)$, while on $]\lambda_2, 1]$, $e_1(\lambda) < e_2(\lambda)$. $\qquad\square$

7 Concluding Remark

It is a natural question to ask, for which type of problems the *single* exponential sum estimates discussed are of most importance. The answer is readily found by a look at the application to the Friedlander–Iwaniec problem we considered: It is the case of exponential sums depending on *several parameters*[12] where such estimates are most meaningful.

In contrast, for exponential sums depending on one parameter only, multiple sum estimations are often more powerful. On the latter, a very thorough exposition was given in Krätzel's monograph [13], however, without incorporating results from the *Discrete Hardy–Littlewood method* which was just being developed at that time.

References

1. F.V. Atkinson, A divisor problem. Q. J. Math. (Oxford) **12**, 193–200 (1941)
2. E. Bombieri, H. Iwaniec, On the order of $\zeta(\frac{1}{2} + it)$. Ann. Sc. Norm. Super. Pisa Cl. Sci. (5), IV. Ser., **13**, 449–472 (1986)
3. M. Drmota, R.F. Tichy, *Sequences, Discrepancies, and Applications*. Lecture Notes in Mathematics, vol. 1651 (Springer, Berlin, 1997)

---

[11] Any confusion with the previous meaning is impossible.

[12] In the above case, these were x and the moduli D_1, D_2, D_3.

4. J.B. Friedlander, H. Iwaniec, Summation formulae for coefficients of L-functions. Can. J. Math. **57**(3), 494–505 (2005)
5. S.W. Graham, G. Kolesnik, *Van der Corput's Method of Exponential Sums*. LMS Lecture Note Series, vol. 126 (Cambridge University Press, Cambridge, 1991)
6. M.N. Huxley, *Area, Lattice Points, and Exponential Sums*. LMS Monographs, New Series, vol. 13 (Oxford University Press, Oxford, 1996)
7. M.N. Huxley, Exponential sums and lattice points III. Proc. Lond. Math. Soc. (3) **87**, 591–609 (2003)
8. M.N. Huxley, Exponential sums and the Riemann zeta-function V. Proc. Lond. Math. Soc. (3) **90**, 1–41 (2005)
9. M.N. Huxley, N. Watt, The number of ideals in a quadratic field. Proc. Indian Acad. Sci. Math. Sci. **104**, 157–165 (1994)
10. M.N. Huxley, N. Watt, The number of ideals in a quadratic field. II. Isr. J. Math. **120**, 125–153 (2000)
11. H. Iwaniec, C.J. Mozzochi, On the divisor and circle problems. J. Number Theory **29**, 60–93 (1988)
12. G. Kolesnik, On the estimation of multiple exponential sums, in *Recent Progress in Analytic Number Theory I* (Academic Press, London, 1981), pp. 231–246
13. E. Krätzel, *Lattice Points* (Deutscher Verlag der Wissenschaften, Berlin, 1988)
14. W. Müller, On the distribution of ideals in cubic number fields. Mh. Math. **106**, 211–219 (1988)
15. W. Müller, W.G. Nowak, Lattice points in domains $|x|^p + |y|^p \leq R^p$. Arch. Math. (Basel) **51**, 55–59 (1988)
16. W.G. Nowak, Higher order derivative tests for exponential sums incorporating the discrete Hardy-Littlewood method. Acta Math. Hungar. **134**, 12–28 (2012)
17. W.G. Nowak, A problem considered by Friedlander & Iwaniec and the discrete Hardy-Littlewood method. Math. Slovaca **67** (2017), to appear
18. B.R. Srinivasan, On the number of Abelian groups of a given order. Acta Arith. **23**, 195–205 (1973)
19. E.C. Titchmarsh, *The Theory of the Riemann Zeta-Function*, 2nd edn. (Clarendon Press, Oxford, 1986), revised by D.R. Heath-Brown
20. J.G. van der Corput, Verschärfung der Abschätzung beim Teilerproblem. Math. Ann. **87**, 39–65 (1922)
21. H. Weyl, Über die Gleichverteilung der Zahlen mod Eins. Math. Ann. **77**, 313–352 (1916)

Orbits of Algebraic Dynamical Systems in Subgroups and Subfields

Alina Ostafe and Igor E. Shparlinski

Abstract We study intersections of orbits in polynomial dynamics with multiplicative subgroups and subfields of arbitrary fields of characteristic zero, as well as with sets of points that are close with respect to the Weil height to division groups of finitely generated groups of $\overline{\mathbb{Q}}^*$.

2010 *Mathematics Subject Classification* Primary 37P05; Secondary 11G25, 11G35, 13P15, 37P25

1 Introduction

1.1 Motivation

Let $f(X) \in \mathbb{K}[X]$ be a polynomial of degree $d \geq 2$ over a field $\mathbb{K}$. We always assume that $\mathbb{K}$ is the field of definition of f and thus the coefficients of f are not contained in any proper subfield of $\mathbb{K}$.

We set $f^{(0)}(X) = X$ and then define the nth iterate of f recursively as $f^{(n)}(X) = f\left(f^{(n-1)}(X)\right)$, $n = 1, 2, \ldots$.

Given a point $w \in \mathbb{K}$ we define its orbit $\mathrm{Orb}_f(w)$ with respect to the polynomial f as the set

$$\mathrm{Orb}_f(w) = \{f^{(n)}(w) \mid n = 0, 1, \ldots\}. \tag{1}$$

Given a reasonably sparse set $\mathcal{S} \subseteq \mathbb{K}$, defined in terms "unrelated" to f, it is natural to assume that the intersection $\mathrm{Orb}_f(w) \cap \mathcal{S}$ is finite. For example, if $\mathcal{S}$ is an orbit of another polynomial this is known as a problem about orbit intersections, which has recently been studied by Ghioca et al. [18], see also [19].

A. Ostafe • I.E. Shparlinski (✉)
School of Mathematics and Statistics, University of New South Wales, Sydney, NSW 2052, Australia
e-mail: alina.ostafe@unsw.edu.au; igor.shparlinski@unsw.edu.au

© Springer International Publishing AG 2017
C. Elsholtz, P. Grabner (eds.), *Number Theory – Diophantine Problems, Uniform Distribution and Applications*, DOI 10.1007/978-3-319-55357-3_18

One can also consider multivariate generalisations of this question, that is, when $S = V$ is an algebraic variety and $\mathcal{F}$ a polynomial system of n polynomials in $\mathbb{K}[X_1, \ldots, X_n]$. In this case, when also $\mathbb{K}$ is of characteristic zero, the question about the finiteness of $\mathrm{Orb}_{\mathcal{F}}(w) \cap S$, $w \in \mathbb{K}^n$, is known as the *dynamical Mordell–Lang conjecture*, see [4, 6, 11, 18, 19, 22, 29, 36] and the references therein. In positive characteristic this conjecture has to be adjusted to replace the finiteness assertion by a more complex description of the set of possible intersections, see also [4, Conjecture 13.2.0.1], and it is still widely open, however we refer to [3, 17] for some recent progress.

It is certainly natural to expect that for a generic polynomial f, in the above cases the intersections $\mathrm{Orb}_f(w) \cap S$ are finite and furthermore, $\mathrm{Orb}_f(w) \cap S = \emptyset$ for all by finitely many initial values $w \in \mathbb{K}$, however, proving this in full generality appears to be difficult. Results of this type are known only in several very special cases such as:

- $S = \mathrm{Orb}_g(z)$ is an orbit of another polynomial g starting at $z \in \mathbb{C}$, see [18, Theorem 1.1];
- $S = \mathbb{K}^m$ is the set of mth powers of an algebraic number field $\mathbb{K}$ for a fixed integer $m \geq 2$, see [11, Theorem 1.3];
- $S = \Gamma$ is an S-unit subgroup of an algebraic number field $\mathbb{K}$ and f is a monic polynomial with coefficients from the ring of S-integers of $\mathbb{K}$, see [22, Theorem 1.7];
- $S = \mathcal{U}$ is the set of all roots of unity and the polynomial and initial points are defined over the cyclotomic closure $\mathbb{K}^c = \mathbb{K}\,(\mathcal{U})$ over an algebraic number field $\mathbb{K}$, see [27];

we refer to [11, 12, 22, 27] for precise notations and formulations (for example, polynomials f of some special shapes must be excluded from these statements), as well as further generalisations.

We also note that the above problems are related and also share similar techniques with the various problems on preperiodic points in special sets. For example, the result of [27] is based on some results and methods of Dvornicich and Zannier [14], who have shown the finiteness of the set of preperiodic points of $f(X) \in \mathbb{K}[X]$ in the cyclotomic closure $\mathbb{K}^c$ over an algebraic number field $\mathbb{K}$ (unless f is of certain explicitly described form).

Finally, we emphasise that we consider orbits (1) as sets rather than as sequences and in particular for preperiodic points w the orbit $\mathrm{Orb}_f(w)$ is finite and thus the finiteness of intersection with any set is trivial.

1.2 Our Results and Goals

Over number fields, we consider intersections of polynomial orbits with sets which can be roughly described as *approximate division groups* of finitely generated groups Γ. These are sets of elements of the form yz where $y^m \in \Gamma$ and z is an element of restricted height, we refer to Sect. 1.4 for precise definitions. We remark that we do not impose that y or z belongs to the ground field $\mathbb{K}$.

Depending on the restrictions on z, we obtain two types of results.

If the height of z is limited by an absolute constant, we obtain a finiteness result, see Corollary 2.1.

For sets with more generous restrictions on z we consider an apparently easier question of estimating the frequency with which orbits fall in these sets. Previously, such a relaxation has also been studied by Bell et al. [3]. The so-called *gap principle* of Benedetto et al. [5] is also of similar spirit, showing sparsity of elements in some orbits that fall on varieties. In particular, it is shown in [3] (see also [4, Theorem 11.1.0.8]) that amongst the first N elements in a non-periodic orbit of an endomorphism $\Phi : \mathbb{P}^s \to \mathbb{P}^s$ of the s-dimensional projective space over a field $\mathbb{K}$ only $O(N/\log N)$ may fall on a fixed irreducible algebraic curve $\mathcal{C} \subseteq \mathbb{P}^s$. Our bounds of Theorems 2.4 and 3.1 are of similar spirit (and unfortunately none of these results give a power saving in the bound on the number of such elements). Among other tools its proof is based on a combinatorial argument that reduces the problem of estimating the cardinality of $\mathrm{Orb}_f(w) \cap \mathcal{S}$ to estimating the number of points on curves with coordinates in the set $\mathcal{S}$. Although there are some similarities with the approach of [3], they appear to be independent. We hope that our arguments can find more applications in other similar questions.

In the case of arbitrary fields of characteristic zero, we also give a similar result for the frequency of intersections of polynomial orbits with finitely generated groups. Finally, we consider the case when the "target" set $\mathcal{S}$ is a proper subfield $\mathbb{L}$ of an arbitrary field $\mathbb{K}$. In this case we obtain a similar result about the frequency of the intersection of orbits and subfields.

One of the goals of these paper is to attract more attention to the study of distribution of polynomial orbits in various sets and show the variety of its problems and techniques their solutions employ. We also pose several open problems in Sect. 6.

1.3 Underlying Techniques

When working over a number field $\mathbb{K}$, for a polynomial $f \in \mathbb{K}[X]$, we prove first that the intersection of $f(\mathbb{K})$ with sets $\mathscr{B}_n(\mathcal{S}, E)$, which contain points in $\overline{\mathbb{Q}}^*$ that with respect to the Weil height are very close to a given subgroup Γ of finite rank, is a set of bounded height, and thus finite. This immediately implies finiteness for the intersection of polynomial orbits with such sets.

For more general sets $\mathscr{C}_n(\mathcal{S}, \varepsilon)$ (see Sect. 1.4 for concrete definitions and notations), and their intersection with orbits, we employ bounds for the cardinality of intersections of such sets with a curve due to [7].

Intersection of varieties in $\mathbb{G}_m^n$ with sets of the type $\mathscr{B}_n(\mathcal{S}, E)$ or $\mathscr{C}_n(\mathcal{S}, \varepsilon)$ fall within two conjectures, the *Mordell–Lang conjecture* on intersection of varieties with finitely generated subgroups and the *Bogomolov conjecture* which is about the discreteness of the set of points of bounded height in a variety. This direction has been extensively studied over several decades, see [1, 2, 7, 8, 10, 15, 23, 24, 26, 30–32, 35] and the references therein, which in particular give precise quantitative results about the intersection of varieties with $\mathscr{B}_n(\mathcal{S}, E)$ or $\mathscr{C}_n(\mathcal{S}, \varepsilon)$ when $\mathcal{S}$ is a finitely generated subgroup Γ or the division group of a subgroup Γ.

In particular, for the case of curves, Liardet [24] proved that unless a curve is very special, it contains only finitely many points with coordinates in the division group of a finitely generated group Γ. In this paper we appeal to recent work of Bérczes et al. [7] (for the case when $\mathcal{S}$ is the division group of a subgroup Γ) who give upper bounds for the cardinality of intersections of curves with $\mathscr{B}_n(\mathcal{S}, E)$ or $\mathscr{C}_n(\mathcal{S}, \varepsilon)$, see Sect. 4.3.

For arbitrary fields $\mathbb{K}$ of characteristic zero and finitely generated subgroups $\Gamma \subseteq \mathbb{K}^*$ we employ the estimates for the number of solutions in Γ to linear equations $a_1 x_1 + \cdots + a_n x_n = 1$, due to Amoroso and Viada [2].

1.4 Notations, Definitions and Conventions

In this section we introduce the notations and definitions needed to be able only to state our main results in the next section. Further notation and definitions are introduced in Sect. 4 where they are used.

As usual, for a field $\mathbb{K}$, we use the notation $\overline{\mathbb{K}}$ for the algebraic closure of $\mathbb{K}$. In this paper we consider only fields of characteristic zero.

The *height* always refers to the *absolute logarithmic Weil height*, we refer to [9, 38] for a background on heights, see also Sect. 4.1.

Following the established tradition, we denote the n-dimensional torus $\mathbb{G}_m^n$ as $(\overline{\mathbb{Q}}^*)^n$ equipped with the group law defined by component-wise multiplication (we note that here the index m is a part of the notation rather than a parameter).

As usual, we say that $f \in \mathbb{K}[X]$ is a monomial if $f(X) = aX^d$ with $a \in \mathbb{K}^*$.

Definition 1.1 We say that a polynomial $F \in \overline{\mathbb{Q}}[X, Y]$ is *special* if it has a factor of the form $aX^m Y^n - b$ or $aX^m - bY^n$ for some $a, b \in \overline{\mathbb{Q}}$ and $m, n \geq 0$. Otherwise, we call F to be *non-special*.

For a finitely generated subgroup $\Gamma \subseteq \mathbb{G}_m^n$, we define the *division group* $\overline{\Gamma}$ by

$$\overline{\Gamma} = \{x \in \mathbb{G}_m^n \mid \exists k \in \mathbb{N} \text{ with } x^k \in \Gamma\}.$$

For $E, \varepsilon \geq 0$ and a set $\mathcal{S} \subseteq \mathbb{G}_m^n$, we define the sets

$$\mathscr{B}_n(\mathcal{S}, E) = \{x \in \mathbb{G}_m^n \mid \exists y, z \in \mathbb{G}_m^n$$
$$\text{with } x = yz, \ y \in \mathcal{S}, \ h(z) \leq E\} \tag{2}$$

and

$$\mathscr{C}_n(\mathcal{S}, \varepsilon) = \left\{ x \in \mathbb{G}_m^n \mid \exists y, z \in \mathbb{G}_m^n \right.$$
$$\left. \text{with } x = yz, \ y \in \mathcal{S}, \ h(z) \le \varepsilon(1 + h(y)) \right\}. \tag{3}$$

We usually write $\mathscr{B}_n(\mathcal{S}, E)$ when the parameter E is allowed to be large, and write $\mathscr{B}_n(\mathcal{S}, \varepsilon)$ when this parameter is rather small (which is always the case with $\mathscr{C}_n(\mathcal{S}, \varepsilon)$). Clearly

$$\mathscr{B}_n(\mathcal{S}, \varepsilon) \subseteq \mathscr{C}_n(\mathcal{S}, \varepsilon).$$

We also drop the subscript n for $n = 1$ and thus write

$$\mathscr{B}(\mathcal{S}, E) = \mathscr{B}_1(\mathcal{S}, E) \qquad \text{and} \qquad \mathscr{C}(\mathcal{S}, \varepsilon) = \mathscr{C}_1(\mathcal{S}, \varepsilon).$$

Furthermore we also write $\mathscr{A}(\mathbb{K}, H)$ for the set of elements in the algebraic number field $\mathbb{K}$ of height at most H, that is,

$$\mathscr{A}(\mathbb{K}, H) = \{ x \in \mathbb{K}^* \mid h(x) \le H \}. \tag{4}$$

Generally, we note that, for a finitely generated subgroup Γ of $\mathbb{G}_m^n$, the sets $\mathscr{B}(\Gamma, E)$ and $\mathscr{B}(\overline{\Gamma}, E)$ are often denoted as Γ_E and $\overline{\Gamma}_E$, respectively, see [2, 7, 30].

In this paper we consider the sets (2) and (3) with $\mathcal{S} = \overline{\Gamma}$ for a finitely generated subgroup Γ of $\mathbb{G}_m^n$. Moreover, in our results we only consider the cases of $n = 1$ and $n = 2$, that is of univariate polynomials and plane curves, respectively.

Let $\mathbb{K}$ be an arbitrary field and $\mathcal{S} \subseteq \mathbb{K}$. For an integer $N \ge 1$, we use $T_w(N, \mathcal{S})$ to denote the number of $n \le N$ with $f^{(n)}(w) \in \mathcal{S}$.

We denote by $\mathbb{N}$ the set of positive integer numbers. Given functions

$$f, g \colon \mathbb{N} \to \mathbb{N},$$

the Landau symbol $f = o(g)$ means that $f(n)/g(n) \to 0$ as $n \to \infty$.

2 Results for Algebraic Number Fields

2.1 Orbits in Sets $\mathscr{B}(\overline{\Gamma}, E)$

Over $\overline{\mathbb{Q}}$ we can say more about intersections of polynomial orbits with division groups. Let $\mathbb{K}$ be an algebraic number field and let $f \in \mathbb{K}[X]$ with $\deg f = d \ge 2$. Let $\Gamma \subseteq \mathbb{K}^*$ be a subgroup of rank r.

We first remark that $f(\mathbb{K}) \cap \overline{\Gamma}$ is a finite set whenever f has at least two distinct roots in $\overline{\mathbb{Q}}$. Indeed, this follows directly from [22, Proposition 1.5, (a)], which relies on Siegel's Theorem. To see this, let $\gamma_1, \ldots, \gamma_r$ be the generators of Γ and S be a finite set of places in $M_{\mathbb{K}}$, including the Archimedean ones, such that $|\gamma_i|_v = 1$ for any $i = 1, \ldots, r$ and any $v \notin S$ (and thus, $|\alpha|_v = 1$ for any $\alpha \in \Gamma$ and $v \notin S$). Now, if $f(x) \in \overline{\Gamma}$ for some $x \in \mathbb{K}$, then this is equivalent with a power $f(x)^m \in \Gamma$, from where we obtain that $|f(x)|_v = 1$ for any $v \notin S$. Thus, $f(\mathbb{K}) \cap \overline{\Gamma} \subset f(\mathbb{K}) \cap R_S^*$, where R_S^* is the ring of S-units in $\mathbb{K}$, and now the finiteness conclusion follows from [22, Proposition 1.5, (a)].

From the above we immediately obtain that for any $w \in \overline{\mathbb{Q}}^*$ the intersection $\mathrm{Orb}_f(w) \cap \overline{\Gamma}$ is finite. Actually, for this conclusion we only need f not to be a monomial aX^d, where $a \in \mathbb{K}^*$, see [22, Proposition 1.6, (a)].

We now extend the above observations to more general intersections $f(\mathbb{K}) \cap \mathscr{B}(\overline{\Gamma}, E)$. However first we study the structure of the intersection $\mathbb{K} \cap \mathscr{B}(\overline{\Gamma}, E)$. Before we formulate the result we recall the notation (4).

Theorem 2.1 *Let $\mathbb{K}$ be an algebraic number field and $\Gamma \subseteq \mathbb{K}^*$ a finitely generated group. Assume that $\{\gamma_1, \ldots, \gamma_r\}$ is a set of generators of Γ, which minimises $h = \max_{i=1,\ldots,r} h\,(\gamma_i)$. Then, for every $E > 0$, we have*

$$\mathbb{K} \cap \mathscr{B}(\overline{\Gamma}, E) \subseteq \{\gamma\eta \mid (\gamma, \eta) \in \Gamma \times \mathscr{A}(\mathbb{K}, H)\},$$

where $H = E + rh$.

Using the *Northcott property* of algebraic number fields (that is, the finiteness of the set of elements of bounded height, and thus, of the set $\mathscr{A}(\mathbb{K}, H)$), we see that the same argument as above implies the following result.

Corollary 2.2 *Let $\mathbb{K}$ be an algebraic number field and let $f \in \mathbb{K}[X]$ have at least two distinct roots in $\overline{\mathbb{Q}}$. Then for every $E > 0$ and a finitely generated group $\Gamma \subseteq \mathbb{K}^*$ the set $f(\mathbb{K}) \cap \mathscr{B}(\overline{\Gamma}, E)$ is finite.*

This immediately leads to the desired result about the finiteness of the elements of orbits in sets $\mathscr{B}(\overline{\Gamma}, E)$. In fact, we consider a more general preimage set

$$g^{-1}\left(\mathscr{B}(\overline{\Gamma}, E)\right) = \{u \in \overline{\mathbb{Q}} : g(u) \in \mathscr{B}(\overline{\Gamma}, E)\}$$

for a polynomial $g \in \mathbb{K}[X]$.

Corollary 2.3 *Let $\mathbb{K}$ be an algebraic number field and let $f, g \in \mathbb{K}[X]$ be such that at least one of the polynomials $g(X)$ and $g(f(X))$ has at least two distinct roots in $\overline{\mathbb{Q}}$. Then for every $E > 0$, a finitely generated group $\Gamma \subseteq \mathbb{K}^*$ and a point $w \in \mathbb{K}$, the set $\mathrm{Orb}_f(w) \cap g^{-1}\left(\mathscr{B}(\overline{\Gamma}, E)\right)$ is finite.*

Note that one can reformulate Corollary 2.3 in a seemingly more general form with the condition that at least one of the polynomials

$$g\left(f^{(\nu)}(X)\right), \qquad \nu = 0, 1, \ldots, \tag{5}$$

has at least two distinct roots in $\overline{\mathbb{Q}}$. This, however, does not extend the class of polynomials, as if $g(X)$ and $g(f(X))$ are monomials then so are all polynomials (5).

It is also easy to see that all underlying results and thus the results of this section also hold for rational functions f, rather than just for polynomials.

2.2 Orbits in Sets $\mathscr{C}(\overline{\Gamma}, \varepsilon)$

The main goal of this section is to give a general result for the frequency of intersections of orbits with the set $\mathscr{C}(\overline{\Gamma}, \varepsilon)$, defined by (3), for a finitely generated subgroup $\Gamma \subseteq \mathbb{G}_m$.

Theorem 2.4 *Let $\mathbb{K}$ be an algebraic number field and let $f \in \mathbb{K}[X]$ with $\deg f = d \geq 2$ not a monomial. Then, for a finitely generated subgroup $\Gamma \subseteq \mathbb{K}^*$ of rank r, a point $w \in \overline{\mathbb{Q}}^*$, and an integer N we have*

$$T_w(N, \mathscr{C}(\overline{\Gamma}, \vartheta)) \leq \frac{(4 \log d + o(1))N}{\log \log \log N}, \qquad as\ N \to \infty,$$

where $\mathscr{C}(\overline{\Gamma}, \vartheta)$ is defined by (3) with

$$\vartheta = (\log N)^{-2}(\log \log N)^{-7r/2-12}, \tag{6}$$

We note that the bound of Theorem 2.4 does not depend on the field of definition of the initial point w.

3 Results for Arbitrary Fields

3.1 Orbits in Finitely Generated Groups

The following result applies to arbitrary fields of characteristic zero.

Theorem 3.1 *If $\mathbb{K}$ is of characteristic zero and $f \in \mathbb{K}[X]$ is not a monomial with $\deg f = d \geq 2$, then for a finitely generated group $\Gamma \subseteq \mathbb{K}^*$ of rank r, a point $w \in \mathbb{K}$, and an integer N we have*

$$T_w(N, \Gamma) \leq \frac{(10 \log d + o(1))N}{\log \log N}, \qquad as\ N \to \infty.$$

3.2 Orbits in Subfields

Here we study $T_w(N, \mathcal{S})$ in the case of subfields $\mathcal{S} = \mathbb{L} \subseteq \mathbb{K}$.

Let $\mathbb{K}$ be a field of characteristic zero. We note that using the result of Ghioca et al. [18, Theorem 1.5] on the finiteness of the orbit intersections over $\mathbb{K}$ one can obtain (under certain natural conditions on $f \in \mathbb{K}[X]$) a result about finiteness of orbit elements that fall in a proper subfield $\mathbb{L}$ of a field $\mathbb{K}$ for which the Galois group $\mathrm{Gal}(\mathbb{K}/\mathbb{L})$ is finite. Indeed, for this one considers the orbits of all conjugates $f_\sigma(X) = \sigma(f(X)))$, $\sigma \in \mathrm{Gal}(\mathbb{K}/\mathbb{L})$, of f, so the event of falling in a subfield corresponds to an orbit intersection.[1] Thus here we concentrate on the case of arbitrary fields $\mathbb{K}$ (for example, finite fields) and also study polynomial images of subfields.

In fact, as before we consider a more general preimage set

$$g^{-1}(\mathbb{L}) = \{u \in \mathbb{K} \ : \ g(u) \in \mathbb{L}\}$$

for a polynomial $g \in \mathbb{K}[X]$.

We need to impose more conditions on the polynomial f. Namely, for a given $g \in \mathbb{K}[X]$ we say that $f \in \mathbb{K}[X]$ *dynamically g-avoids* a subfield $\mathbb{L} \subseteq \mathbb{K}$ if no iterate $g\left(f^{(n)}(X)\right)$ is defined over $\mathbb{L}$.

Theorem 3.2 *Let $\mathbb{K}$ be an arbitrary field and let $g \in \mathbb{K}[X]$ be a monic polynomial. Assume that $f \in \mathbb{K}[X]$ with $\deg f = d \geq 2$ dynamically g-avoids a subfield $\mathbb{L} \subseteq \mathbb{K}$, then for a point $w \in \mathbb{K}$ and an integer N, satisfying $N \leq \#\mathrm{Orb}_f(w)$ in the case when w is preperiodic, we have*

$$T_w\left(N, g^{-1}(\mathbb{L})\right) \leq \frac{(2\log d + o(1))N}{\log N^*}, \qquad as\ N \to \infty,$$

where $N^* = \min\{N, p\}$ if $\mathbb{K}$ is of characteristic $p > 0$ and $N^* = N$ otherwise.

We remark that the condition $N \leq \#\mathrm{Orb}_f(w)$ in the case when w is preperiodic in Theorem 3.2 is only interesting when $\mathbb{K}$ is a finite field of large characteristic.

4 Preliminaries

4.1 Heights of Iterated Polynomials

We introduce first a necessary background on heights. Let $x \in \overline{\mathbb{Q}}$ and $\mathbb{K}$ be a number field such that $x \in \mathbb{K}$. We denote by $M_{\mathbb{K}}$ the set of places of $\mathbb{K}$. In particular

$$M_{\mathbb{Q}} = \{\infty, p \text{ prime}\}$$

---

[1] The authors are very grateful to Michael Zieve for outlining this argument.

consists of one Archimedean valuation ∞ and p-adic valuation, for each prime p. For any $v \in M_{\mathbb{K}}$ extending an absolute value $v_0 \in M_{\mathbb{Q}}$, which we denote by $v \mid v_0$, there exists a (not necessarily unique) embedding $\sigma_v : \mathbb{K} \hookrightarrow \mathbb{C}_{v_0}$ corresponding to v, such that one has $|a|_v = |\sigma_v(a)|_{v_0}$ for any $a \in \mathbb{K}$. We use the notation $v \mid v_0$ also for $v_0 = \infty$, thus the Archimedean absolute values $v \in M_{\mathbb{K}}$ are those with $v \mid \infty$.

We define now the *absolute logarithmic Weil height* of x by

$$h(x) = \sum_{v \in M_{\mathbb{K}}} \max\{0, \log \|x\|_v\},$$

where

$$\|x\|_v = \begin{cases} |x|_v^{[\mathbb{K}_v : \mathbb{R}]/[\mathbb{K}:\mathbb{Q}]} & \text{if } v \mid \infty, \\ |x|_v^{[\mathbb{K}_v : \mathbb{Q}_p]/[\mathbb{K}:\mathbb{Q}]} & \text{if } v \mid p, \end{cases} \tag{7}$$

see [9, Chapter 1] for a background on the valuations and the Weil height. In particular, a remarkable property of the logarithmic Weil height $h(x)$ is that it does not depend on the field $\mathbb{K}$ (hence the name "absolute").

Let $f = \sum_{i=0}^{d} a_i X^i \in \overline{\mathbb{Q}}[X]$ and $\mathbb{K}$ a number field containing all the coefficients a_i, $i = 0, \ldots, d$. We define the *Weil height* of f by

$$h(f) = \sum_{v \in M_{\mathbb{K}}} \log \max_{0 \leq i \leq d} \|a_i\|_v.$$

Taking into account (7), we have the equivalent definition

$$h(f) = \frac{1}{[\mathbb{K} : \mathbb{Q}]} \sum_{v \in M_{\mathbb{K}}} n_v h_{v_0}(\sigma_v(f)), \quad n_v = [\mathbb{K}_v : \mathbb{Q}_{v_0}], \tag{8}$$

and $h_{v_0}(\sigma_v(f)) = \log \max_{0 \leq i \leq d} |\sigma_v(a_i)|_{v_0}$, where $v_0 \in M_{\mathbb{Q}}$, $v \mid v_0$ and σ_v is an embedding of $\mathbb{K}$ in $\mathbb{C}_{v_0}$ corresponding to v.

We need the following bound for the Weil height of polynomial iterates, which follows from [21, Lemma 1.2 (1.c and 2.c)], see also [13, Lemma 3.4], where the bound is obtained for the logarithmic naive height of polynomials over $\mathbb{Z}$ (that is, for $v_0 = \infty$). Although these computations are standard and may be done in other works, for the sake of completeness we provide all the details.

Lemma 4.1 *Let $f \in \overline{\mathbb{Q}}[X]$ of degree d and height h. Then*

$$h(f^{(k)}) \leq h(f) \frac{d^k - 1}{d - 1} + d(d + 1) \frac{d^{k-1} - 1}{d - 1} \log 2.$$

Proof For simplicity, for $v \in M_{\mathbb{K}}$ we denote $f_v = \sigma_v(f)$. We also note that one has $f_v^{(k)} = \sigma_v(f^{(k)})$ for $k \geq 1$.

356   A. Ostafe and I.E. Shparlinski

We see from (8) that

$$h(f^{(k)}) = \frac{1}{[\mathbb{K} : \mathbb{Q}]} \sum_{v \in M_{\mathbb{K}}} n_v h_{v_0}(f_v^{(k)}),$$

and thus it is enough to bound $h_{v_0}(f_v^{(k)})$, where $v_0 \in M_{\mathbb{Q}}$ and $v \mid v_0$.

If $v_0 = \infty$ and $v \in M_{\mathbb{K}}$ such that $v \mid \infty$, then the bound is proved in [13, Lemma 3.4], which in turn is based on [21, Lemma 1.2 (1.c)]. That is, one has

$$h_\infty(f_v^{(k)}) \leq h_\infty(f_v)\frac{d^k - 1}{d - 1} + d(d + 1)\frac{d^{k-1} - 1}{d - 1} \log 2. \tag{9}$$

If $v_0 = p$ a prime and $v \in M_{\mathbb{K}}$ such that $v \mid v_0$, by [21, Lemma 1.2 (2.c)], we have

$$h_p(f_v^{(2)}) \leq h_p(f_v)(d + 1).$$

By induction over $k \geq 1$, using again [21, Lemma 1.2 (2.c)], one immediately obtains

$$h_p(f_v^{(k)}) \leq h_p(f_v)\frac{d^k - 1}{d - 1}. \tag{10}$$

Putting together (9) and (10) in (8), one obtains

$$h(f) = \frac{1}{[\mathbb{K} : \mathbb{Q}]} \left(\sum_{\substack{v \in M_{\mathbb{K}} \\ v \mid p}} n_v h_p(f_v^{(k)}) + \sum_{\substack{v \in M_{\mathbb{K}} \\ v \mid \infty}} n_v h_p(f_\infty^{(k)}) \right)$$

$$= \frac{d^k - 1}{d - 1} \cdot \frac{1}{[\mathbb{K} : \mathbb{Q}]} \sum_{v \in M_{\mathbb{K}}} n_v h_{v_0}(f_v)$$

$$+ d(d + 1)\frac{d^{k-1} - 1}{d - 1} \log 2 \cdot \frac{1}{[\mathbb{K} : \mathbb{Q}]} \sum_{\substack{v \in M_{\mathbb{K}} \\ v \mid \infty}} n_v$$

$$= \frac{d^k - 1}{d - 1}h(f^{(k)}) + d(d + 1)\frac{d^{k-1} - 1}{d - 1} \log 2 \cdot \frac{1}{[\mathbb{K} : \mathbb{Q}]} \sum_{\substack{v \in M_{\mathbb{K}} \\ v \mid \infty}} n_v.$$

Since by Bombieri and Gubler [9, Corollary 1.3.2] one has

$$\sum_{\substack{v \in M_{\mathbb{K}} \\ v \mid \infty}} n_v = [\mathbb{K} : \mathbb{Q}],$$

we conclude the proof. $\qquad\square$

4.2 Counting Points of Bounded Height

To derive a bound for the number of points on curves in $\mathscr{C}_2(\overline{\Gamma}, \varepsilon)$ for some $\varepsilon \geq 0$, see Sect. 4.3, we need the following special case of the general result of Schmidt [34] (taken with $n = 2$), see also [37, Theorem 2.3], which gives an upper bound for the number of points in $\mathbb{G}_m^2$ of bounded degree over a number field $\mathbb{K}$ and of bounded height.

Lemma 4.2 *Let $\mathbb{K}$ be a number field of degree m over $\mathbb{Q}$ and $h \geq 1$ a real number. Then, the number of points $\boldsymbol{x} = (x, y) \in \mathbb{G}_m^2$ with*

$$[\mathbb{K}(\boldsymbol{x}) : \mathbb{K}] = e \qquad and \qquad h(\boldsymbol{x}) \leq h$$

is at most $\exp\left(hme(e + 2) + me(e + 5) + e^2 + 10e + 24\right).$

We note that a slight improvement of the result of Schmidt [34], presented in Lemma 4.2, has been obtained by Widmer in [37, Theorem 2.4] for the one-dimensional case, that is, for points in $\mathbb{P}^1(\overline{\mathbb{Q}})$ of given degree over a number field and of bounded height.

From Lemma 4.2 we immediately have the following corollary.

Corollary 4.3 *Let $\mathbb{K}$ be a number field of degree m over $\mathbb{Q}$ and $h \geq 1$ a real number. Then, the number of points $\boldsymbol{x} = (x, y) \in \mathbb{G}_m^2$ with*

$$[\mathbb{K}(\boldsymbol{x}) : \mathbb{K}] \leq e \qquad and \qquad h(\boldsymbol{x}) \leq h$$

is at most $\exp\left(hme(e + 2) + (me + e + 5)(e + 5)\right).$

Proof The statement follows by summing the bounds for the number of points of given degree i, $i = 1, \ldots, e$, and bounded height given by Lemma 4.2. That is, using that $h \geq 1$, we have

$$\sum_{i=1}^{e} \exp\left(hmi(i + 2) + me(i + 5) + i^2 + 10i + 24\right)$$

$$\leq 2 \exp\left(hme(e + 2) + me(e + 5) + e^2 + 10e + 24\right)$$

$$\leq \exp\left(hme(e + 2) + me(e + 5) + e^2 + 10e + 25\right)$$

$$= \exp\left(hme(e + 2) + (me + e + 5)(e + 5)\right),$$

which concludes the proof. $\qquad\square$

4.3 Intersection of Plane Curves with $\mathscr{C}_2(\overline{\Gamma}, \varepsilon)$

Before stating the results in this section, we need first to introduce more notation. We define the height of $\boldsymbol{x} = (x, y) \in \mathbb{G}_m^2$ by $h(\boldsymbol{x}) = h(x) + h(y)$.

Let Γ be a finitely generated subgroup of $\mathbb{G}_m^2$ of rank $r > 0$.

Let $F \in \overline{\mathbb{Q}}[X, Y]$ be an absolutely irreducible polynomial of degree d and height h, which is not special (see Definition 1.1), and define

$$\Delta = \deg_X F + \deg_Y F.$$

Let $\mathbb{K}$ be the smallest number field containing all coefficients of F and also the group Γ, that is, such that

$$F(X) \in \mathbb{K}[X] \qquad \text{and} \qquad \Gamma \subseteq (\mathbb{K}^*)^2.$$

Let $\mathcal{C} \subseteq \overline{\mathbb{Q}}^2$ be the curve defined by the zero set of the above polynomial F.

The following result is a greatly simplified form of [7, Theorem 2.3] which gives fully explicit constants and also applies to more general groups (including so called S-units).

Lemma 4.4 *Let $\mathbb{K}$, Γ, $\mathcal{C}$, Δ and h be defined as above with $\Delta \geq 2$. Then, there is a constant $c_0(\mathbb{K}, \Gamma)$ depending only on $\mathbb{K}$ and the generators of Γ, such that for every $\boldsymbol{x} \in \mathcal{C} \cap \mathscr{C}_2(\overline{\Gamma}, \zeta)$ with ζ defined by*

$$\zeta^{-1} = c_0(\mathbb{K}, \Gamma) \exp(2\Delta^2) \Delta^{7r+22} (\Delta + h) (\log \Delta)^6, \tag{11}$$

where r is the rank of Γ, we have

$$h(\boldsymbol{x}) \leq c_0(\mathbb{K}, \Gamma) \exp(2\Delta^2) \Delta^{7(r+3)} (\Delta + h) \log \Delta,$$

$$[\mathbb{K}(\boldsymbol{x}) : \mathbb{K}] \leq 2^{50} \Delta (\log \Delta)^6.$$

We note that similar results are known for intersections of curves with sets $\mathscr{B}_2(\mathcal{S}, \varepsilon)$ defined by (2) with $\mathcal{S} = \Gamma$ or $\mathcal{S} = \overline{\Gamma}$. Indeed, for a curve $\mathcal{C} \subset (\overline{\mathbb{Q}})^2$ and $\mathcal{S} = \Gamma$, stronger estimates for the cardinality of $\mathcal{C} \cap \mathscr{B}_2(\Gamma, \varepsilon)$ and $\mathcal{C} \cap \mathscr{C}_2(\Gamma, \varepsilon)$ are obtained in [30, Theorems 1.1 and 1.2] improving on previous results of Remond [32], see also [2, Corollary 1.4] (in these works Γ_ε is used instead $\mathscr{B}_2(\Gamma, \varepsilon)$). When $\mathcal{S} = \overline{\Gamma}$, a similar bound as the one in Lemma 4.4 has been obtained for $\mathcal{C} \cap \mathscr{B}_2(\overline{\Gamma}, \varepsilon)$ in [7, Theorem 2.2], but with a larger value for ε.

Putting now together Lemma 4.4 and Corollary 4.3 we obtain the following bound for the number of points in intersections of curves with $\mathscr{C}_2(\overline{\Gamma}, \varepsilon)$.

Lemma 4.5 *Let $\mathbb{K}$, Γ, $\mathcal{C}$, and Δ be defined as in Lemma 4.4. Then, for ζ defined by* (11), *we have*

$$\#\left(\mathcal{C} \cap \mathscr{C}_2(\overline{\Gamma}, \zeta)\right) \leq \exp\left((h+1) \exp\left((2 + o(1)) \Delta^2\right)\right).$$

Proof The desired result follows by applying the bounds of $h(x)$ and $[\mathbb{K}(x) : \mathbb{K}]$ of Lemma 4.4 in Corollary 4.3. In particular, we apply Corollary 4.3 with $e \leq 2^{50}\Delta(\log \Delta)^6$ and

$$h \leq c_0(\mathbb{K}, \Gamma)\exp(2\Delta^2)\Delta^{7(r+3)}(\Delta + h)\log \Delta,$$

which yields

$$\#\left(\mathcal{C} \cap \mathscr{C}_2(\overline{\Gamma}, \zeta)\right) \leq \exp\left(O\left(\exp(2\Delta^2)\Delta^{7(r+3)+2}(\Delta + h)(\log \Delta)^{13}\right)\right)$$

and thus concludes the proof. $\qquad\square$

4.4 Equations with Variables from Finitely Generated Groups

Let $\mathbb{K}$ be an algebraically closed field of characteristic zero. For fixed coefficients $a_1, \ldots, a_n \in \mathbb{K}^*$ we consider the linear equation

$$a_1 x_1 + \cdots + a_n x_n = 1. \tag{12}$$

We say that a solution $(x_1, \ldots, x_n)$ is *nondegenerate* if no subsum on the left-hand side of (12) vanishes, that is

$$\sum_{i \in \mathcal{I}} a_i x_i \neq 0$$

for every nonempty subset $\mathcal{I} \subseteq \{1, \ldots, n\}$.

For real positive k and ℓ, we define the function

$$A(k, \ell) = (8k)^{4k^4(k+\ell+1)},$$

and recall the following bound from [2, Theorem 6.2], which in turn improves the previous result from [16, Theorem 1.1].

Lemma 4.6 *For any multiplicative subgroup $\Gamma \subseteq \mathbb{K}^*$ of rank r, there are at most $A(n, r)$ nondegenerate solutions $(x_1, \ldots, x_n) \in \Gamma^n$ to (12).*

We now immediately derive the following bound.

Lemma 4.7 *Let $F(X) \in \mathbb{K}[X]$ be a polynomial of degree D which is not a monomial and let $\Gamma \subseteq \mathbb{K}^*$ be a multiplicative subgroup of rank r. Then*

$$\#\{(u, v) \in \Gamma^2 \mid F(u) = v\} < DA(D + 1, r) + D2^{D+1}.$$

Proof Assume that F contains $2 \leq n \leq D+1$ non-zero monomials, and write

$$F(X) = \sum_{i=1}^{n} a_i X^{k_i},$$

where $a_i \neq 0$, $i = 1, \ldots n$.

Then the equation $F(u) = v$ implies

$$\sum_{i=1}^{n} a_i u^{k_i} v^{-1} = 1.$$

We now treat $x_i = u^{k_i} v^{-1}$, $i = 1, \ldots n$, as independent variables and see that unless they form a nondegenerate solution, the variable u must satisfy at least one equation of the form

$$\sum_{i \in \mathcal{I}} a_i u^{k_i} = 0$$

for a nonempty subset $\mathcal{I} \subseteq \{1, \ldots, n\}$. Therefore there exist at most $D(2^{D+1} - 1) < D2^{D+1}$ such values of u after which v is fixed. For other solutions we use Lemma 4.6 with $n \leq D+1$, which gives at most $A(D+1, r)$ values for the vector $\left(u^{k_1} v^{-1}, \ldots, u^{k_n} v^{-1}\right) \in \Gamma^n$. Since $n \geq 2$, using the first two components of each such vector and eliminating v, we obtain a nontrivial equation for u of degree $|k_2 - k_1| \leq D$. The result now follows. $\square$

4.5 Combinatorial Result

We also need the following combinatorial statement which in different forms has been proved and used in a number of works, see [13, 28, 33]. We present it in the form given in [13, Lemma 5.7].

Lemma 4.8 *Let* $2 \leq T < N/2$. *For any sequence*

$$0 \leq n_1 < \cdots < n_T \leq N,$$

there exists $r \leq 2N/T$ *such that* $n_{i+1} - n_i = r$ *for at least* $T(T-1)/4N$ *values of* $i \in \{1, \ldots, T-1\}$.

We note that in [13, Lemma 5.7] the authors consider the values of r in the range $r \leq 2N/(T-1)$ and obtain at least $(T-1)^2/4N$ values of $i \in \{1, \ldots, T-1\}$. However, the slightly improved version of Lemma 4.8 follows directly from the proof of [13, Lemma 5.7], see also the proof of [28, Theorem 20] where these calculations are carried on.

We have the following result which is a straightforward application of Lemma 4.8.

Lemma 4.9 *Let $\mathbb{K}$ be an arbitrary field, $w \in \mathbb{K}$ and let $\mathcal{S} \subseteq \mathbb{K}$ be an arbitrary subset of $\mathbb{K}$. If for some $0 < \tau < 1/2$, we have*

$$T_w(N, \mathcal{S}) = \tau N \geq 2,$$

then there is a non-negative integer $k \leq 2\tau^{-1}$ such that

$$\#\{(u, v) \in \mathcal{S}^2 \mid f^{(k)}(u) = v\} \geq \frac{\tau^2 N}{8}.$$

Proof Let $T = T_w(N, \mathcal{S})$. Let $1 \leq n_1 < \cdots < n_T \leq N$ be all the values such that $f^{(n_i)}(w) \in \mathcal{S}$, $i = 1, \ldots, T$. We denote by $I(h)$ the number of $i = 1, \ldots, T - 1$ with $n_{i+1} - n_i = h$.

From Lemma 4.8 there exists $k \leq 2\tau^{-1}$ such that

$$I(k) \geq \frac{T(T-1)}{4N} = \frac{\tau^2 N}{4}\left(1 - \frac{1}{T}\right) \geq \frac{\tau^2 N}{8}. \tag{13}$$

Let $\mathcal{J}$ be the set of $j \in \{1, \ldots, T - 1\}$ with $n_{j+1} - n_j = k$. Then we have

$$f^{(n_j)}(w) \in \mathcal{S} \quad \text{and} \quad f^{(n_{j+1})}(w) = f^{(k)}\left(f^{(n_j)}(w)\right) \in \mathcal{S}, \qquad j \in \mathcal{J},$$

and thus

$$I(k) \leq \#\{(u, v) \in \mathcal{S}^2 \mid f^{(k)}(u) = v\}.$$

Recalling (13) we conclude the proof. $\qquad\square$

5 Proof of Main Results

5.1 Preambule

The proof of Theorem 2.1 is based on some well-known properties of the Weil height, such as the sub-additivity and homogeneousity, which can be found in [9, 38] and several other standard sources.

The proofs of Theorems 2.4, 3.1 and 3.2 follow a similar scheme. In particular, we always assume that w is a non-preperiodic point and thus all elements of the sequence $f^{(n)}(w)$, $n = 0, 1, \ldots$, are pairwise distinct.

5.2 Proof of Theorem 2.1

Let $y \in \overline{\Gamma}$, that is, there exists a smallest integer $m > 0$ such that $y^m = \gamma_1^{a_1} \cdots \gamma_r^{a_r}$, where $a_1, \ldots, a_r \in \mathbb{Z}$. We define

$$w_i = \lfloor a_i/m \rfloor \qquad \text{and} \qquad b_i = a_i - w_i m, \qquad i = 1, \ldots, r.$$

Then for

$$g = \gamma_1^{w_1} \cdots \gamma_r^{w_r} \in \Gamma \qquad \text{and} \qquad \gamma = \gamma_1^{b_1/m} \cdots \gamma_r^{b_r/m}$$

we see that

$$y = \xi g \gamma,$$

for some root of unity ξ. Recalling that $\mathrm{h}(\xi) = 0$ and using the properties of the height, we obtain

$$\mathrm{h}(\xi\gamma) = \mathrm{h}(\gamma) \leq \sum_{i=1}^{r} \mathrm{h}\left(\gamma_i^{b_i/m}\right) \leq \sum_{i=1}^{r} \frac{b_i}{m} \mathrm{h}(\gamma_i) < \sum_{i=1}^{r} \mathrm{h}(\gamma_i) \leq rh.$$

Now, if $x \in \mathbb{K} \cap \mathscr{B}(\overline{\Gamma}, E)$, that is, $x = yz$ with $y \in \overline{\Gamma}$ and $h(z) \leq E$ then, in the above notation we have $x = gv$, where $v = \xi\gamma z$ satisfies

$$\mathrm{h}(v) \leq \mathrm{h}(\xi\gamma) + \mathrm{h}(z) < rh + E.$$

This concludes the proof.

5.3 Proof of Corollary 2.2

The proof follows the same idea as in the remark at the beginning of Sect. 2.1. By Theorem 2.1, we have

$$f(\mathbb{K}) \cap \mathscr{B}(\overline{\Gamma}, E) \subseteq \{\gamma\eta \mid (\gamma, \eta) \in \Gamma \times \mathscr{A}(\mathbb{K}, H)\},$$

where $\mathscr{A}(\mathbb{K}, H)$ is defined by (4) and $H = E + rh$. Moreover, by the Northcott Theorem, the set $\mathscr{A}(\mathbb{K}, H)$ is finite.

Let $\gamma_1, \ldots, \gamma_r$ be the generators of Γ and let S be a finite set of places in $M_{\mathbb{K}}$, including the Archimedean ones, such that $|\gamma_i|_v = |\eta|_v = 1$ for any $i = 1, \ldots, r$, any $\eta \in \mathscr{A}(\mathbb{K}, H)$ and any $v \notin S$ (and thus, $|\gamma|_v = 1$ for any $\gamma \in \Gamma$ and $v \notin S$).

Now, if $f(x) \in \mathscr{B}(\overline{\Gamma}, E)$ for some $x \in \mathbb{K}$, then there exist $\gamma \in \Gamma$ and $\eta \in \mathscr{A}(\mathbb{K}, H)$ such that $f(x) = \gamma\eta$. Therefore, $|f(x)|_v = 1$ for any $v \notin S$, and thus,

$f(\mathbb{K}) \cap \mathscr{B}(\overline{\Gamma}, E) \subset f(\mathbb{K}) \cap R_S^*$, where R_S^* is the group of S-units in $\mathbb{K}$. Now the finiteness conclusion follows from [22, Proposition 1.5, (a)].

5.4 Proof of Corollary 2.3

Clearly if $\mathrm{Orb}_f(w) \cap g^{-1}\left(\mathscr{B}(\overline{\Gamma}, E)\right)$ is infinite then so is $g(\mathrm{Orb}_f(w)) \cap \mathscr{B}(\overline{\Gamma}, E)$. In turn, this implies that for any integer $\nu \geq 0$ the intersection $g\left(f^{(\nu)}\left(\mathrm{Orb}_f(w)\right)\right) \cap \mathscr{B}(\overline{\Gamma}, E)$ is infinite as well. Using this with $\nu = 0$ and $\nu = 1$ and recalling Corollary 2.2, we obtain the result.

5.5 Proof of Theorem 2.4

Define τ by

$$\tau = T_w(N, \Gamma)/N.$$

Clearly we can assume that

$$\tau \geq \frac{4 \log d}{\log \log \log N} \geq \frac{2}{N}$$

as otherwise there is nothing to prove.

From Lemma 4.9 we have that there exists

$$k \leq 2\tau^{-1} \leq \frac{\log \log \log N}{2 \log d} \tag{14}$$

such that

$$\#\{(u, v) \in \mathscr{C}_2(\overline{\Gamma}, \vartheta)^2 \mid f^{(k)}(u) = v\} \geq \frac{\tau^2 N}{8}. \tag{15}$$

The set $\{(u, v) \in \mathscr{C}(\overline{\Gamma}, \vartheta)^2 \mid f^{(k)}(u) = v\}$ is the intersection of the curve $C_k \subseteq \mathbb{G}_m^2$ defined by the zero set of the polynomial $f^{(k)}(X) - Y = 0$ with the set $\mathscr{C}(\overline{\Gamma}, \vartheta)^2$.

Let ζ be defined by (11) with the parameters $\Delta = d^k + 1$ and $H = h(f^{(k)})$. By (14) we have

$$\Delta = d^k + 1 \leq (\log \log N)^{1/2} + 1$$

and also by Lemma 4.1, we have

$$H \leq h\frac{d^k - 1}{d - 1} + d(d + 1)\frac{d^{k-1} - 1}{d - 1}\log 2 = O\left(\Delta\right)$$

Hence,

$$\begin{aligned}
\zeta^{-1} &= \exp(2\Delta^2 + O(1))\Delta^{7r+23}(\log \Delta)^6 \\
&= O\left((\log N)^2 (\log \log N)^{(7r+23)/2}(\log \log \log N)^6\right),
\end{aligned}$$

provided that N is large enough.

First we notice that the choice of ϑ in (6) ensures that $\vartheta \leq \zeta/2$. Thus, one has

$$\mathscr{C}(\overline{\Gamma}, \vartheta) \times \mathscr{C}(\overline{\Gamma}, \vartheta) \subseteq \mathscr{C}_2(\overline{\Gamma} \times \overline{\Gamma}, \zeta),$$

where the sets are defined by (3) with $n = 1$ for the sets in the cartesian product in the left-hand side, and with $n = 2$ for the set in the right-hand side. Thus, from (15) we obtain

$$\#\left(\mathcal{C}_k \cap \mathscr{C}_2(\overline{\Gamma} \times \overline{\Gamma}, \zeta)\right) \geq \frac{\tau^2 N}{8}.$$

We apply now Lemma 4.5 to upper bound $\#\left(\mathcal{C}_k \cap \mathscr{C}_2(\overline{\Gamma} \times \overline{\Gamma}, \zeta)\right)$. Putting everything together in Lemma 4.5 and taking into account that $k \leq 2\tau^{-1}$ and that $\tau \to 0$ as $N \to \infty$, we obtain

$$\begin{aligned}
N &\leq \exp\left(H \exp\left((2 + o(1))\Delta^2\right)\right) \\
&= \exp\left(\exp\left((2 + o(1))\Delta^2\right)\right) \\
&\leq \exp\left(\exp\left(\exp\left((2\log d + o(1))k\right)\right)\right) \\
&= \exp\left(\exp\left(\exp\left((4\log d + o(1))\tau^{-1}\right)\right)\right),
\end{aligned}$$

from where we get

$$\log \log \log N \leq (4\log d + o(1))\tau^{-1},$$

and thus we conclude the proof.

5.6   Proof of Theorem 3.1

As in the proof of Theorem 2.4, define τ by

$$\tau = T_w(N, \Gamma)/N$$

and assume that $\tau \geq 2/N$.

Since $\deg f^{(k)} = d^k$, combining Lemmas 4.7 and 4.9, we have

$$\frac{\tau^2 N}{8} \leq d^{2\tau^{-1}} A(d^{2\tau^{-1}} + 1, r) + d^{2\tau^{-1}} 2^{d^{2\tau^{-1}}+1}.$$

Thus, if $\tau \to 0$, we obtain

$$N \leq 8\tau^{-2} \left(d^{2\tau^{-1}} A(d^{2\tau^{-1}} + 1, r) + d^{2\tau^{-1}} 2^{d^{2\tau^{-1}}+1} \right). \tag{16}$$

In particular, we see that $\tau \to 0$ as $N \to \infty$, in which case the right-hand side of (16) is of the form $\exp\left(\exp((10 \log d + o(1))\tau^{-1})\right)$ and we obtain the result.

5.7   Proof of Theorem 3.2

Let $\tau = T_w\left(N, g^{-1}(\mathbb{L})\right)/N$. We can assume that $\tau N \geq 2$, as otherwise there is nothing to prove, thus Lemma 4.9 applies. As before, by Lemma 4.9 there exists $k \leq 2\tau^{-1}$ and M pairwise distinct points $(u_m, v_m) \in \mathbb{L}^2$ with $g\left(f^{(k)}(u_m)\right) = v_m$, $m = 1, \ldots, M$, such that

$$M \geq \frac{\tau^2 N}{8}.$$

Note that the points u_m, $m = 1, \ldots, M$, are pairwise distinct due to our assumption on N.

If $\mathbb{K}$ is of characteristic $p > 0$, then we note that if $ed^k \geq p$ then $ed^{2\tau^{-1}} \geq p$ and thus

$$\tau \leq \frac{2 \log d + o(1)}{\log p},$$

which concludes this case.

Otherwise, that is, if either $\mathbb{K}$ is of characteristic zero or $ed^k < p$, we note that if

$$\deg g \, \deg f^{(k)} = ed^k < M$$

then by the Lagrange interpolation we obtain $g\left(f^{(k)}\right) \in \mathbb{L}[X]$, which contradicts the condition on f. Thus

$$d^{2\tau^{-1}} \geq ed^k \geq M \geq \frac{\tau^2 N}{8}.$$

The result now follows.

6   Further Directions

One can almost certainly obtain more explicit versions of Corollary 2.2 and hence of Corollary 2.3. In fact if $\mathbb{K} = \mathbb{Q}$, then under this additional condition the result of Gross and Vincent [20] provides a necessary tool. It is very reasonable to expect that the result of [20], based on a lower bound for linear forms in logarithms, can be extended to arbitrary algebraic number fields. It is also very likely that both Corollaries 2.2 and 2.3 can be extended to the multivariate case.

It is certainly natural to expect that the "sparsity" result of Theorem 2.4 can be replaced by a finiteness result as in Corollary 2.3.

Of interest is also studying the finiteness of the set

$$\{w \in \mathbb{K}\left(\overline{\Gamma}\right) \mid f^{(n)}(w) \in \overline{\Gamma} \text{ for some } n \geq 1\},$$

where Γ is a finitely generated group of $\mathbb{K}^*$. For $\Gamma = \{1\}$, this is the main result of [27], which asserts the finiteness of the set

$$\{w \in \mathbb{K}^c \mid f^{(n)}(w) \in \mathcal{U} \text{ for some } n \geq 1\}$$

for a natural class of polynomials $f \in \mathbb{K}[X]$. This naturally leads to a series of related questions, for example, about the finiteness of the set

$$\{w \in \mathbb{K}^c \mid F(f^{(n)}(w), \zeta) = 0 \text{ for some } \zeta \in \mathcal{U} \text{ and } n \geq 1\},$$

where $F(U, V) \in \mathbb{K}[U, V]$ is a fixed polynomial. One can also ask these and similar questions for groups of the form

$$\mathcal{U}_a = \{u \in \mathbb{C} \mid u^m = a^n \text{ for some non-zero } m, n \in \mathbb{Z}\}$$

for a fixed non-zero algebraic number a (thus $\mathcal{U}_1 = \mathcal{U}$) and more general division groups of finitely generated groups. Several possible approaches to these problems have been discussed in [27], which in particular involve obtaining generalisations of both the results of Dvornicich and Zannier [14] (see also [39]) and Loxton [25] and may require quite significant (but worthwhile) efforts. Chen [12] has studied yet another generalisation of this problem.

As the first step one can try to obtain bounds on the number of elements of these sets in the interval $[1, N]$, similar to the bounds in Sect. 2.

Finally, we note that there are also various results about the sparsity of polynomial orbits in structural sets in finite fields (such as affine spaces, algebraic varieties, subgroups, orbits of another polynomial), see [13, 28, 33] and the references therein. Some of them, however, depend on their counterparts in characteristic zero and thus any progress on the above problems can contribute to this direction as well.

Acknowledgements The authors are grateful to Umberto Zannier for several valuable suggestions, in particular the idea of the proof of Theorem 2.1 appeared from one of these suggestions. The authors would also like to thank Michael Zieve for patient explanation of several issues related to the material of Sect. 3.2 and in particular for outlining the argument about orbits in subfields of number fields.

During the preparation of this paper, A. Ostafe was partially supported by the UNSW Vice Chancellor's Fellowship and I.E. Shparlinski by the Australian Research Council Grant DP140100118.

References

1. I. Aliev, C.J. Smyth, Solving algebraic equations in roots of unity. Forum Math. **24**, 641–665 (2012)
2. F. Amoroso, E. Viada, Small points on subvarieties of a torus. Duke Math. J. **150**, 407–442 (2009)
3. J.P. Bell, D. Ghioca, T.J. Tucker, The dynamical Mordell-Lang problem for Noetherian spaces. Funct. Approx. Comment. Math. **53** (2015), 313–328.
4. J.P. Bell, D. Ghioca, T.J. Tucker, *The Dynamical Mordell–Lang Conjecture*. Mathematical Surveys and Monographs, vol. 210 (American Mathematical Society, Providence, 2016)
5. R.L. Benedetto, D. Ghioca, P. Kurlberg, T.J. Tucker, A gap principle for dynamics. Compos. Math. **146**, 1056–1072 (2010)
6. R.L. Benedetto, D. Ghioca, P. Kurlberg, T.J. Tucker, A case of the dynamical Mordell–Lang conjecture (with an Appendix by U. Zannier). Math. Ann. **352**, 1–26 (2012)
7. A. Berczes, J.-H. Evertse, K. Győry, C. Pontreau, Effective results for points on certain subvarieties of tori. Math. Proc. Camb. Philos. Soc. **147**, 69–94 (2009)
8. F. Beukers, C.J. Smyth, Cyclotomic points on curves. *Number Theory for the Millennium I* (Urbana, Illinois, 2000) (A K Peters, Natick, 2002), pp. 67–85
9. E. Bombieri, W. Gubler, *Heights in Diophantine Geometry* (Cambridge University Press, Cambridge, 2006)
10. E. Bombieri, U. Zannier, Algebraic points on subvarieties of $\mathbb{G}_m^n$. Int. Math. Res. Not. **7**, 333–347 (1995)
11. J. Cahn, R. Jones, J. Spear, Powers in orbits of rational functions: cases of an arithmetic dynamical Mordell–Lang conjecture, Preprint 2015 (see http://arxiv.org/abs/1512.03085)
12. E. Chen, Avoiding algebraic integers of bounded house in orbits of rational functions over cyclotomic closures, Preprint 2016 (see http://arxiv.org/abs/1608.04146)
13. C. D'Andrea, A. Ostafe, I. Shparlinski, M. Sombra, Reduction modulo primes of systems of polynomial equations and algebraic dynamical systems, Preprint 2015 (see http://arxiv.org/abs/1505.05814)
14. R. Dvornicich, U. Zannier, Cyclotomic Diophantine problems (Hilbert irreducibility and invariant sets for polynomial maps). Duke Math. J. **139**, 527–554 (2007)

15. J.-H. Evertse, Points on subvarieties of tori, in *A Panorama of Number Theory or the View from Baker's Garden* (Zürich, 1999) (Cambridge University Press, Cambridge, 2002), pp. 214–230
16. J.-H. Evertse, H.P. Schlickewei, W.M. Schmidt, Linear equations in variables which lie in a multiplicative group. Ann. Math. **155**, 807–836 (2002)
17. D. Ghioca, The dynamical Mordell-Lang conjecture in positive characteristic, Preprint 2016 (see http://arxiv.org/abs/1610.00367)
18. D. Ghioca, T. Tucker, M. Zieve, Intersections of polynomial orbits, and a dynamical Mordell–Lang conjecture. Invent. Math. **171**, 463–483 (2008)
19. D. Ghioca, T. Tucker, M. Zieve, Linear relations between polynomial orbits. Duke Math. J. **161**, 1379–1410 (2012)
20. S.S. Gross, A.F. Vincent, On the factorization of $f(n)$ for $f(x)$ in $\mathbb{Z}[x]$. Int. J. Number Theory **9**, 1225–1236 (2013)
21. T. Krick, L.M. Pardo, M. Sombra, Sharp estimates for the arithmetic Nullstellensatz. Duke Math. J. **109**, 521–598 (2001)
22. H. Krieger, A. Levin, Z. Scherr, T. Tucker, Y. Yasufuku, M.E. Zieve, Uniform boundedness of S-units in arithmetic dynamics. Pac. J. Math. **274**, 97–106 (2015)
23. M. Laurent, Equations diophantiennes exponentielles. Invent. Math. **78**, 299–327 (1984)
24. P. Liardet, Sur une conjecture de Serge Lang. Astérisque **24–25**, 187–210 (1975)
25. J.H. Loxton, On the maximum modulus of cyclotomic integers. Acta Arith. **22**, 69–85 (1972)
26. B. Mazur, Abelian varieties and the Mordell–Lang conjecture. *Model Theory, Algebra, and Geometry*. Mathematical Sciences Research Institute Publications, vol. 39 (Cambridge University Press, Cambridge, 2000), pp. 199–227
27. A. Ostafe, On roots of unity in orbits of rational functions. Proc. Amer. Math. Soc. **145**, 1927–1936 (2017)
28. A. Ostafe, Polynomial values in affine subspaces of finite fields. J. Anal. Math. (to appear)
29. A. Ostafe, M. Sha, On the quantitative dynamical Mordell–Lang conjecture. J. Number Theory **156**, 161–182 (2015)
30. C. Pontreau, A Mordell–Lang plus Bogolomov type result for curves in $\mathbb{G}_m^2$. Monatsh. Math. **157**, 267–281 (2009)
31. B. Poonen, Mordell–Lang plus Bogomolov. Invent. Math. **137**, 413–425 (1999)
32. G. Rémond, Sur les sous-variétés des tores. Compos. Math. **134**, 337–366 (2002)
33. O. Roche-Newton, I.E. Shparlinski, Polynomial values in subfields and affine subspaces of finite fields. Q. J. Math. **66**, 693–706 (2015)
34. W.M. Schmidt, Northcott's theorem on heights I. A general estimate. Monatsh. Math. **15**, 169–181 (1993)
35. W.M. Schmidt, Heights of points on subvarieties of $\mathbb{G}_m^n$, in *Number Theory* (Paris, 1993–1994). London Mathematical Society Lecture Note Series, vol. 235 (Cambridge University Press, 1996), pp. 157–187
36. J.H. Silverman, B. Viray, On a uniform bound for the number of exceptional linear subvarieties in the dynamical Mordell–Lang conjecture. Math. Res. Lett. **20**, 547–566 (2013)
37. M. Widmer, Asymptotically counting points of bounded height. Ph.D. thesis, Universität Basel (2007)
38. U. Zannier, *Lecture notes on Diophantine analysis*, Appunti. Scuola Normale Superiore di Pisa (Nuova Serie) [Lecture Notes. Scuola Normale Superiore di Pisa (New Series)], vol. 8 (Edizioni della Normale, Pisa, 2009)
39. U. Zannier, Hilbert irreducibility above algebraic groups. Duke Math. J. **153**, 397–425 (2010)

Patterns of Primes in Arithmetic Progressions

János Pintz

Dedicated to the 60th birthday of Robert F. Tichy

Abstract After the proof of Zhang about the existence of infinitely many bounded gaps between consecutive primes the author showed the existence of a bounded d such that there are arbitrarily long arithmetic progressions of primes with the property that $p' = p+d$ is the prime following p for each element of the progression. This was a common generalization of the results of Zhang and Green-Tao. In the present work it is shown that for every m we have a bounded m-tuple of primes such that this configuration (i.e. the integer translates of this m-tuple) appear as arbitrarily long arithmetic progressions in the sequence of all primes. In fact we show that this is true for a positive proportion of all m-tuples. This is a common generalization of the celebrated works of Green-Tao and Maynard/Tao.

1 Introduction

In their ground-breaking work Green and Tao [5] proved the existence of infinitely many k-term arithmetic progressions in the sequence of primes for every integer $k > 0$. I showed a conditional strengthening of it [10] according to which if the primes have a distribution level $\vartheta > 1/2$ (for the definition of the distribution level see (1) below), then there exists a constant $C(\vartheta)$ such that we have a positive even $d \leqslant C(\vartheta)$ with the property that $0 < d \leqslant C(\vartheta)$ and for every k there exist infinitely many arithmetic progressions $\{p_i^*\}_{i=1}^k$ of length k with $p_i^* \in \mathcal{P}$ ($\mathcal{P}$ denotes the set of primes) such that $p_i^* + d$ is a prime too, in particular, the prime following p_i^*. After the proof of Zhang [13], proving the unconditional existence of infinitely many bounded gaps between primes (this was proved earlier in our work [2] under the condition that primes have a distribution level $\vartheta > 1/2$) I showed this without any unproved hypotheses [11].

J. Pintz (✉)
Alfréd Rényi Institute of Mathematics, Hungarian Academy of Sciences, Reáltanoda u. 13–15,
H-1053 Budapest, Hungary
e-mail: pintz.janos@renyi.mta.hu

© Springer International Publishing AG 2017
C. Elsholtz, P. Grabner (eds.), *Number Theory – Diophantine Problems,
Uniform Distribution and Applications*, DOI 10.1007/978-3-319-55357-3_19

We say that θ is a distribution level of the primes if

$$\sum_{\substack{q \leq x^\theta}} \max_{\substack{a \\ (a,q)=1}} \left| \pi(x,q,a) - \frac{\pi(x)}{\varphi(q)} \right| \ll_A \frac{x}{(\log x)^A} \tag{1}$$

holds for any $A > 0$ where the $\ll$ symbol of Vinogradov means that $f(x) = O(g(x))$ is abbreviated by $f(x) \ll g(x)$.

In his recent work Maynard [8] gave a simpler and more efficient proof of Zhang's theorem. In particular he gave an unconditional proof of a weaker version of Dickson's conjecture [1] which we abbreviate as Conjecture DHL since Hardy and Littlewood formulated a stronger quantitative version of it 20 years later [7].

Conjecture DHL (Prime k-Tuples Conjecture) Let $\mathcal{H} = \{h_1, \ldots, h_k\}$ be admissible, which means that for every prime p there exists an integer a_p such that for any i $a_p \not\equiv h_i \pmod{p}$. Then there are infinitely many integers n such that all of $n+h_1, \ldots, n+h_k$ are primes.

The weaker version showed by Maynard (and simultaneously and independently by T. Tao (unpublished)) was that Conjecture $\mathrm{DHL}(k, k_0)$ (formulated below) holds for $k \gg k_0^2 e^{4k_0}$.

Conjecture $DHL(k, k_0)$ If $\mathcal{H}$ is admissible of size k, then there are infinitely many integers n such that $\{n + h_i\}_{i=1}^k$ contains at least k_0 primes.

A brief argument, given by Maynard [8] (see Theorem 1.2 of his work) shows that if there exists a $C(k_0)$ such that $\mathrm{DHL}(k, k_0)$ holds for $k \geq C(k_0)$, then a positive proportion of all admissible m-tuples satisfy the prime m-tuple conjecture for every m (for the exact formulation see Theorem 1.2 of [8]).

The purpose of the present work is to show a common generalization of the result of Maynard (and Tao) and that of Green–Tao.

Theorem 1 *Let $m > 0$ and $\mathcal{A} = \{a_1, \ldots, a_r\}$ be a set of r distinct integers with r sufficiently large depending on m. Let $N(\mathcal{A})$ denote the number of integer m-tuples $\{h_1, \ldots, h_m\} \subseteq A$ such that there exist for every ℓ infinitely many ℓ-term arithmetic progressions of integers $\{n_i\}_{i=1}^\ell$ where $n_i + h_j$ is prime for each pair i, j. Then*

$$N(\mathcal{A}) \gg_m \#\{(h_1, \ldots, h_m) \in \mathcal{A}\} \gg_m |A|^m = r^m. \tag{2}$$

This is an unconditional generalization of the result in [10].

A function field analogue of our results was recently proved in a somewhat similar way by Parshall [9].

2 Preparation: First Part of the Proof of Theorem 2

The arguments in the last three paragraphs of Section 4 of [8] can be applied here practically without any change and so, similarly to Theorems 1.1 and 1.2 of [8], our Theorem 1 will also follow in essentially the same way from (the weaker)

Theorem 2 *Let m be a positive integer, $\mathcal{H} = \{h_1, \ldots, h_k\}$ be an admissible set of k distinct non-negative integers $h_i \leq H$, $k = \lceil Cm^2 e^{4m} \rceil$ with a sufficiently large absolute constant C. Then there exists an m-element subset*

$$\{h_1', h_2', \ldots, h_m'\} \subseteq \mathcal{H} \tag{3}$$

such that for every positive integer ℓ we have infinitely many ℓ-element non-trivial arithmetic progressions of integers n_i such that $n_i + h_j' \in \mathcal{P}$ for $1 \leq i \leq \ell$, $1 \leq j \leq m$, further $n_i + h_j'$ is always the jth prime following n_i.

Remark 1

 (i) For $\ell = m = 2$ this is Zhang's theorem,
 (ii) for $\ell = 2$, m arbitrary this is the Maynard–Tao theorem,
 (iii) for $m = 1$, ℓ arbitrary this is the Green–Tao theorem,
 (iv) for $m = 2$, ℓ arbitrary this was proved under the condition that primes have a distribution level $\theta > 1/2$ in [10], unconditionally (using Zhang's method) in [11].

Remark 2 In fact we prove the following stronger result:

Theorem 3 *There is some C, such that for all k_0 and all $k > Ck_0^2 e^{4k_0}$ there is some $c > 0$, such that for all admissible tuples $\{h_1, \ldots, h_k\}$ the number $N(x)$ of integers $n \leq x$, such that $n + h_i$ is n^c-pseudo prime, and among these k integers there are at least k_0 primes, satisfies $N(x) \gg \frac{x}{\log^k x}$.*

In order to show our Theorem 2 we will follow the scheme of [8]. We therefore emphasize just a few notations here, but we will use everywhere Maynard's notation throughout our work. Similarly to his work, k will be a fixed integer, $\mathcal{H} = \{h_1, \ldots, h_k\} \subseteq [0, H]$ a fixed admissible set. Any constants implied by the $\ll$ and 0 notations may depend on k and H. N will denote a large integer and asymptotics will be understood as $N \to \infty$. Most variables will be natural numbers, p (with or without subscripts) will denote always primes, $[a, b]$ the least common multiple of $[a, b]$ (however, sometimes the closed interval $[a, b]$). $\theta > 0$ will denote a distribution of primes, $R = N^{\theta/2-\varepsilon}$ with a fixed but arbitrarily small $\varepsilon > 0$. We will weight the integers with a non-negative weight w_n which will be zero unless n lies in a fixed residue class $\nu_0 \pmod{W}$ where $W = \prod_{p \leq D_0} p$. D_0 tends in [8] slowly to infinity with N. His choice is actually $D_0 = \log \log \log N$. However, it is sufficient to choose

$$D_0 = C^*(k), \tag{4}$$

with a sufficiently large constant $C^*(k)$, depending on k.

The proof runs similarly in this case as well just we lose the asymptotics then, but the dependence on D_0 is explicitly given in [8]. The weights w_n are defined in

(2.4) of [8] as

$$w_n = \left(\sum_{d_i \mid n + h_i \; \forall i} \lambda_{d_1,\dots,d_k} \right)^2. \tag{5}$$

The choice of $\lambda_{d_1,\dots,d_k}$ will be through the choice of other parameters $y_{r_1,\dots,r_k}$ by the aid of the identity

$$\lambda_{d_1,\dots,d_k} = \left(\prod_{i=1}^{k} \mu(d_i) d_i \right) \sum_{\substack{r_1,\dots,r_k \\ d_i \mid r_i \; \forall i \\ (r_i, W) = 1}} \frac{\mu\left(\prod_{i=1}^{k} r_i \right)^2}{\prod_{i=1}^{k} \varphi(r_i)} y_{r_1,\dots,r_k} \tag{6}$$

whenever $\left(\prod_{i=1}^{k} d_i, W \right) = 1$ and $\lambda_{d_1,\dots,d_r} = 0$ otherwise. Here $y_{r_1,\dots,r_k}$ will be defined by the aid of a piecewise differentiable function F,

$$y_{r_1,\dots,r_k} = F\left(\frac{\log r_1}{\log R}, \dots, \frac{\log r_k}{\log R} \right) \tag{7}$$

where F will be real valued, supported on

$$R_k = \left\{ (x_1,\dots,x_k) \in [0,1]^k : \sum_{i=1}^{k} x_i \le 1 \right\}. \tag{8}$$

All this is in complete agreement with the notation of Proposition 1 and (6.3) of [8].

Our proof will also make use of the main pillars of Maynard's proof, his Propositions 1–3, which we quote now with the above notations as

Proposition 1′ *With the above notation let*

$$S_1 := \sum_{\substack{n \\ N \le n < 2N \\ n \equiv v_0 \;(\mathrm{mod}\; W)}} w_n, \quad S_2 := \sum_{\substack{n \\ N \le n < 2N \\ n \equiv v_0 \;(\mathrm{mod}\; W)}} \left(w_n \sum_{i=1}^{k} \chi_{\mathcal{P}}(n + h_i) \right), \tag{9}$$

where $\chi_{\mathcal{P}}(n)$ denotes the characteristic function of the primes. Then we have as $N \to \infty$

$$S_1 = \frac{\left(1 + O\left(\frac{1}{D_0} \right) \right) \varphi(W)^k N (\log R)^k}{W^{k+1}} I_k(F), \tag{10}$$

$$S_2 = \frac{\left(1 + O\left(\frac{1}{D_0} \right) \right) \varphi(W)^k N (\log R)^{k+1}}{W^{k+1}} \sum_{j=1}^{k} J_k^{(j)}(F), \tag{11}$$

provided $I_k(F) \neq 0$ and $J_k^{(j)}(F) \neq 0$ for each j, where

$$I_k(F) = \int_0^1 \cdots \int_0^1 F(t_1, \ldots, t_k)^2 dt_1 \ldots dt_k, \tag{12}$$

$$J_k^{(j)}(F) = \int_0^1 \cdots \int_0^1 \left(\int_0^1 F(t_1, \ldots, t_k) dt_j \right)^2 dt_1 \ldots dt_{j-1} dt_{j+1} \ldots dt_k. \tag{13}$$

Proposition 2' *Let $\mathcal{S}_k$ denote the set of piecewise differentiable functions with the earlier given properties, including $I_k(F) \neq 0$ and $J_k^{(j)}(F) \neq 0$ for $1 \leq j \leq k$. Let*

$$M_k = \sup \frac{\sum_{j=1}^k J_k^{(j)}(F)}{I_k(F)}, \quad r_k = \left\lceil \frac{\theta M_k}{2} \right\rceil \tag{14}$$

and let $\mathcal{H}$ be a fixed admissible sequence $\mathcal{H} = \{h_1, \ldots, h_k\}$ of size k. Then there are infinitely many integers n such that at least r_k of the $n + h_i$ $(1 \leq i \leq k)$ are simultaneously primes.

Proposition 3' $M_{105} > 4$ *and* $M_k > \log k - 2 \log \log k - 2$ *for* $k > k_0$.

Remark In the proof Maynard will use for every k an explicitly given function $F = F_k$ satisfying the above inequality. Therefore the additional dependence on F will be actually a dependence on k.

The main idea (beyond the original proof of Maynard–Tao) is that in the weighted sum S_1 in (9) all those weights w_n for numbers $n \in [N, 2N]$ are in total negligible for which any of the $n + h_i$ terms $(1 \leq i \leq k)$ has a small prime factor p (i.e., with a sufficiently small $c_1(k)$ depending on k, $p \mid n + h_i$, $p < n^{c_1(k)}$).

To make it more precise let $c_1(k)$ be a sufficiently small fixed constant (to be determined later and fixed for the rest of the work). Let $P^-(n)$ be the smallest prime factor of n. Then we have

Lemma 1 *We have*

$$S_1^- = \sum_{\substack{N \leq n < 2N \\ n \equiv v_0 \;(\mathrm{mod}\; W) \\ P^-\left(\prod_{i=1}^k (n+h_i) \right) < n^{c_1(k)}}} w_n \ll_{k,H} \frac{c_1(k) \log N}{\log R} S_1. \tag{15}$$

Since $R = N^{\frac{\theta}{2} - \varepsilon}$, S_1^-/S_1 will be arbitrarily small if $c_1(k)$ is chosen sufficiently small. The proof of Lemma 1 will be postponed to Sect. 3. This means that during the whole proof we can neglect those numbers n for which $P^-\left(\prod_{i=1}^{k}(n+h_i)\right) < n^{c_1(k)}$ and it is sufficient to deal with numbers n with $n + h_i$ being almost primes for each $i = 1, 2, \ldots, k$ (by which we mean that $n + h_i$ has only prime factors at least $n^{c_1(k)}$). A trivial consequence of this fact is that for such numbers n $\prod_{i=1}^{k}(n + h_i)$ has a bounded number of prime factors. Consequently we have for these numbers n by (5.9) and (6.3) of [8]

$$w_n \ll \lambda_{\max}^2 \ll y_{\max}^2 (\log R)^{2k} \ll (\log R)^{2k} \ll (\log R)^{2k} \tag{16}$$

with the convention that the constants implied by the $\ll$ and O constants can depend on k and both $c_1(k)$ and $F = F_k$ will only depend on k.

The essence of Maynard's proof is that (see (4.1)–(4.4) of [8])

$$S_2 > \left(\left(\frac{\theta}{2} - \varepsilon\right)(M_k - \varepsilon) + O\left(\frac{1}{D_0}\right)\right) S_1 \tag{17}$$

which directly implies the existence of infinitely many values n such that there are at least

$$r_k = \left\lceil \frac{\theta M_k}{2} \right\rceil \tag{18}$$

primes among $n + h_i$ $(1 \leqslant i \leqslant k)$.

Let us denote, in analogy with (9)

$$S_1^+ := \sum_{\substack{n \\ N \leqslant n < 2N \\ n \equiv v_0 \ (\mathrm{mod}\ W) \\ P^-\left(\prod_{i=1}^{k}(n+h_i)\right) \geqslant n^{c_1(k)}}} w_n, \quad S_2^+ := \sum_{\substack{n \\ N \leqslant n < 2N \\ n \equiv v_0 \ (\mathrm{mod}\ W) \\ P^-\left(\prod(n+h_i)\right) \geqslant n^{c_1(k)}}} w_n \left(\sum_{i=1}^{k} \chi_{\mathcal{P}}(n+h_i)\right). \tag{19}$$

Then Lemma 1, i.e. (15) implies together with (17) that (if $c_1(k)$ and ε are chosen sufficiently small, D_0 sufficiently large, then)

$$S_2^+ > \left(\left(\frac{\theta}{2} - \varepsilon\right)(M_k - \varepsilon) + O(c_1(k)) + O\left(\frac{1}{D_0}\right) + o(1)\right) S_1, \tag{20}$$

which implies the existence of a large number of n values in $[N, 2N)$, $n \equiv v_0$ (mod W) with at least r_k primes among them and additionally almost primes with $P^-(n + h_i) > n^{c_1(k)}$ in all other components $i \in [1, k]$.

Together with (16) this implies

$$S_1^* := \sum_{\substack{n \\ N \leq n < 2N \\ n \equiv \nu_0 \ (\text{mod } W) \\ P^-\left(\prod_{i=1}^{k}(n+h_i)\right)>n^{c_1(k)} \\ \#\{i; n+h_i \in \mathcal{P}\} \geq r_k}} 1 \gg \frac{S_1}{(\log R)^{2k}} = \frac{\left(1 + O\left(\frac{1}{D_0}\right)\right) \varphi(W)^k N I_k(F)}{W^{k+1}(\log R)^k}. \tag{21}$$

Since $D_0 = C^*(k)$ we have $\varphi(W)^k / W^{k+1} \geq C'(k)$. Thus a positive proportion (depending on k) of the integers $n \in [N, 2N)$ with $n \equiv \nu_0 \ (\text{mod } W)$ and $P^-\left(\prod_{i=1}^{k}(n + h_i)\right) > n^{c_1(k)}$ contain at least r_k primes among $n + h_i$ $(1 \leq i \leq k)$. This follows from (21) and

$$\sum_{\substack{N \leq n < 2N, \, n \equiv \nu_0 \ (\text{mod } W) \\ P^-\left(\prod_{i=1}^{k}(n+h_i)\right)>n^{c_1(k)}}} 1 \ll \frac{N}{\log^k N} \tag{22}$$

where the implied constant in the $\ll$ symbol depends only on k, H and $c_1(k)$, therefore only on k, finally. (22) is a consequence of Selberg's sieve (see, for example, Theorem 5.1 of [6] or Theorem 2 in § 2.2.2 of [4]).

If Lemma 1 will be proved (see Sect. 3), then Theorem 2 will follow from Theorem 5 of [10] which we quote here as

Main Lemma *Let k be an arbitrary positive integer and $\mathcal{H} = \{h_1, \ldots, h_k\}$ be an admissible k-tuple. If the set $\mathcal{N}(\mathcal{H})$ satisfies with constants $c_1(k)$, $c_2(k)$*

$$\mathcal{N}(\mathcal{H}) \subseteq \left\{ n; P^-\left(\prod_{i=1}^{k}(n + h_i)\right) \geq n^{c_1(k)} \right\} \tag{23}$$

and

$$\#\{n \leq X, n \in \mathcal{N}(\mathcal{H})\} \geq \frac{c_2(k)X}{\log^k X} \tag{24}$$

for $X > X_0$, then $N(\mathcal{H})$ contains ℓ-term arithmetic progressions for every ℓ.

In order to see that the extra condition that the given prime pattern occurs also for consecutive primes we have to work in the following way. For any given $\mathcal{H} = \{h_1, \ldots, h_k\}$ with $k = \lceil Cm^2 \log m \rceil$ we choose an m-element subset $\mathcal{H}' = \{h_1', \ldots, h_m'\} \subseteq \mathcal{H}$ with minimal diameter $h_m' - h_1'$ such that with some constants

$c_1'(k), c_2'(k) > 0$ the relations (23)–(24), more exactly

$$\#\left\{ n \leq X;\ P^-\left(\prod_{i=1}^{k}(n + h_i) \right) \geq n^{c_1'(k)},\ n + h_i' \in \mathcal{P}\ (1 \leq i \leq m) \right\} \geq \frac{c_2'(k)X}{\log^k X} \qquad (25)$$

should hold for $X > X_0$.

By the condition that $\mathcal{H}'$ has minimal diameter we can delete from our set $\mathcal{N}(\mathcal{H})$ those n's for which there exists any $h_i \in \mathcal{H} \setminus \mathcal{H}'$, $h_1' < h_i < h_m'$ such that beyond (25) also $n + h_i \in \mathcal{P}$ would hold.

On the other hand, we can also neglect those $n \in \mathcal{N}(\mathcal{H})$ for which with a given $h \in [1, H]$, $h \notin \mathcal{H}_k$ we would have additionally $n + h \in \mathcal{P}$ since the total number of such $h \in [1, H]$ is by (22) at most

$$O_k\left(\frac{NH}{\log^{k+1} N} \right) = o\left(\frac{N}{\log^k N} \right) \qquad (26)$$

since our original H in Theorem 2 was fixed.

We note that the above way of specifying the m-element sets $\mathcal{H}_m'$ for which we have arbitrarily long (finite) arithmetic progressions of n's such that $n + h_i'$ ($1 \leq i \leq m$) would be a given bounded pattern of *consecutive* primes does not change the validity of the argument of Maynard (see Theorem 1.2 of [8]) which shows that the above is true for a positive proportion of all m-element sets (the proportion depends on m).

3   Proof of Lemma 1: End of the Proof of Theorem 2

The proof of Lemma 1 will be a trivial consequence of the following

Lemma 2 *The following relation holds for any prime $D_0 < p < N^{c_1}$ and all $i \in [1, \ldots, k]$:*

$$S_{1,p}^* := \sum_{\substack{N \leq n < 2N \\ n \equiv v_0\ (\text{mod } W) \\ p \mid n + h_i}} w_n \ll_{F,H,k} \frac{\log p}{p \log R} \sum_{\substack{N \leq n < 2N \\ n \equiv v_0\ (\text{mod } W)}} w_n = \frac{\log p}{p \log R} S_1. \qquad (27)$$

Proof It is clear that it is enough to show this for $i = 1$, for example. During the proof we will use the analogue of Lemma 6 of [3] for the special case $k = 1$, $\delta = p \in \mathcal{P}$ and for squarefree n with

$$f(n) = n \quad f_1(n) = \mu * f(n) = \prod_{p \mid n}(p - 1) = \varphi(n) \qquad (28)$$

which is as follows:

$$T_p := \sum_{d,e} \frac{\lambda_d \lambda_e}{[d,e,p]/p} = \sum_{\substack{r \\ p \nmid r}} \frac{\mu^2(r)}{\varphi(r)} (y_r - y_{rp})^2. \tag{29}$$

This form appears as the last displayed equation on page 85 of Selberg [12] or Eq. (1.9) on page 287 of Greaves [4]. We note the general starting condition that similarly to [8] the numbers $W, [d_1, e_1], \ldots, [d_k, e_k]$ will be always coprime to each other.

Writing $n + h_1 = pm$ we see that we have for any $\varepsilon > 0$ and denoting $\sum^*$ for the conditions $n \in [N, 2N)$, $n \equiv v_0 \pmod{W}$; $d_i, e_i \mid n + h_i$ $(2 \leqslant i \leqslant k)$

$$S^*_{1,p} = \sum_1 + \sum_2 + O(R^{2+\varepsilon}) \tag{30}$$

where

$$\sum_1 = \sideset{}{^*}\sum_{\substack{p \nmid [d_1, e_1] \\ d_1 \mid m, e_1 \mid m}} \lambda_{d_1,\ldots,d_k} \lambda_{e_1,\ldots,e_k}, \tag{31}$$

$$\sum_2 = \sideset{}{^*}\sum_{\substack{p \mid [d_1, e_1] \\ d_1 \mid pm, e_1 \mid pm}} \lambda_{d_1,\ldots,d_k} \lambda_{e_1,\ldots,e_k}. \tag{32}$$

Distinguishing further in $\sum_2$ according to $p^2 \mid d_1 e_1$ or not we obtain from (31) and (32) for any $\varepsilon > 0$

$$\sum_1 = \frac{N}{pW} \sideset{}{^*}\sum_{p \nmid [d_1, e_1]} \frac{\lambda_{d_1,\ldots,d_k} \lambda_{e_1,\ldots,e_k}}{\prod_{i=1}^{k} [d_i, e_i]} + O(R^{2+\varepsilon}) \tag{33}$$

and

$$\sum_2 = \frac{N}{pW} \left\{ \left(\sideset{}{^*}\sum_{d_1 = pd'_1, p \nmid e_1} \frac{\lambda_{pd'_1,\ldots,d_k} \lambda_{e_1,\ldots,e_k}}{[d'_1, e_1] \prod_{i=2}^{k} [d_i, e_i]} + \sideset{}{^*}\sum_{e_1 = pe'_1, p \nmid d_1} \frac{\lambda_{d_1,\ldots,d_k} \lambda_{pe'_1,\ldots,e_k}}{[d_1, e'_1] \prod_{i=2}^{k} [d_i, e_i]} \right) \right.$$
$$\left. + \sideset{}{^*}\sum_{d_1 = pd'_1, e_1 = pe'_1} \frac{\lambda_{pd'_1,\ldots,d_k} \lambda_{e'_1 p,\ldots,e_k}}{[d'_1, e'_1] \prod_{i=1}^{k} [d_i, e_i]} \right\} + O(R^{2+\varepsilon}). \tag{34}$$

Consequently we have

$$S_{1,p}^* = \frac{N}{pW} \sum{}^* \frac{\lambda_{d_1,\ldots,d_k}\lambda_{e_1,\ldots,e_k}}{\frac{[d_1,e_1,p]}{p} \prod_{i=2}^{k}[d_ie_i]} + O(R^{2+\varepsilon}). \tag{35}$$

Let us denote the sum in (35) analogously to (29) by $T_{p,1}$. Then, similarly to (29) we obtain using additionally the argument of Section 5 of [8]

$$T_{p,1} = \sum_{u_1,\ldots,u_k} \frac{\prod_{i=1}^{k} \mu^2(u_i)}{\prod_{i=1}^{k} \varphi(u_i)} \left(y_{u_i,\ldots,u_k} - y_{u_1p,u_2,\ldots,u_k}\right)^2. \tag{36}$$

However by the choice (6.3) of [8] we have (since F is piecewise differentiable)

$$\left(y_{u_1,\ldots,u_k} - y_{u_1p,u_2,\ldots,u_k}\right)^2 = F\left(\frac{\log u_1}{\log R},\ldots,\frac{\log u_k}{\log R}\right)^2 - F\left(\frac{\log u_1 + \log p}{\log R},\ldots,\frac{\log u_k}{\log R}\right)^2 \tag{37}$$

$$\ll_F \frac{\log p}{\log R},$$

since F depends only on k, and hence the constant implied by the $\ll$ symbol may depend on k. Hence we have by Proposition (4.1) of [8]

$$T_{p,1} \ll \frac{N}{W} \cdot \frac{\log p}{p \log R} \sum_{(u_1,W)=1} \frac{\prod_{i=1}^{k}(\mu^2(u_i))}{\prod_{i=1}^{k} \varphi(u_i)} \ll \frac{\log p}{p \log R} \cdot S_1 \tag{38}$$

which proves Lemma 2 and thereby Lemma 1 and Theorem 2.

Acknowledgements The author thanks to the referee for his valuable suggestions. The author was supported by National Research, Development and Innovation Office, NKFIH, OTKA NK 104181, K100291 and ERC-AdG. 321104.

References

1. L.E. Dickson, A new extension of Dirichlet's theorem on prime numbers. Messenger Math. (2) **33**, 155–161 (1904)
2. D.A. Goldston, J. Pintz, C.Y. Yıldırım, Primes in tuples I. Ann. Math. **170**, 819–862 (2009)
3. D.A. Goldston, S.W. Graham, J. Pintz, C.Y. Yıldırım, Small gaps between products of two primes. Proc. Lond. Math. Soc. (3) **98**(3), 741–774 (2009)
4. G. Greaves, *Sieves in Number Theory* (Springer, Berlin, 2001)
5. B. Green, T. Tao, The primes contain arbitrarily long arithmetic progressions. Ann. Math. (2) **167**(2), 481–547 (2008)
6. H. Halberstam, H.-E. Richert, *Sieve Methods* (Academic Press, New York, 1974)
7. G.H. Hardy, J.E. Littlewood, Some Problems of 'Partitio Numerorum'. III. On the expression of a number as a sum of primes. Acta Math. **44**, 1–70 (1923)
8. J. Maynard, Small gaps between primes, Ann. Math. (2) **181**(1), 383–413 (2015)
9. H. Parshall, Small gaps between configurations of prime polynomials. J. Number Theory **162**, 35–53 (2016)
10. J. Pintz, Are there arbitrarily long arithmetic progressions of twin primes? in *An irregular mind. Szemerédi is 70*. Bolyai Society Mathematical Studies, vol. 21 (Springer, New York, 2010)
11. J. Pintz, Polignac numbers, conjectures of Erdős on gaps between primes, arithmetic progressions in primes, and the bounded gap conjecture, in *From Arithmetic to Zeta-Functions, Number Theory in Memory of Wolfgang Schwarz*, ed. by J. Sander, J. Steuding, R. Steuding (Springer, 2016), pp. 367–384
12. A. Selberg, *Collected Papers*, vol. II (Springer, Berlin, 1991)
13. Y. Zhang, Bounded gaps between primes. Ann. Math. (2) **179**(2), 1121–1174 (2014)

On Simple Linear Recurrences

Andrzej Schinzel

To Professor Robert Tichy on his 60th anniversary

Abstract It is proved that every simple linear recurrence defined over a number field K, that has zeros modulo almost all prime ideals of K, takes the value 0 for a certain integer index. A similar theorem does not hold, in general, for simple linear recurrences of order $n > 3$. The case $n = 3$ is studied, but not decided.

AMS Classification 2010. 11B37, 11D61

I have proved [4] that for every essentially ternary integral recurrence u_n that the companion polynomial has a double zero there exists an integer $D > 0$ such that u_n contains terms divisible by m for any integer m prime to D independently of solubility of the equation $u_n = 0$ (a linear recurrence is of order essentially k, if it is of order k and not of order $k - 1$). For simple linear recurrences the situation is different. According to Skolem's conjecture [5] (see [2]) every simple linear recurrence u_n over a number field K that for every $m > 0$ contains terms divisible by m contains 0 (possibly with a negative index). We shall prove

Theorem 1 *Let K be a number field and $u_n \in K$ a simple binary linear recurrence. If for almost all (in the sense of density) prime ideals $\mathfrak{p}$ of K, the congruence*

$$u_n \equiv 0 \pmod{\mathfrak{p}} \tag{1}$$

is soluble for integers n, then the equation

$$u_n = 0 \tag{2}$$

is soluble for integers n.

A. Schinzel (✉)
Institute of Mathematics, Polish Academy of Sciences, Śniadeckich 8, 00-656 Warsaw, Poland
e-mail: schinzel@impan.pl

© Springer International Publishing AG 2017
C. Elsholtz, P. Grabner (eds.), *Number Theory – Diophantine Problems,
Uniform Distribution and Applications*, DOI 10.1007/978-3-319-55357-3_20

For $K = \mathbb{Q}$ Theorem 1 was proved by Somer [6]. In the sequel "almost all" is used in the sense of density.

Theorem 2 *Let K be a number field and $u_n \in K$ a simple essentially ternary linear recurrence with the companion polynomial*

$$(z - 1)(z - \alpha_1)(z - \alpha_2),$$

where $\alpha_1^2 = \alpha_2^x$ ($x \in \{0, 1, 2\}$). If for almost all prime ideals $\mathfrak{p}$ of K the congruence

$$u_n \equiv 0 \pmod{\mathfrak{p}} \tag{3}$$

is soluble for integers n, then the equation

$$u_n = 0 \tag{4}$$

is soluble for integers n.

Theorem 3 *Let $u_n \in \mathbb{Q}$ be a simple essentially ternary recurrence with the companion polynomial $(z - 1)(z - \alpha_1)(z - \alpha_2)$, where $\alpha_1^3 = \alpha_2^x$ ($x \in \{0, 1, 2, 3\}$). If for almost all primes p the congruence*

$$u_n \equiv 0 \pmod{p} \tag{5}$$

is soluble for integers n, then the Eq. (4) is soluble for integers n.

The following extension of Theorem 2 holds.

Let K be a number field and $u_n \in K$ a simple essentially ternary linear recurrence with the companion polynomial $(z - \alpha_1)(z - \alpha_2)(z - \alpha_3)$, where $\alpha_1^{x_1}\alpha_2^{x_2}\alpha_3^{x_3} = 1$, $x_1 + x_2 + x_3 = 0$, $0 < |x_1| + |x_2| + |x_3| \leq 4$. If for almost all prime ideals $\mathfrak{p}$ of K the congruence (3) is soluble for integers n, then the Eq. (4) is soluble for integers n.

While the proof of this extension reduces easily to the proof of Theorem 2, the proof of a similar extension of Theorem 3 does not reduce to the proof of Theorem 3 and requires unknown information about reducibility of sixnomials over the rational field.

Theorem 4 *There exist a real quadratic field K and $u_n \in K$ a simple essentially ternary linear recurrence with the companion polynomial $(z - 1)(z - \alpha)(z - \alpha^3)$, such that the congruence (3) is soluble for all prime ideals $\mathfrak{p}$ of K, but the Eq. (4) is insoluble.*

Theorem 5 *For every $k \geq 4$ there exists a simple linear integral recurrence u_n of order essentially k such that the congruence*

$$u_n \equiv 0 \pmod{p} \tag{6}$$

is soluble for all primes p, but the equation

$$u_n = 0 \tag{7}$$

is insoluble.

The main problem left open in the paper is the question, whether for simple ternary linear recurrences u_n over $\mathbb{Q}$, the solubility of the congruence (5) for almost all primes p implies the solubility of the Eq. (4). Thanks are due to two anonymous referees for correcting many mistakes.

Notation ζ_q is a primitive root of unity of order q.

Proof of Theorem 1 We have $u_n = c_1\alpha_1^n + c_2\alpha_2^n$, where the companion polynomial $(z - \alpha_1)(z - \alpha_2) \in K[z]$ and $\alpha_1 \neq \alpha_2$. If $c_1 = c_2 = 0$, or for $i = 1$ or 2, $c_i = \alpha_{3-i} = 0$, we have identically $u_n = 0$, thus the Eq. (2) is soluble. If for $i = 1$ or 2, $c_i\alpha_i \neq 0$, $c_{3-i}\alpha_{3-i} = 0$, then the congruence (1) is soluble only for $\mathfrak{p}$ dividing c_i and the numerator or the denominator of α_i. Otherwise, $c_1c_2\alpha_1\alpha_2 \neq 0$ and taking $\alpha = \alpha_1/\alpha_2$, $\beta = -c_2/c_1$ we infer from solubility of (1) that the congruence $\alpha^n \equiv \beta \pmod{\mathfrak{p}}$ is soluble for almost all prime ideals. By Theorem 2 of [1] with $k = 1$, K replaced by $K(\alpha)$ we infer that the equation $\alpha^n = \beta$ is soluble for integers n. This in turn implies solubility of (2). $\qquad\square$

Proof of Theorem 2 We have $u_n = c_0 + c_1\alpha_1^n + c_2\alpha_2^n$, $c_0c_1c_2 \neq 0$. We shall consider successively $x = 0, 1$, or 2. If $\alpha_1^2 = 1$, since u_n is simple, $\alpha_1 = -1$, $u_{2n} = c_0 + c_1 + c_2\alpha_2^{2n}$, $u_{2n+1} = c_0 - c_1 + c_2\alpha_2^{2n+1}$. Solubility of (3) implies solubility of the congruence $f(\alpha^x) \equiv 0 \pmod{\mathfrak{p}}$, where $f(z) = (c_2z + c_0 + c_1)(c_2\alpha_2z + c_0 - c_1)$, $\alpha = \alpha_2^2$. By Theorem 5 of [2] with $k = 1$, K replaced by $K(\alpha_2)$ we infer that the equation $f(\alpha^n) = 0$ is soluble for integers n. This in turn implies solubility of (4).

If $\alpha_1^2 = \alpha_2$ solubility of (3) implies solubility of the congruence $f_1(\alpha_1^n) \equiv 0 \pmod{\mathfrak{p}}$, where $f_1(z) = c_0 + c_1z + c_2z^2$. We again apply Theorem 5 of [2] with $k = 1$, K replaced by $K(\alpha_1)$.

If $\alpha_1^2 = \alpha_2^2$, then since u_n is simple, we have $\alpha_2 = -\alpha_1$, $u_{2n} = c_0 + c_1\alpha_1^{2n} + c_2\alpha_1^{2n}$, $u_{2n+1} = c_0 + c_1\alpha_1^{2n+1} - c_2\alpha_1^{2n+1}$. Solubility of (3) implies solubility of the congruence $f_2(\alpha^x) \equiv 0 \pmod{\mathfrak{p}}$, where $f_2(z) = (c_0 + c_1z + c_2z)(c_0 + c_1\alpha_1z - c_2\alpha_1z)$, $\alpha = \alpha_1^2$. We again apply Theorem 5 of [2] with $k = 1$, K replaced by $K(\alpha_1)$. $\qquad\square$

For the proof of Theorem 3 we need two lemmas.

Lemma 1 *Let K be a number field, $\alpha \in K^*$, $f \in K[z]$ be monic of degree three. The congruence*

$$f(\alpha^n) \equiv 0 \pmod{\mathfrak{p}} \tag{8}$$

is soluble for almost all prime ideals $\mathfrak{p}$ of K, if and only if either the equation $f(\alpha^n) = 0$ is soluble in integers n or for a β in the splitting field of f

$$\alpha = \beta^2, \quad f(z) = (z - \beta^{2r_1+1})(z + \beta^{2r_2})(z + \beta^{2r_3+1}), \quad r_i \in \mathbb{Z}.$$

If the last condition is satisfied, then the congruence (8) is soluble for all prime ideals $\mathfrak{p}$ of K, for which β is a $\mathfrak{p}$-adic unit and

$$either\ f(z) = (z^2 - \alpha^{2r_1+1})(z + \alpha^{r_2}) \quad or\ \beta \in K. \tag{9}$$

Proof The lemma is contained in Theorem 1 of [3] or its proof, except for the last sentence. To prove this we infer from $f \in K[z]$, $\beta^2 \in K$ that

$$\beta(\beta^{2r_1} - \beta^{2r_3}) = \beta^{2r_1+1} - \beta^{2r_3+1} \in K,$$

hence (9) holds. $\qquad\square$

Lemma 2 *If $x \in \{0, 1, 2, 3\}$; $\alpha_1^3 = \alpha_2^x$; $1, \alpha_1, \alpha_2$ distinct and $(z-\alpha_1)(z-\alpha_2) \in \mathbb{Q}[z]$, then either $x \in \{1, 2\}$, $\alpha_1 \in \mathbb{Q}^*$, $\alpha_2 \in \mathbb{Q}^*$, or $x = 0$, $\alpha_1 = \zeta_3$, $\alpha_2 = \zeta_3^2$, or $x = 1$, $\alpha_1 = \zeta_4$, $\alpha_2 = \zeta_4^3$, or $x = 3$, $\alpha_1 = \zeta_3^i r$, $\alpha_2 = \zeta_3^{2i} r (i \in \{1, 2\}, r \in \mathbb{Q}^*)$.*

Proof We shall consider successively $x = 0, 1, 2, 3$. If $\alpha_1^3 = 1$, since we have $\alpha_1 \neq 1$, $\alpha_1 = \zeta_3$ and, since $\alpha_1\alpha_2 \in \mathbb{Q}$, $\alpha_2 = \zeta_3^2 r$ $(r \in \mathbb{Q})$. Since $\alpha_1 + \alpha_2 \in \mathbb{Q}$, $\alpha_2 = \zeta_3^2$. If $\alpha_1^3 = \alpha_2$, since $\alpha_1, \alpha_2 \in \mathbb{Q}$, we have $\alpha_1^4 \in \mathbb{Q}$, $\alpha_1 + \alpha_1^3 = s \in \mathbb{Q}$ and taking the traces of both sides of last formula, we obtain that either $\alpha_1 \in \mathbb{Q}^*$, $\alpha_2 \in \mathbb{Q}^*$ or $s = 0$, $\alpha_1 + \alpha_1^3 = 0$, $\alpha_1 = \zeta_4$, $\alpha_2 = \zeta_4^3$. If $\alpha_1^3 = \alpha_2^2$, since $\alpha_1\alpha_2 \in \mathbb{Q}$, $\alpha_1 + \alpha_2 \in \mathbb{Q}$, we have $\alpha_1^5 \in \mathbb{Q}$, $\alpha_2^5 \in \mathbb{Q}$, $\alpha_1 + \alpha_2 = s \in \mathbb{Q}$ and taking traces of both sides of the last formula we obtain that either $\alpha_1 \in \mathbb{Q}^*$, $\alpha_2 \in \mathbb{Q}^*$, or $s = 0$, $\alpha_2 = -\alpha_1$, $\alpha_1^3 = \alpha_1^2$, $\alpha_1 = 0, 1$, which contradicts the assumption that we have $\alpha_1 \neq \alpha_2$, $\alpha_2 = \zeta_3^i \alpha_1$ $(i = 1, 2)$ and, since $\alpha_1 + \alpha_2 \in \mathbb{Q}$, $\zeta_3^{2i}\alpha_1 \in \mathbb{Q}$; $\alpha_1 = \zeta_3^i r$ $(r \in \mathbb{Q})$, $\alpha_2 = \zeta_3^{2i} r$. $\qquad\square$

Proof of Theorem 3 We have $u_n = c_0 + c_1\alpha_1^n + c_2\alpha_2^n \in \mathbb{Q}$,

$$c_0 c_1 c_2 \neq 0. \tag{10}$$

By Lemma 2 either $x \in \{1, 2\}$, $\alpha_1 \in \mathbb{Q}^*$, $\alpha_2 \in \mathbb{Q}^*$, or $x = 0$, $\alpha_1 = \zeta_3$, $\alpha_2 = \zeta_3^2$, or $x = 1$, $\alpha_1 = \zeta_4$, $\alpha_2 = \zeta_4^3$, or $x = 3$, $\alpha_1 = \zeta_3^i r$, $\alpha_2 = \zeta_3^{2i} r$ $(i \in \{1, 2\}, r \in \mathbb{Q}^*)$. We shall consider successively $x = 0, 1, 2, 3$. If $x = 0$, $\alpha_1 = \zeta_3$, $\alpha_2 = \zeta_3^2$, we have $u_{3n} = c_0 + c_1 + c_2$, $u_{3n+1} = c_0 + c_1\zeta_3 + c_2\zeta_3^2$, $u_{3n+2} = c_0 + c_1\zeta_3^2 + c_2\zeta_3$ and solubility of (5) for almost all primes p implies that

$$(c_0 + c_1 + c_2)(c_0 + c_1\zeta_3 + c_2\zeta_3^2)(c_0 + c_1\zeta_3^2 + c_2\zeta_3) = 0,$$

which implies solubility of (4).

If $x = 1$, $\alpha_1 = \zeta_4$, $\alpha_2 = \zeta_4^3$, we have $u_{4n} = c_0 + c_1 + c_2$, $u_{4n+1} = c_0 + c_1\zeta_4 + c_2\zeta_4^3$, $u_{4n+2} = c_0 - c_1 - c_2$, $u_{4n+3} = c_0 + c_1\zeta_4^3 + c_2\zeta_4$ and solubility of (5) for almost all primes p implies that

$$(c_0 + c_1 + c_2)(c_0 + c_1\zeta_4 + c_2\zeta_4^3)(c_0 - c_1 - c_2)(c_0 + c_1\zeta_4^3 + c_2\zeta_4) = 0,$$

which implies solubility of (4).

If $x = 1$, $\alpha_1 \in \mathbb{Q}^*$, $\alpha_2 \in \mathbb{Q}^*$, we have $c_i \in \mathbb{Q}^*$ $(0 \le i \le 2)$, thus applying Lemma 1 with $f(z) = z^3 + \frac{c_1}{c_2}z + \frac{c_0}{c_2}$, $K = \mathbb{Q}$, $\alpha = \alpha_1$ we infer from Lemma 1 that either

$$f(\alpha_1^n) = 0 \tag{11}$$

is soluble, or

$$f(z) = (z^2 - \alpha_1^{2r_1+1})(z + \alpha_1^{r_2}), \tag{12}$$

or

$$\alpha_1 = \beta^2, \ \beta \in \mathbb{Q}^*, \ f(z) = (z - \beta^{2r_1+1})(z + \beta^{2r_2})(z + \beta^{2r_3+1}),$$
$$r_i \in \mathbb{Z}, \ r_1 \ne r_3. \tag{13}$$

If (11) holds, then $u_n = 0$, while if (12) holds, then $\frac{c_0}{c_2} = 0$, contrary to (10). If (13) holds, then

$$c_2\beta^{6r_1+3} + c_1\beta^{2r_1+1} + c_0 = 0,$$
$$-c_2\beta^{6r_2} - c_1\beta^{2r_2} + c_0 = 0,$$
$$-c_2\beta^{6r_3+3} - c_1\beta^{2r_3+1} + c_0 = 0.$$

We have by (10)

$$\begin{vmatrix} \beta^{6r_1+3} & \beta^{2r_1+1} & 1 \\ -\beta^{6r_2} & -\beta^{2r_2} & 1 \\ -\beta^{6r_3+3} & -\beta^{2r_3+1} & 1 \end{vmatrix} = 0,$$

hence

$$-\beta^{6r_1+2r_2+3} + \beta^{6r_2+2r_3+1} - \beta^{6r_3+2r_1+4}$$
$$-\beta^{6r_3+2r_2+3} + \beta^{6r_1+2r_3+4} + \beta^{6r_2+2r_1+1} = 0.$$

Since $r_1 \ne r_3$, $\beta \ne 0$ is a zero of a monic polynomial with integral coefficients and the constant term ± 1 (this requires comparing the exponents in the above equation). Since $\beta \in \mathbb{Q}$, we have $\beta = \pm 1$, $\alpha_1 = 1$, $\alpha_2 = 1$, contrary to the assumption that u_n is simple.

If $x = 2$, $\alpha_1 \in \mathbb{Q}^*$, $\alpha_2 \in \mathbb{Q}^*$, then the numbers $\alpha_3 = \frac{\alpha_1}{\alpha_2}$, $\alpha_4 = \frac{1}{\alpha_2}$ satisfy $\alpha_3 \in \mathbb{Q}^*$, $\alpha_4 \in \mathbb{Q}^*$, $\alpha_4 = \alpha_3^3$ and this case reduces to the former.

If $x = 3$, $\alpha_1 = \zeta_3^i r$, $\alpha_2 = \zeta_3^{2i} r$ ($i \in \{1, 2\}$, $r \in \mathbb{Q}^*$), it follows from $u_n \in \mathbb{Q}$ that $c_0 \in \mathbb{Q}$, c_1, c_2 are conjugates in the field $\mathbb{Q}(\zeta_3)$ and

$$u_{3n} = c_0 + c_1 r^{3n} + c_2 r^{3n}, \quad u_{3n+1} = c_0 + c_1 \zeta_3^i r^{3n+1} + c_2 \zeta_3^{2i} r^{3n+1},$$

$$u_{3n+2} = c_0 + c_1 \zeta_3^{2i} r^{3n+2} + c_2 \zeta_3^{i} r^{3n+2}.$$

Solubility of (5) implies solubility of the congruence

$$f(r^{3n}) \equiv 0 \pmod{\mathfrak{p}}, \tag{14}$$

where

$$f(z) = (c_0 + c_1 z + c_2 z)(c_0 + c_1 \zeta_3^i rz + c_2 \zeta_3^{2i} rz)(c_0 + c_1 \zeta_3^{2i} r^2 z + c_2 \zeta_3^i r^2 z) \in \mathbb{Q}[z].$$

If $\deg f \leq 2$, then Theorem 5 of [2] is applicable. If $\deg f = 3$, then by Lemma 1 with $\alpha = r^3$ either

$$f(r^{3n}) = 0 \tag{15}$$

is soluble, or $c = (c_1 + c_2)(c_1 \zeta_3^i + c_2 \zeta_3^{2i})(c_1 \zeta_3^{2i} + c_2 \zeta_3^i) \neq 0$ (c is the leading coefficient of f) and

$$c^{-1} r^{-3} f(z) = (z^2 - r^{6r_1+3})(z + r^{3r_2}), \quad r_i \in \mathbb{Z}, \tag{16}$$

or $r^3 = \beta^2$, $\beta \in \mathbb{Q}^*$ and

$$c^{-1} r^{-3} f(z) = (z - \beta^{2r_1+1})(z + \beta^{2r_2})(z + \beta^{2r_3+1}), \quad r_i \in \mathbb{Z}. \tag{17}$$

If (15) holds, (4) is solvable, if (16) holds, $c_0 = 0$, contrary to (10). Finally, if (17) holds, then $r = \gamma^2$, $\beta = \gamma^3$, $\gamma \in \mathbb{Q}^*$ and either

$$c_0 + c_1 \gamma^{6r_1+3} + c_2 \gamma^{6r_1+3} = 0, \quad c_0 - c_1 \zeta_3^i \gamma^{6r_2+2} - c_2 \zeta_3^{2i} \gamma^{6r_2+2} = 0, \tag{18}$$
$$c_0 - c_1 \zeta_3^{2i} \gamma^{6r_3+7} - c_2 \zeta_3^i \gamma^{6r_3+7} = 0,$$

or

$$c_0 + c_1 \gamma^{6r_1+3} + c_2 \gamma^{6r_1+3} = 0, \quad c_0 - c_1 \zeta_3^i \gamma^{6r_3+5} - c_2 \zeta_3^{2i} \gamma^{6r_3+5} = 0, \tag{19}$$
$$c_0 - c_1 \zeta_3^{2i} \gamma^{6r_2+4} - c_2 \zeta_3^i \gamma^{6r_2+4} = 0,$$

or

$$c_0 - c_1 \gamma^{6r_2} - c_2 \gamma^{6r_2} = 0, \quad c_0 + c_1 \zeta_3^i \gamma^{6r_1+5} + c_2 \zeta_3^{2i} \gamma^{6r_1+5} = 0, \tag{20}$$
$$c_0 - c_1 \zeta_3^{2i} \gamma^{6r_3+7} - c_2 \zeta_3^i \gamma^{6r_3+7} = 0,$$

or

$$c_0 - c_1\gamma^{6r_2} - c_2\gamma^{6r_2} = 0, \quad c_0 - c_1\zeta_3^i\gamma^{6r_3+5} - c_2\zeta_3^{2i}\gamma^{6r_3+5} = 0, \qquad (21)$$
$$c_0 + c_1\zeta_3^{2i}\gamma^{6r_1+7} + c_2\zeta_3^i\gamma^{6r_1+7} = 0,$$

or

$$c_0 - c_1\gamma^{6r_3+3} - c_2\gamma^{6r_3+3} = 0, \quad c_0 + c_1\zeta_3^i\gamma^{6r_1+5} + c_2\zeta_3^{2i}\gamma^{6r_1+5} = 0, \qquad (22)$$
$$c_0 - c_1\zeta_3^{2i}\gamma^{6r_2+4} - c_2\zeta_3^i\gamma^{6r_2+4} = 0,$$

or

$$c_0 - c_1\gamma^{6r_3+3} - c_2\gamma^{6r_3+3} = 0, \quad c_0 - c_1\zeta_3^i\gamma^{6r_2+2} - c_2\zeta_3^{2i}\gamma^{6r_2+2} = 0, \qquad (23)$$
$$c_0 + c_1\zeta_3^{2i}\gamma^{6r_1+7} + c_2\zeta_3^i\gamma^{6r_1+7} = 0.$$

However, the cases (21), (22), (23) are obtained from the cases (20), (19), (18), respectively, by replacing γ by $-\gamma$, and interchanging r_1 and r_3. Therefore, it remains only to consider the cases (18), (19), (20).

From (18) and (10) we obtain

$$\begin{vmatrix} 1 & \gamma^{6r_1+3} & \gamma^{6r_1+3} \\ 1 & -\zeta_3^i\gamma^{6r_2+2} & -\zeta_3^{2i}\gamma^{6r_2+2} \\ 1 & -\zeta_3^{2i}\gamma^{6r_3+7} & -\zeta_3^i\gamma^{6r_3+7} \end{vmatrix} = 0,$$

hence

$$\zeta_3^{2i}\gamma^{6r_2+6r_3+9} - \zeta_3^{2i}\gamma^{6r_1+6r_3+10} - \zeta_3^{2i}\gamma^{6r_1+6r_2+5}$$
$$+\zeta_3^i\gamma^{6r_1+6r_2+5} - \zeta_3^i\gamma^{6r_2+6r_3+9} + \zeta_3^i\gamma^{6r_1+6r_3+10}$$
$$= (\zeta_3^{2i} - \zeta_3^i)(\gamma^{6r_2+6r_3+9} - \gamma^{6r_1+6r_3+10} - \gamma^{6r_1+6r_2+5}),$$

and, since $\gamma \in \mathbb{Q}^*$, $\gamma = \pm 1$, which is impossible.

From (19) and (10) we obtain

$$\begin{vmatrix} 1 & \gamma^{6r_1+3} & \gamma^{6r_1+3} \\ 1 & -\zeta_3^i\gamma^{6r_3+5} & -\zeta_3^{2i}\gamma^{6r_3+5} \\ 1 & -\zeta_3^{2i}\gamma^{6r_2+4} & -\zeta_3^i\gamma^{6r_2+4} \end{vmatrix} = 0,$$

hence

$$\zeta_3^{2i}\gamma^{6r_2+6r_3+9} - \zeta_3^{2i}\gamma^{6r_1+6r_2+7} - \zeta_3^{2i}\gamma^{6r_1+6r_3+8}$$

$$+\zeta_3^{i}\gamma^{6r_1+6r_3+8} - \zeta_3^{i}\gamma^{6r_2+6r_3+9} + \zeta_3^{i}\gamma^{6r_1+6r_2+7}$$

$$= (\zeta_3^{2i} - \zeta_3^{i})(\gamma^{6r_2+6r_3+9} - \gamma^{6r_1+6r_2+7} - \gamma^{6r_1+6r_3+8}),$$

and, since $\gamma \in \mathbb{Q}^*$, $\gamma = \pm 1$, which is impossible.

From (20) and (10) we obtain

$$\begin{vmatrix} 1 & -\gamma^{6r_2} & -\gamma^{6r_2} \\ 1 & \zeta_3^{i}\gamma^{6r_1+5} & \zeta_3^{2i}\gamma^{6r_1+5} \\ 1 & -\zeta_3^{2i}\gamma^{6r_3+7} & -\zeta_3^{i}\gamma^{6r_3+7} \end{vmatrix} = 0,$$

hence

$$-\zeta_3^{2i}\gamma^{6r_1+6r_3+12} + \zeta_3^{2i}\gamma^{6r_2+6r_3+7} - \zeta_3^{2i}\gamma^{6r_1+6r_2+5}$$

$$+\zeta_3^{i}\gamma^{6r_1+6r_2+5} + \zeta_3^{i}\gamma^{6r_1+6r_3+12} - \zeta_3^{i}\gamma^{6r_2+6r_3+7}$$

$$= (\zeta_3^{i} - \zeta_3^{2i})(\gamma^{6r_1+6r_3+12} - \gamma^{6r_2+6r_3+7} + \gamma^{6r_1+6r_2+5}),$$

and, since $\gamma \in \mathbb{Q}^*$, $\gamma = \pm 1$, which is impossible. $\qquad\square$

Proof of Theorem 4 Let $K = \mathbb{Q}(\sqrt{5})$, $\alpha = \frac{3+\sqrt{5}}{2}$, $u_n = -1 - 2\alpha^n + \alpha^{3n}$. The companion polynomial of u_n is $(z - 1)(z - \alpha)(z - \alpha^3)$. The congruence (3) is equivalent to $f(\alpha^n) \equiv 0 \pmod{\mathfrak{p}}$, where $f(z) = -1 - 2z + z^3 = (z + 1)(z - \beta)(z + \beta^{-1})$ and $\beta = \frac{1+\sqrt{5}}{2}$. By Lemma 1 with $r_1 = r_2 = 0$, $r_3 = -1$, the congruence (3) is soluble for almost all (in fact for all, since β is a unit) prime ideals $\mathfrak{p}$ of K, but since β is not a root of unity, (4) is not soluble in integers. $\qquad\square$

Proof of Theorem 5 Take $u_n = (4^{2n+1} - 1)(4^n + 1)^{k-3}$. The congruence (6) is equivalent for $p > 2$ to $f(4^n) \equiv 0 \pmod{p}$, where $f(z) = (z + 1)(z^2 - \frac{1}{4})$. By Lemma 1 with $K = \mathbb{Q}$, $\alpha = 4$, $\beta = 2$, $r_1 = r_3 = -1$, $r_2 = 0$, (6) is soluble for almost all (in fact for all) primes p, but (7) is not soluble in integers n.

It remains to prove that u_n is of order essentially k. We have

$$u_n = (4^{2n+1} - 1)\sum_{i=0}^{k-3}\binom{k-3}{i}4^{in}$$

$$= -1 - (k-3)4^n + \sum_{i=0}^{k-3}4^{(i+2)n}\left(4\binom{k-3}{i} - \binom{k-3}{i+2}\right).$$

If for $i \leq k - 3$: $4\binom{k-3}{i} = \binom{k-3}{i+2}$, then $i \leq k - 5$, $k^2 - k(2i + 7) - 3i^2 - 5i + 4 = 0$ and

$$(4i + 6)^2 > (2k - 2i - 7)^2 = 16i^2 + 48i + 33 > (4i + 5)^2,$$

which is impossible. Therefore, the companion polynomial of u_n is

$$(x - 1)(x - 4) \cdots (x - 4^{k-1}) = x^k - c_1 x^{k-1} - \cdots - c_k, \quad \text{where } c_k \neq 0,$$

and $u_n = c_1 u_{n-1} + c_2 u_{n-2} + \cdots + c_k u_{n-k}$ is of order essentially k. $\qquad\square$

References

1. A. Schinzel, On power residues and exponential congruences. Acta Arith. **27**, 397–420 (1975); Selecta, vol. 2, 915–938
2. A. Schinzel, Abelian binomials, power residues and exponential congruences. Acta Arith. **32**, 245–274 (1977); Corrigenda and addenda, ibid. 36 (1980), 101–104; Selecta, vol. 2, 939–970
3. A. Schinzel, On the congruence $u_n \equiv c \pmod{p}$, where u_n is a recurring sequence of the second order. Acta Acad. Paedagog. Agriensis Sect. Mat. (N.S.) **30**, 147–165 (2003)
4. A. Schinzel, On ternary integral recurrences, Bull. Pol. Acad. Sci. Math. **63**, 19–23 (2015)
5. T. Skolem, Anwendung exponentieller Kongruenzen zum Beweis der Unlösbarkeit gewisser diophantischer Gleichungen. Vid. Akad. Avh. Oslo I 12, 1–16 (1937)
6. L. Somer, Which second-order linear integral recurrences have almost all primes as divisors? Fibonacci Q. **17**, 111–116 (1979)

Corrections to [3]

p. 148,	line 10:	before *root* insert *primitive*,
	line -1:	for 0 read $\frac{m}{2}$,
p. 155,	line -12:	for h read h_2,
p. 158,	line 4:	for 12 read 13,
	line -6:	for 13 read 12,
	line -5:	for 4 twice read 2,
	line -2:	for $ad_2 - \varepsilon_2 \frac{a}{2}$ read $\frac{ad_2}{e} - \varepsilon_2 \frac{w}{4}$,
p. 159,	line 1:	for ad_3 read $\frac{ad_3}{e}$,
	line 5:	for 4 twice read 2,
p. 164,	line -14:	after *take* insert α *real*,
	formula (52):	for $\alpha^2 z + 1$ read $\alpha^2 z + \lambda_2^2 \alpha^{-2} \lambda_1^2$.

Equivalence of the Logarithmically Averaged Chowla and Sarnak Conjectures

Terence Tao

Abstract Let λ denote the Liouville function. The Chowla conjecture asserts that

$$\sum_{n \leq X} \lambda(a_1 n + b_1)\lambda(a_2 n + b_2) \ldots \lambda(a_k n + b_k) = o_{X \to \infty}(X)$$

for any fixed natural numbers $a_1, a_2, \ldots, a_k$ and non-negative integer $b_1, b_2, \ldots, b_k$ with $a_i b_j - a_j b_i \neq 0$ for all $1 \leq i < j \leq k$, and any $X \geq 1$. This conjecture is open for $k \geq 2$. As is well known, this conjecture implies the conjecture of Sarnak that

$$\sum_{n \leq X} \lambda(n)f(n) = o_{X \to \infty}(X)$$

whenever $f \colon \mathbb{N} \to \mathbb{C}$ is a fixed deterministic sequence and $X \geq 1$. In this paper, we consider the weaker logarithmically averaged versions of these conjectures, namely that

$$\sum_{X/\omega \leq n \leq X} \frac{\lambda(a_1 n + b_1)\lambda(a_2 n + b_2) \ldots \lambda(a_k n + b_k)}{n} = o_{\omega \to \infty}(\log \omega)$$

and

$$\sum_{X/\omega \leq n \leq X} \frac{\lambda(n)f(n)}{n} = o_{\omega \to \infty}(\log \omega)$$

under the same hypotheses on $a_1, \ldots, a_k, b_1, \ldots, b_k$ and f, and for any $2 \leq \omega \leq X$. Our main result is that these latter two conjectures are logically equivalent to each other, as well as to the "local Gowers uniformity" of the Liouville function. The main tools used here are the entropy decrement argument of the author used recently to establish the $k = 2$ case of the logarithmically averaged Chowla conjecture, as

T. Tao (✉)
Department of Mathematics, UCLA, 405 Hilgard Ave, Los Angeles, CA 90095, USA
e-mail: tao@math.ucla.edu

© Springer International Publishing AG 2017

C. Elsholtz, P. Grabner (eds.), *Number Theory – Diophantine Problems, Uniform Distribution and Applications*, DOI 10.1007/978-3-319-55357-3_21

well as the inverse conjecture for the Gowers norms, obtained by Green, Ziegler, and the author.

1   Introduction

Let λ denote the Liouville function, thus λ is the completely multiplicative function such that $\lambda(p) = -1$ for all primes p. We have the following well-known conjecture of Chowla [6]:

Conjecture 1.1 (Chowla Conjecture)   Let $k \geqslant 1$, let $a_1, \ldots, a_k$ be natural numbers and let $b_1, \ldots, b_k$ be distinct nonnegative integers such that $a_i b_j - a_j b_i \neq 0$ for $1 \leqslant i < j \leqslant k$. Then

$$\sum_{n \leqslant X} \lambda(a_1 n + b_1) \ldots \lambda(a_k n + b_k) = o_{X \to \infty}(X)$$

for all $X \geqslant 1$. (See Sect. 1.1 below for our asymptotic notation conventions.)

Note that the bound of $o_{X \to \infty}(X)$ improves slightly over the trivial bound of $O(X)$. The conjectures discussed later in this introduction will also similarly claim a slight improvement (of "little-o" type) over the corresponding trivial bound.

The $k = 1$ case of the Chowla conjecture is equivalent to the prime number theorem. The higher k cases are open, although there are a number of partial results available if one allows for some averaging in the $b_1, \ldots, b_k$ parameters, or if one wishes to obtain an upper bound in magnitude of the form $(1 - \varepsilon + o(1))X$ rather than $o(X)$; see [5, 13, 21, 43, 44] for some results in this direction. A routine application of the identity $\mu(n) = \sum_{d^2 \mid n} \mu(d)\lambda(\frac{n}{d^2})$ (or the inverse identity $\lambda(n) = \sum_{d^2 \mid n} \mu(\frac{n}{d^2})$) allows one to replace the Liouville function λ in Conjecture 1.1 by the Möbius function μ if desired; see, e.g., [29, §6] for a closely related argument. See also [33, 34, 45] for some results on the related topic of sign patterns for the Liouville function.

In [49, 50], Sarnak introduced the following related conjecture. Recall that a *topological dynamical system* (Y, T) is a compact metric space Y with a homeomorphism $T: Y \to Y$, and the *topological entropy* $h(Y, T)$ of such a system is defined as

$$h(Y, T) := \lim_{\varepsilon \to 0} \limsup_{n \to \infty} \frac{1}{n} \log N(\varepsilon, n)$$

where $N(\varepsilon, n)$ is the largest number of ε-separated points in Y using the metric $d_n : Y \times Y \to \mathbb{R}^+$ defined by

$$d_n(x, y) := \max_{0 \leqslant i \leqslant n} d(T^i x, T^i y).$$

A sequence $f: \mathbb{Z} \to \mathbb{C}$ is said to be *deterministic* if it is of the form

$$f(n) = F(T^n x_0)$$

for all n and some topological dynamical system (Y, T) of zero topological entropy $h(Y, T) = 0$, a base point $x_0 \in Y$, and a continuous function $F: Y \to \mathbb{C}$.

Conjecture 1.2 (Sarnak Conjecture) Let $f: \mathbb{N} \to \mathbb{C}$ be a deterministic sequence. Then

$$\sum_{n \leqslant X} \lambda(n) f(n) = o_{X \to \infty}(X)$$

for all $X \geqslant 1$.

Both Conjectures 1.1 and 1.2 can be viewed as instances of the "Möbius pseudorandomness principle" (see, e.g., [37, §13]). In [49] it was observed that Conjecture 1.2 was implied by Conjecture 1.1; see [52] for some proofs of this implication. The Sarnak conjecture has been verified for many particular instances of zero entropy topological dynamical systems [2–4, 7–11, 14–16, 18, 19, 25, 29, 32, 38–42, 46, 47, 51, 56]; for further variants of the Sarnak conjecture, see [12, 36].

Recently in [54], we introduced the following logarithmically averaged version of Conjecture 1.1:

Conjecture 1.3 (Logarithmically Averaged Chowla Conjecture) Let $k \geqslant 1$, let $a_1, \ldots, a_k$ be natural numbers, and let $b_1, \ldots, b_k$ be distinct nonnegative integers such that $a_i b_j - a_j b_i \neq 0$ for $1 \leqslant i < j \leqslant k$. Then one has

$$\sum_{X/\omega \leqslant n \leqslant X} \frac{\lambda(a_1 n + b_1) \ldots \lambda(a_k n + b_k)}{n} = o_{\omega \to \infty}(\log \omega) \tag{1}$$

for all $2 \leqslant \omega \leqslant X$.

We bound ω from below by 2 rather than 1 to avoid the minor inconvenience of $\log \omega$ vanishing. A standard averaging argument shows that Conjecture 1.1 implies Conjecture 1.3 for any fixed choice of k. Conversely, if we could prove Conjecture 1.3 for $\omega > 1$ fixed and an error term of $o_{X \to \infty}(1)$ instead of $o_{\omega \to \infty}(\log \omega)$, one could establish Conjecture 1.1 by a summation by parts argument. We leave the details of these (routine) arguments to the interested reader.

By introducing the *entropy decrement argument*, we were able to establish the $k = 2$ case of Conjecture 1.3 in [54]; using this result (or more precisely, a generalisation of this result in which λ is replaced by a more general bounded completely multiplicative function, in the spirit of the Elliott conjecture [17]), we were able to affirmatively settle the Erdős discrepancy problem [53].

One can of course restrict this conjecture to the model case $a_1 = \cdots = a_k = 1$:

Conjecture 1.4 (Logarithmically Averaged Chowla Conjecture, Special Case) Let $k \geqslant 1$, and let $h_1 < \cdots < h_k$ be distinct nonnegative integers. Then

$$\sum_{X/\omega \leqslant n \leqslant X} \frac{\lambda(n+h_1)\ldots\lambda(n+h_k)}{n} = o_{\omega \to \infty}(\log \omega) \tag{2}$$

for all $2 \leqslant \omega \leqslant X$.

By using this conjecture to compute the asymptotics of logarithmic averages of $\prod_{i=1}^{k}(1 - \epsilon_i \lambda(n+i))$ for various signs $\epsilon_1, \ldots, \epsilon_k \in \{-1, +1\}$, we see that Conjecture 1.4 implies that every sign pattern $(\epsilon_1, \ldots, \epsilon_k) \in \{-1, +1\}^k$ occurs infinitely often within the Liouville sequence $\lambda(1), \lambda(2), \ldots$. This latter claim is currently only established for $k = 3$; see [34, 45]; even the weaker claim that the Liouville sequence is not deterministic, which was also conjectured by Sarnak [49], remains open.

We also have a logarithmically averaged version of the Sarnak conjecture:

Conjecture 1.5 (Logarithmically Averaged Sarnak Conjecture) Let $f: \mathbb{N} \to \mathbb{C}$ be a deterministic sequence. Then

$$\sum_{X/\omega \leqslant n \leqslant X} \frac{\lambda(n)f(n)}{n} = o_{\omega \to \infty}(\log \omega) \tag{3}$$

for all $2 \leqslant \omega \leqslant X$.

We introduce two further conjectures which will be relevant in the proof of our main theorem. Recall that for any finitely supported function $f: \mathbb{Z} \to \mathbb{C}$ and any $d \geqslant 1$, the *Gowers uniformity norm* $\|f\|_{U^d(\mathbb{Z})}$, first introduced in [23, 24], is defined by the formula

$$\|f\|_{U^d(\mathbb{Z})} := \left(\sum_{x, h_1, \ldots, h_d \in \mathbb{Z}} \prod_{\vec{\omega} \in \{0,1\}^d} \mathcal{C}^{|\vec{\omega}|} f(x + \omega_1 h_1 + \cdots + \omega_d h_d) \right)^{1/2^d},$$

where $\vec{\omega} = (\omega_1, \ldots, \omega_d)$, $|\vec{\omega}| := \omega_1 + \cdots + \omega_d$, and $\mathcal{C}: z \mapsto \bar{z}$ is the complex conjugation operator. One can verify that $\|f\|_{U^d(\mathbb{Z})}$ is well defined as a non negative real. Given a non-empty discrete interval I in the integers $\mathbb{Z}$, we define the local Gowers norm $\|f\|_{U^d(I)}$ by the formula

$$\|f\|_{U^d(I)} := \|f 1_I\|_{U^d(\mathbb{Z})} / \|1_I\|_{U^d(\mathbb{Z})}$$

where 1_I is the indicator function of I. We then form the following conjecture:

Conjecture 1.6 (Logarithmically Averaged Local Gowers Uniformity of Liouville Conjecture) Let $d \geqslant 1$. Then one has

$$\sum_{X/\omega \leqslant n \leqslant X} \frac{\|\lambda\|_{U^d([n,n+H]\cap\mathbb{Z})}}{n} = o_{H\to\infty}(\log\omega) \qquad (4)$$

for all $2 \leqslant H \leqslant \omega \leqslant X$.

The constraint $H \leqslant \omega$ is mainly for aesthetic convenience (otherwise one would have to replace the $o_{H\to\infty}(\log\omega)$ term on the right-hand side with $o_{H\to\infty}(\log\omega) + o_{\omega\to\infty}(\log\omega))$; in any event, the conjecture is strongest and most interesting in the regime where H is small compared with X. The $d = 1$ form of this conjecture follows from the recent breakthrough work of Matomäki and Radziwiłł [43], but the $d > 1$ cases remain open. However, when one considers the regime where ω is fixed and H is large, the results in [29, 31] give the claim (4) when $H \geqslant X$, and when $d = 2$ the results of [58] extend this to $H \geqslant X^{5/8+\varepsilon}$ for any fixed $\varepsilon > 0$.

The Gowers norms are known to be connected to a special type of deterministic sequence, namely the *nilsequences*, through the *inverse conjecture for the Gowers norms*, proven in [31] after building on prior work in [23, 24, 26, 30]. As we shall see later in this paper, this result shows that Conjecture 1.6 can be placed in the following equivalent form. Recall that an *s-step nilmanifold* is a manifold of the form G/Γ where G is a connected, simply connected nilpotent Lie group of step s, and Γ is a cocompact discrete subgroup of G. We can give such a manifold a smooth Riemannian metric for the purpose of defining concepts such as a Lipschitz function on G/Γ; we will not specify the exact choice of this metric as any two such metrics are equivalent. The topological dynamical systems $(G/\Gamma, x \mapsto gx)$ for $g \in G$ are known as *nilsystems*, and sequences of the form $n \mapsto F(g^n x_0)$ for some continuous $F: G/\Gamma \to \mathbb{C}$, group element $g \in G$, and base point $x_0 \in G/\Gamma$ are known as (basic) *nilsequences*. It is not difficult to show that nilsystems have zero topological entropy, and hence all nilsequences are deterministic.

Conjecture 1.7 (Logarithmically Averaged Local Liouville-Nilsequences Conjecture) Let $s \geqslant 0$. Let G/Γ be an s-step nilmanifold, let $F: G/\Gamma \to \mathbb{C}$ be Lipschitz continuous, and let $x_0 \in G/\Gamma$. Then

$$\sum_{X/\omega \leqslant n \leqslant X} \frac{\sup_{g\in G} |\sum_{h=1}^{H} \lambda(n+h)F(g^h x_0)|}{n} = o_{H\to\infty}(H\log\omega). \qquad (5)$$

for all $2 \leqslant H \leqslant \omega \leqslant X$.

Note carefully that the supremum in g here is *inside* the summation in n. Analogously with the preceding conjecture, the $s = 0$ case of this conjecture was established in [43], but the $s \geqslant 1$ cases remain open. As with Conjecture 1.6, in the regime where ω is fixed and H is large, the results in [29] give the above claim for $H \geqslant X$, and when $s = 1$ the results of Zhan [58] extend this to $H \geqslant X^{5/8+\varepsilon}$.

A variant of the $s = 1$ case of Conjecture 1.7, in which the supremum in g, x_0 is placed outside the summation in n, but ω can be taken to be independent of x, was established in [44].

We are now ready to state the main result of this paper.

Theorem 1.8 *Conjectures 1.3–1.7 are equivalent.*

Remark 1.9 An inspection of the arguments in this paper reveals that all of the equivalences in this theorem continue to hold if we enforce a fixed functional relationship between ω and X. For instance, choosing the relationship $X = \omega$, we can show the equivalence of the logarithmically averaged Chowla conjecture

$$\sum_{n \leqslant X} \frac{\lambda(n + h_1) \ldots \lambda(n + h_k)}{n} = o_{X \to \infty}(\log X)$$

for all fixed distinct natural numbers $h_1, \ldots, h_k$, with the logarithmically averaged Sarnak conjecture

$$\sum_{n \leqslant X} \frac{\lambda(n)f(n)}{n} = o_{X \to \infty}(\log X)$$

for all fixed deterministic sequences f.

We summarise the key implications in this theorem as follows (see Fig. 1):

- The implication of Conjecture 1.4 from Conjecture 1.3 is trivial.
- The implication of Conjecture 1.5 from Conjecture 1.4 was essentially already observed in [49], but for the convenience of the reader we give a self-contained derivation in Sect. 2.

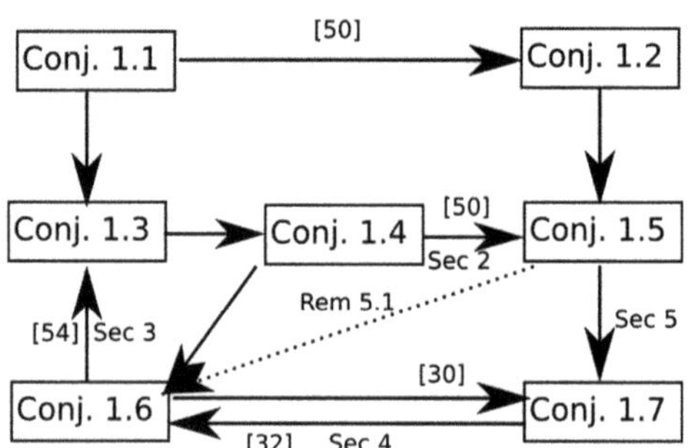

Fig. 1 Logical implications between conjectures, annotated by the reference or section where the implication (or some minor variant of that implication) is essentially proven. Implications without any annotation are trivial. The *dotted arrow* refers to the potential implication sketched in Remark 5.1. One could enlarge this diagram by adding non-logarithmically averaged versions of Conjectures 1.4, 1.6, 1.7; we leave this task to the interested reader

- The derivation of Conjecture 1.3 from Conjecture 1.6 follows from adapting the entropy decrement argument in [54], and is given in Sect. 3.
- The derivation of Conjecture 1.6 from Conjecture 1.7 follows from the inverse conjecture for the Gowers norms [31, Theorem 1.1], and is given in Sect. 4. (The converse implication is proven similarly using the converse [26, Proposition 12.6] to the inverse conjecture, which is much easier to prove.)
- Finally, the derivation of Conjecture 1.7 from Conjecture 1.5 follows from an estimation of the metric entropy of the space of nilsequences of controlled complexity, and is morally (though not quite) a consequence of the zero-entropy nature of nilsystems; we detail this in Sect. 5.

Remark 1.10 Most of the arguments in this paper should extend if one replaces the Liouville function by a more general bounded multiplicative function; the main obstruction to this is that one would now need some sort of "higher order restriction theorem for the primes" in the entropy decrement step (used to deduce Conjecture 1.3 from Conjecture 1.6), generalising the "linear restriction theorem" used in [54, Lemma 3.7]. We will not pursue this matter here (but see Remark 3.5 below).

Remark 1.11 The implication of Conjecture 1.3 from Conjecture 1.6 is the only part of the argument that requires the logarithmic averaging; all of the other implications are valid if Conjectures 1.3–1.7 are replaced by their non-logarithmically averaged counterparts (such as Conjecture 1.1 or Conjecture 1.2).

Remark 1.12 In addition to the above implications, there is also an easy way to deduce Conjecture 1.6 from Conjecture 1.4. Indeed, from expanding out the Gowers norms and interchanging summations, we see from Conjecture 1.4 that

$$\sum_{X/\omega \leqslant n \leqslant X} \frac{\|\lambda\|^{2^d}_{U^d([n,n+H]\cap\mathbb{Z})}}{n} = o_{H\to\infty}(\log\omega)$$

if H is sufficiently slowly growing as a function of ω, which by Hölder's inequality gives Conjecture 1.6 in the case when H is sufficiently slowly growing; one can then use the Gowers–Cauchy–Schwarz inequality [24] to control the Gowers norms for large values of H in terms of Gowers norms for small values of H, giving Conjecture 1.6 in general. We leave the details to the interested reader. See also Remark 5.1 for another possible implication that avoids the use of the (difficult) inverse conjecture for the Gowers norms.

1.1 Notation

We adopt the usual asymptotic notation of $A \ll B$, $B \gg A$, or $A = O(B)$ to denote the assertion that $|A| \leqslant CB$ for some constant C. If we need C to depend on an additional parameter, we will denote this by subscripts, e.g. $A = O_\varepsilon(B)$ denotes the bound $|A| \leqslant C_\varepsilon B$ for some C_ε depending on ε.

In all of our results, there will be a number of asymptotic parameters such as X, ω, H, as well as "fixed" quantities (such as $k, f, d, a_1, \ldots, a_k, b_1, \ldots, b_k$) that do not depend on the asymptotic parameters; the distinction should be clear from context. (In particular, in each of the conjectures stated in the introduction, the variables introduced before the word "Then" are fixed, and the variables appearing afterwards are asymptotic parameters.) Given an asymptotic parameter such as X, we use $A = o_{X \to \infty}(B)$ to denote the bound $|A| \leqslant c(X)B$ where $c(X)$ depends only on X and fixed quantities and goes to zero as $X \to \infty$ (subject to whatever restrictions are in place on the asymptotic parameters, such as $1 \leqslant H \leqslant \omega \leqslant X$).

If E is a statement, we use 1_E to denote the indicator, thus $1_E = 1$ when E is true and $1_E = 0$ when E is false, and $1_A(x) = 1_{x \in A}$ for any set A and point x.

Given a finite set S, we use $|S|$ to denote its cardinality.

For any real number α, we write $e(\alpha) := e^{2\pi i \alpha}$; this quantity lies in the unit circle $S^1 := \{z \in \mathbb{C} : |z| = 1\}$. By abuse of notation, we can also define $e(\alpha)$ when α lies in the additive unit circle $\mathbb{R}/\mathbb{Z}$.

All sums and products will be over the natural numbers $\mathbb{N} = \{1, 2, \ldots\}$ unless otherwise specified, with the exception of sums and products over p which is always understood to be prime.

We use $d|n$ to denote the assertion that d divides n, and $n\ (d)$ to denote the residue class of n modulo d.

We will frequently use probabilistic notation such as the expectation $\mathbb{E}\mathbf{X}$ of a random variable $\mathbf{X}$ or a probability $\mathbb{P}(E)$ of an event E. We will use boldface symbols such as $\mathbf{X}, \mathbf{Y}$, or $\mathbf{n}$ to refer to random variables.

A particularly important random variable for us will be the following. Suppose we are given some parameters $2 \leqslant \omega \leqslant X$. We then define $\mathbf{n}$ to be the random natural number $\{n : X/\omega \leqslant n \leqslant X\}$ drawn with probability distribution

$$\mathbb{P}(\mathbf{n} = n) := \frac{1/n}{\sum_{X/\omega \leqslant n \leqslant X} 1/n}.$$

Since $\sum_{X/\omega \leqslant n \leqslant X} 1/n$ is comparable to $\log \omega$, we can rewrite many of the logarithmically averaged claims conjectured in the introduction in probabilistic notation. Specifically, the bound (1) may be rewritten as

$$\mathbb{E}\lambda(a_1\mathbf{n} + b_1) \ldots \lambda(a_k\mathbf{n} + b_k) = o_{\omega \to \infty}(1), \tag{6}$$

and similarly (2) may be rewritten as

$$\mathbb{E}\lambda(\mathbf{n} + h_1) \ldots \lambda(\mathbf{n} + h_k) = o_{\omega \to \infty}(1). \tag{7}$$

Continuing in this vein, (3) is equivalent to

$$\mathbb{E}\lambda(\mathbf{n})f(\mathbf{n}) = o_{\omega \to \infty}(1), \tag{8}$$

Eq. (4) is equivalent to

$$\mathbb{E}\|\lambda\|_{U^d([\mathbf{n},\mathbf{n}+H]\cap\mathbb{Z})} = o_{H\to\infty}(1), \tag{9}$$

and (5) is equivalent to

$$\mathbb{E}\sup_{g\in G}\left|\sum_{h=1}^{H}\lambda(\mathbf{n}+h)F(g^{\mathbf{n}}x_0)\right| = o_{H\to\infty}(H). \tag{10}$$

We will rely heavily on the following approximate affine invariance of the random variable $\mathbf{n}$:

Lemma 1.13 (Approximate Affine Invariance) *Let q be a natural number, and let r be an integer. Suppose that ω is sufficiently large depending on q, r. Then for any complex-valued random variable $F(\mathbf{n})$ depending on $\mathbf{n}$ and bounded in magnitude by $O(1)$, one has*

$$\mathbb{E}F(\mathbf{n})1_{\mathbf{n}=r\ (q)} = \frac{1}{q}\mathbb{E}F(q\mathbf{n}+r) + o_{\omega\to\infty}(1).$$

Proof See [54, Lemma 2.5]. (The statement there involved additional parameters H_+, A intermediate between q, r and ω, but it is easy to see that one can delete these parameters from the statement and proof of that lemma.) $\qquad\square$

Specialising this lemma to the case $q = 1$, we obtain the approximate translation invariance

$$\mathbb{E}F(\mathbf{n}) = \mathbb{E}F(\mathbf{n}+r) + o_{\omega\to\infty}(1) \tag{11}$$

when ω is sufficiently large depending on r. This translation invariance will be sufficient for establishing the implications in Sects. 2 and 4, but the argument in Sect. 3 requires the full affine invariance from Lemma 1.13, which is only available in the logarithmically averaged setting.

2 From Chowla to Sarnak

In this section we deduce Conjecture 1.5 from Conjecture 1.4. Our arguments are an adaptation of those in [52].

Fix a topological dynamical system (Y, T) of zero topological entropy, a base point $x_0 \in Y$, and a continuous function $F: Y \to \mathbb{C}$. We allow all implied constants in the asymptotic notation to depend on these quantities. We introduce the following parameters:

- We let $\varepsilon > 0$ be a quantity that is sufficiently small (depending on the fixed quantities $(Y, T), x_0, F$).
- Then, we let H be a quantity that is sufficiently large depending on ε (and the fixed quantities).
- Finally, we let $2 \leqslant \omega \leqslant X$ be quantities with ω sufficiently large depending on ε, H (and the fixed quantities).

Let $\mathbf{n}$ be as in the previous section. Using the form (8) of Conjecture 1.5, we see that it will suffice to establish the bound

$$\mathbb{E}\lambda(\mathbf{n})F(T^{\mathbf{n}}x_0) \ll \varepsilon$$

under the above assumptions on $\varepsilon, H, \omega, X$.

From approximate translation invariance (11), we have

$$\mathbb{E}\lambda(\mathbf{n} + h)F(T^{\mathbf{n}+h}x_0) = \mathbb{E}\lambda(\mathbf{n})F(T^{\mathbf{n}}x_0) + o_{\omega\to\infty}(1)$$

for any $1 \leqslant h \leqslant H$, so in particular upon averaging in h we obtain

$$\mathbb{E}\frac{1}{H}\sum_{h=1}^{H}\lambda(\mathbf{n} + h)F(T^{\mathbf{n}+h}x_0) = \mathbb{E}\lambda(\mathbf{n})F(T^{\mathbf{n}}x_0) + o_{\omega\to\infty}(1).$$

Thus it will suffice to show that

$$\frac{1}{H}\sum_{h=1}^{H}\lambda(\mathbf{n} + h)F(T^{\mathbf{n}+h}x_0) \ll \varepsilon$$

with probability $1 - O(\varepsilon)$, since this expression is already bounded by $O(1)$.

As F is uniformly continuous, there exists $\delta > 0$ depending on ε, F such that $|F(x) - F(y)| \leqslant \varepsilon$ whenever $d(x, y) \leqslant \delta$. As (Y, T) has zero entropy, we see (if H is large enough) that we can cover Y by $O(\exp(\varepsilon^3 H))$ balls of radius δ in the d_H metric. That is to say, we can find points $x_1, \ldots, x_m \in Y$ with $m \ll \exp(\varepsilon^3 H)$ such that for each $y \in Y$, there exists $1 \leqslant i \leqslant m$ such that

$$d(T^h x_i, T^h y) \leqslant \delta$$

for all $1 \leqslant h \leqslant H$. Applying this with y replaced by $T^{\mathbf{n}}x_0$, we conclude that there exists a random variable $1 \leqslant \mathbf{i} \leqslant m$ such that

$$d(T^h x_{\mathbf{i}}, T^{\mathbf{n}+h}x_0) \leqslant \delta$$

for all $1 \leqslant h \leqslant H$, and in particular

$$\frac{1}{H}\sum_{h=1}^{H}\lambda(\mathbf{n} + h)F(T^{\mathbf{n}+h}x_0) = \frac{1}{H}\sum_{h=1}^{H}\lambda(\mathbf{n} + h)F(T^h x_{\mathbf{i}}) + O(\varepsilon).$$

Thus it will suffice to show that

$$\left| \sum_{h=1}^{H} \lambda(\mathbf{n} + h) F(T^h x_{\mathbf{i}}) \right| \leqslant \varepsilon H$$

with probability $1 - O(\varepsilon)$. Since there are only $O(\exp(\varepsilon^3 H))$ choices for $\mathbf{i}$, it suffices by the union bound to show that

$$\left| \sum_{h=1}^{H} \lambda(\mathbf{n} + h) F(T^h x_i) \right| \leqslant \varepsilon H$$

with probability $1 - O(\exp(c\varepsilon^2 H))$ for some fixed $c > 0$ and all (deterministic) $i = 1, \ldots, m$.

Let $k \leqslant H/2$ be a natural number to be chosen later. By the Chebyshev inequality, we have

$$\mathbb{P}\left(\left| \sum_{h=1}^{H} \lambda(\mathbf{n} + h) F(T^h x_i) \right| > \varepsilon H \right) \leqslant (\varepsilon H)^{-2k} \mathbb{E} \left| \sum_{h=1}^{H} \lambda(\mathbf{n} + h) F(T^h x_i) \right|^{2k}. \tag{12}$$

On the other hand, from Conjecture 1.4 (in the form (7)), we have

$$\mathbb{E}\lambda(\mathbf{n} + h_1) \ldots \lambda(\mathbf{n} + h_{2k}) = o_{\omega \to \infty}(1) \tag{13}$$

for any $1 \leqslant h_1 < \cdots < h_{2k} \leqslant H$, since ω is assumed sufficiently large depending on H.

Expanding out the expression inside the expectation in (12), we obtain H^{2k} terms, most of which are $o_{\omega \to \infty}(1)$ thanks to (13). The cumulative contribution of all such terms to (12) is still $o_{\omega \to \infty}$, since ω is assumed large depending on H (and hence on k). The only terms which are not of this form are terms in which each factor of $\lambda(\mathbf{n}+h)$ occurs at least twice (so in particular at most k different values of h appear). Crude counting shows that there are at most $k^{2k} \binom{H}{k} = O(Hk)^k$ such terms, each of which contributes at most $O(1)$ to the above sum, and hence

$$\mathbb{P}\left(\left| \sum_{h=1}^{H} \lambda(\mathbf{n} + h) F(T^h x_i) \right| > \varepsilon H \right) \ll (\varepsilon H)^{-2k} O(Hk)^k + o_{\omega \to \infty}(1).$$

Choosing k to be a small multiple of $\varepsilon^2 H$ (rounded to the nearest integer), we obtain the claim.

3 The Entropy Decrement Argument

In this section we use the entropy decrement argument from [54], together with some Cauchy–Schwarz type manipulations similar to that used in [22, 57], as well as known results on linear equations on primes [28], to deduce Conjecture 1.3 from Conjecture 1.6.

We first make some easy reductions in Conjecture 1.3. Firstly, we may assume $k > 2$, since the $k \leqslant 2$ case was already established in [54]. Next, if we set $a := a_1 \ldots a_k$, then $\lambda(a_i n + b_i)$ is a constant multiple of $\lambda(an + b_i')$, where $b_i' := a_1 \ldots a_{i-1} b_i a_{i+1} \ldots a_k$. Thus (replacing a_i, b_i with a, b_i' for each i) we may assume without loss of generality that $a_1 = \cdots = a_k = a$, in which case the condition $a_i b_j - a_j b_i \neq 0$ now simplifies to the requirement that the $b_1, \ldots, b_k$ are distinct.

Henceforth $k, a, b_1, \ldots, b_k$ are considered fixed. We allow all implied constants in the argument below to depend on $k, a, b_1, \ldots, b_k$. We select some further quantities:

- First, we let $\varepsilon > 0$ be a quantity that is sufficiently small depending on $k, a, b_1, \ldots, b_k$.
- Then, we select a natural number w that is sufficiently large depending on $k, a, b_1, \ldots, b_k, \varepsilon$.
- Then, we select a natural number H_- that is sufficiently large depending on $k, a, b_1, \ldots, b_k, \varepsilon, w$.
- Then, we select a natural number H_+ that is sufficiently large depending on $k, a, b_1, \ldots, b_k, \varepsilon, w, H_-$.
- Finally, we let ω, X be quantities such that $2 \leqslant \omega \leqslant X$ such that ω is sufficiently large depending on $k, a, b_1, \ldots, b_k, \varepsilon, w, H_-, H_+$.

The reader may find it convenient to keep the hierarchy

$$1 \ll \frac{1}{\varepsilon} \ll w \ll H_- \ll H_+ \ll \omega \leqslant X$$

in mind in the arguments which follow.

Using the form (6), it will now suffice to establish the bound

$$\mathbb{E} \prod_{i=1}^{k} \lambda(a\mathbf{n} + b_i) \ll \varepsilon.$$

Using approximate translation invariance (11), we may assume without loss of generality that $b_1 = 0$.

Assume for the sake of contradiction that the claim failed, thus

$$\left| \mathbb{E} \prod_{i=1}^{k} \lambda(a\mathbf{n} + b_i) \right| \gg \varepsilon. \tag{14}$$

We now use Lemma 1.13 to convert the single average in (14) to a double average, as in [54, Proposition 2.6]:

Proposition 3.1 *Suppose that* (14) *holds. Let* $H_- \leqslant H \leqslant H_+$, *and let* $\mathcal{P}_H$ *denote the set of primes between* $\frac{\varepsilon^2}{2}H$ *and* $\varepsilon^2 H$. *Then*

$$\left| \mathbb{E} \sum_{p \in \mathcal{P}_H} \sum_j 1_{a\mathbf{n}+j=0 \; (ap)} \prod_{i=1}^{k} \lambda(a\mathbf{n} + j + pb_i) 1_{[1,H]}(j + pb_i) \right| \gg \varepsilon \frac{H}{\log H}.$$

Proof Write

$$Q := \mathbb{E} 1_{\mathbf{n}=0 \; (a)} \prod_{i=1}^{k} \lambda(\mathbf{n} + b_i),$$

then (14) and Lemma 1.13 imply that $|Q| \gg \varepsilon$. For any prime p, we have $\lambda(p) = -1$, and hence from the complete multiplicativity of the Liouville function we have the identity

$$1_{\mathbf{n}=0 \; (a)} \prod_{i=1}^{k} \lambda(\mathbf{n} + b_i) = (-1)^k 1_{p\mathbf{n}=0 \; (ap)} \prod_{i=1}^{k} \lambda(p\mathbf{n} + pb_i)$$

and thus

$$\mathbb{E} 1_{p\mathbf{n}=0 \; (ap)} \prod_{i=1}^{k} \lambda(p\mathbf{n} + pb_i) = (-1)^k Q.$$

Applying Lemma 1.13 and noting that $1_{\mathbf{n}=0 \; (ap)} 1_{\mathbf{n}=0 \; (p)} = 1_{\mathbf{n}=0 \; (ap)}$, we conclude that

$$\mathbb{E} 1_{\mathbf{n}=0 \; (ap)} \prod_{i=1}^{k} \lambda(\mathbf{n} + pb_i) = (-1)^k \frac{Q}{p} + o_{\omega \to \infty}(1).$$

for any prime $p \leqslant H$. Shifting $\mathbf{n}$ by j using another application of Lemma 1.13, we conclude that

$$\mathbb{E} 1_{\mathbf{n}+j=0 \; (ap)} \prod_{i=1}^{k} \lambda(\mathbf{n} + j + pb_i) = (-1)^k \frac{Q}{p} + o_{\omega \to \infty}(1).$$

for any prime $p \leqslant H$ and any $1 \leqslant j \leqslant H$. Summing in j, we conclude (recalling that ω is assumed large compared with H_+ and hence H)

$$\mathbb{E} \sum_{j=1}^{H} 1_{\mathbf{n}+j=0 \; (ap)} \prod_{i=1}^{k} \lambda(\mathbf{n} + j + pb_i) = (-1)^k \frac{HQ}{p} + o_{\omega \to \infty}(1).$$

If we now introduce the quantity

$$R(s) = R_p(s) := \mathbb{E} \sum_{j=1}^{H} 1_{\mathbf{n}+j=0 \ (ap)} \prod_{i=1}^{k} \lambda(\mathbf{n}+j+pb_i) 1_{\mathbf{n}=s \ (a)}$$

for $s \in \mathbb{Z}/a\mathbb{Z}$, we therefore have

$$\sum_{s \in \mathbb{Z}/a\mathbb{Z}} R(s) = (-1)^k \frac{HQ}{p} + o_{\omega \to \infty}(1). \tag{15}$$

On the other hand, applying Lemma 1.13 with $\mathbf{n}$ shifted to $\mathbf{n}+1$, and then shifting j by one, we have

$$R(s+1) := \mathbb{E} \sum_{j=2}^{H+1} 1_{\mathbf{n}+j=0 \ (ap)} \prod_{i=1}^{k} \lambda(\mathbf{n}+j+pb_i) 1_{\mathbf{n}=s \ (a)}.$$

The difference between $\sum_{j=2}^{H+1} 1_{\mathbf{n}+j=0 \ (ap)} \prod_{i=1}^{k} \lambda(\mathbf{n} + j + pb_i) 1_{\mathbf{n}=s \ (a)}$ and $\sum_{j=1}^{H} 1_{\mathbf{n}+j=0 \ (ap)} \lambda(\mathbf{n}+j+pb_1) \ldots \lambda(\mathbf{n}+j+pb_k) 1_{\mathbf{n}=s \ (a)}$ is zero with probability $1 - O(1/p)$, and $O(1)$ on the remaining event. Absorbing the $o_{\omega \to \infty}(1)$ error into the $O(1/p)$ error, we conclude that

$$R(s+1) = R(s) + O\left(\frac{1}{p}\right)$$

for all $s \in \mathbb{Z}/a\mathbb{Z}$, so R fluctuates by at most $O(a/p)$. Combining this with (15), we conclude in particular that

$$R(0) = (-1)^k \frac{HQ}{ap} + O\left(\frac{a}{p}\right).$$

Summing over $\mathcal{P}_H$, we conclude that

$$\mathbb{E} \sum_{j=1}^{H} \sum_{p \in \mathcal{P}_H} 1_{\mathbf{n}+j=0 \ (ap)} \prod_{i=1}^{k} \lambda(\mathbf{n}+j+pb_i) 1_{\mathbf{n}=0 \ (a)}$$
$$= \left((-1)^k \frac{HQ}{a} + O(a)\right) \sum_{p \in \mathcal{P}_H} \frac{1}{p}$$

and hence by the prime number theorem and the lower bound $|Q| \gg \varepsilon$, we have

$$\left| \mathbb{E} \sum_{j=1}^{H} \sum_{p \in \mathcal{P}_H} 1_{\mathbf{n}+j=0 \ (ap)} \prod_{i=1}^{k} \lambda(\mathbf{n}+j+pb_i) 1_{\mathbf{n}=0 \ (a)} \right| \gg \varepsilon \frac{H}{a \log H}.$$

Applying Lemma 1.13, we obtain

$$\left| \mathbb{E} \sum_{j=1}^{H} \sum_{p \in \mathcal{P}_H} 1_{a\mathbf{n}+j=0 \ (ap)} \prod_{i=1}^{k} \lambda(a\mathbf{n} + j + pb_1) \right| \gg \varepsilon \frac{H}{\log H}.$$

If one of the $j + pb_i$ lies outside of $[1, H]$, then j lies in either $[1, B\varepsilon^2 H]$ or $[(1 - B\varepsilon^2)H, H]$, where $B := \max(|b_1|, \ldots, |b_k|)$. The contribution of these values of j can be easily estimated to be $O(\frac{\varepsilon^2 BH}{\log H})$, which is negligible since ε was assumed small. Discarding these contributions, we obtain the proposition. $\qquad\square$

We rewrite the conclusion of Proposition 3.1 as

$$|\mathbb{E}F(\mathbf{X}_H, \mathbf{Y}_H)| \gg \varepsilon \frac{H}{\log H} \tag{16}$$

where $\mathbf{X}_H$ is the discrete random variable

$$\mathbf{X}_H := (\lambda(a\mathbf{n} + j))_{j=1,\ldots,H}$$

(taking values in $\{-1, +1\}^H$), $\mathbf{Y}_H$ is the discrete random variable

$$\mathbf{Y}_H := \mathbf{n} \ (P_H)$$

(taking values in $\mathbb{Z}/P_H\mathbb{Z}$) with $P_H := \prod_{p \in H} p$, and $F: \{-1, +1\}^H \times \mathbb{Z}/P_H\mathbb{Z} \to \mathbb{R}$ is the function

$$F((x_j)_{j=1,\ldots,H}, y \ (P_H)) := \sum_{p \in \mathcal{P}_H} \sum_{j} 1_{ay+j=0 \ (ap)} \prod_{i=1}^{k} x_{j+pb_1} \tag{17}$$

with the convention that $x_j = 0$ for $j \notin [1, H]$.

Crucially, we can locate a scale H in which $\mathbf{X}_H$ and $\mathbf{Y}_H$ have a weak independence property:

Proposition 3.2 (Entropy Decrement Argument) *There exists a natural number H between H_- and H_+ which is a multiple of a, such that*

$$\mathbb{I}(\mathbf{X}_H, \mathbf{Y}_H) \leq \frac{H}{\log H \log \log \log H},$$

where $\mathbb{I}(\mathbf{X}_H, \mathbf{Y}_H)$ denotes the mutual information between $\mathbf{X}_H$ and $\mathbf{Y}_H$ (see [54, §3] for a definition).

Proof See [54, Lemma 3.2]. $\qquad\square$

Let H be as in the above proposition. Repeating the derivation of [54, (3.16)] (using in particular the Hoeffding concentration inequality [35]) almost verbatim, we may now conclude from (16) that

$$\left| \mathbb{E} \frac{1}{P_H} \sum_{y \in \mathbb{Z}/P_H \mathbb{Z}} F(\mathbf{X}_H, y) \right| \gg \varepsilon \frac{H}{\log H}.$$

But from the Chinese remainder theorem and (17), the left-hand side can be written as

$$\left| \mathbb{E} \sum_{p \in \mathcal{P}_H} \frac{1}{p} \sum_{j=0 \ (a)} \prod_{i=1}^{k} \lambda(a\mathbf{n} + j + pb_i) 1_{[1,H]}(j + pb_i) \right|.$$

Writing $1 = \frac{\log p}{\log H} + O_\varepsilon(\frac{1}{\log H})$ and discarding the error term by the triangle inequality and prime number theorem, we thus have

$$\left| \mathbb{E} \sum_{p \in \mathcal{P}_H} \frac{\log p}{p} \sum_{j=0 \ (a)} \prod_{i=1}^{k} \lambda(a\mathbf{n} + j + pb_i) 1_{[1,H]}(j + pb_i) \right| \gg \varepsilon H.$$

If we let Λ denote the von Mangoldt function, we thus have

$$\left| \mathbb{E} \sum_{\frac{\varepsilon^2}{2} H \leq m \leq \varepsilon^2 H} \frac{\Lambda(m)}{m} \sum_{j=0 \ (a)} \prod_{i=1}^{k} \lambda(a\mathbf{n} + j + pb_i) 1_{[1,H]}(j + pb_i) \right| \gg \varepsilon H, \qquad (18)$$

since the contribution of those m which are powers of primes, rather than primes, is easily seen to be negligible.

It is now convenient to use the "W-trick" from [27]. We recall the parameter w introduced (but not yet used) at the beginning of the argument. We set

$$W := \prod_{p \leq w} p$$

and observe that the contribution to (18) of those m that share a common factor with W is negligible. Discarding these terms and applying the pigeonhole principle, we conclude the existence of a natural number $1 \leq r \leq W$ coprime with W, such that

$$\left| \mathbb{E} \sum_{\frac{\varepsilon^2}{2} H \leq m \leq \varepsilon^2 H} \frac{\Lambda(m) 1_{m=r \ (W)}}{m} \sum_{j=0 \ (a)} \prod_{i=1}^{k} \lambda(a\mathbf{n} + j + mb_i) 1_{[1,H]}(j + mb_i) \right|$$

$$\gg \varepsilon \frac{H}{\phi(W)},$$

where $\phi(W)$ is the Euler totient function of W. Making the substitution $m = Wm' + r$, and discarding some negligible error terms, we conclude that

$$\left| \mathbb{E} \sum_{\frac{\varepsilon^2}{2} \frac{H}{W} \leq m \leq \varepsilon^2 \frac{H}{W}} \frac{\Lambda(Wm + r)}{Wm} \sum_{j=0}^{k} \prod_{(a)\ i=1}^{k} \lambda(a\mathbf{n} + j + (Wm + r)b_i) 1_{[1,H]}(j + (Wm + r)b_i) \right|$$

$$\gg \varepsilon \frac{H}{\phi(W)},$$

so if we define

$$\Lambda_{W,r}(m) := \frac{\phi(W)}{W} \Lambda(Wm + r)$$

then

$$\left| \mathbb{E} \sum_{\frac{\varepsilon^2}{2} \frac{H}{W} \leq m \leq \varepsilon^2 \frac{H}{W}} \frac{\Lambda_{W,r}(m)}{m} \sum_{j=0}^{k} \prod_{(a)\ i=1}^{k} \lambda(a\mathbf{n} + j + (Wm + r)b_i) 1_{[1,H]}(j + (Wm + r)b_i) \right|$$

$$\gg \varepsilon H.$$

$$(19)$$

We now replace $\Lambda_{W,r}$ by 1. Manipulations of this form have appeared in [22, 57]; we will use an argument somewhat similar to that in [22]:

Proposition 3.3 (Elimination of von Mangoldt Weight) *We have*

$$\mathbb{E} \sum_{\frac{\varepsilon^2}{2} \frac{H}{W} \leq m \leq \varepsilon^2 \frac{H}{W}} \frac{\Lambda_{W,r}(m) - 1}{m} \sum_{j=0}^{k} \prod_{(a)\ i=1}^{k} \lambda(a\mathbf{n} + j + (Wm + r)b_i) 1_{[1,H]}(j + (Wm + r)b_i)$$

$$= o_{w \to \infty}(H).$$

Proof By the triangle inequality, it suffices to show the deterministic estimate

$$\sum_{\frac{\varepsilon^2}{2} \frac{H}{W} \leq m \leq \varepsilon^2 \frac{H}{W}} \frac{\Lambda_{W,r}(m) - 1}{m} \sum_{j} \prod_{i=1}^{k} f_i(j + (Wm + r)b_i) = o_{w \to \infty}(H)$$

for any functions $f_1, \ldots, f_k : \mathbb{Z} \to [-1, 1]$ supported on $[1, H]$ (note that the constraint $j = 0$ (a) can be absorbed into (say) the f_1 factor). By shifting each f_i by rb_i (and restricting back to $[1, H]$ at the cost of a negligible error), we may replace each term $f_i(j + (Wm + r)b_i)$ here by $f_i(j + Wmb_i)$.

By embedding $[1, H]$ into $\mathbb{Z}/2H\mathbb{Z}$ and extending functions by zero, it suffices to show that

$$\mathbb{E}_{j,m\in\mathbb{Z}/2H\mathbb{Z}} c_m \prod_{i=1}^{k} f_i(j + Wmb_i) = o_{w\to\infty}(1)$$

for any functions $f_1, \ldots, f_k \colon \mathbb{Z}/2H\mathbb{Z} \to [-1, 1]$, where $c_m := \frac{\Lambda_{W,r}(m)-1}{m}$ if m is an integer between $\frac{\varepsilon^2}{2}\frac{H}{W}$ and $\varepsilon^2 \frac{H}{W}$ (identified with an element of $\mathbb{Z}/2H\mathbb{Z}$), and $c_m = 0$ otherwise, and we use the averaging notation $\mathbb{E}_{n\in A} f(n) := \frac{1}{|A|} \sum_{n\in A} f(n)$. Making the substitution $m = m_1 + \cdots + m_k$ and $j = n - Wm_1 b_1 - \cdots - Wm_k b_k$, we reduce to showing that

$$\mathbb{E}_{n,m_1,\ldots,m_k\in\mathbb{Z}/2H\mathbb{Z}} c_{m_1+\cdots+m_k} \prod_{i=1}^{k} F_i(n, m_1, \ldots, m_k) = o_{w\to\infty}(1),$$

where $F_i \colon \mathbb{Z}^{k+1} \to [-1, 1]$ is the function

$$F_i(n, m_1, \ldots, m_k) := f_i\left(n + \sum_{j=1}^{k} Wm_j(b_i - b_j)\right).$$

Observe that for each $i = 1, \ldots, k$, F_i does not depend on the m_i variable. Applying the triangle inequality in n and the Cauchy–Schwarz inequality k times (as in [28, (B.7)]), we see that it suffices to show that

$$\mathbb{E}_{m_1^{(0)},\ldots,m_k^{(0)},m_1^{(1)},\ldots,m_k^{(1)}\in\mathbb{Z}/2H\mathbb{Z}} \prod_{\vec{\omega}\in\{0,1\}^k} c_{\sum_{i=1}^{k} m_i^{(\omega_i)}} = o_{w\to\infty}(1)$$

where $\vec{\omega} = (\omega_1, \ldots, \omega_k)$. Writing $h_i := m_i^{(1)} - m_i^{(0)}$ and $x := m_1^{(0)} + \cdots + m_k^{(0)}$, we can rewrite the left-hand side as

$$\mathbb{E}_{x,h_1,\ldots,h_k\in\mathbb{Z}/2H\mathbb{Z}} \prod_{\vec{\omega}\in\{0,1\}^k} c_{x+\vec{\omega}\cdot\vec{h}},$$

where $\vec{\omega} \cdot \vec{h} := \omega_1 h_1 + \cdots + \omega_k h_k$, so by definition of c_m, it suffices to show that

$$\sum_{x,h_1,\ldots,h_k\in\mathbb{Z}} \prod_{\vec{\omega}\in\{0,1\}^k} 1_{\frac{\varepsilon^2}{2}\frac{H}{W}\leq x+\vec{\omega}\cdot\vec{h}\leq\varepsilon^2\frac{H}{W}} \frac{\Lambda_{W,r}(x + \vec{\omega}\cdot\vec{h}) - 1}{x + \vec{\omega}\cdot\vec{h}} = o_{w\to\infty}(H^{k+1}).$$

Using a Riemann sum approximation, it suffices to show that

$$\sum_{x\in I, h_1\in J_1,\ldots,h_k\in J_k} \prod_{\vec{\omega}\in\{0,1\}^k} (\Lambda_{W,r}(x + \vec{\omega}\cdot\vec{h}) - 1) = o_{w\to\infty}((H\log^{-10} H)^{k+1})$$

for all intervals $I, J_1, \ldots, J_k \subset [1, H]$ of length $H \log^{-10} H$ (say). But this follows from the results in [28], or more precisely from the localised estimate in [20, (A.9)].

$\square$

From (19), the above proposition, and the triangle inequality, we have

$$\left| \mathbb{E} \sum_{\frac{\varepsilon^2}{2} \frac{H}{W} \leq m \leq \varepsilon^2 \frac{H}{W}} \frac{1}{m} \sum_{j=0 \ (a)} \prod_{i=1}^{k} \lambda(a\mathbf{n} + j + (Wm + r)b_i) 1_{[1,H]}(j + (Wm + r)b_i) \right| \gg \varepsilon H.$$

Since the expression inside the summation is $O(H)$, we conclude that with probability $\gg_\varepsilon 1$, one has

$$\left| \sum_{\frac{\varepsilon^2}{2} \frac{H}{W} \leq m \leq \varepsilon^2 \frac{H}{W}} \frac{1}{m} \sum_{j=0 \ (a)} \prod_{i=1}^{k} \lambda(a\mathbf{n} + j + (Wm + r)b_i) 1_{[1,H]}(j + (Wm + r)b_i) \right| \gg_\varepsilon H. \tag{20}$$

Let us condition to this event. Using our hypothesis that Conjecture 1.6 holds (in the form (9)), together with Markov's inequality, we see that with conditional probability $1 - o_{H \to \infty}(1)$ one also has

$$\|\lambda\|_{U^{k-1}([a\mathbf{n}, a\mathbf{n}+H] \cap \mathbb{Z})} = o_{H \to \infty}(1), \tag{21}$$

and we condition to this event also.

Replacing m by $Wm + r$, and dropping some negligible boundary terms, we see from (20) that

$$\left| \sum_{\frac{\varepsilon^2}{2} H \leq m \leq \varepsilon^2 H} \frac{1_{m=r \ (W)}}{m} \sum_{j=0 \ (a)} \prod_{i=1}^{k} \lambda(a\mathbf{n} + j + mb_i) 1_{[1,H]}(j + mb_i) \right| \gg_{\varepsilon, W} H.$$

Since $m = r \ (W)$, and W is a multiple of a, we can write $1_{j=0 \ (a)}$ as $1_{a\mathbf{n}+j+mb_k = rb_k \ (a)}$. As $b_1 = 0$, we may thus write the above estimate in the form

$$\left| \sum_{\frac{\varepsilon^2}{2} H \leq m \leq \varepsilon^2 H} \frac{1_{m=r \ (W)}}{m} \sum_{j} f_1(j) f_2(j + mb_2) \ldots f_k(j + mb_k) \right| \gg_{\varepsilon, W} H$$

for some ($\mathbf{n}$-dependent) functions $f_1, f_2, \ldots, f_k \colon \mathbb{Z} \to [-1, 1]$ supported on $[1, H]$, with $f_1(j) := \lambda(a\mathbf{n} + j) 1_{[1,H]}(j)$ (the precise values of $f_2, \ldots, f_k$ will not be relevant). Note from (21) that

$$\|f_1\|_{U^{k-1}([1,H] \cap \mathbb{Z})} = o_{H \to \infty}(1). \tag{22}$$

We now dispose of the m weights. Note that the quantity $f_1(j)f_2(j + mb_2)\ldots f_k(j + mb_k)$ is only non-vanishing when $m = O(H)$, so we may embed the m variable in (say) $\mathbb{Z}/HW\mathbb{Z}$. We can Fourier expand $m \mapsto 1_{m=r\ (W)}$ into a linear combination of exponential phases $m \mapsto e(sm/W)$ with $s = 1, \ldots, W$ and coefficients of size $O(1)$. Similarly, using a standard Fourier expansion (e.g., using[1] Fejér kernels), one can approximate $m \mapsto 1_{\frac{\varepsilon^2}{2}H \leq m \leq \varepsilon^2 H}\frac{1}{m}$ on $\mathbb{Z}/HW\mathbb{Z}$ by a linear combination of $O_{\varepsilon,\delta}(1)$ exponential phases $m \mapsto e(sm/HW)$ with $s = 1, \ldots, H$ and coefficients $O_{\varepsilon,\delta,W}(1/H)$, plus an error whose $\ell^1(\mathbb{Z}/HW\mathbb{Z})$ norm in m is at most δ, for any given $\delta > 0$. Applying these expansions for $\delta > 0$ sufficiently small depending on ε, W, and using the pigeonhole principle, we conclude that

$$\left| \sum_m e(sm/HW) \sum_j f_1(j)f_2(j + mb_2)\ldots f_k(j + mb_k) \right| \gg_{\varepsilon,W} H^2$$

for some integer s, where we now revert to m as taking values in $\mathbb{Z}$ rather than $\mathbb{Z}/HW\mathbb{Z}$. To deal with the phase $e(sm/HW)$, we write m as a linear combination of $j + mb_{k-1}$ and $j + mb_k$, and conclude (using our assumption $k > 2$) that

$$\left| \sum_m \sum_j f_1(j)f_2'(j + mb_2)\ldots f_k'(j + mb_k) \right| \gg_{\varepsilon,W} H^2$$

for some functions $f_2', \ldots, f_k' : \mathbb{Z} \to \mathbb{C}$ supported on $[1, H]$ and bounded in magnitude by 1. But from the "generalised von Neumann inequality" (see, e.g., [55, Lemma 11.4], after embedding $[1, H]$ in a cyclic group $\mathbb{Z}/p\mathbb{Z}$ of some prime p between $2H$ and $4H$, say) we have

$$\left| \sum_m \sum_j f_1(j)f_2'(j + mb_2)\ldots f_k'(j + mb_k) \right| \ll \|f_1\|_{U^{k-1}([1,H]\cap\mathbb{Z})}$$

giving a contradiction to (22). This concludes the derivation of Conjecture 1.3 from Conjecture 1.6.

Remark 3.4 An inspection of the above argument shows that if one wishes to establish Conjecture 1.3 for a specific choice of $k \geq 3$, then it would suffice to establish Conjecture 1.6 for $d = k - 1$. In particular, the first open case $k = 3$ of Conjecture 1.3 would follow from a non-trivial bound on the local U^2 norms of the Liouville function.

---

[1] Alternatively, one can perform a Fourier series expansion of $1_{\frac{\varepsilon^2}{2W} \leq x \leq \frac{\varepsilon^2}{W}}\frac{1}{x}$ on the unit circle.

Remark 3.5 In the spirit of the Elliott conjecture [17] (see also [44] for a correction to that conjecture), one could more generally consider estimates of the form

$$\sum_{x/\omega \leqslant n \leqslant x} \frac{g_1(n + h_1) \dots g_k(n + h_k)}{n} = o_{\omega \to \infty}(\log \omega)$$

for bounded completely multiplicative functions $g_1, \dots, g_k$. The weight $\Lambda(m)$ appearing in the above analysis would now be replaced by $\Lambda g_1 \dots g_k(m)$, and so the results on linear equations in primes used in Proposition 3.1 are no longer available. Nevertheless, one should still be able to deploy a "transference principle" to approximate the weight $\Lambda g_1 \dots g_k$ by a small number of "structured" functions (such as nilsequences), which should still allow one to derive a suitable generalisation of Conjecture 1.3 for the $g_1, \dots, g_k$ from Conjecture 1.6 (possibly after increasing d to k instead of $k + 1$), in the spirit of [54, Theorem 1.3] (which used a "restriction theorem for the primes" as a proxy for the transference principle). We will not pursue this matter here.

4 Applying the Inverse Conjecture for the Gowers Norms

In this section we show how Conjecture 1.6 can be deduced from Conjecture 1.7.

Let $d \geqslant 1$, let $\varepsilon > 0$ be sufficiently small depending on d, and let $2 \leqslant H \leqslant \omega \leqslant X$ be such that H is sufficiently large depending on d, ε. We allow implied constants to depend on d. Using the formulation (9), our goal is now to show that

$$\mathbb{E}\|\lambda\|_{U^d([\mathbf{n},\mathbf{n}+H]\cap\mathbb{Z})} \ll \varepsilon.$$

Suppose this claim failed, then we must have

$$\|\lambda\|_{U^d([\mathbf{n},\mathbf{n}+H]\cap\mathbb{Z})} \gg \varepsilon \tag{23}$$

with probability $\gg \varepsilon$.

Suppose that we are in the event that (23) holds. Then, by the inverse conjecture for the Gowers norms [31, Theorem 1.3], there exists a $d - 1$-step (random) nilmanifold $\mathbf{G}/\mathbf{\Gamma}$ from a finite list $\mathcal{M}_{d-1,\varepsilon}$ (each of which is equipped with a smooth Riemannian metric), and a (random) function $\mathbf{F}\colon \mathbf{G}/\mathbf{\Gamma} \to \mathbb{C}$ with Lipschitz constant $O_\varepsilon(1)$ and a random group element $\mathbf{g} \in G$ and random base point $\mathbf{x}_0 \in \mathbf{G}/\mathbf{\Gamma} \to \mathbb{C}$ such that

$$\left|\sum_{h=1}^{H} \lambda(\mathbf{n} + h)\mathbf{F}(\mathbf{g}^h\mathbf{x}_0)\right| \gg_\varepsilon 1. \tag{24}$$

By the pigeonhole principle, one can find a *deterministic $d-1$-step* nilmanifold G/Γ such that $\mathbf{G}/\mathbf{\Gamma}$ is equal to G/Γ with probability $\gg_\varepsilon 1$. We condition to this event. Next, we fix a deterministic base point x_0 in G/Γ. For the random base point $\mathbf{x}_0$, we can write $\mathbf{x}_0 = \mathbf{g}_1 x_0$ for some bounded element $\mathbf{g}_1 \in G$. We can then write

$$\mathbf{F}(\mathbf{g}^h \mathbf{x}_0) = \mathbf{F}(\mathbf{g}_1(\mathbf{g}_1^{-1}\mathbf{g}\mathbf{g}_1)^h x_0).$$

Replacing $\mathbf{g}$ by $\mathbf{g}_1^{-1}\mathbf{g}\mathbf{g}_1$ and $\mathbf{F}$ by the function $x \mapsto \mathbf{F}(\mathbf{g}_1 x)$, we see that we may assume without loss of generality that $\mathbf{x}_0 = x_0$. Finally, by the Arzela–Ascoli theorem, the class of Lipschitz functions from G/Γ to $\mathbb{C}$ of Lipschitz constant $O_\varepsilon(1)$ is totally bounded in the uniform topology. Thus, we can restrict the range of possible values of the random function $\mathbf{F}$ to a finite collection of $O_\varepsilon(1)$ deterministic Lipschitz functions without significantly affecting (24). By the pigeonhole principle, we can thus find a deterministic Lipschitz function $F: G/\Gamma \to \mathbb{C}$ such that

$$\left| \sum_{h=1}^{H} \lambda(\mathbf{n}+h)F(\mathbf{g}^h x_0) \right| \gg_\varepsilon 1$$

with probability $\gg_\varepsilon 1$. In particular,

$$\sup_{g \in G} \left| \sum_{h=1}^{H} \lambda(\mathbf{n}+h)F(g^h x_0) \right| \gg_\varepsilon 1$$

with probability $\gg_\varepsilon 1$, which implies that

$$\mathbb{E} \sup_{g \in G} \left| \sum_{h=1}^{H} \lambda(\mathbf{n}+h)F(g^h x_0) \right| \gg_\varepsilon 1.$$

But this contradicts Conjecture 1.7 (in the form (10)).

Remark 4.1 An inspection of the above argument shows that in order to prove Conjecture 1.6 for a specific choice of $d \geqslant 2$, it suffices to establish Conjecture 1.7 with $s = d - 1$. Combining this with Remark 3.4, we see that to establish Conjecture 1.3 for a specific choice of $k \geqslant 3$, it suffices to establish Conjecture 1.7 with $s = d - 1$. In particular, and after performing a Fourier expansion of 1-step nilsequences $n \mapsto F(g^n x_0)$, we see that to prove the $k = 3$ case of Conjecture 1.3, it will suffice to establish the bound

$$\sum_{X/\omega \leqslant n \leqslant X} \frac{\sup_{\alpha \in \mathbb{R}/\mathbb{Z}} |\sum_{h=1}^{H} \lambda(n+h)e(h\alpha)|}{n} = o_{H \to \infty}(H \log \omega).$$

for all $1 \leqslant H \leqslant \omega \leqslant x$. Bounds of this form are available for very large values of H; for instance, the estimates in [58] give this bound when $\omega > 1$ is fixed and $H \geqslant$

$x^{5/8+\varepsilon}$ for any fixed $\varepsilon > 0$. In [44] a weaker version of this estimate was established in which $\omega > 1$ is fixed and the supremum in α was outside the summation in n.

Remark 4.2 One can reverse the above arguments, using [26, Proposition 12.6] in place of [31, Theorem 1.3], to show directly that Conjecture 1.6 implies Conjecture 1.7; we leave the details of this implication to the interested reader. This implication of course already follows from the arguments used to prove other components of Theorem 1.8 in this paper, but this alternate argument is also valid in the absence of logarithmic averaging.

5 Constructing a Deterministic Sequence

In this section we show that Conjecture 1.7 follows from Conjecture 1.5.

Let $s, G/\Gamma$, x_0, F be as in Conjecture 1.7; we allow all implied constants to depend on these quantities. By splitting in to real and imaginary parts we may take F to be real-valued. Let $\varepsilon > 0$. Our task is to show that

$$\sum_{X/\omega \leqslant n \leqslant X} \frac{\sup_{g \in G} |\sum_{h=1}^{H} \lambda(n+h)F(g^h x_0)|}{n} \ll \varepsilon H \log \omega$$

whenever $1 \leqslant H \leqslant \omega \leqslant X$, and H is sufficiently large depending on ε. From[2] [29, Theorem 1.1], we see that

$$\sup_{g \in G} \left| \sum_{h=1}^{H} \lambda(n+h)F(g^h x_0) \right| = o_{H \to \infty}(H)$$

whenever $n \leqslant H \log H$ (say); in fact the results in [29] allow one to improve upon the trivial bound of $O(H)$ by an arbitrary fixed power of $\log H$. Thus the net contribution of the case $n \leqslant H \log H$ to (25) is negligible, so we may restrict to the case $n > H \log H$. In this regime, one has $\frac{1}{n+h} = \frac{1}{n} + O(\frac{1}{\log H}\frac{1}{n})$; the contribution of the error term is negligible (cf. (11)), so it suffices to show that

$$\sum_{H \log H, X/\omega \leqslant n \leqslant X} \sup_{g \in G} \left| \sum_{h=1}^{H} \frac{\lambda(n+h)}{n+h} F(g^h x_0) \right| \ll \varepsilon H \log \omega$$

---

[2]This result is stated for the Möbius function in place of the Liouville function, but the arguments extend to the Liouville case; see [29, §6].

It will suffice to just establish the positive part

$$\sum_{H \log H, X/\omega \leqslant n \leqslant X} \sup_{g \in G} \max\left(\sum_{h=1}^{H} \frac{\lambda(n+h)}{n+h} F(g^h x_0), 0\right) \ll \varepsilon H \log \omega \qquad (25)$$

of this estimate, since the full estimate then follows by applying (25) for both F and $-F$ and using the triangle inequality.

Suppose for contradiction that the bound (25) failed. Then we can find sequences H_i, ω_i, X_i with

$$1 \leqslant H_i \leqslant \omega_i \leqslant X_i$$

and $H_i \to \infty$ as $i \to \infty$, such that

$$\sum_{H_i \log H_i, X_i/\omega_i \leqslant n \leqslant X_i} \sup_{g \in G} \max\left(\sum_{h=1}^{H_i} \frac{\lambda(n+h)}{n+h} F(g^h x_0), 0\right) \gg \varepsilon H_i \log \omega_i. \qquad (26)$$

By sparsifying the sequences H_i, ω_i, X_i we may assume that

$$H_{i+1} \geqslant 100 X_i \qquad (27)$$

(say) for all i.

The quantity $\sup_{g \in G} |\sum_{h=1}^{H_i} \frac{\lambda(n+h)}{n+h} F(g^h x_0)|$ is bounded above by $O(H_i/n)$. Thus we can find a set S_i of numbers n with $H_i \log H_i, X_i/\omega_i \leqslant n \leqslant X_i$ such that

$$\sum_{n \in S_i} \frac{1}{n} \gg \varepsilon \log \omega_i$$

and such that

$$\sup_{g \in G} \sum_{h=1}^{H_i} \frac{\lambda(n+h)}{n+h} F(g^h x_0) \gg \varepsilon \frac{H_i}{n} \qquad (28)$$

for all $n \in S_i$, since the contribution to the left-hand side (26) of those n for which (28) fails can be made to be significantly smaller than the right-hand side of (26) by choosing the implicit constants appropriately.

By a greedy algorithm, we can then find a subset S_i' of S_i that is H_i-separated (that is to say, $|n - m| \geqslant H_i$ for any distinct $n, m \in S_i'$) such that

$$\sum_{n \in S_i'} \frac{1}{n} \gg \frac{\varepsilon}{H_i} \log \omega_i. \qquad (29)$$

For each $n \in S_i'$, we can find a group element $g_n \in G$ such that

$$\sum_{h=1}^{H_i} \frac{\lambda(n+h)}{n+h} F(g_n^h x_0) \gg \varepsilon \frac{H_i}{n}. \tag{30}$$

If we now set $f \colon \mathbb{Z} \to \mathbb{R}$ to be the function defined by setting

$$f(n+h) := F(g_n^h x_0)$$

whenever $n \in S_i'$ and $1 \leqslant h \leqslant H_i$ is an integer for some i, and $f(m) = 0$ for all other m, we see that f is well defined because all the intervals $\{n+1, \ldots, n+H_i\}$ with $n \in S_i'$ and $i \geqslant 1$ are disjoint, thanks to (27) and the H_i-separation of the S_i'.

Summing (30) over all $n \in S_i'$ and using (29), we conclude that

$$\sum_{H_i \log H_i, X_i/\omega_i \leqslant n \leqslant 2X_i} \frac{\lambda(n)}{n} f(n) \gg \varepsilon^2 \log \omega_i.$$

On the other hand, if f is deterministic, then Conjecture 1.5 gives

$$\sum_{H_i \log H_i, X_i/\omega_i \leqslant n \leqslant 2X_i} \frac{\lambda(n)}{n} f(n) = o_{\omega_i \to \infty}(\log \omega_i)$$

(one can divide here into two cases, depending on whether $\log \frac{2X_i}{H_i \log H_i}$ is smaller than (say) $\sqrt{\log \omega_i}$ or not). Thus it will suffice to show that the sequence f is deterministic.

Since F is bounded, f takes values in a compact interval $[-C, C]$. Consider the compact space

$$[-C, C]^{\mathbb{Z}} = \{(y_n)_{n \in \mathbb{Z}} : y_n \in [-C, C] \forall n \in \mathbb{Z}\}$$

which we endow with the shift

$$T(y_n)_{n \in \mathbb{Z}} := (y_{n+1})_{n \in \mathbb{Z}}$$

and metric

$$d((x_n)_{n \in \mathbb{Z}}, (y_n)_{n \in \mathbb{Z}}) := \sup_{n \in \mathbb{Z}} 2^{-|n|} |x_n - y_n|.$$

We can identify f with a point $y_0 := (f(n))_{n \in \mathbb{Z}}$ in $[-C, C]^{\mathbb{Z}}$. We let $Y = \overline{\{T^n y_0 : n \in \mathbb{Z}\}}$ be the orbit closure of y_0 in $[-C, C]^{\mathbb{Z}}$, then (Y, T) is a topological dynamical system. If we let $F_0 \colon Y \to \mathbb{R}$ be the function

$$F_0((y_n)_{n \in \mathbb{Z}}) := y_0,$$

then F_0 is continuous and

$$f(n) = F_0(T^n y_0)$$

for all $n \in \mathbb{Z}$. Thus, to show that f is deterministic, it suffices to show that (Y, T) has zero topological entropy. That is to say, for any fixed $\varepsilon > 0$ and any sufficiently large N, we should be able to cover Y by at most $\exp(O(\varepsilon N))$ balls of radius $O(\varepsilon)$ in the metric

$$d_N(x, y) := \max_{0 \leqslant i \leqslant N} d(T^i x, T^i y)$$

or equivalently

$$d_N((x_n)_{n \in \mathbb{Z}}, (y_n)_{n \in \mathbb{Z}}) = \sup_{n \in \mathbb{Z}} 2^{-\max(-n, 0, n-N)} |x_n - y_n|.$$

Observe that if two sequences $(x_n)_{n \in \mathbb{Z}}, (y_n)_{n \in \mathbb{Z}}$ are such that $x_n = y_n + O(\varepsilon)$ for all $-N \leqslant n \leqslant 2N$, then (for N sufficiently large depending on ε) we have $d_N((x_n)_{n \in \mathbb{Z}}, (y_n)_{n \in \mathbb{Z}}) = O(\varepsilon)$. Thus it suffices to find a collection $\mathcal{S}_{\varepsilon,N}$ of finite sequences $(x_h)_{-N \leqslant h \leqslant 2N}$ of cardinality $\exp(O(\varepsilon N))$ with the property that for every $n \in \mathbb{Z}$, there exists a sequence $(x_h)_{-N \leqslant h \leqslant 2N}$ in $\mathcal{S}_{\varepsilon,N}$ such that

$$f(n + h) = x_h + O(\varepsilon)$$

for all $-N \leqslant h \leqslant 2N$.

Observe that if we can prove this claim for a given value of N, then we automatically obtain the claim for any larger $N' \geqslant N$ (with a slightly worse implicit constant), by covering the interval $[-N', 2N']$ by $O(N'/N)$ translates of $[-N, 2N]$. In particular, it will suffice to verify the claim with $N = \lfloor H_{i_0}/10 \rfloor$ for i_0 sufficiently large depending on ε.

We may remove from consideration those n for which $|n| \leqslant 2N$, since these cases can be accommodated simply by adding the sequences $(f(n + h))_{-N \leqslant h \leqslant 2N}$ for $|n| \leqslant 2N$ to $\mathcal{S}_{\varepsilon,N}$, which only increases the cardinality of that family by a negligible amount. If $n < -2N$, then one has $f(n + h) = 0$ for all $-N \leqslant h \leqslant 2N$, and this case can be accommodated by adding the zero sequence $(0)_{-N \leqslant h \leqslant 2N}$ to $\mathcal{S}_{\varepsilon,N}$. Thus we may assume that $n > 2N$.

Recall that the function f is only supported on the union of the intervals $\{m + 1, \ldots, m + H_i\}$ with $i \geqslant 1$ and $m \in S_i'$, so in particular $H_i \log H_i \leqslant m \leqslant X_i$. Since $n > 2N$, such an interval can only intersect the interval $\{n - N, \ldots, n + 2N\}$ if one has

$$H_i \log H_i \ll n \ll X_i;$$

in particular there is at most one choice of i in which this can occur. Since $n \geqslant 2N = 2\lfloor H_{i_0}/10 \rfloor$, we conclude from (27) that $i \geqslant i_0$, so in particular $H_i \geqslant 10N$.

In particular, each interval $\{n - N, \ldots, n + 2N\}$ meets at most two of the intervals $\{m + 1, \ldots, m + H_i\}$. It will now suffice to exhibit a set $\mathcal{S}'_{\varepsilon,N}$ of finite sequences $(x_n)_{-N \leqslant h \leqslant 2N}$ of cardinality $O(\exp(O(\varepsilon N)))$ with the property that for any $i \geqslant i_0$, any $m \in S'_i$, and sub-interval $\{n - N, \ldots, n + 2N\}$ of $\{m + 1, \ldots, m + H_i\}$, there exists a sequence $(x_n)_{-N \leqslant h \leqslant 2N}$ in $\mathcal{S}'_{\varepsilon,N}$ for which

$$f(n + h) = x_h + O(\varepsilon)$$

for all $-N \leqslant h \leqslant 2N$. Indeed, one can now set $\mathcal{S}_{\varepsilon,N}$ to be the collection of all sequences $(x_n)_{-N \leqslant h \leqslant 2N}$ formed by concatenating at most two subsequences of sequences in $\mathcal{S}'_{\varepsilon,N}$, together with some blocks of zeroes; the cardinality of $\mathcal{S}_{\varepsilon,N}$ is $O(N^{O(1)}|\mathcal{S}'_{\varepsilon,N}|^2)$, which will be at most $\exp(O(\varepsilon N))$ if N is large enough.

It remains to exhibit $\mathcal{S}'_{\varepsilon,N}$. If n, m are as above, then

$$f(n + h) = F(g_m^{n+h-m} x_0)$$

for $-N \leqslant h \leqslant 2N$. In particular, there exists a polynomial sequence $g_n \colon \mathbb{Z} \to G$, that is to say a sequence of the form

$$g_n(h) = g_{n,0} g_{n,1}^h g_{n,2}^{\binom{h}{2}} \cdots g_{n,s}^{\binom{h}{s}}$$

where $g_{n,i} \in G_i$ for $i = 0, \ldots, s$, and $G = G_0 = G_1 \geqslant G_2 \geqslant \cdots \geqslant G_s$ is the lower central series of G, such that

$$f(n + h) = F(g_n(h)\Gamma)$$

for $-N \leqslant h \leqslant 2N$. Currently we have $g_{n,0} = g_m^{n-m}$, $g_{n,1} = g_m$, and all other coefficients trivial; however, we shall shortly consider more general polynomial sequences in which the higher coefficients $g_{n,2}, \ldots, g_{n,s}$ are allowed to be non-trivial.

The coefficients $g_{n,i}$ of an arbitrary polynomial sequence g_n can be unbounded. However, any such sequence g_n may be factorised as

$$g_n = \tilde{g}_n \gamma_n$$

where $\tilde{g}_n$ is a polynomial sequence with coefficients taking values in a compact set (depending only on G, Γ) and γ_n is a polynomial sequence with coefficients in Γ; see [31, Lemma C.1] for a proof. In particular, we have $g_n(h)\Gamma = \tilde{g}_n(h)\Gamma$ for any h, and hence

$$f(n + h) = F\left(\tilde{g}_{n,0} \tilde{g}_{n,1}^h \tilde{g}_{n,2}^{\binom{h}{2}} \cdots \tilde{g}_{n,s}^{\binom{h}{s}} \Gamma\right)$$

for all $-N \leqslant h \leqslant 2N$ and some coefficients $\tilde{g}_{n,0}, \ldots, \tilde{g}_{n,s}$ in some fixed compact subset K of G.

Let A be a large constant depending on G, Γ to be chosen later. From many applications of the Baker–Campbell–Hausdorff formula (which is a polynomial formula in a connected, simply connected nilpotent Lie group such as G), we see that if we modify each of the coefficients $\tilde{g}_{n,0}, \ldots, \tilde{g}_{n,s}$ by at most $O(N^{-A})$ (after endowing G with some smooth left-invariant Riemannian metric), then the quantities $\tilde{g}_{n,0} \tilde{g}_{n,1}^h \tilde{g}_{n,2}^{\binom{h}{2}} \cdots \tilde{g}_{n,s}^{\binom{h}{s}}$ for $-N \leqslant h \leqslant 2N$ only change by $O(N^{-A+O(1)})$ in the G metric. In particular, if we select a maximal N^{-A}-separated net Σ of K, and let $g'_{n,i}$ be $\tilde{g}_{n,i}$ rounded to the nearest element of Σ (breaking ties arbitrarily), then from the Lipschitz nature of F we have

$$f(n+h) = F\left(g'_{n,0}(g'_{n,1})^h (g'_{n,2})^{\binom{h}{2}} \cdots (g'_{n,s})^{\binom{h}{s}} \Gamma\right) + O\left(N^{-A+O(1)}\right).$$

If we choose A large enough, then the error term $O(N^{-A+O(1)})$ is $O(\varepsilon)$. If we now set $\mathcal{S}'_{\varepsilon,N}$ to be the collection of all sequences of the form

$$\left(F\left(g_0 g_1^h g_2^{\binom{h}{2}} \cdots g_s^{\binom{h}{s}} \Gamma\right)\right)_{-N \leqslant h \leqslant 2N}$$

with $g_0, \ldots, g_s \in \Gamma$, then $\mathcal{S}'_{\varepsilon,N}$ has cardinality $O(N^{O(A)}) = O(\exp(O(\varepsilon N)))$ for N large enough, and the claim follows.

Remark 5.1 The main fact that was used in the above argument is that the collection of nilsequences $n \mapsto F(g^n x_0)$, where F is a Lipschitz function on a nilmanifold G/Γ of "bounded complexity," $g \in G$, and $x_0 \in G/\Gamma$, has "uniform zero entropy" in the sense that[3] for any $\varepsilon > 0$ and any N sufficiently large depending on ε, the set of sequences formed from evaluating an arbitrary nilsequence in this collection at N consecutive values has a metric entropy of $O(\exp(O(\varepsilon N)))$ at scale $\varepsilon > 0$. This is stronger than asserting that each individual nilsystem $(G/\Gamma, x \mapsto gx)$, $g \in G$ has zero entropy, as one needs to control the metric entropy of the set of sequences arising from *arbitrary* shifts g, rather than just one shift at a time. On the other hand, if all one is interested in is deducing Conjecture 1.6 from Conjecture 1.5, it is likely that one does not need the full strength of the inverse conjecture in [31], and in particular one does not need to introduce the notion of a nilmanifold or nilsequence at all. Instead, one can rely on "soft" inverse theorems in which the role of nilsequences are replaced by those of dual functions (see, e.g., [55, §11.4]), in which case the task is basically reduced to establishing that the collection of dual functions also has "uniform zero entropy" in a certain sense. This in turn should be provable using some sort of random sampling argument to show that the dual function of a given function f is almost completely controlled by the values of f at some sparse random subset of the domain. We will, however, not attempt to formalise these arguments here.

---

[3]See also [1, Lemma B.9] for a closely related claim.

Acknowledgements The author is supported by NSF grant DMS-1266164 and by a Simons Investigator Award. The author also thanks Ben Green and Peter Sarnak for comments and encouragement, and Christian Elsholtz, Yi Ji Gao, and the anonymous referees for corrections and suggestions.

References

1. B. Bhattacharya, S. Ganguly, X. Shao, Y. Zhao, Upper tails for arithmetic progressions in a random set, preprint. arXiv:1605.02994
2. J. Bourgain, Möbius-Walsh correlation bounds and an estimate of Mauduit and Rivat. J. Anal. Math. **119**, 147–163 (2013)
3. J. Bourgain, On the correlation of the Moebius function with rank-one systems. J. Anal. Math. **120**, 105–130 (2013)
4. J. Bourgain, P. Sarnak, T. Ziegler, Disjointness of Moebius from horocycle flows, in *From Fourier Analysis and Number Theory to Radon Transforms and Geometry*. Developments in Mathematics, vol. 28 (Springer, New York, 2013), pp. 67–83
5. F. Cassaigne, S. Ferenczi, R. Mauduit, A. Sárközy, On finite pseudorandom binary sequences. III. The Liouville function. I. Acta Arith. **87**(4), 367–390 (1999)
6. S. Chowla, *The Riemann Hypothesis and Hilbert's Tenth Problem* (Gordon and Breach, New York, 1965)
7. C. Dartyge, G. Tenenbaum, Sommes des chiffres de multiples d'entiers. Ann. Inst. Fourier (Grenoble) **55**, 2423–2474 (2005)
8. H. Davenport, On some infinite series involving arithmetical functions (II). Q. J. Math. **8**, 313–320 (1937)
9. J.-M. Deshouillers, M. Drmota, C. Müllner, Automatic Sequences generated by synchronizing automata fulfill the Sarnak conjecture. Stud. Math. **231**, 83–95 (2015)
10. T. Downarowicz, S. Kasjan, Odometers and Toeplitz subshifts revisited in the context of Sarnak's conjecture. Stud. Math. **229**(1), 45–72 (2015)
11. M. Drmota, Subsequences of automatic sequences and uniform distribution, in *Uniform Distribution and Quasi-Monte Carlo Methods*. Radon Series on Computational and Applied Mathematics, vol. 15 (De Gruyter, Berlin, 2014), pp. 87–104
12. T. Eisner, A polynomial version of Sarnak's conjecture. C. R. Math. Acad. Sci. Paris **353**(7), 569–572 (2015)
13. E. El Abdalaoui, X. Ye, A cubic nonconventional ergodic average with Möbius and Liouville weight, preprint, arXiv:1504.00950
14. E.H. El Abdalaoui, M. Lemańczyk, T. de la Rue, On spectral disjointness of powers for rank-one transformations and Möbius orthogonality. J. Funct. Anal. **266**, 284–317 (2014)
15. E.H. El Abdalaoui, S. Kasjan, M. Lemańczyk, 0–1 sequences of the Thue-Morse type and Sarnak's conjecture. Proc. Am. Math. Soc. **144**, 161–176 (2016)
16. E.H. El Abdalaoui, M. Lemańczyk, T. de la Rue, Automorphisms with quasi-discrete spectrum, multiplicative functions and average orthogonality along short intervals, preprint, arXiv:1507.04132
17. P.D.T.A. Elliott, On the correlation of multiplicative functions. Notas Soc. Mat. Chile (Notas de la Sociedad de Matemática de Chile) **11**, 1–11 (1992)
18. S. Ferenczi, C. Mauduit, On Sarnak's conjecture and Veech's question for interval exchanges, preprint
19. S. Ferenczi, J. Kułaga-Przymus, M. Lemańczyk, C. Mauduit, Substitutions and Möbius disjointness, in *Ergodic Theory, Dynamical Systems, and the Continuing Influence of John C. Oxtoby*. Contemporary Mathematics, vol. 678 (American Mathematical Society, Providence, 2016), pp. 151–173

20. K. Ford, B. Green, S. Konyagin, T. Tao, Large gaps between consecutive prime numbers. Ann. Math. (2) **183**(3), 935–974 (2016)
21. N. Frantzikinakis, B. Host, Asymptotics for multilinear averages of multiplicative functions. Math. Proc. Camb. Philos. Soc. **161**(1), 87–101 (2016)
22. N. Frantzikinakis, B. Host, B. Kra, Multiple recurrence and convergence for sequences related to the prime numbers. J. Reine Angew. Math. **611**, 131–144 (2007)
23. W.T. Gowers, A new proof of Szemerédi's theorem for arithmetic progressions of length four. Geom. Funct. Anal. **8**(3), 529–551 (1998)
24. W.T. Gowers, A new proof of Szemerédi's theorem. Geom. Funct. Anal. **11**(3), 465–588 (2001)
25. B. Green, On (not) computing the Möbius function using bounded depth circuits. Combin. Probab. Comput. **21**, 942–951 (2012)
26. B. Green, T. Tao, An inverse theorem for the Gowers $U^3(G)$ norm. Proc. Edinb. Math. Soc. **51**, 73–153 (2008)
27. B. Green, T. Tao, The primes contain arbitrarily long arithmetic progressions. Ann. Math. (2) **167**(2), 481–547 (2008)
28. B. Green, T. Tao, Linear equations in primes. Ann. Math. (2) **171**(3), 1753–1850 (2010)
29. B. Green, T. Tao, The Möbius function is strongly orthogonal to nilsequences. Ann. Math. (2) **175**(2), 541–566 (2012)
30. B. Green, T. Tao, T. Ziegler, An inverse theorem for the Gowers U^4-norm. Glasg. Math. J. **53**(1), 1–50 (2011)
31. B. Green, T. Tao, T. Ziegler, An inverse theorem for the Gowers $U^{s+1}[N]$-norm. Ann. Math. (2) **176**(2), 1231–1372 (2012)
32. G. Hanna, Sur les occurrences des mots dans les nombres premiers. Acta Arith. **178**(1), 15–42 (2017)
33. G. Harman, J. Pintz, D. Wolke, A note on the Möbius and Liouville functions. Stud. Sci. Math. Hung. **20**(1–4), 295–299 (1985)
34. A. Hildebrand, On consecutive values of the Liouville function. Enseign. Math. (2) **32**(3–4), 219–226 (1986)
35. W. Hoeffding, Probability inequalities for sums of bounded random variables. J. Am. Stat. Assoc. **58**, 13–30 (1963)
36. W. Huang, Z. Lian, S. Shao, X. Ye, Sequences from zero entropy noncommutative toral automorphisms and Sarnak Conjecture, preprint, arXiv:1510.06022
37. H. Iwaniec, E. Kowalski, *Analytic Number Theory*. American Mathematical Society Colloquium Publications, vol. 53 (American Mathematical Society, Providence, RI, 2004)
38. D. Karagulyan, On Möbius orthogonality for interval maps of zero entropy and orientation-preserving circle homeomorphisms. Ark. Mat. (2015). doi:10.1007/s11512-014-0208-5
39. J. Kułaga-Przymus, M. Lemańczyk, The Moebius function and continuous extensions of rotations. Monatsh. Math. **178**(4), 553–582 (2015)
40. J. Liu, P. Sarnak, The Möbius function and distal flows. Duke Math. J. **164**, 1353–1399 (2015)
41. C. Mauduit, J. Rivat, Sur un problème de Gelfond: la somme des chiffres des nombres premiers. Ann. Math. **171**, 1591–1646 (2010)
42. C. Mauduit, J. Rivat, Prime numbers along Rudin-Shapiro sequences. J. Eur. Math. Soc. **17**(10), 2595–2642 (2015)
43. K. Matomäki, M. Radziwiłł, Multiplicative functions in short intervals. Ann. Math. (2) **183**(3), 1015–1056 (2016)
44. K. Matomäki, M. Radziwiłł, T. Tao, An averaged form of Chowla's conjecture. Algebra Number Theory **9**, 2167–2196 (2015)
45. K. Matomäki, M. Radziwiłł, T. Tao, Sign patterns of the Möbius and Liouville functions. Forum Math. Sigma **4**(e14), 44 pp. (2016)
46. C. Müllner, Automatic sequences fulfill the full Sarnak conjecture, preprint, arXiv:1602.03042
47. R. Peckner, Möbius disjointness for homogeneous dynamics, preprint, arXiv:1506.07778
48. T. Rue, M. Lemańczyk, J. Kułaga-Przymus, E.H. Abdalaoui, The Chowla and the Sarnak conjectures from ergodic theory point of view. Discrete Contin. Dyn. Syst. **37**(6), 2899–2944 (2017)

49. P. Sarnak, Three lectures on the Möbius Function randomness and dynamics (2010), publications.ias.edu/sarnak/paper/506
50. P. Sarnak, Mobius randomness and dynamics. Not. S. Afr. Math. Soc. **43**(2), 89–97 (2012)
51. P. Sarnak, A. Ubis, The horocycle flow at prime times. J. Math. Pures Appl. **103**, 575–618 (2015)
52. T. Tao, The Chowla conjecture and the Sarnak conjecture (2012), terrytao.wordpress.com/2012/10/14
53. T. Tao, The Erdős discrepancy problem. Discrete Anal. **1**, 26 pp. (2016)
54. T. Tao, The logarithmically averaged Chowla and Elliott conjectures for two-point correlations, preprint, arXiv:1509.05422
55. T. Tao, V. Vu, *Additive Combinatorics* (Cambridge University Press, Cambridge, 2006)
56. W.A. Veech, Möbius orthogonality for generalized Morse-Kakutani flows, preprint
57. T. Wooley, T. Ziegler, Multiple recurrence and convergence along the primes. Am. J. Math. **134**(6), 1705–1732 (2012)
58. T. Zhan, On the representation of large odd integer as a sum of three almost equal primes. Acta Math. Sinica (N.S.) **7**(3), 259–272 (1991)

Discrepancy Bounds for β-adic Halton Sequences

Jörg M. Thuswaldner

Dedicated to Professor Robert F. Tichy on the occasion of his 60th birthday

Abstract Van der Corput and Halton sequences are well-known low-discrepancy sequences. Almost 20 years ago Ninomiya defined analogues of van der Corput sequences for β-numeration and proved that they also form low-discrepancy sequences if β is a Pisot number. Only very recently Robert Tichy and his co-authors succeeded in proving that β-adic Halton sequences are equidistributed for certain parameters $\boldsymbol{\beta} = (\beta_1, \ldots, \beta_s)$ using methods from ergodic theory. In the present paper we continue this research and give discrepancy estimates for β-adic Halton sequences for which the components β_i are m-bonacci numbers. Our methods are quite different and use dynamical and geometric properties of Rauzy fractals that allow to relate β-adic Halton sequences to rotations on high dimensional tori. The discrepancies of these rotations can then be estimated by classical methods relying on W.M. Schmidt's Subspace Theorem.

2010 *Mathematics Subject Classification* Primary: 11K38, 11B83; Secondary: 11A63

1 Introduction

Given $q \in \mathbb{N}$ with $q \geq 2$, each integer $n \geq 0$ admits a unique *q-ary expansion* $n = \sum_{j=0}^{L} \varepsilon_j(n) q^j$ with $\varepsilon_j(n) \in \{0, \ldots, q-1\}$ and $\varepsilon_L(n) \neq 0$ for $L \neq 0$. Using this expansion we can define the so-called *van der Corput sequence*

$$v_q(n) = \sum_{j=0}^{L} \varepsilon_j(n) q^{-j-1} \in [0, 1) \qquad (n \geq 0).$$

J.M. Thuswaldner (✉)
Chair of Mathematics and Statistics, University of Leoben, Franz-Josef-Strasse 18, A-8700 Leoben, Austria
e-mail: joerg.thuswaldner@unileoben.ac.at

© Springer International Publishing AG 2017
C. Elsholtz, P. Grabner (eds.), *Number Theory – Diophantine Problems, Uniform Distribution and Applications*, DOI 10.1007/978-3-319-55357-3_22

As mentioned, for instance, in Kuipers and Niederreiter [31, Chap. 2, Sect. 3] this sequence has optimal equidistribution properties modulo $[0, 1)$ and, hence, is a so-called *low-discrepancy sequence*. A generalization to higher dimensions $s \in \mathbb{N}$ is provided by the *Halton sequence* which is defined for each parameter vector $q = (q_1, \ldots, q_s)$, $q_i \geq 2$, by

$$h_q(n) = (v_{q_1}(n), \ldots, v_{q_s}(n)) \qquad (n \geq 0).$$

Halton sequences admit strong equidistribution properties modulo $[0, 1)^s$ for parameter vectors q with pairwise relatively prime entries (see [31, p. 129] or [23]).

These concepts can be carried over to so-called *linear recurrent number systems* (see, e.g., [22, 43]) and β-*expansions* (cf. [2, 42, 45]). As we start out from linear recurrent number systems, we briefly recall the definition of these objects. For $d \in \mathbb{N}$ let

$$G_{k+d} = a_1 G_{k+d-1} + a_2 G_{k+d-2} + \cdots + a_d G_k \qquad (k \geq 0) \tag{1}$$

be a linear recurrence with integral coefficients $a_1, \ldots, a_d$ and integral initial values $G_0, \ldots, G_{d-1}$. Suppose the coefficients satisfy $a_1 \geq a_2 \geq \cdots \geq a_d \geq 1$ ($a_1 > 1$ is needed for $d = 1$) and the conditions

$$G_0 = 1, \quad G_k = a_1 G_{k-1} + \cdots + a_k G_0 + 1 \quad \text{for} \quad k \in \{1, \ldots, d - 1\} \tag{2}$$

are in force. Then we can expand each $n \in \mathbb{N}$ uniquely by a *greedy algorithm* as

$$n = \sum_{j=0}^{\infty} \varepsilon_j(n) G_j, \tag{3}$$

where the digit string $\ldots \varepsilon_1(n)\varepsilon_0(n)$ satisfies

$$\varepsilon_k(n) \ldots \varepsilon_1(n)\varepsilon_0(n)0^{\infty} < (a_1 \ldots a_{d-1}(a_d - 1))^{\infty} \tag{4}$$

in lexicographic order for each $k \in \mathbb{N}$ (see [43] for details).

Let β be the dominant root of the characteristic equation of (1). Then the β-*adic van der Corput sequence* is defined by

$$V_\beta(n) = \sum_{j \geq 0} \varepsilon_j(n)\beta^{-j-1}. \tag{5}$$

One of the problems in the β-adic case is that the asymmetry of the language of possible digit strings $\ldots \varepsilon_1(n)\varepsilon_0(n)$ entails that reflecting expansions on the decimal point destroys the equidistribution properties (see [53, Sect. 2]). However, symmetric languages still lead to equidistribution of $V_\beta(n)$ modulo 1. For instance, as observed in Barat and Grabner [7, Sect. 4], β-adic van der Corput sequences are equidistributed modulo 1 for recurrences of the types

$$G_{k+d} = a(G_{k+d-1} + \cdots + G_k) \quad (d \geq 2, \, a \in \mathbb{N}) \qquad \text{and}$$

$$G_{k+2} = (a+1)G_{k+1} + aG_k \qquad (a \in \mathbb{N}). \tag{6}$$

Ninomiya [41] came up with a slightly different definition of β-adic van der Corput sequences that are equidistributed modulo 1 for every Pisot number β. In particular, instead of reflecting expansions on the decimal point he reorders them w.r.t. the reverse lexicographic order. However, as mentioned in [41, Acknowledgements], if the linear recurrence is of the form (6) then Ninomiya's sequence agrees with the one defined in (5). Generalizing Ninomiya's construction, Steiner [53] defines and studies van der Corput sequences for abstract numeration systems in the sense of Lecomte and Rigo [32, 33].

Analogously to the classical Halton sequence we define its $\boldsymbol{\beta}$-adic variant by

$$H_{\boldsymbol{\beta}}(n) = (V_{\beta_1}(n), \dots, V_{\beta_s}(n)) \qquad (\boldsymbol{\beta} = (\beta_1, \dots, \beta_s)).$$

The first result on equidistribution properties of $\boldsymbol{\beta}$-adic Halton sequences is due to Robert Tichy and his co-authors (see Hofer et al. [27]) and reads as follows. Let $(G_k^{(1)}), \dots, (G_k^{(s)})$ be linear recurrent sequences of the form (6) (indeed, they could exhibit a slightly larger class, see [27, Lemma 1]), and let $\beta_1, \dots, \beta_s$ be the dominant roots of the associated characteristic equations. Then, under suitable assumptions on the algebraic independence of the elements β_i, the sequence $H_{\boldsymbol{\beta}}(n)$ is equidistributed modulo $[0, 1)^s$ for $\boldsymbol{\beta} = (\beta_1, \dots, \beta_s)$. This result is proved by methods from ergodic theory and provides no information on the discrepancy (see also the generalizations proved in [30]). Recently, Drmota [13] considered "hybrid" Halton sequences of vectors containing classical van der Corput sequences plus one component which is equal to the van der Corput sequence $V_\varphi(n)$, with φ being the golden ratio (corresponding to the Fibonacci sequence). His approach is different and he is able to give good bounds on the discrepancy that are close to optimality.

To state some of the mentioned results more precisely we introduce some notation. For $s \in \mathbb{N}$ and $A \subset [0, 1)^s$ we denote by $\mathbf{1}_A$ the characteristic function of A. For a given sequence $(\mathbf{y}_n)_{n \geq 0}$ in $[0, 1)^s$ we define the *(star) discrepancy* by

$$D_N((\mathbf{y}_n)_{n \geq 0}) = \sup_{0 < \omega_1, \dots, \omega_s \leq 1} \left| \frac{1}{N} \sum_{n=0}^{N-1} \mathbf{1}_{[0,\omega_1) \times \cdots \times [0,\omega_s)}(\mathbf{y}_n) - \omega_1 \cdots \omega_s \right|.$$

It is well known that $D_N((\mathbf{y}_n)_{n \geq 0})$ tends to 0 for $N \to \infty$ if and only if $(\mathbf{y}_n)_{n \geq 0}$ is equidistributed modulo $[0, 1)^s$. Moreover, this quantity provides a gauge for the quality of the equidistribution of a sequence. For more on discrepancy we refer to [14, 31].

Using this notion the above-mentioned results read as follows: for the van der Corput sequence it is well known that $D_N((v_q(n))_{n\geq0}) \ll \log N/N$ and for the s-dimensional Halton sequence we have $D_N((h_q(n))_{n\geq0}) \ll (\log N)^s/N$ (see, e.g., [31]; in [25] it is shown that this result extends to Halton sequences defined in terms of number systems with rational bases in the sense of [4]). In the β-adic case Ninomiya [41] shows that $D_N((V_\beta(n))_{n\geq0}) \ll \log N/N$ as well, while for the β-adic Halton sequences $H_\beta(n)$ no such estimates seem to be known. Concerning the hybrid case Drmota [13] shows that for each $\varepsilon > 0$ one has $D_N((v_{q_1}(n), \ldots, v_{q_s}(n), V_\varphi(n))_{n\geq0}) \ll N^{\varepsilon-1}$.

A large number of other variants of van der Corput and Halton sequences have been studied in the literature and for many of them strong discrepancy bounds can be derived (see, e.g., [12, 18, 19, 21, 25, 26, 38]).

Our aim is to give estimates for the discrepancy of β-adic Halton sequences that are defined in terms of m-bonacci numbers. To be more precise, for $m \geq 2$ define the *m-bonacci sequence* $(F_k^{(m)})_{k\geq0}$ by

$$F_k^{(m)} = 2^k, \qquad 0 \leq k \leq m - 1,$$
$$F_k^{(m)} = \sum_{j=1}^{m} F_{k-j}^{(m)}, \ k \geq m. \tag{7}$$

It is easy to see that $(F_k^{(m)})_{k\geq0}$ satisfies the conditions in (2) for $m \geq 2$ and, hence, the m-bonacci sequence admits a unique "greedy" expansion (3) for each $n \in \mathbb{N}$. In what follows, we denote the dominant root of the characteristic equation of the recurrence of $(F_k^{(m)})$ by φ_m and call it the *m-bonacci number*. Choose pairwise distinct integers $m_1, \ldots, m_s$ greater than 1, and let $\beta_i = \varphi_{m_i}$ for $i \in \{1, \ldots, s\}$ and $\boldsymbol{\beta} = (\beta_1, \ldots, \beta_s)$. Our main result will establish a discrepancy estimate for the sequence $H_\beta(n)$ for these choices of vectors $\boldsymbol{\beta}$ (see Theorem 5.1).

2 Preliminaries

In [13, Lemma 6] the discrepancy of the van der Corput sequence $V_\varphi(n)$ associated with the golden ratio φ was expressed in terms of local discrepancies of certain rotations. To this matter, relations between Ostrowski expansions and irrational rotations as discussed in [24, Sect. 3.2] are used. As anticipated in [13, Sect. 1], this doesn't generalize to higher degree recurrences and should be replaced by "a distribution analysis of linear sequences (modulo 1) in Rauzy fractal type sets." This is what we do for our class of m-bonacci recurrences. Before we can define these objects we need some notation.

Let $m \geq 2$ be fixed. A *substitution* σ over a finite alphabet $\mathcal{A} = \{1, \ldots, m\}$ is an endomorphism over the free monoid $\mathcal{A}^*$ of finite words over $\mathcal{A}$ (equipped with the operation of concatenation). The *incidence matrix* B_σ of σ is the $m \times m$ matrix given by $B_\sigma = (|\sigma(j)|_i)_{i,j \in \mathcal{A}}$. Here, for a word $w \in \mathcal{A}^*$ we denote the number of occurrences of the letter $i \in \mathcal{A}$ in w by $|w|_i$. Moreover, we will use $|w|$ for the number of letters of w. The *abelianization map* $\mathbf{l} : \mathcal{A}^* \to \mathbb{N}^m$ is defined by $\mathbf{l}(w) = (|w|_1, \ldots, |w|_m)$. It satisfies the relation $\mathbf{l}\sigma(w) = B_\sigma \mathbf{l}(w)$ for each $w \in \mathcal{A}^*$. There is a natural extension of a substitution σ to the space of infinite words $\mathcal{A}^\mathbb{N}$, which is equipped with the product topology of the discrete topology on $\mathcal{A}$. If we assume that $\sigma(1)$ starts with the letter 1 (what we will do throughout this paper), it is easy to see that the limit

$$u = u_1 u_2 \ldots = \lim_{n \to \infty} \sigma^n(11 \ldots) \tag{8}$$

exists and satisfies $\sigma(u) = u$.

A substitution σ is called a *unit Pisot substitution* if the characteristic polynomial of B_σ is the minimal polynomial of a Pisot unit. This class of substitutions is well studied (see, e.g., [5, 9, 29] for their dynamical and geometric properties). We wish to associate a *Rauzy fractal* with a unit Pisot substitution σ. To this matter let $\mathbf{u}$ and $\mathbf{v}$ be the normalized right and left eigenvector of B_σ associated with the dominant Pisot eigenvalue of B_σ, respectively. Then the orthogonal space $\mathbf{v}^\perp$ is the contractive invariant hyperspace of the matrix B_σ. We define π_c to be the projection of $\mathbb{R}^m$ along $\mathbf{u}$ to this contracting hyperspace. If $u = u_1 u_2 \ldots$ is given by (8), the *Rauzy fractal* $\mathcal{R}$ associated with the substitution σ is defined by

$$\mathcal{R} = \overline{\{\pi_c \mathbf{l}(u_1 u_2 \ldots u_n) \; : \; n \in \mathbb{N}\}}.$$

Obviously we have that $\mathcal{R}$ is a subset of $\mathbf{v}^\perp$ which can be viewed as the union of the *subtiles*

$$\mathcal{R}(i) = \overline{\{\pi_c \mathbf{l}(u_1 u_2 \ldots u_n) \; : \; u_{n+1} = i, n \in \mathbb{N}\}} \qquad (i \in \mathcal{A}).$$

These sets first occur in [44] and have been extensively studied since then (cf., e.g., [5, 6, 9, 29]). The set $\mathcal{R}(i)$, $i \in \mathcal{A}$, is a compact subset of $\mathbf{v}^\perp$ which is the closure of its interior and whose boundary has zero measure (see for instance [9]).

Under certain conditions (such as the *super coincidence condition* introduced in [29]) the subtiles $\mathcal{R}(i)$ are essentially disjoint and $\mathcal{R}$ forms a fundamental domain of the lattice $\mathcal{L} = \langle \pi_c(\mathbf{e}_1 - \mathbf{e}_2), \ldots, \pi_c(\mathbf{e}_1 - \mathbf{e}_m) \rangle_\mathbb{Z}$ and, hence, tiles $\mathbf{v}^\perp$ by translates taken from this lattice (see [29, Theorems 1.2 and 1.3]). Here $\mathbf{e}_i$ $(1 \le i \le m)$ are the standard basis vectors of $\mathbb{R}^m$.

To state a set equation for the subtiles $\mathcal{R}(i)$ we need the following *prefix–suffix graph* G_σ (cf. [11, Sect. 3]). The set of vertices of this labeled directed graph is given by the set of letters $i \in \mathcal{A}$ and there is an edge from i to j labeled by $(p, i, s) \in \mathcal{A}^* \times \mathcal{A} \times \mathcal{A}^*$ if $\sigma(j) = pis$. This edge will be written as $i \xrightarrow{(p,i,s)} j$ or just as $i \xrightarrow{p} j$, since (p, i, s) is completely determined by j and p.

Using this graph we can view the subtiles $\mathcal{R}(i)$ as the unique solution of the *graph directed iterated function system* (in the sense of Mauldin and Williams [34])

$$\mathcal{R}(i) = \bigcup_{\substack{(p,i,s) \\ i \xrightarrow{\quad} j}} B_\sigma R(j) + \pi_c \mathbf{l}(p) \qquad (i \in \mathcal{A}). \tag{9}$$

Here the union, which is extended over all edges in G_σ starting at the vertex i is always essentially disjoint. This set equation goes back to Sirvent and Wang [51].

We now define the class of substitutions we will need for our purposes. Let $m \geq 2$ be fixed. To the linear recurrent sequence $(F_k^{(m)})$ given by (7) we associate the substitution

$$\sigma_m(i) = \begin{cases} 1(i+1) & \text{for } i < m, \\ 1 & \text{for } i = m, \end{cases}$$

and denote its incidence matrix by $B_m = B_{\sigma_m}$. By induction we see that

$$F_k^{(m)} = |\sigma_m^k(1)| \qquad (k \geq 0). \tag{10}$$

It is well known that σ_m is a unit Pisot substitution for each $m \geq 2$ (see, e.g., [8, Sect. 2.3]). Thus we may associate a Rauzy fractal $\mathcal{R}_m = \bigcup_{i \in \mathcal{A}} \mathcal{R}_m(i)$ with it. It has the following properties.

Lemma 2.1 *Let $m \geq 2$. For each $k \in \mathbb{N}$ define the collection*

$$\mathcal{S}_k^{(m)} = \mathcal{S}_k = \left\{ B_\sigma^k \mathcal{R}_m(i_k) + \pi_c \mathbf{l}\left(\sigma_m^{k-1}(p_{k-1}) \dots \sigma_m(p_1)p_0\right) \; : \; i_0 \xrightarrow{p_0} \cdots \xrightarrow{p_{k-1}} i_k \in G_{\sigma_m} \right\},$$

where the range reaches over all walks of length k in G_{σ_m}. Then the elements of $\mathcal{S}_k$ overlap only on their boundaries,

$$\mathcal{R}_m = \bigcup_{S \in \mathcal{S}_k} S, \tag{11}$$

and $\mathcal{R}_m$ forms a fundamental domain of the lattice $\mathcal{L} = \mathcal{L}_m$.

The elements of $\mathcal{S}_k$ are called the *level k subtiles* of $\mathcal{R}_m$.

Proof It follows from [29, Example 3.1 and Theorem 1.2] that the subtiles $\mathcal{R}_m(i)$, $i \in \mathcal{A}$, are essentially disjoint and that their union $\mathcal{R}_m$ is a fundamental domain of the lattice $\mathcal{L} = \mathcal{L}_m$ in $\mathbf{v}^\perp = \mathbf{v}_m^\perp$ (see also [1]). Using the disjointness of the subtiles and iterating (9) for k times we get (11), where the union has no essential overlaps. $\qquad \square$

The prefix–suffix graph G_{σ_m} of σ_m is easy to calculate and a drawing of it is provided in Fig. 1.

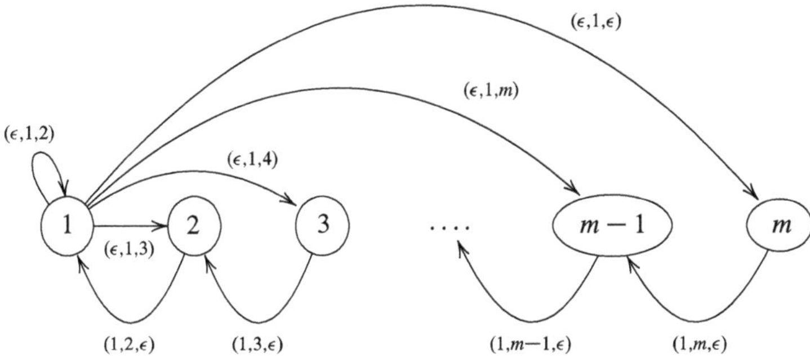

Fig. 1 The prefix–suffix graph G_{σ_m} of σ_m (ϵ denotes the empty word)

We equip $\mathbf{v}_m^{\perp}$ with the Lebesgue measure $\lambda_{\mathbf{v}_m}$ that is normalized in a way that a fundamental mesh of $\mathcal{L}_m$ has measure 1. Using [29, Eq. (3.3)] it is easy to see that

$$\lambda_{\mathbf{v}}(\mathcal{R}_m(i)) = \varphi_m^{-i} \qquad (1 \leq i \leq m).\tag{12}$$

In what follows, whenever convenient we will write $\mathcal{L} = \mathcal{L}_m$, $\mathbf{v} = \mathbf{v}_m$, $\pi_c = \pi_{c,m}$, etc. to emphasize for which m we use the object in question.

3 β-adic van der Corput Sequences and Torus Rotations

The main result of this section relates the φ_m-adic van der Corput sequence to a rotation on an $(m-1)$-dimensional torus. In its statement and in the sequel we will write $|A|$ for the Lebesgue measure of a measurable set $A \subset \mathbb{R}$.

Proposition 3.1 *Fix $m \geq 2$, let $(F_j^{(m)})_{j \geq 0}$ be the m-bonacci sequence with dominant root φ_m, and choose $k \in \mathbb{N}$ arbitrary. For $n \in \mathbb{N}$ let*

$$n = \sum_{j \geq 0} \varepsilon_j(n) F_j^{(m)}\tag{13}$$

be the expansion of n in the linear recurrent number system associated with $(F_j^{(m)})$. Let $r \in \{0, \ldots, m-1\}$ be given in a way that $\varepsilon_{k-r-1}(n) = 0$, $\varepsilon_{k-r}(n) = \cdots = \varepsilon_{k-1}(n) = 1$. Then the following assertions hold.

(1) For

$$\mu_k = \sum_{j=0}^{k-1} \varepsilon_{k-1-j}(n) \varphi_m^j$$

we have that

$$V_{\varphi_m}(n) \in \left[\frac{\mu_k}{\varphi_m^k}, \frac{\mu_k + \varphi_m^r - \sum_{i=0}^{r-1} \varphi_m^i}{\varphi_m^k}\right).$$

(2) *For*

$$v_k = \sum_{j=0}^{k-1} \varepsilon_j(n) F_j^{(m)}$$

we have that

$$n\pi_c(\mathbf{e}_1) \in v_k \pi_c(\mathbf{e}_1) + B_m^k \bigcup_{i=1}^{m-r} \mathcal{R}_m(i) + \mathcal{L}_m.$$

The measures of the occurring sets agree, i.e., we have

$$\left|\left[\frac{\mu_k}{\varphi_m^k}, \frac{\mu_k + \varphi_m^r - \sum_{i=0}^{r-1} \varphi_m^i}{\varphi_m^k}\right)\right| = \lambda_v \left(v_k \pi_c(\mathbf{e}_1) + B_m^k \bigcup_{i=1}^{m-r} \mathcal{R}_m(i)\right). \tag{14}$$

Proof To show item (1) just note that according to (4) (see also [43, Theorem 1]) the language of possible digit strings $\ldots \varepsilon_2(n)\varepsilon_1(n)\varepsilon_0(n)$ in (13) is given by the set of finite words in $\{0, 1\}^*$ in which blocks of m consecutive occurrences of 1 are forbidden (here we cut leading zeros to get finite sequences). The assertion then follows by easy calculations using the fact that $\varphi_m^m = \varphi_m^{m-1} + \cdots + \varphi_m + 1$.

To prove item (2) first consider the labels of a walk in the prefix–suffix graph G_{σ_m}. According to Fig. 1 we see that the prefixes of these labels are either ϵ (the empty word) or 1. Moreover, each sequence of prefixes can occur in the labeling of a walk subject to the condition that it contains no block of m prefixes 1. Looking at the lengths of the prefixes, this reflects the restriction on the language of digit strings for the linear recurrent sequence $(F_j^{(m)})$. More precisely, if n is given as in (13), then there is a (unique) walk $w : i_0 \xrightarrow{p_0} i_1 \xrightarrow{p_1} \cdots$ in the prefix–suffix graph such that $p_j = 1^{\varepsilon_j(n)}$ $(j \in \mathbb{N})$ and vice versa. For this walk w, using (10) we get

$$n = \sum_{j \geq 0} \varepsilon_j(n) F_j^{(m)} = \sum_{j \geq 0} \varepsilon_j(n)|\sigma_m^j(1)| = \sum_{j \geq 0} |\sigma_m^j(1^{\varepsilon_j(n)})| = \sum_{j \geq 0} |\sigma_m^j(p_j)|$$
$$= |\ldots \sigma_m^2(p_2)\sigma_m(p_1)p_0| \tag{15}$$

(note that since the digits $\varepsilon_j(n)$ are eventually 0, the prefixes p_j are eventually empty so that we can truncate the sums to finite sums in this formula). Thus $\ldots \sigma_m^2(p_2)\sigma_m(p_1)p_0$ is a word of length n and since by the definition of $\mathcal{L}_m$ we have

$$\pi_c \mathbf{l}(v_1 \ldots v_n) = \pi_c \sum_{i=1}^{n} \mathbf{e}_{v_i} \equiv n\pi_c(\mathbf{e}_1) \pmod{\mathcal{L}_m}$$

for each word $v_1 \ldots v_n \in \mathcal{A}^n$ we gain

$$n\pi_c(\mathbf{e}_1) \equiv \pi_c\mathbf{1}\Big(\ldots \sigma_m^2(p_2)\sigma_m(p_1)p_0\Big) \pmod{\mathcal{L}_m}. \tag{16}$$

In the same way we derive that

$$v_k\pi_c(\mathbf{e}_1) \equiv \pi_c\mathbf{1}\Big(\sigma_m^{k-1}(p_{k-1}) \ldots \sigma_m(p_1)p_0\Big) \pmod{\mathcal{L}_m}. \tag{17}$$

We need to identify the level k subtile of $\mathcal{R}_m$ containing a representative of $n\pi_c(\mathbf{e}_1)$ modulo $\mathcal{L}_m$. Since B_m acts as a uniform contraction on $\mathbf{v}^\perp$ we have $B_m^\ell \mathcal{R}_m(i) \to \{\mathbf{0}\}$ for $\ell \to \infty$ in Hausdorff metric. Thus, iterating the set equation in Eq. (9) *ad infinitum* for each $j_0 \in \mathcal{A}$ and taking the union over $j_0 \in \mathcal{A}$ we get

$$\mathcal{R}_m = \bigcup_{j_0 \xrightarrow{q_0} j_1 \xrightarrow{q_1} \cdots} \pi_c\mathbf{1}\Big(\ldots \sigma_m^2(q_2)\sigma_m(q_1)q_0\Big), \tag{18}$$

where the union runs over all infinite walks of G_{σ_m}. The walk w corresponding to the expansion (15) of n starts with $w_k : i_0 \xrightarrow{p_0} i_1 \xrightarrow{p_1} \cdots \xrightarrow{p_{k-1}} i_k$ and occurs in the range of the union in (18). Thus we restrict this union to the infinite walks starting with w_k. Doing this we produce the subtile of level k corresponding to the first k digits of n. In particular, we get again from (9) that

$$
\begin{aligned}
& B_m^k \mathcal{R}_m(i_k) + \pi_c\mathbf{1}\Big(\sigma_m^{k-1}(p_{k-1}) \ldots \sigma_m(p_1)p_0\Big) \\
& \quad = \bigcup_{i_k=j_k \xrightarrow{q_k} j_{k+1} \xrightarrow{q_{k+1}} \cdots} \pi_c\mathbf{1}\Big(\ldots \sigma_m^{k+1}(q_{k+1})\sigma_m^k(q_k)\sigma_m^{k-1}(p_{k-1}) \ldots \sigma_m(p_1)p_0\Big),
\end{aligned}
\tag{19}
$$

and the element $\pi_c\mathbf{1}\big(\ldots \sigma_m^2(p_2)\sigma_m(p_1)p_0\big)$ is contained in this union for some $i_k \in \mathcal{A}$. We inspect the prefix–suffix graph G_{σ_m} in Fig. 1 to specify the possible values of i_k. Since $p_{k-r-1} = \varepsilon_{k-r-1}(n) = 0$ and $p_{k-r} = \varepsilon_{k-r}(n) = \cdots = p_{k-1} = \varepsilon_{k-1}(n) = 1$ we see that $i_k \in \{1, \ldots, m-r\}$, as the other vertices cannot be accessed by a walk ending with a block of prefixes 1 of length r in G_{σ_m}. Thus

$$\pi_c\mathbf{1}\Big(\ldots \sigma_m^2(p_2)\sigma_m(p_1)p_0\Big) \in B_m^k \mathcal{R}_m(i_k) + \pi_c\mathbf{1}\Big(\sigma_m^{k-1}(p_{k-1}) \ldots \sigma_m(p_1)p_0\Big) \tag{20}$$

holds for some $i_k \in \{1, \ldots, m-r\}$. The result in item (2) now follows by inserting (16) and (17) in (20).

To show (14) we use (12) together with the fact that $\lambda_\mathbf{v}(B_m A) = \varphi_m^{-1}\lambda_\mathbf{v}(A)$ holds for each measurable $A \subset \mathbf{v}^\perp$. $\qquad\square$

Remark 3.2 The representation of n in terms of lengths of prefixes under powers of σ_m in (15) is called the *Dumont–Thomas expansion* of n. Dumont–Thomas expansions are introduced in [15, 16]. As mentioned in Rigo and Steiner [46, p. 284], an abstract numeration system in the sense of [32, 33] corresponding to a primitive automaton leads to Dumont–Thomas numeration for a primitive substitution.

We will need the following tiling result of the intervals in Proposition 3.1 (1).

Lemma 3.3 *Fix $m \geq 2$ and $k \in \mathbb{N}$. To each $n < F_k^{(m)}$ associate $r = r(n)$ and $\mu_k = \mu_k(n) = \sum_{j=0}^{k-1} \varepsilon_{k-1-j}(n)\varphi_m^j$ as in Proposition 3.1. Then the collection of intervals*

$$C_k^{(m)} = C_k = \left\{ \left[\frac{\mu_k(n)}{\varphi_m^k}, \frac{\mu_k(n) + \varphi_m^{r(n)} - \sum_{i=0}^{r(n)-1} \varphi_m^i}{\varphi_m^k} \right) \; : \; 0 \leq n < F_k^{(m)} \right\}$$

forms a partition of the interval $[0, 1)$.

Proof This follows easily by direct calculation using $\varphi_m^m = \varphi_m^{m-1} + \cdots + \varphi_m + 1$. □

Since by Lemma 2.1 the subtiles of level k of $\mathcal{R}_m$ overlap on the boundary, we will need a slightly shifted rotation that always hits the interior of the subtiles.

Corollary 3.4 *Fix $m \geq 2$, let $(F_j^{(m)})_{j \geq 0}$ be the m-bonacci sequence with dominant root φ_m, and choose $k, N \in \mathbb{N}$ arbitrary. For $0 \leq n < N$ let*

$$n = \sum_{j \geq 0} \varepsilon_j(n) F_j^{(m)}$$

be the expansion of n in the linear recurrent number system associated with $(F_j^{(m)})$. Let $r \in \{0, \ldots, m-1\}$ be given in a way that $\varepsilon_{k-r-1}(n) = 0$, $\varepsilon_{k-r}(n) = \cdots = \varepsilon_{k-1}(n) = 1$. Then there is a constant $\alpha = \alpha(k, N) \in \mathbf{v}^\perp$ depending only on k and N such that

$$n\pi_c(\mathbf{e}_1) + \alpha \in \nu_k \pi_c(\mathbf{e}_1) + \bigcup_{i=1}^{m-r} \mathrm{int}\left(B_m^k \mathcal{R}_m(i) \right) + \mathcal{L}_m \quad (0 \leq n < N)$$

with ν_k as in Proposition 3.1.

Proof Let L be defined in a way that $F_{L-1}^{(m)} \leq N - 1 < F_L^{(m)}$ and $M = \max(k, L)$. Then, arguing as in (15), the expansion of each $n < N$ can be written as

$$n = \sum_{j=0}^{M-1} \varepsilon_j(n) F_j^{(m)} = \sum_{j=0}^{M-1} |\sigma_m^j(p_j)|$$

for some walk labelled by $p_0, \ldots, p_{M-1}$ in the prefix–suffix graph G_{σ_m}. From Proposition 3.1 we know that

$$n\pi_c(\mathbf{e}_1) \equiv \pi_c \mathbf{1}\left(\sigma_m^{M-1}(p_{M-1}) \ldots \sigma_m(p_1)p_0 \right) \in \nu_k \pi_c \mathbf{e}_1 + B_m^k \bigcup_{i=1}^{m-r} \mathcal{R}_m(i) + \mathcal{L}_m.$$

Since $r \in \{0, \ldots, m - 1\}$, $i = 1$ is always in the range of the union on the right-hand side, by the set equation in Eq. (9) this right-hand side contains all elements $\pi_c \mathbf{1}(\ldots \sigma_m^{M-1}(p_{M-1}) \ldots \sigma_m(p_1)p_0)$ corresponding to a walk w in the prefix–suffix automaton starting with $i_0 \xrightarrow{p_0} \cdots \xrightarrow{p_{M-2}} i_{M-1} \xrightarrow{p_{M-1}} 1$. However, by (16) and (19) this is equivalent to

$$n\pi_c(\mathbf{e}_1) + B_m^M \mathcal{R}_m(1) \subset v_k \pi_c \mathbf{e}_1 + B_m^k \bigcup_{i=1}^{m-r} \mathcal{R}_m(i) + \mathcal{L}_m \qquad (0 \le n < N).$$

Since $B_m^M \mathcal{R}_m(1)$ has positive measure, while the boundaries of $\mathcal{R}_m(i)$, $i \in \mathcal{A}$, have zero measure we can choose $\alpha \in B_m^M \mathcal{R}_m(1)$ in a way that $n\pi_c(\mathbf{e}_1) + \alpha$ avoids the boundaries of the subtiles for every $n \in \{0, \ldots, N - 1\}$ and the result follows. $\qquad \square$

4 Discrepancy Estimates

Using the results of the previous section we will express the discrepancy of β-adic Halton sequences in terms of local discrepancies of rotations. We first state a generalization of [13, Lemma 6] on the discrepancy of φ_m-adic van der Corput sequences that is of interest in its own right.

Lemma 4.1 *Fix $m \ge 2$, let $N \in \mathbb{N}$ be given, and choose L in a way that $F_{L-1}^{(m)} \le N - 1 < F_L^{(m)}$. Then*

$$D_N((V_{\varphi_m}(n))_{n \ge 0}) \ll \frac{1}{\varphi_m^L} + \sum_{1 \le k \le L} \delta_k,$$

where

$$\delta_k = \sup_{S \in \mathcal{S}_k} \left| \frac{1}{N} \sum_{n=0}^{N-1} \mathbf{1}_S\left(n\pi_c(\mathbf{e}_1) + \alpha_k \bmod \mathcal{L}_m\right) - \lambda_{\mathbf{v}}(S) \right|$$

with $\alpha_k = \alpha_k(N)$ chosen as in Corollary 3.4.

Proof Choose $x \in [0, 1)$ arbitrary. We can construct a family $I_1, \ldots, I_L, I^*$ of intervals such that $I_k = \emptyset$ or $I_k \in \mathcal{C}_k$ and $|I^*| < \varphi_m^{-L}$ in a way that

$$I_1 \cup \cdots \cup I_L \cup I^* = [0, x) \quad \text{with no overlaps.} \tag{21}$$

Indeed, as $\mathcal{C}_L$ is a partition of $[0, 1)$ by Lemma 3.3 we can choose $\varepsilon_{L-1} \ldots \varepsilon_0 \in \{0, 1\}^L$ containing no block of m consecutive 1s in a way that $\mu_L = \sum_{j=0}^{L-1} \varepsilon_{L-1-j}\varphi^j$ satisfies $\mu_L \varphi_m^{-L} \le x < (\mu_L + 1)\varphi_m^{-L}$. Then let $I^* = [\mu_L \varphi_m^{-L}, x)$. For the definition of I_k let $\xi_k = 0 \cdot F_{k-1}^{(m)} + \varepsilon_{k-2} F_{k-2}^{(m)} + \cdots + \varepsilon_0 F_0^{(m)}$ and use the notation of Lemma 3.3.

Then, since the $(k-1)$-st digit of ξ_k is 0 we have $r(\xi_k) = 0$ and we set

$$I_k = \begin{cases} \left[\dfrac{\mu_k(\xi_k)}{\varphi_m^k}, \dfrac{\mu_k(\xi_k)+\varphi_m^{r(\xi_k)}-\sum_{i=0}^{r(\xi_k)-1}\varphi_m^i}{\varphi_m^k}\right) = \left[\dfrac{\mu_k(\xi_k)}{\varphi_m^k}, \dfrac{\mu_k(\xi_k)+1}{\varphi_m^k}\right), & \varepsilon_{k-1} = 1, \\ \emptyset, & \varepsilon_{k-1} = 0. \end{cases}$$

Clearly, we have $I_k \in \mathcal{C}_k$ whenever it is nonempty and with these choices it is easy to check that (21) is true.

Set $y_n = V_{\varphi_m}(n)$ for convenience. In order to estimate $D_N((y_n)_{n\geq 0})$ it is sufficient to deal with intervals of the form I_k and I^* since by the triangle inequality we have

$$\left|\frac{1}{N}\sum_{n=0}^{N-1}\mathbf{1}_{[0,x)}(y_n) - x\right| \leq \sum_{k=1}^{L}\left|\frac{1}{N}\sum_{n=0}^{N-1}\mathbf{1}_{I_k}(y_n) - |I_k|\right| + \left|\frac{1}{N}\sum_{n=0}^{N-1}\mathbf{1}_{I^*}(y_n) - |I^*|\right|. \tag{22}$$

On the right-hand side it obviously suffices to take the sum over all values of k corresponding to nonempty sets I_k because we get no contribution otherwise. Since in this case $I_k \in \mathcal{C}_k$ we can apply Proposition 3.1 and Corollary 3.4. Indeed, since $\mathcal{C}_k$ is a partition of $[0,1)$ by Lemma 3.3, Proposition 3.1 yields

$$y_n \in I_k \quad\Longrightarrow\quad n\pi_c(\mathbf{e}_1) \in \xi_k\pi_c(\mathbf{e}_1) + B_m^k\bigcup_{i=1}^{m}\mathcal{R}_m(i) + \mathcal{L}_m.$$

The right-hand side is a union of m level k subtiles and since the elements of $\mathcal{S}_k$ overlap on their boundaries by Lemma 2.1, the converse implication is not true. Thus we have to use Corollary 3.4 which implies the existence of $\alpha_k \in \mathbf{v}^\perp$ such that

$$y_n \in I_k \quad\Longrightarrow\quad n\pi_c(\mathbf{e}_1)+\alpha_k \in \xi_k\pi_c(\mathbf{e}_1)+\bigcup_{i=1}^{m}\mathrm{int}(B_m^k\mathcal{R}_m(i))+\mathcal{L}_m. \tag{23}$$

Since the interiors of the elements of $\mathcal{S}_k$ are pairwise disjoint by Lemma 2.1, the implication in (23) can be reversed and it becomes an equivalence. By this equivalence, for each I_k there is a union of m elements $S_1,\ldots,S_m \in \mathcal{S}_k$ (viz. $S_i = B_m^k\mathcal{R}_m(i) + \xi_k\pi_c(\mathbf{e}_1)$) such that

$$\sum_{n=0}^{N-1}\mathbf{1}_{I_k}(y_n) = \sum_{n=0}^{N-1}\mathbf{1}_{S_1\cup\cdots\cup S_m}\left(n\pi_c(\mathbf{e}_1) + \alpha_k \mod \mathcal{L}_m\right).$$

Thus, observing that (14) yields $|I_k| = \lambda_{\mathbf{v}}(S_1 \cup \cdots \cup S_m)$ and applying the triangle inequality again we gain for each $k \in \{1,\ldots,L\}$

$$\left|\frac{1}{N}\sum_{n=0}^{N-1}\mathbf{1}_{I_k}(y_n) - |I_k|\right| \leq \sum_{i=1}^{m}\left|\frac{1}{N}\sum_{n=0}^{N-1}\mathbf{1}_{S_i}\left(n\pi_c(\mathbf{e}_1) + \alpha_k \mod \mathcal{L}_m\right) - \lambda_{\mathbf{v}}(S_i)\right| \tag{24}$$

(α_k depends only on k and N). To treat the interval I^* put $J = [\mu_L \varphi_m^{-L}, (\mu_L + 1)\varphi_m^{-L})$ and note that $J \supset I^*$ is an element of $\mathcal{C}_L$. We have

$$\sum_{n=0}^{N-1} \mathbf{1}_{I^*}(y_n) \le \sum_{n=0}^{N-1} \mathbf{1}_J(y_n) \le \left| \sum_{n=0}^{N-1} \mathbf{1}_J(y_n) - |J| \right| + |J| \tag{25}$$

and, hence,

$$\left| \sum_{n=0}^{N-1} \mathbf{1}_{I^*}(y_n) - |I^*| \right| \le \left| \sum_{n=0}^{N-1} \mathbf{1}_J(y_n) - |J| \right| + |J| + |I^*|. \tag{26}$$

Since $J \in \mathcal{C}_L$ we may apply Proposition 3.1 and Corollary 3.4 as above and since $|I^*| < |J| = \varphi_m^{-L}$ we gain

$$\left| \sum_{n=0}^{N-1} \mathbf{1}_{I^*}(y_n) - |I^*| \right| \le \sum_{i=1}^{m} \left| \frac{1}{N} \sum_{n=0}^{N-1} \mathbf{1}_{S_i}(n\pi_c(\mathbf{e}_1) + \alpha_L \mod \mathcal{L}_m) - \lambda_{\mathbf{v}}(S_i) \right| + \frac{2}{\varphi_m^L} \tag{27}$$

for some $S_1, \ldots, S_m \in \mathcal{S}_L$. Now we insert (24) and (27) in (22) and take the supremum over all $x \in [0, 1)$. This yields the result. $\qquad \square$

Lemma 4.1 implies Ninomiya's [41] low-discrepancy estimate for the φ_m-adic van der Corput sequence since the level k subtiles of the Rauzy fractals are *bounded remainder sets* for the rotation by $\pi_c(\mathbf{e}_i)$ (see [10, Theorem 1]). Moreover, its proof paves the way for the proof of the following proposition.

Proposition 4.2 *Fix $m_1, \ldots, m_s \ge 2$, let $N \in \mathbb{N}$ be given, and choose $L_1, \ldots, L_s$ in a way that $F_{L_j-1}^{(m_j)} \le N - 1 < F_{L_j}^{(m_j)}$. Then for $\boldsymbol{\beta} = (\beta_1, \ldots, \beta_s) = (\varphi_{m_1}, \ldots, \varphi_{m_s})$*

$$D_N((H_{\boldsymbol{\beta}}(n))_{n \ge 0}) \ll \sum_{i=1}^{s} \frac{1}{\beta_i^{L_i}} + \sum_{1 \le k_1 \le L_1} \cdots \sum_{1 \le k_s \le L_s} \delta_{k_1, \ldots, k_s}, \tag{28}$$

where

$$\delta_{k_1, \ldots, k_s} = \sup_{S_1 \in \mathcal{S}_{k_1}^{(m_1)}, \ldots, S_s \in \mathcal{S}_{k_s}^{(m_s)}} \left| \frac{1}{N} \sum_{n=0}^{N-1} \prod_{i=1}^{s} \mathbf{1}_{S_i}\left(n\pi_{c,m_i}(\mathbf{e}_1) + \alpha_{k_i} \mod \mathcal{L}_{m_i} \right) - \prod_{i=1}^{s} \lambda_{\mathbf{v}_{m_i}}(S_i) \right|.$$

Here $\alpha_{k_i} = \alpha_{k_i}(N)$ is chosen as in Corollary 3.4.

Proof We start with an arbitrary box $[0, x_1) \times \cdots \times [0, x_s) \subset [0, 1)^s$. Each of the intervals $[0, x_i)$ can be partitioned into subintervals $I_1^i, \ldots, I_{L_i}^i, I^{*,i}$ with $I_k^i \in \mathcal{C}_k^{(m_i)}$ or empty and $|I^{*,i}| \le \beta_i^{-L_i}$ (see the proof of Lemma 4.1). Using the triangle inequality we obtain, setting $\mathbf{y}_n = H_\beta(n)$ and $I^i_{L+1} = I^{*,i}$ for convenience,

$$\left| \frac{1}{N} \sum_{n=0}^{N-1} \mathbf{1}_{[0,x_1)\times\cdots\times[0,x_s)}(\mathbf{y}_n) - x_1\cdots x_s \right| \leq \sum_{k_1=1}^{L_1+1}\cdots\sum_{k_s=1}^{L_s+1}\left| \frac{1}{N}\sum_{n=0}^{N-1}\mathbf{1}_{I^1_{k_1}\times\cdots\times I^s_{k_s}}(\mathbf{y}_n) - |I^1_{k_1}|\cdots|I^s_{k_s}| \right|.$$

For each box $I^1_{k_1}\times\cdots\times I^s_{k_s}$ with $k_1 \leq L_1,\ldots,k_s \leq L_s$ we can argue in the same way as in the proof of Lemma 4.1 to see that

$$\left| \frac{1}{N}\sum_{n=0}^{N-1}\mathbf{1}_{I^1_{k_1}\times\cdots\times I^s_{k_s}}(\mathbf{y}_n) - |I^1_{k_1}|\cdots|I^s_{k_s}| \right|$$

$$\leq \sum_{j_1=1}^{m_1}\cdots\sum_{j_s=1}^{m_s}\left| \frac{1}{N}\sum_{n=0}^{N-1}\prod_{i=1}^{s}\mathbf{1}_{I^i_{k_i}}(V_{\beta_i}(n)) - \prod_{i=1}^{s}\lambda_{\mathbf{v}_{m_i}}(S_{ij_i}) \right|$$

$$= \sum_{j_1=1}^{m_1}\cdots\sum_{j_s=1}^{m_s}\left| \frac{1}{N}\sum_{n=0}^{N-1}\prod_{i=1}^{s}\mathbf{1}_{S_{ij_i}}\left(n\pi_{c,m_i}(\mathbf{e}_1) + \alpha_{k_i} \mod \mathcal{L}_{m_i}\right) - \prod_{i=1}^{s}\lambda_{\mathbf{v}_{m_i}}(S_{ij_i}) \right|$$

for certain $S_{ij} \in \mathcal{S}^{(m_i)}_{k_i}$ $(1 \leq i \leq s,\ 1 \leq j \leq m_i)$. If some of the k_i equal $L_i + 1$, then blowing up the according intervals $I^i_{L+1} = I^{*,i}$ as in the proof of Lemma 4.1 we can treat them in the same way as the intervals $I^i_{k_i}$ with $k_i \leq L_i$ causing an overall error bounded by $\sum_{i=1}^{s}\frac{1}{\beta_i^{L_i}}$. Taking the supremum over all boxes $[0,x_1)\times\cdots\times[0,x_s) \subset [0,1)^s$ gives the result.
$\qquad\square$

5 The Main Result

We are now in a position to state and prove our main result, a discrepancy estimate for β-adic Halton sequences with m-bonacci arguments.

Theorem 5.1 *Let $m_1,\ldots,m_s$ be pairwise distinct integers greater than or equal to 2 and set $\beta = (\varphi_{m_1},\ldots,\varphi_{m_s})$. If $\{1,\varphi_{m_1},\ldots,\varphi_{m_1}^{m_1-1},\ldots,\varphi_{m_s},\ldots,\varphi_{m_s}^{m_s-1}\}$ is linearly independent over $\mathbb{Q}$, then the discrepancy of the β-adic Halton sequence $H_\beta(n)$ satisfies*

$$D_N((H_\beta(n))_{n\geq 0}) \ll N^{\frac{\max\{d_i-(m_i-1)\,:\,1\leq i\leq s\}}{(m_1-1)+\cdots+(m_s-1)}+\varepsilon} \tag{29}$$

for each $\varepsilon > 0$. Here $d_i = \dim_B(\partial\mathcal{R}_{m_i})$, which is strictly smaller than $m_i - 1$, denotes the box counting dimension of the boundary of the Rauzy fractal $\mathcal{R}_{m_i}$, $1 \leq i \leq s$.

Remark 5.2 The box counting dimension of $\partial\mathcal{R}_m$ can be calculated explicitly in terms of the so-called *boundary graph* (see, for instance, [50, Theorem 4.4] for a formula or [20, Theorem 3.1] for an estimate).

A sufficient condition for the set $\{1, \varphi_{m_1}, \ldots, \varphi_{m_1}^{m_1-1}, \ldots, \varphi_{m_s}, \ldots, \varphi_{m_s}^{m_s-1}\}$ to be linearly independent over $\mathbb{Q}$ is that the degree of the extension $\mathbb{Q}(\varphi_{m_1}, \ldots, \varphi_{m_s}) : \mathbb{Q}$ satisfies $[\mathbb{Q}(\varphi_{m_1}, \ldots, \varphi_{m_s}) : \mathbb{Q}] = m_1 \cdots m_s$. This condition holds, for instance, if the integers $m_1, \ldots, m_s$ or the discriminants of the number fields $\mathbb{Q}(\varphi_{m_1}), \ldots, \mathbb{Q}(\varphi_{m_s})$ are pairwise relatively prime (see, e.g. Mordell [37]).

We set the stage for the proof of Theorem 5.1 by establishing a series of preparatory results. First we provide a technical lemma which shows how to conjugate the rotation by the vector $\pi_c(\mathbf{e}_1)$ on the fundamental domain of the lattice $\mathcal{L}_m$ in $\mathbf{v}_m^\perp$ to a rotation on the standard torus $\mathbb{R}^{m-1}/\mathbb{Z}^{m-1} \simeq [0, 1)^{m-1}$.

Lemma 5.3 *Let $m \geq 2$ be given. Then*

$$\pi_c(\mathbf{e}_1) = \sum_{i=2}^{m} \pi_c(\mathbf{e}_1 - \mathbf{e}_i)\varphi_m^{-i}, \tag{30}$$

i.e., the rotation $n\pi_c(\mathbf{e}_1) + \alpha \mod \mathcal{L}_m$, $n \in \mathbb{N}$, is conjugate to the rotation

$$n\left(\varphi_m^{-2}, \ldots, \varphi_m^{-m}\right) + Q_m\alpha \mod \mathbb{Z}^{m-1}, \qquad n \in \mathbb{N},$$

by a linear conjugacy Q_m.

Proof The identity in (30) can be verified by a (somewhat tedious) direct computation. Indeed, it is easy to see that each coordinate of (30) is an element of $\mathbb{Q}(\varphi_m)$. Using this it suffices to compare the coefficients of φ_m^i for each $0 \leq i \leq m - 1$ in each coordinate. The conjugacy assertion follows immediately from (30). $\qquad\square$

Choose $m_1, \ldots, m_s \geq 2$. In what follows we denote by

$$Q = Q_{m_1} \times \cdots \times Q_{m_s} \tag{31}$$

the conjugacy between the rotation

$$R(n, \boldsymbol{\alpha}) = \prod_{i=1}^{s}(n\pi_{c,m_i}(\mathbf{e}_1) + \alpha_i \mod \mathcal{L}_{m_i}) \tag{32}$$

with offset $\boldsymbol{\alpha} = (\alpha_1, \ldots, \alpha_s)$ on the $(m_1 - 1) + \cdots + (m_s - 1)$ dimensional torus $\prod_{i=1}^{s}(\mathbf{v}_{m_i}^\perp/\mathcal{L}_{m_i})$ and the rotation

$$QR(n, \boldsymbol{\alpha}) = \prod_{i=1}^{s}\left(n\left(\varphi_{m_i}^{-2}, \ldots, \varphi_{m_i}^{-m_i}\right) + Q\alpha_i \mod \mathbb{Z}^{(m_i-1)}\right) \tag{33}$$

on the standard torus $[0, 1)^{(m_1-1)+\cdots+(m_s-1)}$. This conjugacy exists by Lemma 5.3
In what follows we will need to study properties of products of the form

$$P = P(k_1, \ldots, k_s) = Q\left(B_{m_1}^{k_1}\mathcal{R}_{m_1} \times \cdots \times B_{m_s}^{k_s}\mathcal{R}_{m_s}\right) \in \mathbb{R}^{(m_1-1)+\cdots+(m_s-1)}$$

with Q as in (31). We will need the *box counting dimension* of ∂P, whose value doesn't depend on $k_1, \ldots, k_s$. It is defined by (see Falconer [17, Sect. 3.1])

$$D = \dim_{\mathrm{B}}(\partial P) = \lim_{\ell \to \infty} \frac{\log N(\partial P, \ell)}{\ell \log 2},$$

where $N(\partial P, \ell)$ is the number of boxes of side length $2^{-\ell}$ in $\mathbb{R}^{(m_1-1)+\cdots+(m_s-1)}$ arranged in a grid centered at 0 having nonempty intersection with ∂P. The boundary ∂P is given by the union of cartesian products of the shape ($1 \le i \le s$)

$$Q\left(B_{m_1}^{k_1} \mathcal{R}_{m_1} \times \cdots \times B_{m_{i-1}}^{k_{i-1}} \mathcal{R}_{m_{i-1}} \times \partial B_{m_i}^{k_i} \mathcal{R}_{m_i} \times B_{m_{i+1}}^{k_{i+1}} \mathcal{R}_{m_{i+1}} \times \cdots \times B_{m_s}^{k_s} \mathcal{R}_{m_s}\right)$$

and, hence, its box counting dimension satisfies (cf. e.g. [17, Product Formula 7.5])

$$D \le \max \left\{ \sum_{j=1, j \ne i}^{s} (m_j - 1) + \dim_{\mathrm{B}}(\mathcal{R}_{m_i}) \ : \ 1 \le i \le s \right\}.$$

As mentioned above, in [20] an explicitly computable bound for $\dim_{\mathrm{B}}(\mathcal{R}_m)$ is given. Since $\mathcal{R}_m$ admits a tiling w.r.t. the lattice $\mathcal{L}_m$ by Lemma 2.1, it follows from [50, Theorem 4.1] that this bound is nontrivial for each $m \ge 2$, and, hence, we have

$$D < (m_1 - 1) + \cdots + (m_s - 1). \tag{34}$$

We now turn to a covering property for products of Rauzy fractals.

Lemma 5.4 *Let $m_1, \ldots, m_s \ge 2$ and $k_1, \ldots, k_s \ge 0$, and set*

$$P = Q(B_{m_1}^{k_1} \mathcal{R}_{m_1} \times \cdots \times B_{m_s}^{k_s} \mathcal{R}_{m_s}).$$

Let $D = \dim_{\mathrm{B}}(\partial P)$ be the box counting dimension of ∂P and fix $\varepsilon > 0$. Then for each $M \in \mathbb{N}$ we can cover P by boxes $U_1, \ldots, U_r, V_1, \ldots, V_{r'} \subset \mathbb{R}^{(m_1-1)+\cdots+(m_s-1)}$ with $r, r' \ll 2^{M(D+\varepsilon)}$ in the following way.

- *Each box U_i has empty intersection with the complement of P ($1 \le i \le r$).*
- *Each box V_i intersects the boundary of P ($1 \le i \le r'$) and*

$$\sum_{i=1}^{r'} \mu(V_i) \ll 2^{M(D-(m_1-1)-\cdots-(m_s-1)+\varepsilon)},$$

where μ denotes the Lebesgue measure on $\mathbb{R}^{(m_1-1)+\cdots+(m_s-1)}$.

Proof Cover P by the collections $\mathcal{K}_\ell$ of boxes of side length $2^{-\ell}$ for each $\ell \in \{1, \ldots, M\}$ arranged in a grid centered at 0. So for $\ell > 1$ each box in $\mathcal{K}_\ell$ is contained in a unique larger box of $\mathcal{K}_{\ell-1}$ which is called the *parent* of this box.

Choose the boxes U_i inductively as follows. First take all elements of $\mathcal{K}_1$ that have empty intersection with the complement of P. For $\ell > 1$ take all elements of $\mathcal{K}_\ell$ that have empty intersection with the complement of P and which has not been covered so far. Thus $\{U_1, \ldots, U_r\}$ contains, apart from elements of $\mathcal{K}_1$, only elements of $\mathcal{K}_\ell$ whose parents intersect the boundary of P. By the definition of the box counting dimension, the number z_ℓ of such elements satisfies $z_\ell \ll 2^{\ell(D+\varepsilon)}$. Summing up z_ℓ over all $1 \le \ell \le M$ we obtain the bound for r asserted in the statement.

Since the sets V_i can be chosen to be boxes from $\mathcal{K}_M$ in the claimed way by the definition of the box counting dimension the lemma is proved. $\qquad\square$

In the last preparatory lemma we recall a classical discrepancy estimate for rationally independent rotations on the torus. We will use the following notation (see, e.g., Niederreiter [39, 40]). We say that $\boldsymbol{\gamma} \in \mathbb{R}^s$ is of *finite type* η_0 if $\eta_0 \in \mathbb{R}$ is the infimum of all $\eta \in \mathbb{R}$ for which there exists a positive constant $c = c(\eta, \boldsymbol{\gamma})$ such that

$$\left(\prod_{j=1}^{s} \max(|h_j|, 1) \right)^{\eta} \, ||(h_1, \ldots, h_s) \cdot \boldsymbol{\gamma}|| \ge c$$

holds for all $(h_1, \ldots, h_s) \in \mathbb{Z}^s \setminus \{\boldsymbol{0}\}$.

Lemma 5.5 *Let* $\boldsymbol{\gamma} = (\gamma_1, \ldots, \gamma_s) \in \mathbb{R}^s$ *with algebraic numbers* $\gamma_1, \ldots, \gamma_s$ *and assume that* $\{1, \gamma_1, \ldots, \gamma_s\}$ *is linearly independent over* $\mathbb{Q}$. *Then for each* $\varepsilon > 0$ *we have*

$$D_N((n\boldsymbol{\gamma} \bmod [0, 1)^s)_{n \ge 0}) \ll N^{\varepsilon - 1}.$$

Proof Using a classical result by Schmidt [48, Theorem 2] we see that a vector $(\gamma_1, \ldots, \gamma_s)$ of real algebraic numbers for which $\{1, \gamma_1, \ldots, \gamma_s\}$ is linearly independent over $\mathbb{Q}$ is of finite type 1. By Kuipers and Niederreiter [31, p. 132, Exercise 3.17] (or [14, Theorem 1.80]) this implies the result. $\qquad\square$

After these preparations we turn to the proof of Theorem 5.1.

Proof of Theorem 5.1 In view of Proposition 4.2 we have to estimate the quantities $\delta_{k_1, \ldots, k_s}$. First note that by the definition of the linear conjugacy Q in (31), setting $\boldsymbol{\alpha} = (\alpha_{k_1}, \ldots, \alpha_{k_s})$, we can write

$$\delta_{k_1, \ldots, k_s} = \sup_{S_1 \in \mathcal{S}_{k_1}^{(m_1)}, \ldots, S_s \in \mathcal{S}_{k_s}^{(m_s)}} \left| \frac{1}{N} \sum_{n=0}^{N-1} \mathbf{1}_{S_1 \times \cdots \times S_s}(R(n, \boldsymbol{\alpha})) - \prod_{i=1}^{s} \lambda_{v_{m_i}}(S_i) \right|$$

$$= \sup_{S_1 \in \mathcal{S}_{k_1}^{(m_1)}, \ldots, S_s \in \mathcal{S}_{k_s}^{(m_s)}} \left| \frac{1}{N} \sum_{n=0}^{N-1} \mathbf{1}_{Q(S_1 \times \cdots \times S_s)}(QR(n, \boldsymbol{\alpha})) - \mu(Q(S_1 \times \cdots \times S_s)) \right|,$$

with $R(n) = R(n, \boldsymbol{\alpha})$ as defined in (32), $QR(n, \boldsymbol{\alpha})$ as in (33), and μ the Lebesgue measure on $\mathbb{R}^{(m_1-1)+\cdots+(m_s-1)}$. Since $\{1, \varphi_{m_1}, \ldots, \varphi_{m_1}^{m_1-1}, \ldots, \varphi_{m_s}, \ldots, \varphi_{m_s}^{m_s-1}\}$ is linearly independent over $\mathbb{Q}$, due to (33) the rotation $QR(n)$ satisfies the assumptions of Lemma 5.5 (the offset $\boldsymbol{\alpha}$ doesn't change the discrepancy significantly, see [14, Lemma 1.7]), and we have

$$D_N((QR(n))_{n\geq0}) \ll N^{\varepsilon/3-1} \tag{35}$$

for $\varepsilon > 0$ chosen as in the statement of the theorem.

We will now estimate the quantities $\delta_{k_1,\ldots,k_s}$ in terms of this discrepancy. To this end choose $M = \lfloor \frac{\log_2 N}{(m_1-1)+\cdots+(m_s-1)} \rfloor$ and cover $Q(S_1 \times \cdots \times S_s)$ for $S_1 \times \cdots \times S_s \in \mathcal{S}_{k_1}^{(m_1)} \times \cdots \times \mathcal{S}_{k_s}^{(m_s)}$ by boxes $\{U_1, \ldots, U_r\}$ and $\{V_1, \ldots, V_{r'}\}$ as specified in Lemma 5.4. Set now $\mathbf{S} = S_1 \times \cdots \times S_s$. Then, using the triangle inequality and arguing in the same way as in (25) and (26), we get

$$\left| \frac{1}{N} \sum_{n=0}^{N-1} \mathbf{1}_{Q\mathbf{S}}(QR(n)) - \mu(Q\mathbf{S}) \right| \leq \sum_{j=0}^{r} \left| \frac{1}{N} \sum_{n=0}^{N-1} \mathbf{1}_{U_j}(QR(n)) - \mu(U_j) \right|$$

$$+ \sum_{j=0}^{r'} \left| \frac{1}{N} \sum_{n=0}^{N-1} \mathbf{1}_{V_j \cap Q\mathbf{S}}(QR(n)) - \mu(V_j \cap Q\mathbf{S}) \right|$$

$$\leq \sum_{j=0}^{r} \left| \frac{1}{N} \sum_{n=0}^{N-1} \mathbf{1}_{U_j}(QR(n)) - \mu(U_j) \right|$$

$$+ \sum_{j=0}^{r'} \left(\left| \frac{1}{N} \sum_{n=0}^{N-1} \mathbf{1}_{V_j}(QR(n)) - \mu(V_j) \right| + 2\mu(V_j) \right).$$

Since U_j and V_j are boxes, the moduli on the right-hand side can be estimated by the discrepancy of $QR(n)$ (again the fact that the boxes are not located at the origin doesn't cause significant difference by Drmota and Tichy [14, Lemma 1.7]). Hence, taking the supremum over all $\mathbf{S} \in \mathcal{S}_{k_1}^{(m_1)} \times \cdots \times \mathcal{S}_{k_s}^{(m_s)}$ and keeping in mind that by Lemma 5.4 we have the estimates $r, r' \ll 2^{M(D+\varepsilon/3)}$ and

$$\sum_{j=0}^{r'} \mu(V_j) \ll 2^{M(D-(m_1-1)-\cdots-(m_s-1)+\varepsilon/3)}$$

this yields

$$\delta_{k_1,\ldots,k_s} \ll 2^{M(D+\varepsilon/3)} D_N((R(n))_{n\geq0}) + 2^{M(D-(m_1-1)-\cdots-(m_s-1)+\varepsilon/3)}.$$

By (35) we finally end up with

$$\delta_{k_1,\dots,k_s} \ll N^{\frac{D+\varepsilon/3}{(m_1-1)+\cdots+(m_s-1)}-1+\frac{\varepsilon}{3}} + N^{\frac{D-(m_1-1)+\cdots+(m_s-1)}{(m_1-1)+\cdots+(m_s-1)}+\frac{\varepsilon}{3}} \ll N^{\frac{\max\{d_i-(m_i-1)\,:\,1\le i\le s\}}{(m_1-1)+\cdots+(m_s-1)}+\frac{2\varepsilon}{3}}.$$

Inserting this in Proposition 4.2 proves (29); indeed, the sums over $k_1,\dots,k_s$ occurring in (28) contribute only logarithmic factors that are absorbed by $N^{\varepsilon/3}$, and $\sum_{i=1}^s \beta_i^{-L_i} \ll N^{-1}$.

The claim that $\dim_{\mathrm{B}}(\partial \mathcal{R}_{m_i}) < m_i - 1$ has already been treated in the paragraph preceding (34). $\qquad\qquad\square$

We conclude this section with an easy example.

Example 5.6 Consider the golden mean φ_2 and the dominant root φ_3 of the *tribonacci polynomial* $X^3 - X^2 - X - 1$. The Rauzy fractal $\mathcal{R}_2$ is an interval, hence, $\dim_{\mathrm{B}}\partial\mathcal{R}_2 = 0$. For $\mathcal{R}_3$ we know from [28] that $\dim_{\mathrm{B}}\partial\mathcal{R}_3 = 1.09336\ldots$ (in fact, in [28] the Hausdorff dimension is determined, however, for the set $\partial\mathcal{R}_3$ the Hausdorff dimension is the same as the box counting dimension because the restriction of B_3 to $\mathbf{v}_3^{\perp}$ is a similarity transformation). Since $\mathbb{Q}(\varphi_2, \varphi_3)$ has degree 6 over $\mathbb{Q}$ the linear independence assumption in Theorem 5.1 is satisfied and we obtain (choosing $\varepsilon > 0$ sufficiently small)

$$D_N((H_{(\varphi_2,\varphi_3)}(n))_{n\ge 0}) \ll N^{-0.30221}.$$

6 Final Remarks on Possible Further Research

There are several directions for further research on this topic. The first task would be to generalize Theorem 5.1 to the full class of linear recurrences given in (6) or in [27]. To do this, several obstacles have to be mastered. The reason is that the dominant root β of the characteristic polynomial of the recurrences in (6) is a Pisot number but in general not a unit. Again one can associate substitutions to these linear recurrences, however, the Rauzy fractals no longer live in Euclidean space but in an open subring of the adèle ring $\mathbb{A}_{\mathbb{Q}(\beta)}$. The theory of these fractals is well developed (see, e.g., [3, 35, 36, 49]), and with some more technical effort they should relate the Halton sequences in question with certain rotations on these subrings. To estimate the discrepancy of these rotations generalizations of the Erdős–Turán–Koksma inequality (cf. [14, Theorem 1.21] for the Euclidean version) and Schlickewei's p-adic subspace theorem (cf. [47]) could be of use. We want to come back to this in a forthcoming paper.

To be more general one could define Halton sequences for the substitutive case (based on Dumont–Thomas numeration in the sense of Steiner [53]), where even more examples with symmetric languages should come up.

Beyond that it would be interesting to get results on Halton sequences related to β-expansions with asymmetric languages. In this case it is not so clear how to proceed and one needs to deal with the reverse language in some way to define the appropriate Rauzy fractals in order to derive the rotation related to the Halton sequence in question. Also generalizations of Drmota's hybrid case (see [13]) deserve interest.

Since the discrepancy estimates in Theorem 5.1 are certainly not optimal it would be of great interest to gain a better understanding of the distribution properties of β-adic Halton sequences that would lead to improved discrepancy estimates and to the characterization of bounded remainder sets (see [52] for bounded remainder sets for β-adic van der Corput sequences).

Acknowledgements Supported by projects I1136 and P27050 granted by the Austrian Science Fund (FWF)

References

1. S. Akiyama, On the boundary of self affine tilings generated by Pisot numbers. J. Math. Soc. Jpn. **54**(2), 283–308 (2002)
2. S. Akiyama, Pisot number system and its dual tiling, in *Physics and Theoretical Computer Science*. NATO Science for Peace and Security Series D: Information and Communication Security, vol. 7 (IOS, Amsterdam, 2007), pp. 133–154
3. S. Akiyama, G. Barat, V. Berthé, A. Siegel, Boundary of central tiles associated with Pisot beta-numeration and purely periodic expansions. Monatsh. Math. **155**(3–4), 377–419 (2008)
4. S. Akiyama, C. Frougny, J. Sakarovitch, Powers of rationals modulo 1 and rational base number systems. Isr. J. Math. **168**, 53–91 (2008)
5. S. Akiyama, M. Barge, V. Berthé, J.-Y. Lee, A. Siegel, On the Pisot substitution conjecture, in *Mathematics of Aperiodic Order*. Progress in Mathematics, vol. 309 (Birkhäuser/Springer, Basel, 2015), pp. 33–72
6. P. Arnoux, S. Ito, Pisot substitutions and Rauzy fractals. Bull. Belg. Math. Soc. Simon Stevin, **8**(2), 181–207 (2001); Journées Montoises d'Informatique Théorique (Marne-la-Vallée, 2000)
7. G. Barat, P.J. Grabner, Distribution properties of G-additive functions. J. Number Theory **60**(1), 103–123 (1996)
8. V. Berthé, A. Siegel, Tilings associated with beta-numeration and substitutions. Integers **5**(3), 1–46 (2005), #A02
9. V. Berthé, A. Siegel, J. Thuswaldner, Substitutions, Rauzy fractals and tilings, in *Combinatorics, Automata and Number Theory*. Encyclopedia of Mathematics and its Applications, vol. 135 (Cambridge University Press, Cambridge, 2010), pp. 248–323
10. V. Berthé, W. Steiner, J.M. Thuswaldner, Geometry, dynamics and arithmetic of S-adic shifts (preprint, 2016)
11. V. Canterini, A. Siegel, Automate des préfixes-suffixes associé à une substitution primitive. J. Théor. Nombres Bordeaux **13**(2), 353–369 (2001)
12. I. Carbone, Discrepancy of LS-sequences of partitions and points. Ann. Mat. Pura Appl. (4) **191**(4), 819–844 (2012)
13. M. Drmota, The discrepancy of generalized van-der-Corput–Halton sequences. Ind. Math. **26**(5), 748–759 (2015)
14. M. Drmota, R.F. Tichy, *Sequences, Discrepancies and Applications*. Lecture Notes in Mathematics, vol. 1651 (Springer, Berlin, 1997)
15. J.-M. Dumont, A. Thomas, Systemes de numeration et fonctions fractales relatifs aux substitutions. Theor. Comput. Sci. **65**(2), 153–169 (1989)

16. J.-M. Dumont, A. Thomas, Digital sum moments and substitutions. Acta Arith. **64**(3), 205–225 (1993)
17. K.J. Falconer, *Fractal Geometry* (Wiley, Chichester, 1990)
18. H. Faure, C. Lemieux, Improved Halton sequences and discrepancy bounds. Monte Carlo Methods Appl. **16**(3–4), 231–250 (2010)
19. H. Faure, P. Kritzer, F. Pillichshammer, From van der Corput to modern constructions of sequences for quasi-Monte Carlo rules. Ind. Math. **26**(5), 760–822 (2015)
20. D.-J. Feng, M. Furukado, S. Ito, J. Wu, Pisot substitutions and the Hausdorff dimension of boundaries of atomic surfaces. Tsukuba J. Math. **30**(1), 195–223 (2006)
21. T. Fujita, S. Ito, S. Ninomiya, The generalized van der Corput sequence and its application to numerical integration. Sūrikaisekikenkyūsho Kōkyūroku **1240**, 114–124 (2001); 5th Workshop on Stochastic Numerics (Japanese) (Kyoto, 2001)
22. P.J. Grabner, R.F. Tichy, Contributions to digit expansions with respect to linear recurrences. J. Number Theory **36**(2), 160–169 (1990)
23. J.H. Halton, On the efficiency of certain quasi-random sequences of points in evaluating multi-dimensional integrals. Numer. Math. **2**, 84–90 (1960)
24. A. Haynes, Equivalence classes of codimension-one cut-and-project nets. Ergodic Theory Dyn. Syst. **36**(3), 816–831 (2016)
25. R. Hofer, Halton-type sequences to rational bases in the ring of rational integers and in the ring of polynomials over a finite field. Math. Comput. Simul. (to appear)
26. R. Hofer, P. Kritzer, G. Larcher, F. Pillichshammer, Distribution properties of generalized van der Corput-Halton sequences and their subsequences. Int. J. Number Theory **5**(4), 719–746 (2009)
27. M. Hofer, M.R. Iacò, R. Tichy, Ergodic properties of β-adic Halton sequences. Ergodic Theory Dyn. Syst. **35**(3), 895–909 (2015)
28. S. Ito, M. Kimura, On Rauzy fractal. Jpn. J. Ind. Appl. Math. **8**(3), 461–486 (1991)
29. S. Ito, H. Rao, Atomic surfaces, tilings and coincidence. I. Irreducible case. Isr. J. Math. **153**, 129–155 (2006)
30. A. Jassova, P. Lertchoosakul, R. Nair, On variants of the Halton sequence. Monatsh. Math. **180**, 743–764 (2016)
31. L. Kuipers, H. Niederreiter, *Uniform Distribution of Sequences* (Wiley-Interscience [Wiley], New York/London/Sydney, 1974); Pure and Applied Mathematics
32. P. Lecomte, M. Rigo, Numeration systems on a regular language. Theory Comput. Syst. **34**(1), 27–44 (2001)
33. P. Lecomte, M. Rigo, On the representation of real numbers using regular languages. Theory Comput. Syst. **35**(1), 13–38 (2002)
34. R.D. Mauldin, S.C. Williams, Hausdorff dimension in graph directed constructions. Trans. Am. Math. Soc. **309**(2), 811–829 (1988)
35. M. Minervino, W. Steiner, Tilings for Pisot beta numeration. Ind. Math. **25**(4), 745–773 (2014)
36. M. Minervino, J. Thuswaldner, The geometry of non-unit Pisot substitutions. Ann. Inst. Fourier (Grenoble) **64**(4), 1373–1417 (2014)
37. L.J. Mordell, On the linear independence of algebraic numbers. Pac. J. Math. **3**, 625–630 (1953)
38. M. Mori, M. Mori, Dynamical system generated by algebraic method and low discrepancy sequences. Monte Carlo Methods Appl. **18**(4), 327–351 (2012)
39. H. Niederreiter, Application of Diophantine approximations to numerical integration, in *Diophantine Approximation and Its Applications (Proceedings of a Conference, Washington, DC, 1972)* (Academic, New York, 1973), pp. 129–199
40. H. Niederreiter, A discrepancy bound for hybrid sequences involving digital explicit inversive pseudorandom numbers. Unif. Distrib. Theory **5**(1), 53–63 (2010)
41. S. Ninomiya, Constructing a new class of low-discrepancy sequences by using the β-adic transformation. Math. Comput. Simul. **47**(2–5), 403–418 (1998); IMACS Seminar on Monte Carlo Methods (Brussels, 1997)

42. W. Parry, On the β-expansions of real numbers. Acta Math. Acad. Sci. Hungar. **11**, 401–416 (1960)
43. A. Pethő, R.F. Tichy, On digit expansions with respect to linear recurrences. J. Number Theory **33**(2), 243–256 (1989)
44. G. Rauzy, Nombres algébriques et substitutions. Bull. Soc. Math. France **110**(2), 147–178 (1982)
45. A. Rényi, Representations for real numbers and their ergodic properties. Acta Math. Acad. Sci. Hungar. **8**, 477–493 (1957)
46. M. Rigo, W. Steiner, Abstract β-expansions and ultimately periodic representations. J. Théor. Nombres Bordeaux **17**(1), 283–299 (2005)
47. H.P. Schlickewei, The p-adic Thue-Siegel-Roth-Schmidt theorem. Arch. Math. (Basel) **29**(3), 267–270 (1977)
48. W.M. Schmidt, Simultaneous approximation to algebraic numbers by rationals. Acta Math. **125**, 189–201 (1970)
49. A. Siegel, Représentation des systèmes dynamiques substitutifs non unimodulaires. Ergodic Theory Dyn. Syst. **23**(4), 1247–1273 (2003)
50. A. Siegel, J.M. Thuswaldner, Topological properties of Rauzy fractals. Mém. Soc. Math. Fr. (N.S.) **118**, 1–144 (2009)
51. V.F. Sirvent, Y. Wang, Self-affine tiling via substitution dynamical systems and Rauzy fractals. Pac. J. Math. **206**(2), 465–485 (2002)
52. W. Steiner, Regularities of the distribution of β-adic van der Corput sequences. Monatsh. Math. **149**(1), 67–81 (2006)
53. W. Steiner, Regularities of the distribution of abstract van der Corput sequences. Unif. Distrib. Theory **4**(2), 81–100 (2009)